Mathematical Morphology and its Applications to Image and Signal Processing

Computational Imaging and Vision

Volume 12

Mathematical Morphology and its Applications to Image and Signal Processing

edited by

Henk J. A. M. Heijmans

Centre for Mathematics and Computer Science (CWI), Amsterdam, The Netherlands

and

Jos B. T. M. Roerdink

Department of Mathematics and Computing Science, University of Groningen, Groningen, The Netherlands

KLUWER ACADEMIC PUBLISHERS

DORDRECHT / BOSTON / LONDON

A C.I.P. Catalogue record for this book is available from the Library of Congress.

ISBN 978-90-481-5053-3

Published by Kluwer Academic Publishers,
P.O. Box 17, 3300 AA Dordrecht, The Netherlands.

Sold and distributed in North, Central and South America
by Kluwer Academic Publishers,
101 Philip Drive, Norwell, MA 02061, U.S.A.

In all other countries, sold and distributed
by Kluwer Academic Publishers,
P.O. Box 322, 3300 AH Dordrecht, The Netherlands.

Printed on acid-free paper

Contents

Shape Analysis and Partial Differential Equations

Connected Operators

Segmentation

Random Models and Statistical Techniques

Algorithms

Applications

Preface

This volume contains the papers presented at the fourth International Symposium on Mathematical Morphology and its Applications to Image and Signal Processing (ISMM'98) held in Amsterdam, The Netherlands, June 3–5, 1998. The first ISMM meeting took place in 1993 in Barcelona. The second and third meeting were held in Fontainebleau (1994) and Atlanta (1996). Their main goal is to provide a forum for researchers in image and signal processing, pattern recognition, and computer vision, for the exchange of ideas and techniques in mathematical morphology and related fields.

These proceedings contain 5 invited and 46 accepted papers. We want to thank all members of the programme committee for their invaluable assistance in the reviewing process (using a double blind procedure) of all submitted papers. The geographic spectrum of the authors comprises more than 15 countries all over the world. We succeeded in supplying five conference grants for high-quality contributions by researchers from Eastern Europe. A special note concerns the creation of computer art with mathematical morphology by J.S. Ploem, now retired from the University of Leiden, who was one of the earliest persons in the Netherlands to apply morphological techniques in cytology. A selection of his artwork was on exhibition during the symposium.

The organisation of ISMM'98 and the realisation of these proceedings before the beginning of the meeting would not have been possible without the help of many individuals and organisations. First of all we would like to express our gratitude to all authors who submitted their papers or accepted our invitation. We want to thank the Centre for Mathematics and Computer Science (CWI) in Amsterdam, in particular Frans Snijders, for providing us with all necessary facilities. Thanks are also due to CWI and SION (Dutch science foundation for research in computer science) for financial support. Finally, special thanks go to Lieke Schultze who took care of the ISMM'98 secretariat and who is responsible for most of the local arrangements.

Henk Heijmans
Jos Roerdink

Introduction

Mathematical Morphology is a methodology for the quantitative analysis of spatial structures based on concepts of shape and form. It has a strong mathematical flavour in its use of concepts from set theory, topology, lattice algebra, random functions, etc. These theoretical developments are driven by practical problems of very diverse nature. The actual analysis of the structures under study is performed on digital images, necessitating the use of powerful algorithms and hardware for processing and analysis of images on digital computers.

The present book consists of the invited and contributed papers presented at the ISMM'98 symposium. The choice of invited papers was largely motivated by the editors' aim to stimulate the interchange of ideas between the morphological community and people working in (closely) related areas, who use methods based on random models and statistics, partial differential equations, curve evolution, scale space, etc.

The contributions are classified under a number of broad themes, which are briefly introduced hereafter. Note that several papers have aspects belonging to two or more of these themes.

Theory. This concerns representation of morphological operators, multivariate morphology, lattice theory and shape description.

Topology and Geometry. Central issues are metrics, convexity and connectivity, as well as their digitization.

Shape Analysis and Partial Differential Equations. This is an area of active research, not only within mathematical morphology. Partial differential equations are used for image iterative filtering, curve evolution and image segmentation. Models (e.g. of anatomic objects from medical images) are extracted based on skeletons or related primitives which enable a compact representation.

Connected Operators. Connected operators are operators that act on the level of the flat zones (homogeneous regions) of an image rather than on the level of individual pixels. In this section, several approaches are described for constructing connected operators with applications to segmentation.

Segmentation. Apart from the classical watershed algorithm, this section contains papers on colour segmentation and segmentation of non-image data.

Random Models and Statistical Techniques. These methods are used for the design of optimal filters, inference of random fields, image restoration, as well as the description of grey value images by means of random sets and the construction of new image metrics.

Algorithms. Special algorithms are used for finding optimal sequential decompositions of morphological operators, for function interpolation, motion compensation, detection of linear features, opening transforms and skeletonization of compressed binary pictures.

Applications. The range of applications of mathematical morphology is very large. Included in this volume are papers on character extraction, analysis of interferometric phase-maps, water depth determination, road network extraction, radar imaging, and the creation of computer graphics with mathematical morphology, to name a few.

We hope that the papers in this book will contribute ideas which will be useful to people working in mathematical morphology and other areas of signal and image processing, pattern recognition and computer vision.

HENK HEIJMANS
JOS ROERDINK

NUMERICAL AND APPROXIMATION TOOLS BASED ON MATHEMATICAL MORPHOLOGY

ANTONY T. POPOV
St. Kliment University of Sofia, Faculty of Mathematics and Informatics,
5, James Bourchier Blvd., BG-1126 Sofia, BULGARIA
e-mail: atpopov@fmi.uni-sofia.bg, fax (+359 2) 687 180

Abstract. This work shows that complex interval operations used in scientific computations and approximation tasks can be regarded as morphological ones on an appropriate chosen complete lattice. Thus these operations can be generalized on more complicated objects, like for instance fuzzy numbers. It is demonstrated also that the distance measuring the likelihood of the graphs of two functions, usually referred as 'Hausdorff distance' between functions and widely used in function approximation tasks and signal processing, can be calculated efficiently using grey-tone morphological operations.

Key words: Interval arithmetic operations, Complete lattice, Morphological operations, Hausdorff approximations, Baire functions

1. Introduction

Mathematical morphology has been created as an approach based on concepts like transformation groups. The basic morphological operations — erosions and dilations — are introduced by Matheron [7] and Serra [10] to examine the structure of binary images which can be represented mathematically as sets. Dilations and erosions over sets are represented as Minkowski addition and subtraction by means of a fixed set called structuring element. Morphological operations have been extended to grey-scale images modelled mathematically as functions. Serra [11] and Heijmans [3] have shown that morphological operations can be extended to any complete lattice.

Interval based algorithms are used to handle uncertain and ambiguous data in an easy and efficent way tracking systematical, aleatory and rounding errors. They have been applied successfuly in numerical approximation, dynamical systems analysis, computational geometry tasks etc. In our work we show that the operations of interval algebra can be expressed by morphological operations on an appropriately chosen lattice over the set of intervals on the real line. It is demonstrated also that the distance measuring the likelihood of the graphs of two functions used in [9], usually referred as 'Hausdorff distance' between functions can be calculated efficiently using grey-tone morphological operations. Generalizations of the Hausdorff distance in functional spaces are widely used in numerical approximation tasks [9] and signal processing [2].

In this work we use the same notions and notations about complete lattices and the morphological operations on them as in [3]. For instance, let $\mathcal{L}$ be a complete lattice with a supremum generating family l, and let T be an Abelian group of automorphisms of $\mathcal{L}$ acting transitively over l. The elements of T are denoted by τ_x, namely for any $x \in l$ $\tau_x(o) = x$, where o is a fixed element of l interpreted as an origin. Then also, we can consider a symmetry in $\mathcal{L}$ as $\check{A} = \bigvee_{a \in l(A)} \tau_a^{-1}(o)$. Evidently $\check{a} = \tau_a^{-1}(o) = \tau_{\check{a}}(o)$ for any $a \in l$. If A is an arbitrary element of the lattice $\mathcal{L}$ let us denote by $l(A) = \{a \in l \mid a \leq A\}$ the supremum-generating set of A. Following [3] we define the operations

$$\delta_A = \bigvee_{a \in l(A)} \tau_a \tag{1}$$

and

$$\varepsilon_A = \bigwedge_{a \in l(A)} \tau_a^{-1} = \bigwedge_{a \in l(A)} \tau_{\check{a}} \tag{2}$$

which form an adjunction. δ_A and ε_A are T-invariant operators called *dilation and erosion by the structuring element A*.

2. Additive interval operations

Consider the lattice $\mathcal{L}$ with elements the subsets of a linear space M. First, we shall regard the so called classical, or *binary* morphological operations defined in this lattice. Order relation in $\mathcal{L}$ is the ordinary set inclusion, supremum operation is the union, and infimum is the intersection operation. The minimal element of the lattice is the empty set, while the maximal element is M itself. Let the commutative group of automorphisms be the group of translations, i.e. $\tau_b(A) = \{x \mid x = a + b, a \in A\}$, and let the supremum generating family be the family of one-point subsets of M, namely $l = \{\{x\} \mid x \in M\}$. Then $\delta_A(B) = \delta_B(A) = A \oplus B$ is the standard Minkowski addition, $A \oplus B = \{a + b \mid a \in A, b \in B\}$ and $\varepsilon_A(B) = B \ominus A$ is the Minkowski subtraction operation introduced by Hadwiger [7]:

$$A \ominus B = \bigcap_{b \in B} \tau_{-b}(A) = \{x \mid \tau_x(B) \subseteq A\}. \tag{3}$$

Then closing and opening of A by B are defined as

$$A \bullet B = (A \oplus B) \ominus B \supseteq A, \tag{4}$$

$$A \circ B = (A \ominus B) \oplus B = \{x \in A \mid \exists t : x \in \tau_t(A) \subseteq B\}. \tag{5}$$

Let us denote by X^c the complement of the set X, i.e. $X^c = M \setminus X$. Then the following simple equalities hold [10]:

$$(A \oplus B)^c = A^c \ominus (-B), \quad (A \circ B)^c = A^c \bullet (-B), \tag{6}$$

where by $-B$ we denote the centrally symmetric set of B. Having in mind (6), and since every dilation fixes the minimal element of the lattice and every erosion fixes its maximal element, for any non-degenerate element A of $\mathcal{L}$ we have that

$$A \oplus \emptyset = \emptyset, \ M \ominus A = M, \ \emptyset \ominus A = \emptyset, \ A \ominus \emptyset = M. \tag{7}$$

Following [5] we introduce the following notations: Let $I(\mathbf{R})$ be the family of all closed finite intervals of the real line. For completeness we may assume that the empty set $\emptyset$ is an element of $I(\mathbf{R})$. But in our work we shall deal only with proper intervals, i.e. such that their left endpoint is less or equal to their right endpoint. The case when any of the operands is the empty interval is covered by (7). The non-empty finite intervals from $I(\mathbf{R})$ will be denoted by Roman capitals, namely $A = [a^-, a^+]$, i.e. a^- is the left endpoint of the interval A and a^+ is its right endpoint. By $\omega(A)$ we denote the length of the interval A, $\omega(A) = a^+ - a^-$.

Consider the outer and inner interval additive operations as defined by S. Markov in [5, 6]:

$$A + B = [a^- + b^-, a^+ + b^+],\ A +^- B = [a^{-s} + b^s, a^s + b^{-s}],$$
$$A - B = [a^- - b^+, a^+ - b^-],\ A -^- B = [a^{-s} - b^{-s}, a^s - b^s],$$

where

$$s = \begin{cases} +, & \omega(A) \geq \omega(B), \\ -, & \omega(A) < \omega(B). \end{cases}$$

Here and further on, a^s with $s \in \{+,-\}$ denotes certain endpoint of the interval A — the left one if $s = -$ and the right one if $s = +$. We define the 'product' st for $s,t \in \{+,-\}$ by $++ = -- = +$, $+- = -+ = -$, i.e. $a^{++} = a^{--} = a^+$ and $a^{+-} = a^{-+} = a^-$.

As proved in [8] the outer and inner interval operations in $I(\mathbf{R})$ are related to dilations and erosions as follows:

$$A + B = A \oplus B = \delta_A(B) = \delta_B(A), \tag{8}$$

$$A +^- B = A \ominus (-B) \cup B \ominus (-A) = \varepsilon_{-B}(A) \cup \varepsilon_{-A}(B). \tag{9}$$

It is easy to demonstrate that for the outer and inner subtraction we have the following relations:

$$A - B = A + (-B) = A \oplus (-B), \tag{10}$$
$$A -^- B = A +^- (-B) = A \ominus B \cup -(B \ominus A). \tag{11}$$

If $\mathcal{L}$ is a complete lattice with a supremum generating family l, and T is an Abelian group of automorphisms of $\mathcal{L}$ acting transitively over l, having in mind (9) we can introduce an analog of the inner interval addition, which we will call *inner morphological operation*, $\eta : \mathcal{L} \times \mathcal{L} \mapsto \mathcal{L}$. Namely

$$\eta(A, B) = \varepsilon_{\check{B}}(A) \vee \varepsilon_{\check{A}}(B).$$

Since $\varepsilon_{\check{B}}(A) = \bigwedge_{-z \in l(\check{B})}(A) = \bigwedge_{y \in l(B)}(A) \leq \bigvee_{y \in l(B)}(A) = \delta_B(A)$, then we obtain evidently that

$$\eta(A, B) = \eta(B, A) \leq \delta_B(A) = \delta_A(B). \tag{12}$$

3. Multiplicative Interval Operations

In many applications such as locating the roots of a polynomial with interval coefficients, it is desirable to spread also the multiplicative operations over the set of real intervals. Let $\mathcal{L}$ be the set of all closed finite and infinite real intervals different from the singleton $[0,0]$. The order relation is the set inclusion, while the supremum is defined as the closed convex hull of the union, and the infimum is defined as intersection:

$$\bigvee_i H_i = \mathrm{co}\,(\mathrm{cl}\,(\bigcup_i H_i)) = [\inf h_i^-, \sup h_i^+]$$

$$\bigwedge_i H_i = \bigcap_i H_i = [\sup h_i^-, \inf h_i^+].$$

A supremum generating family is

$$l = \{\{x\} = [x,x] \mid x \in \mathbf{R},\, x \neq 0\},$$

therefore for any $A \in \mathcal{L}$ $l(A) = \{\{x\} \mid x \in A,\ x \neq 0\}$. There is an Abelian group of automorphisms

$$T = \{\tau_h \mid h \in \mathbf{R} \setminus \{0\},\, \tau_h(\{x\}) = \{xh\}\},$$

which acts transitively on l and $\tau_h(A) = [\min(a^-h, a^+h), \max(a^-h, a^+h)]$.

Then we can define dilation and erosion operations by the structuring element A using expressions (1) and (2):

$$\delta_A(B) = \bigvee_{a \in l(A)} \tau_a(B),$$

$$\varepsilon_A(B) = \bigwedge_{a \in l(A)} \tau_{1/a}(B)\ .$$

Define $\mathcal{Z}$ as the family of closed intervals containing 0. If $A \notin \mathcal{Z}$

$$\sigma(A) = \begin{cases} +, & \text{if } 0 < a^-; \\ -, & \text{if } a^+ < 0. \end{cases}$$

Outer multiplication in $I(\mathbf{R})$ (see [5]) is defined by

$$A \times B = \begin{cases} [a^{-\sigma(B)}b^{-\sigma(A)}, a^{\sigma(B)}b^{\sigma(A)}], & \text{for } A, B \in \mathcal{L} \setminus \mathcal{Z}, \\ [a^{\sigma(A)}b^{-\sigma(A)}, a^{\sigma(A)}b^{\sigma(A)}], & \text{for } A \in \mathcal{L} \setminus \mathcal{Z}, B \in \mathcal{Z}, \\ [a^{-\sigma(B)}b^{\sigma(B)}, a^{\sigma(B)}b^{\sigma(B)}], & \text{for } B \in \mathcal{L} \setminus \mathcal{Z}, A \in \mathcal{Z}, \\ [\min(a^-b^+, a^+b^-), \max(a^-b^-, a^+b^+)], & \text{for } A, B \in \mathcal{Z}, \end{cases}$$

The following two propositions are proved in [8]:

Proposition 3.1 *For the dilation defined above we have that* $\delta_B(A) = \delta_A(B) = A \times B$.

Proposition 3.2 *For intervals B such that $b^+b^- \neq 0$ we have that*

$$\varepsilon_B(A) = \begin{cases} \left[\frac{a^{-\sigma(B)}}{b^{-\sigma(A)}}, \frac{a^{\sigma(B)}}{b^{\sigma(A)}}\right], & \text{for } A, B \in \mathcal{L}\setminus\mathcal{Z}, \\ \left[\frac{a^{-\sigma(B)}}{b^{\sigma(B)}}, \frac{a^{\sigma(B)}}{b^{\sigma(B)}}\right], & \text{for } B \in \mathcal{L}\setminus\mathcal{Z}, A \in \mathcal{Z} \\ \emptyset, & \text{for } A \in \mathcal{L}\setminus\mathcal{Z}, B \in \mathcal{Z}, \\ \left[\max\left(\frac{a^+}{b^-}, \frac{a^-}{b^+}\right), \min\left(\frac{a^+}{b^+}, \frac{a^-}{b^-}\right)\right], & \text{for } A, B \in \mathcal{Z}. \end{cases}$$

Let us consider the inner multiplication operation "$\times^-$" defined as [5] :

$$A \times^- B = \begin{cases} [a^{\sigma(B)\epsilon}b^{-\sigma(A)\epsilon}, a^{-\sigma(B)\epsilon}b^{\sigma(A)\epsilon}], & \epsilon = \psi(A,B),\ A, B \in \mathcal{L}\setminus\mathcal{Z}, \\ [a^{-\sigma(A)}b^{-\sigma(A)}, a^{-\sigma(A)}b^{\sigma(A)}], & A \in \mathcal{L}\setminus\mathcal{Z},\ B \in \mathcal{Z} \\ [a^{-\sigma(B)}b^{-\sigma(B)}, a^{\sigma(B)}b^{-\sigma(B)}], & A \in \mathcal{Z}, B \in \mathcal{L}\setminus\mathcal{Z}, \\ [\max(a^-b^+, a^+b^-), \min(a^-b^-, a^+b^+)], & A, B \in \mathcal{Z}. \end{cases}$$

Here $\psi(A,B) = +$ if $\chi(A) \geq \chi(B)$ and $\psi(A,B) = -$ otherwise, where $\chi(A) = a^-/a^+$ if $a^- + a^+ \geq 0$ and $\chi(A) = a^+/a^-$ if $a^- + a^+ < 0$.

Denoting $\frac{1}{B} = \left[\frac{1}{b^+}, \frac{1}{b^-}\right]$, (which in this case is namely the symmetric element of B in $\mathcal{L}$) we obtain:

Corollary 3.3 $A \times^- B = \varepsilon_{\frac{1}{A}}(B) \cup \varepsilon_{\frac{1}{B}}(A)$ *for any intervals* $A, B \in \mathcal{L}\setminus\mathcal{Z}$.

Proof. From Proposition 3.2 we have that

$$\varepsilon_{\frac{1}{B}}(A) = [a^{-\sigma(B)}b^{\sigma(A)}, a^{\sigma(B)}b^{-\sigma(A)}],$$

$$\varepsilon_{\frac{1}{A}}(B) = [a^{\sigma(B)}b^{-\sigma(A)}, a^{-\sigma(B)}b^{\sigma(A)}],$$

Consider first the case when $\sigma(B) = \sigma(A) = +$. Then $\varepsilon_{\frac{1}{B}}(A) = [a^-b^+, a^+b^-]$ and $\varepsilon_{\frac{1}{A}}(B) = [a^+b^-, a^-b^+]$. Then $\varepsilon_{\frac{1}{A}}(B) = \emptyset$ if and only if $a^+b^- > a^-b^+$, which holds if and only if $\chi(A) < \chi(B)$, i.e. $\psi(A,B) = -$. Therefore $\varepsilon_{\frac{1}{A}}(B) \cup \varepsilon_{\frac{1}{B}}(A) = [a^-b^+, a^+b^-] = A \times^- B$. If $\chi(A) > \chi(B)$ then $\psi(A,B) = +$, and $a^+b^- > a^-b^+$. This means that $\varepsilon_{\frac{1}{B}}(A) = \emptyset$ and $\varepsilon_{\frac{1}{A}}(B) \cup \varepsilon_{\frac{1}{B}}(A) = [a^+b^-, a^-b^+] = A \times^- B$. In the case when $\chi(A) = \chi(B)$, by definition $\psi(A,B) = +$, and both intervals $\varepsilon_{\frac{1}{A}}(B)$ and $\varepsilon_{\frac{1}{B}}(A)$ degenerate into same one-point interval — in this case the statement is trivial. Thus the proposition is proved in the case when $\sigma(B) = \sigma(A) = +$.

The cases $\sigma(B) = \sigma(A) = -$ and $\sigma(B) = -\sigma(A)$ are regarded in the same way. □

Thus we demonstrate that the inner interval multiplication is an inner morphological operation. In [8] a consideration about complex circular and rectangular interval operations (as defined in [1]) is presented and new inner complex operations are defined, which are used in solving algebraic equations with real interval coefficients.

4. Computing the Hausdorff distance between functions by grey-tone morphological operations

In the previous sections we considered the space of closed finite intervals of the real line, while here we deal with interval valued functions.

As mentioned in [9], in many approximation problems it is useful to work with the so called 'Hausdorff distance' between functions. The main profit is that we approximate the original function by a simpler one (piecewise polygons, Bezier functions, splines etc.) which graph shape is close to the original graph shape. We shall consider only functions with bounded domain. But before setting the main result we need some additional notions and notations.

Let consider the grey-tone morphological operations in $\mathrm{Fun}(M)$ – the family of all functions mapping the linear space M to the compactified real line $\bar{\mathbf{R}} = \mathbf{R} \cup \{\infty, -\infty\}$. The basic morphological operations – erosions and dilations – are defined by

$$(\delta_g(f))(x) = (f \oplus g)(x) = \sup_{x \in M} (f(x-h) + g(h)), \tag{13}$$

$$(\varepsilon_g(f))(x) = (f \ominus g)(x) = \inf_{x \in M} (f(x+h) - g(h)), \tag{14}$$

supposing that $s + t = -\infty$ if $s = -\infty$ or $t = -\infty$, and $s - t = \infty$ if $s = \infty$ or $t = -\infty$. Then every translational - invariant dilation is given by (13), and every translational - invariant erosion – by (14). These operations are named *grey-tone morphological operations* [10] and are widely used in the analysis of grey - tone images. For the construction of these operations using the complete lattice approach (expressions (1) and (2)) see [3].

In [3] it is shown that there exists a simple relation between binary and grey-tone operations. So, let define the set named *umbra* $U \subseteq M \times \bar{\mathbf{R}}$ with the property: the point $(x, t) \in U$ if and only if for every $s < t$ it follows that $(x, s) \in U$. Given a function $f(x)$ from $\mathrm{Fun}(M)$, we can define an umbra

$$\Gamma(f) = \{(x,t) \in M \times \bar{\mathbf{R}} |\, t \leq f(x)\}.$$

Conversely, given an umbra U, we can construct the function

$$f_U(x) = \sup\{t \in \mathbf{R} | (x,t) \in U\},$$

such that $\Gamma(f_U)$ is namely the umbra set U. We can define also an *upper closure* of the set $A \subseteq M \times \bar{\mathbf{R}}$ as the minimal umbra set containing A, which will be denoted by $\Upsilon(A)$, i.e. $\Upsilon(A)$ is the intersection of all umbrae containing A. It is evident that the $\Upsilon()$ operation is a closing. In [3] also the class of pre-umbra sets is considered: The set H is a pre-umbra if for every point $(x, t) \in H$ and for every $s < t$ the inclusion $(x, s) \in H$ holds. It is clear that the Minkowski sum of two umbrae is a pre-umbra. Also, if H is a pre-umbra, then $\mathrm{cl}\,(\Upsilon(H)) = \mathrm{cl}\,(H)$. The last equality follows from the fact that H consists of closed and non-closed rays and therefore $\Upsilon(H)$ consists of closed rays. Then the equality follows immediately from the fact that the sequence of real numbers $\{x_n\}$ converges just when $\{x_n - \frac{1}{n}\}$ converges and vice versa and they have a common limit. Also, if A is a topological space and A is closed in $M \times \bar{\mathbf{R}}$, then $\Upsilon(A)$ is closed in $M \times \bar{\mathbf{R}}$ as well.

In [3] the following theorem is proved:

Theorem 4.1 $\Gamma(f \oplus g) = \Upsilon(\Gamma(f) \oplus \Gamma(g))$, $\Gamma(f \ominus g) = \Gamma(f) \ominus \Gamma(g)$.

Let now $M = \mathbf{R}^n$ and let us consider the *closed graph of the function* $f : \mathbf{R}^n \mapsto \bar{\mathbf{R}}$ denoted by $\Phi(f)$ (see [9]) – it is defined as an interval valued function $(\Phi(f))(x) = [(I(f))(x), (S(f))(x)]$, where $(I(f))(x)$ and $(S(f))(x)$ are the upper and lower Baire functions of f respectively [9]:

$$(I(f))(x) = \lim_{\delta \to +0} \inf\{f(t) | t \in B_\delta(x) \cap \operatorname{dom}(f)\} = \liminf_{t \to x} f(t),$$
$$(S(f))(x) = \lim_{\delta \to +0} \sup\{f(t) | t \in B_\delta(x) \cap \operatorname{dom}(f)\} = \limsup_{t \to x} f(t).$$

Here by $\operatorname{dom}(f)$ we denote the domain of f, i.e. $\operatorname{dom}(f) = \{x | \infty > f(x) > -\infty\}$. Here and henceforth $B_r^m(x)$ denotes the closed ball in the m-dimensional space (disk in two dimensional space) with centre x and radius r. Sometimes we miss the upper index when there is no confusion with dimensions.

The closed graph can be considered also as a closed set in $\mathbf{R}^n \times \bar{\mathbf{R}}$.

Lemma 4.2 *If* $f : \mathbf{R}^n \mapsto \bar{\mathbf{R}}$, *then* $\operatorname{cl}(\Gamma(f)) = \Upsilon(\Phi(f))$.

Proof. Let $y \in \Upsilon(\Phi(f))$. Without lack of generality we may assume that $y = (S(f))(x)$ for any x. We shall show that $(x, y) \in \operatorname{cl}(\Gamma(f))$.

Consider a sequence of closed balls $B_{\delta_i}(x)$, $\lim_{i\to\infty} \delta_i = 0$. Let us define the numbers $S_i = \sup\{f(t) | t \in B_{\delta_i}(x) \cap \operatorname{dom}(f)\}$. Then there exist points $x_i \in B_{\delta_i}(x) \cap \operatorname{dom}(f)$ such that $S_i \geq f(x_i) \geq S_i - \frac{1}{n}$. Therefore $\lim_{i\to\infty} x_i = x$ and $\lim_{i\to\infty} f(x_i) = (S(f))(x) = y$. Since the points $(x_i, f(x_i))$ are from $\Gamma(f)$, then $(x, y) \in \operatorname{cl}(\Gamma(f))$ and $\Upsilon(\Phi(f)) \subseteq \operatorname{cl}(\Gamma(f))$.

Conversely, let us now choose a point $(x, y) \in \Gamma(f)$and suppose for simplicity that $y = f(x)$. Then $(I(f))(x) \leq f(x) \leq (S(f))(x)$ which means that $\Gamma(f) \subseteq \Upsilon(\Phi(f))$. From the upper semicontinuity of the upper Baire function (see [9]) it follows that $\Upsilon(\Phi(f))$ is closed. Therefore $\operatorname{cl}(\Upsilon(\Gamma(f))) \subseteq \Upsilon(\Phi(f))$, which proves the lemma. □

Corollary 4.3 $\Gamma(S(f)) = \operatorname{cl}(\Gamma(f)) = \Upsilon(\Phi(f))$.

In [9] Sendov defines the 'Hausdorff distance' between the functions f and g as $\mathrm{d}(f, d) = \operatorname{dist}(\Phi(f), \Phi(g))$, where dist is the ordinary Hausdorff distance between sets: If P and Q are non-empty compacts in $\mathbf{R}^n$, then

$$\operatorname{dist}(P, Q) \;=\; \inf\{\varepsilon \,|\, Q \subseteq P \oplus B_\varepsilon(0)\,,\, P \subseteq Q \oplus B_\varepsilon(0)\}. \tag{15}$$

In this work we demonstrate that the distance d can be easily expressed by grey-tone morphological operations, which enables its simpler computation using only the values of the upper Baire function.

Sendov's definition of Hausdorff distance between functions is more general and more precise than that studied in [2] and correctly treats the function discontinuities. For discrete signals both definitions lead to the same result.

Let $x \in \mathbf{R}^n$. For every $\varepsilon > 0$ we define the function $b_\varepsilon(x) = \max\{y | (x, y) \in B_\varepsilon^{n+1}(x)\}$ with domain the ball $B_\varepsilon^n(0)$.

Theorem 4.4 $\mathrm{d}(f,g) = \sup\{\varepsilon | S(f) \leq S(g) \oplus b_\varepsilon, S(g) \leq S(f) \oplus b_\varepsilon\}$.

Proof. From Theorem 4.1 it follows that $\Gamma(S(f)) \oplus b_\varepsilon) = \Upsilon(\Gamma(S(f)) \oplus \Gamma(b_\varepsilon))$ From simple geometrical reasons it follows also that $\Gamma(S(f)) \oplus \Gamma(b_\varepsilon) = \Gamma(S(f)) \oplus B_\varepsilon^{n+1}(0)$. Since $\Gamma(S(f))$ is closed (see Lemma 4.2), and $B_\varepsilon^{n+1}(0)$ is compact, then their Minkowski sum is topologically closed. Also, the sum of two umbrae is a pre-umbra, but when a set is a topologically closed pre-umbra, it is an umbra. Therefore $\Gamma(S(f) \oplus b_\varepsilon) = \Gamma(S(f)) \oplus B_\varepsilon(0)$. For the two functions u and v we have $u \leq v$ just when $\Gamma(u) \subseteq \Gamma(v)$. Then from the Corollary of Lemma 4.2, $\Gamma(S(f)) = \Upsilon(\Phi(f))$. Therefore the inequality $S(g) \leq S(f) \oplus b_\varepsilon$ is equivalent to the inclusion $\Upsilon(\Phi(g)) \subseteq \Upsilon(\Phi(f)) \oplus B_\varepsilon(0)$ which is satisfied evidently when $\Phi(g) \subseteq \Upsilon(\Phi(f)) \oplus B_\varepsilon(0)$. Then the proof of our theorem follows immediately from the definition of Hausdorff distance between sets. □

5. Conclusion and future research

The theorem which is proved in the last section gives the opportunity for easy computation of the distance between functions, especially in the case of discrete signals. It has been shown that the interval based computations have strong relations with the morphological operations. Using these relations, we can generalize the basic interval operations to more complex objects. Further we shall direct our efforts to express directed interval arithmetic in the framework of morphological operations and to generalize the interval algebra, esp. multiplicative operations for fuzzy numbers, L-fuzzy sets etc.

References

1. G. Alefeld, J. Herzberger. ***Introduction to interval computations***, Academic Press, New York 1983.
2. E. R. Dougherty. Application of the Hausdorff metric in gray-scale mathematical morphology via truncated umbrae. ***J. Visual Comm. Image Representation*** 2 (2): 177 – 187, 1991.
3. H. J. A. M. Heijmans. ***Morphological image operators,*** Academic Press, Boston 1994.
4. V. Kreinovich, V. Nesterov and N. A. Zheludeva. Interval methods that are guaranteed to underestimate (and the resulting new justification of Kaucher arithmetic. ***Reliable Computing*** 2 (2): 119 – 124, 1996.
5. S. M. Markov. Extended interval arithmetic. ***Compt. rend. Acad. Bulg. Sci.*** 30 (9):1239–1242, 1977.
6. S. M. Markov. On the algebra of intervals and convex bodies. ***J. UCS***, 1998, to appear.
7. G. Matheron. ***Random sets and integral geometry***, Wiley, New York 1975.
8. A. T. Popov. A relation between morphological and interval operations, ***Reliable Computing***, 4 (2), 1998.
9. Bl. Sendov. ***Hausdorff approximations***, Kluwer, Dordrecht 1990.
10. J. Serra. ***Image analysis and mathematical morphology***, Academic Press, London 1982.
11. J. Serra. Mathematical morphology for complete lattices. In J. Serra, editor, ***Image analysis and mathematical morphology***, vol. 2, Academic Press, London 1988.

TOGGLE MAPPINGS AND SOME RELATED TRANSFORMATIONS: A STUDY OF CONTRAST ENHANCEMENT

IVAN R. TEROL VILLALOBOS
CIDETEQ, S.C. Parque Tecnologico Quéretaro, Sanfandila-Pedro Escobedo, C.P. 76700-APDO 064, Querétaro, Mexico. email: famter@ciateq.mx

Abstract. This paper deals with the notion of toggle mappings. Toggle mappings have been proposed by Serra in a framework theory for morphological contrast operators. In this paper, we discuss some related nonincreasing transformations that are built using gradient criteria. These transformations can be considered toggle mappings, then we study them from a contrast point of view. By studying their stability, it is shown that the activity of these mappings exhibit a fixed zone growth. In particular, we discuss a family of sequential nonincreasing transformations. We focus on two important issues: contrast enhancement and image homotopy modification.

Key words: Toggle mappings, morphological gradient, morphological slope filters, contrast enhancement, geodesic transformations.

1. Introduction

Morphological contrast enhancement was studied by Meyer and Serra [1] in order to propose a framework theory for morphological contrast operators. A consequence of this work was the proposition of toggle mappings [2]. The interest in this paper deals with the contrast enhancement. Because of this, a class of filters proposed with the name of morphological slope filters (MSF) [3], that can be considered toggle mappings, is studied. They are built using morphological [4] gradient criteria which means that these filters are nonincreasing transformations. Because these filters modify the gradient on an image in order to contrast it, the gradient homotopy modification [5] is studied. In particular, we study a family of sequential MSF. This family enables us to obtain intermediate results that permit to control better the output image [6]. The stability of MSF is studied and relate them to the toggle mappings. We also study a generalization using a geodesic approach.

2. Toggle Mappings and Some Related Transformations

2.1. Toggle mappings.

The main idea of toggle mappings relies on the comparison of a real-valued function f to two patterns and to choose at each point x the closest value between the original function and the two patterns. The original idea was proposed by Kramer and Bruckner [7] (see section 3). Meyer and Serra used it to propose a framework theory for morphological contrast operators. As expressed in Serra's work [2], a mapping W which may modify f only when $f(x)$ lies inside the interval $[\varsigma(f_x), \eta(f_x)]$, provides us with two contrasted versions of f ($\varsigma(f_x)$ being clearer than f and $\eta(f_x)$ being darker). The idea of taking more than two primitives, and of adding conditions suggest the concept of toggle mappings expressed by he following definition.

Definition 1 Let F be the class of the functions $E \to \overline{R}$ ($\overline{Z}$ in this work) and F' that of the mappings F→F. Given a family Ψ_i of elements of F', one calls toggle mapping of primitives Ψ_i to any mapping W such that:

- At each point x, W_x equals one of the $\Psi_{i,x}$.

- The criterion that affects one Ψ_i to W_x at a given point x depends only on the primitives Ψ_i on the value W_x and on possible constants.

- In particular, if at point x, one of the Ψ_i coincides with the identity mapping I_x, then: $W_x = I_x = \Psi_x$.

The set of mappings F' is equipped with the following relationship:

$$W_1, W_2 \in F'; W_1 \succ W_2 \Leftrightarrow \begin{cases} I \wedge W_1 > I \wedge W_2 \\ I \vee W_1 < I \vee W_2 \end{cases}$$

The study reported in [1,2] revealed that, when toggle mappings are built, idempotence assumption is wished because it ensures that the enhancement process is well controlled. However, it is possible to weaken the idempotence assumption, by replacing it by some modality which controls the behavior under iteration. Two possible cases have been studied in [1]: a) One jump growth, and b) Fixed zone growth.

2.2. Nonincreasing filters using gradient criteria.

In [3] a new class of morphological nonincreasing filters was proposed. The idea to build them are the notions of morphological gradient criteria and idempotent mappings. Zones of weak contrast (weak gradient) are attenuated and zones of great contrast remain unchanged. This means, zones of great contrast on the image are retained (without changing the gray- level values) and enhanced by attenuating the other zones. Consider the definitions of internal and external gradients [4].

$$gradi_B(f(x)) = f(x) - \epsilon_B(f(x)) \qquad grads_B f(x) = \delta_B f(x) - f(x)$$

δ_B and ϵ_B, are, respectively, the morphological dilation and erosion transformations, B the elementary structuring element. From these equations and from the notion of toggle mappings, we define two operators,

$$\xi_\phi^{\epsilon_n}(f(x)) = \begin{cases} \epsilon_B(f(x)) & \text{if } gradi_B(f(x)) \leq \phi \\ f(x) & \text{otherwise} \end{cases}$$

$$\xi_\phi^{\delta_n}(f(x)) = \begin{cases} \delta_B(f(x)) & \text{if } grads_B(f(x)) \leq \phi \\ f(x) & \text{otherwise} \end{cases} \tag{1}$$

In here, we will working with eqns. (1), by iterating these operators. At the nth step when idempotence is reached,

$$\xi_\phi^{\epsilon_n} = \xi_\phi^{\epsilon_1}(\xi_\phi^{\epsilon_{n-1}}) \quad \text{and} \quad \xi_\phi^{\delta_n} = \xi_\phi^{\delta_1}(\xi_\phi^{\delta_{n-1}}) \tag{2}$$

Invariably, we will write $\xi_\phi^{\epsilon_\infty}$ or $\xi_\phi^{\epsilon_n}$ at the nth step when idempotence is reached. The filters $\xi_\phi^{\epsilon_\infty}$ and $\xi_\phi^{\delta_\infty}$ are anti-extensive and extensive transformations respectively, they are idempotent transformations at the nth step, but they are not increasing. Generally, we will present results for the $\xi_\phi^{\epsilon_\infty}$ filter case, but similar results can be expressed for the filter $\xi_\phi^{\delta_\infty}$.

3. Fixed Zone Growth: Stability of MSF.

When morphological erosion and dilation are used as primitives to build toggle mappings, the risk to degenerate the image by iteration of these operators can appear. In our case, both basic transformations are used to build toggle mappings but in a separated way. This option enables us to have some control of the output image. Let us to analyze the stability problem by studying the case of fixed zone growth.

Definition 2 The activity of a mapping Ψ is said to exhibit a fixed zone growth when

$$\Psi \cdot \Psi \succ \Psi \quad \textit{and} \quad \Psi_x = I_x \Rightarrow (\Psi \cdot \Psi)_x = I_x$$

This definition means that the successive iterated values at point x, namely Ψ_x^1, Ψ_x^2, $\ldots$, Ψ_x^n may only increase (or decrease), or stop. If Ψ does not modify the function value at point x, it remains unchanged as long as the operator is iterated. An example which shows that non-idempotent toggles generate jumps and oscillations is the Kramer and Bruckner transformation. This transformation is given by the following relationship.

$$W^{\delta\epsilon}(f(x)) = \begin{cases} \epsilon_{\lambda B}(f(x)), & \text{if } gradi_B(f(x)) \leq grads_B(f(x)) \\ \delta_{\lambda B}(f(x)), & \text{otherwise} \end{cases} \tag{3}$$

When expressing it by gradient criteria the interpretation becomes more complex. At each point x, the erosion transformation is selected if $gradi_B(f(x)) \leq grads_B(f(x))$ and the dilation transformation otherwise. In other words, we compare the gradient definitions to select the smallest gradient value between both definitions. The interpretation of this contrast operator is problematic because the supports of two gradient operators (i.e. the two sets of points of the domain of definition where these functions are strictly positive) are relatively disjoint. In contrast to this transformation which includes both gradient definitions, more consistent interpretations are obtained, working separately with both gradient definitions. In this case, the successive iterated values at point x, namely $\xi_{\phi,x}^{\epsilon_1}$, $\xi_{\phi,x}^{\epsilon_2}$, $\ldots$, $\xi_{\phi,x}^{\epsilon_n}$ decrease or stop. If $\xi_\phi^{\epsilon_1}$ does not modify the function value at point x, it remains unchanged as long as the operator is iterated. Suppose that at point x, $\xi_\phi^{\epsilon_1}(f(x)) = f(x)$ $(gradi_B(f(x)) > \phi)$. Because in a neighborhood B_x, using equation (1), the points $y \in B_x$ can only take the eroded or the original function value, then $\epsilon_B(\xi_\phi^{\epsilon_1}(f(x))) \leq \epsilon_B(f(x))$ and $gradi_B(\xi_\phi^{\epsilon_1}(f(x))) \geq gradi_B(f(x)) > \phi$. The function value f(x) remains unchanged, $\xi_\phi^{\epsilon_1}(\xi_\phi^{\epsilon_1}(f(x))) = \xi_\phi^{\epsilon_2}(f(x)) = f(x)$ if $gradi_B(f(x)) > \phi$. Then, we can define a set S_ϕ (great contrast points), associated to this operator, such that for all $x \in S_\phi \subset D_f$, the function value f(x) remains unchanged for all iteration.

Property 1 Let $x \in S_\phi \subset D_f$ be the points such that, $\xi_\phi^{\epsilon_1}(f(x)) = f(x)$ with $gradi_B(f(x)) > \phi$ then $\xi_\phi^{\epsilon_\infty}(f(x)) = f(x)$ $\forall x \in S_\phi$.

Also a set of weak contrast points can be defined that enable us to ensure that at each iteration, the function value of a weak contrast point decrease or stop.

Property 2 For all points x belongs to the complement of great contrast points set S_ϕ^C (the set of weak contrast points), we have that

$$\xi_\phi^{\epsilon_{k+1}}(f(x)) = \epsilon_B(\xi_\phi^{\epsilon_k}(f(x))).$$

At the first iteration of this operator, all points are classified in two categories and remain in them at each iteration. Similar comments can be expressed for $\xi_\phi^{\delta_n}$. Because at first step the points of image are classified (S_ϕ and S_ϕ^C), it is easy to show that for two given parameters $\phi_2 > \phi_1$, there exists inclusion relation.

Property 3 For two given parameters ϕ_1 and ϕ_2 such that $\phi_1 < \phi_2$, the support S_{ϕ_2} is included in support S_{ϕ_1}.

By thresholding the gradient image between $\phi_1 + 1$ and 255 to obtain S_{ϕ_1}, and between $\phi_2 + 1$ and 255 to obtain S_{ϕ_2} it is possible to observe this relation. Also, there are inclusion relations between the MSF depending on the ordering relation $\leq$ of the ϕ value. That means, for ϕ_1,ϕ_2 with $\phi_1 \leq \phi_2$,

$$\xi_{\phi_1}^{\epsilon_\infty} \geq \xi_{\phi_2}^{\epsilon_\infty} \qquad \text{and} \qquad \xi_{\phi_1}^{\delta_\infty} \leq \xi_{\phi_2}^{\delta_\infty} \tag{4}$$

Then, by iterating non idempotent mappings, two idempotent mappings are built. The behavior under iteration of the fixed zones growth, enables us to control the stability.

4. A Family of Sequential MSF.

For the filtering of the input image, we employ a family of MSF in a sequential way (see [6]). This family of filters depends on a family of parameters ϕ_i with $i \in S = \{1, 2, \ldots, m\}$ and $\phi_j < \phi_k; j < k$. Consider two given parameters ϕ_1, ϕ_2 such that $\phi_1 < \phi_2$. Property (3) and equation (4) show that filtering with parameter ϕ_2 is more discriminative than that with parameter ϕ_1. In order to obtain intermediate results the filters can be applied sequentially $\xi_{\phi_2}^{\epsilon_n}(\xi_{\phi_1}^{\epsilon_n}(f))$. For the sake of simplicity, let us to use the notation $\xi_{\phi_2}^{\epsilon_n}(\xi_{\phi_1}^{\epsilon_n}) = \xi_{\phi_1,\phi_2}^{\epsilon_n}$. Image in Fig. 1(d) has been obtained by applying the sequential filter $\xi_{\phi_1,\ldots,\phi_m}^{\epsilon_n}$ on the image in Fig. 1(a), using m=6 and $\phi_1 = 6, \phi_2 = 8, \phi_3 = 10, \phi_4 = 12, \phi_5 = 14, \phi_6 = 16$. By applying the filter $\xi_{\phi_j}^{\epsilon_n}$, with parameter ϕ_j, before applying $\xi_{\phi_k}^{\epsilon_n}$ with $\phi_j < \phi_k$, we change the contrast in some points of the original function in order to keep these ones when we apply $\xi_{\phi_k}^{\epsilon_n}$. Compare these filtered images with those in Figs. 1(b) and 1(c) corresponding to filters $\xi_{\phi_1}^{\epsilon_n}$ and $\xi_{\phi_6}^{\epsilon_n}$ with $\phi_1 = 6$, $\phi_6 = 16$. Then, the choice of a sequential family of filters leaves us a certain range of freedom. For $\phi_1 < \phi_2$, we have that, $\xi_{\phi_2}^{\epsilon_n} \leq \xi_{\phi_1,\phi_2}^{\epsilon_n} \leq \xi_{\phi_1}^{\epsilon_n}$ and $\xi_{\phi_1}^{\epsilon_n}(\xi_{\phi_2}^{\epsilon_n}) \leq \xi_{\phi_2,\phi_1}^{\epsilon_n} = \xi_{\phi_2}^{\epsilon_n}$. Moreover, let $\{\phi_i\}$ be a family of parameters with $i \in S = \{1, 2, \ldots, m\}$ and such that $\phi_j \leq \phi_k$ for $j < k$,

$$\xi_{\phi_m}^{\epsilon_n} \leq \xi_{\phi_1,\ldots,\phi_m}^{\epsilon_n} \leq \xi_{\phi_1}^{\epsilon_n} \tag{5}$$

By applying a family of filters with parameters ϕ_i between ϕ_1 and ϕ_m, the output image obtained by $\xi_{\phi_1,\ldots,\phi_m}^{\epsilon_n}$, contains some great contrast regions of the levels

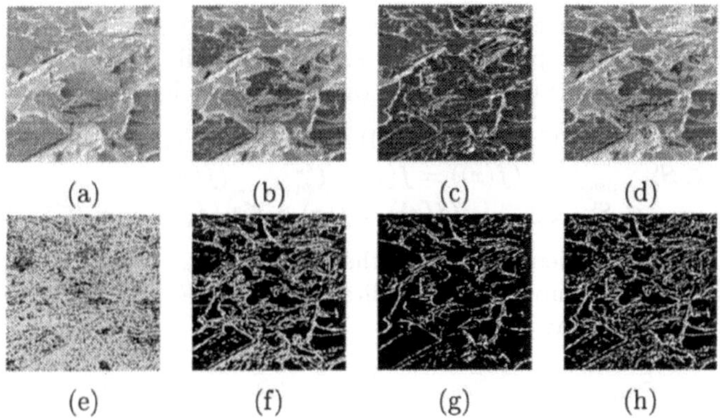

Fig. 1. (a) Original image, (b) filtered image $\xi^{\epsilon_n}_{\phi}$, $\phi = 6$, (c) filtered image $\xi^{\epsilon_n}_{\phi}$, $\phi = 16$, (d) filtered image $\xi^{\epsilon_n}_{\phi_1,\dots,\phi_6}$, $\phi = 6,..,\phi = 16$, (e) Gradient support of image (a), (f) Gradient support of image (b), (g) Gradient support of image (c), (h) Gradient support of image (d).

$\phi_1, \dots, \phi_m$ employed for its computation. This can be observed from the following,

$$\xi^{\epsilon_n}_{\phi_m} \le \xi^{\epsilon_n}_{\phi_{m-1},\phi_m} \le \xi^{\epsilon_n}_{\phi_{m-1}}, \dots, \le \xi^{\epsilon_n}_{\phi_2} \le \xi^{\epsilon_n}_{\phi_1,\phi_2} \le \xi^{\epsilon_n}_{\phi_1} \tag{6}$$

$$\xi^{\epsilon_n}_{\phi_1,\phi_2,\dots,\phi_m} \ge \xi^{\epsilon_n}_{\phi_2,\dots,\phi_m} \ge, \dots, \ge \xi^{\epsilon_n}_{\phi_{m-1},\phi_m} \ge \xi^{\epsilon_n}_{\phi_m} \tag{7}$$

$\xi^{\epsilon_n}_{\phi_{i-1},\phi_i}(f)$ preserves some great contrast regions from f that are eliminated by $\xi^{\epsilon_n}_{\phi_i}(f)$. On the other hand, if $\xi^{\epsilon_n}_{\phi_1,\phi_2,\dots,\phi_m}$ contains some great contrast regions of the levels $\phi_1, \phi_2, \dots, \phi_m$ employed for the sequential family, then $\xi^{\epsilon_n}_{\phi_2,\dots,\phi_m}$ does not contain details from $\xi^{\epsilon_n}_{\phi_1}$ and similar for $\xi^{\epsilon_n}_{\phi_k,\dots,\phi_m}$ that does not contain details from all $\xi^{\epsilon_n}_{\phi_j}$ with $\phi_j \le \phi_k$.

5. Homotopy Modification using MSF.

Eroding or dilating a region image where the contrast is low and leaves it unchanged where contrast is high will flatten low-contrast region, thereby enhancing contrast and modifying the gradient homotopy.

Property 4 Let ϕ be a given parameter and suppose that the idempotence is reached at the nth step. Then,

$$\forall x \in S_\phi \;\; gradi_B(\xi^{\epsilon_n}_{\phi}(f(x))) > \phi \;\text{ and }\; \forall x \in S^C_\phi \;\; gradi_B(\xi^{\epsilon_n}_{\phi}(f(x))) = 0$$

Moreover, using sequential MSF, the output image obtained by $\xi^{\epsilon_n}_{\phi_1,\dots,\phi_m}$, contains some great contrast regions of the levels $\phi_1, \phi_2, \dots, \phi_m$ employed for its computation. A low-contrast region at level ϕ_k of the family can be transformed in a high contrast region with regard to ϕ_m. Then, the intensity gray-level at level ϕ_k of the gradient of the output image can be amplified. Concerning the support of the gradient image, from property (4), in a filtered image $\xi^{\epsilon_n}_{\phi}$ the set defined by the support of the gradient is the same than that of points of great contrast S_ϕ. Using properties

(1), (2) and (3), we can express the next relations between the supports of the great and weak contrast points. For a sequential family of parameters ϕ_i with $i \in S = \{1, 2, \dots, m\}$ and such that $\phi_j < \phi_k$ for $j < k$ we have that,

$$\forall x \in S_{\phi_m} \subset S_{\phi_1,\dots,\phi_m} \subset S_{\phi_1}; \; \xi^{\epsilon_n}_{\phi_m}(f(x)) = \xi^{\epsilon_n}_{\phi_1,\dots,\phi_m}(f(x)) = \xi^{\epsilon_n}_{\phi_1}(f(x)) = f(x)$$
$$\forall x \in S_{\phi_1} \subset S^C_{\phi_1,\dots,\phi_m}; \; \xi^{\epsilon_n}_{\phi_1}(f(x)) = f(x) > \xi^{\epsilon_n}_{\phi_1,\dots,\phi_m}(f(x))$$
$$\forall x \in S_{\phi_1,\dots,\phi_m} \subset S^C_{\phi_m} \; \xi^{\epsilon_n}_{\phi_1,\dots,\phi_m}(f(x)) = f(x) > \xi^{\epsilon_n}_{\phi_m}(f(x)) = \epsilon^1_B(\xi^{\epsilon_{n-1}}_{\phi_m}(f(x)))$$

In Fig. 1(e) the gradient support of the original image is shown. Figs. 1(f) to 1(h) show the gradient supports of the filtered images (Figs. 1(b) to 1(d)). Using equation (7), we have that,

$$S_{\phi_1,\phi_2,\dots,\phi_m} \supseteq S_{\phi_2,\dots,\phi_m} \supseteq, \dots, \supseteq S_{\phi_{m-1},\phi_m} \supseteq S_{\phi_m}$$

where $S_{\phi_k,\phi_{k+1},\dots,\phi_m}$ is the support of great contrast points (or the gradient support) of $\xi^{\epsilon_n}_{\phi_k,\phi_{k+1},\dots,\phi_m}$. Similar relations can be expressed for equations (5) and (6). By observing the gradient supports of filtered images in Figs. 1(f) and 1(g) ($\xi^{\epsilon_n}_{\phi_1}$ and $\xi^{\epsilon_n}_{\phi_6}$, $\phi_1 = 6$, $\phi_6 = 16$), notice that some contours are preserved, especially for the first filtered image. The image in Fig. 1(h) shows well why sequential filters depending of a family ϕ_i with $i \in S = \{1, 2, \dots, m\}$ gives more interesting results. Some contours from the level $\phi_1 = 6$ of the family have been preserved after applying the filter $\xi^{\epsilon_n}_{\phi_1,\dots\phi_m}$. This inclusion relation between gradient supports can be used in order to minimize the effect of degradations for great values of the ϕ parameter. For example, using the gradient support image in Fig. 1(g) as contour marks, some contours from the gradient support image in Fig. 1(h) can be recovered. Now, let us to study the gray level histogram of the gradient images computed only in the gradient support. Property (4) shows that the gray-level histogram of the gradient of filtered images present well-defined zones. Observe the histograms in Fig. 2(a), obtained from the gradients of original image and filtered images. Note in Fig. 2(a), that for the gradient of images, Figs. 1(c) and 1(d) (curves G4 and G3), the gray-level number is zero for all histogram value less than $\phi_6 = 16$. The histogram curves for $gradi_B(\xi^{\epsilon_n}_{\phi_6})$ and $gradi_B(\xi^{\epsilon_n}_{\phi_1,\dots,\phi_m})$ (G4 and G3) begin at a gray level greater than 16. However, the histogram curve for $gradi_B(\xi^{\epsilon_n}_{\phi_6})$ shows that many contours have been eliminated of the image. When comparing the two histogram curves obtained from the gradient of the images in Figs. 1(b) and 1(d) (G2 and G3), notice that they essentially have the same behavior, apart from a horizontal shift. This shift (x-axis of the intensity gray level) is proportional to the ϕ value ($max\{\phi_i\}$ for a sequential family). This means that both filtered images have more or less similar zones but with a greater contrast for the sequential filtered image.

6. A Geodesic Approach.

From the analysis presented in Section (3) a geodesic mask can be defined by means of the set of great contrast points S_ϕ (or its complement). Then, by thresholding the gradient image, we obtain the geodesic mask S^C_ϕ, where the original image will be transformed by the erosion or dilation transformations. Moreover, for sequential MSF with parameters ϕ_i, $i \in S = \{1, 2, \dots, m\}$, a similar thresholding procedure can be realized at each level of the family. This main remark enables us to use a

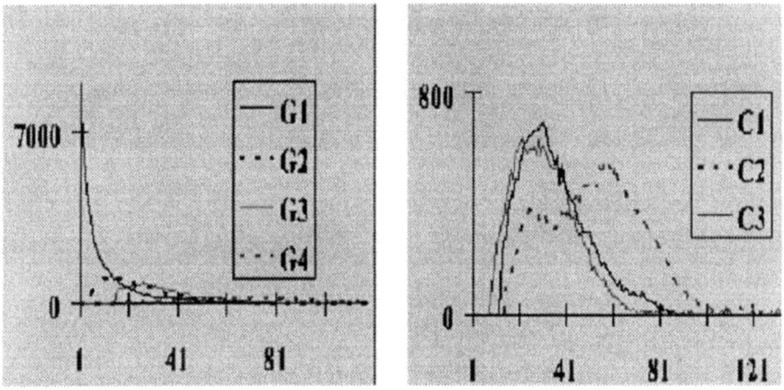

Fig. 2. Grey-Level histograms

geodesic mask for the generalization of eqns. (1) and (2). In fact, other gradient definitions or a filtered gradient image can be used to obtain the geodesic mask and to build other filters. This is not directly possible from equations (1). Consider the following relationship

$$\xi_{\phi}^{\delta_\lambda}(f(x)) = \begin{cases} \delta_{\lambda B}(f(x)), & \text{if } gradi_{\lambda B}(f(x)) \leq \phi \\ f(x), & \text{otherwise} \end{cases} \tag{8}$$

where λ is an homothetic parameter. This transformation has a different behavior than that of expression (1) which uses an elementary structuring element. For $\lambda > 1$ eqn.(8) defines a conditional transformation and not a geodesic one. Consider two points y and y' on an image, such that $(f(y') > f(y))$. The function value f(y') cannot propagate to the point y if there is not a path of weak contrast points linking y and y'. When $\lambda > 1$, a function value can propagate even if there is not a path of weak contrast points. Figs. 3(a) and 3(b) show the original and the filtered image using $\xi_{\phi}^{\delta_n}$ with $\phi = 9$. Fig. 3(c) illustrates the filtered image using $\xi_{\phi}^{\delta\lambda_n}$ with $\lambda = 3$ and $\phi = 14$. By comparing the filtered image $\xi_{\phi}^{\delta_n}$ in Fig. 3(b) to the filtered image $\xi_{\phi}^{\delta\lambda_n}$ in figure 3(c) we observe a greater contrast enhancement in $\xi_{\phi}^{\delta\lambda_n}$ than in $\xi_{\phi}^{\delta_n}$. In this case $(\xi_{\phi}^{\delta\lambda_n})$, some zones change their gray level values to another value, coming from a neighbor region, even if there is not a path of weak contrast points linking these zones. Now, let us briefly study the directional gradient case. The structuring element B is a segment of length l in a given direction α. Filters (eqns. (1) and (2)) are built using linear dilations (or erosions) in order to calculate directional gradients. In this example zones of weak contrast are attenuated using an elementary structuring element. To protect the edges from the noise, a filtering process in a perpendicular direction for each directional gradient is applied. This enables us to retain the most significant contours at each direction. Finally, the sup of directional gradients is computed. Image in Fig. 3(d) has been filtered using this procedure. A sequential family of these linear transformations with parameters ϕ_i (9,10,11,12,13) was used. When comparing the quality of filtered images in Figs.

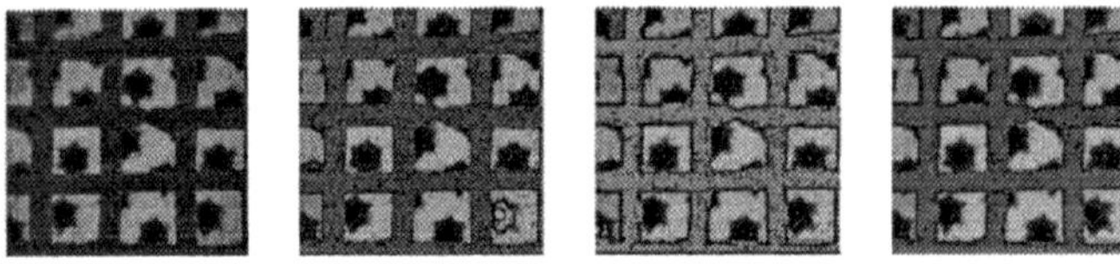

(a) (b) (c) (d)

Fig. 3. (a) Original image, (b) Filtered image $\xi_{\phi_1}^{\delta_n}$, (c) Filtered image $\xi_{\phi}^{\delta\lambda_n}$, (d) Filtered image $\xi_{\phi_1,\ldots,\phi_m}^{\delta_n}$

3(b),3(c) with that in Fig. 3(d), observe that the filtered image $\xi_{\phi}^{\delta_n}$ with $\phi = 9$ is well contrasted. However, some zones are merged. This fact leads to a loss of quality. In contrast, filtered image in Fig. 3(d) present well-defined contrast and well-defined zones. Now, let us to analyze the histogram curves of the gradients in Fig.2(b), computed from images in Figs. 3(b),3(d) (C1 and C3 respectively). When comparing these histograms, notice that they have the same behavior. However, the histogram C3 shows that more contours are retained with a greater contrast in image Fig. 3(d) than that of image in Fig. 3(b) (C1). By comparing the histogram of the filtered image in Fig. 3(c) (C2), we also observe a greater contrast enhancement than those of images in Fig. 3(b) and 3(d).

7. Conclusion

In this paper, we realize a study of contrast enhancement using a class of morphological nonincreasing filters (MSF) that can be considered toggle mappings. The main results are the gradient homotopy modification using sequential MSF and a geodesic approach that generalizes our results. A stability study is also presented and we show that the activity of these mappings exhibit a fixed zone growth.

Acknowledgements

The author would like to thank Diego Rodrigo. T. G. for much encouragement. This work was partially funded by the government agency CONACyT.

References

1. F. Meyer, J. Serra. Contrasts and activity lattice. *Signal Processing*, 16:303–317, 1989.
2. J. Serra. Toggle mappings. *Technical Report CMM School of Mines*,N–18/88/MM, 1989.
3. I. R. Terol-Villalobos. Nonincreasing filters using morphological gradient criteria. *Optical Engineering*, 35(11):3172–3182, 1996.
4. J. F. Rivest et al. Morphological gradients. *J. of Electronic Imaging*, 2:326–336, 1993.
5. F. Meyer, S. Beucher. Morphological Segmentation. *J. Visual Comm. Image Represent.*, 1:21–46, 1990.
6. I. R. Terol-Villalobos, J. A. Cruz-Mandujano. Contrast enhancement and image segmentation using a class of morphological nonincreasing filters. *To appear in J. of Electronic Imaging.*
7. H. P. Kramer, J. B. Bruckner. Iteration of non-linear transformation for enhancement of digital image. *Pattern Recognition* 7:53–58, 1975.

REPRESENTATION OF GRAY-SCALE WINDOWED OPERATORS

JUNIOR BARRERA AND EDWARD R. DOUGHERTY

Departamento de Ciência da Computação do IME/USP
Instituto de Matemática e Estatística. Universidade de São Paulo.

The Texas Center for Applied Technology
Department of Electrical Engineering. Texas A & M University.

Abstract. The automatic design of morphological operators may be modeled as a statistical optimization process in a space of operators. A limitation of this approach is that it may lead to extremely hard computational problems. An alternative to overcome this difficult is to fix a reduced family of useful operators as the search space. The restriction of the family of operators considered by a spatial constraint has been useful for designing binary image operators, however, it is not enough for designing gray-scale image operators. In this paper, we present a gray-scale constraint, that added to the spatial constraint, permits to control the size of the search space considered. Besides, we study characterizations and representations of gray-scale image operators that satisfy the spatial and gray-scale constraints.

Key words: gray-scale windowed operators, morphological representations

1. Introduction

A fundamental result on decomposition of operators in Mathematical Morphology was given by Matheron, in his classical book on Random Sets [11]. Matheron proved that any increasing and translation invariant (t.i.) set operator could be decomposed as the union of erosions. The ideas under this theorem originated an active branch of research in Mathematical Morphology. Matheron 's original theorem was refined, with the introduction of the notion of basis [10], and generalized to non necessarily increasing t.i set operators [3], to locally defined t.i. set operators [5], to non necessarily t.i. set operators [1] and to generic complete lattice operators [4].

The status of these operator decompositions changed from purely theoretical to effective computational tools with the work of Dougherty [6] [7], who approached the design of morphological operators by a statistical optimization model, where the search space is a family of operators represented by some decomposition structure.

A limitation of this approach is that it may lead to extremely hard computational problems. A way to overcome this difficult is the introduction of prior information. The simplest way for modeling prior information is to fix a reduced family of useful operators as the search space.

The family of *W-operators* is composed by image operators that are both spatialy locally defined (i.e., depend just on observations of the image through a fixed window)

and spatialy translation invariant. The optimization in the space of W-operators has been applied with success in the design of binary image operators. However, the same spatial constraint has not been enough for designing effective gray-scale image operators.

A W-operator is characterized by its *characteristic function* (i.e., a mapping defined from gray-scale window images to an integer range), that is, for each W-operator there exists a corresponding characteristic function and conversely.

When a characteristic function has finite domain and range, it is called a *computational function.* The cardinality of the family of W-operators characterized by computational functions is l^{m^n}, where l and m are, respectively, the cardinality of the input and output image ranges and n is the cardinality of the window considered. In the binary case, $l = m = 2$. Typically, in the gray-scale case, $l = m = 256$.

In this paper, we introduce a constraint for gray-scale image operators: the *gray-scale window.* This constraint is just an application of the notion of local definition on the image range. Under the spatial constraint, the operators depend just on observations of the images via a fixed window, similarly, under the gray-scale constraint, the operators depend just on observations via a fixed gray-scale integer interval, the gray-scale window. This constraint may be also stated on characteristic functions.

The family of *K-characteristic functions* is composed by the characteristic functions that satisfy both the gray-scale window and the gray-scale translation invariant constraints. The sub-family of the W-operators that are characterized by K-characteristic functions is called the family of *WK-operators.*

A K-characteristic function is characterized by a computational function, which has as range the gray-scale window itself. Therefore, denoting by k the cardinality of the gray-scale window, the cardinality of the family of WK-operators is k^{k^n}.

Comparing the cardinality of the families of W-operators and WK-operators, we note that they depend, respectively, on the spatial window and on both the spatial and gray-scale window. In other words, in the first case we have just the parameter n to control (l and m being fixed by the images considered), while in the second one we have the parameters n and k (with no dependence of the images considered).

Following this introduction, Section 2 presents a characterization of the W-operators in terms of characteristic functions. Section 3 gives a general decomposition for characteristic functions. Section 4 states the family of K-characteristic functions and studies their characterization. Section 5 presents a decomposition for WK-operators. Finally, Section 6 discuss the main contributions of this paper and some future steps of this research.

2. Characterization of W-operators

Let E be a non empty set that is an Abelian group with respect to a binary operation denoted by $+$. Elements of E will be denoted by x and z.

The *translation* of a subset W of E by the vector x is the subset given by

$$W_x = \{z \in E : z - x \in W\}.$$

Let L and M be two subsets of $\mathbb{Z}$. Elements of M will be denoted by y and t. The operations of *maximum* and *minimum* on subsets of M will be denoted, respectively,

$\vee$ and $\wedge$.

An *image* is a mapping from E to L or from E to M. Images will be denoted by f and g. The set of all images from E to L (respec., M) will be denoted $Fun[E, L]$ (respec., $Fun[E, M]$).

For any $f \in Fun[E, L]$ and $x \in E$, the translation of f by x is the function f_x given by, for any $z \in E$, $f_x(z) = f(z - x)$.

An *image operator* is a mapping from $Fun[E, L]$ to $Fun[E, M]$. Image operators will be denoted by the capital Greek letter Ψ.

Let W be a finite subset of E. An *image window* is a mapping from W to L. Image windows will be denoted by a, b and u. The set of all images window will be denoted $Fun[W, L]$.

For any $f \in Fun[E, L]$, the *restriction* to W of f is the image window f/W given by, for any $z \in W$, $(f/W)(z) = f(z)$.

The *restriction class* of f to W, denoted $\mathcal{F}_{f/W}$, is the family of images whose restriction to W gives f/W, that is,

$$\mathcal{F}_{f/W} = \{g \in Fun[E, L] : f/W = g/W\}$$

An image operator Ψ is called *spatially locally defined* in the window W if and only if (iff), for any $f \in Fun[E, L]$ and $x \in E$,

$$\Psi(f)(x) = \Psi(g)(x), \forall g \in \mathcal{F}_{f_{-x}/W}.$$

An image operator Ψ is called *spatially translation invariant* iff, for any $z \in E$,

$$\Psi(f_z) = \Psi(f)_z.$$

An operator that is both spatially locally defined in W and spatially translation invariant is called a *W-operator.*

A *characteristic function* is a mapping from $Fun[W, L]$ to M. Characteristic functions will be denoted by lower case Greek letters $\alpha, \beta, \gamma, \ldots$ When L and M are finite sets the characteristic functions will be also called *computational functions.*

An important property of the W-operators is that they can be characterized uniquely by the characteristic functions, that is, there exists a bijection between the set of W-operators and the set of characteristic functions [8] [9].

The characterization of a W-operator Ψ by its characteristic function ψ is given by, for any $f \in Fun[E, L]$ and $x \in E$,

$$\Psi(f)(x) = \psi(f_{-x}/W).$$

3. Sup-representation of characteristic functions

The results of the last section imply that the problem of representing W-operators reduces to the problem of representing their corresponding characteristic functions. In the following, we give a sup-representation that holds for any characteristic function.

Let $a, b \in Fun[W, L]$. We will say that a is *less than or equal to* b, denoted $a \leq b$, iff $a(x) \leq b(x), \forall x \in W$.

Let $a, b \in Fun[W, L]$ such that $a \leq b$. The subset of $Fun[W, L]$, given by

$$[a, b] = \{u \in Fun[W, L] : a \leq u \leq b\},$$

is called an *interval* of extremities a and b.

Let $\lambda_{a,b}$ denote the characteristic function from $Fun[W, L]$ to $\{0, 1\}$ given by, for any $u \in Fun[W, L]$,

$$\lambda_{a,b}(u) = \begin{cases} 1 \text{ if } u \in [a, b] \\ 0 \text{ otherwise} \end{cases}.$$

$\lambda_{a,b}$ is called a *sup-generating characteristic function.*

For any characteristic function ψ from $Fun[W, L]$ to M and for any $y \in M$, the *kernel* of ψ at y is the set given by

$$\mathcal{K}(\psi)(y) = \{u \in Fun[W, L] : y \leq \psi(u)\}.$$

Theorem 1 *If ψ is a characteristic function from $Fun[W, L]$ to M, then it can be represented by, for any $u \in Fun[W, L]$,*

$$\psi(u) = \bigvee \{y \in M : \bigvee \{\lambda_{a,b}(u) : [a, b] \subseteq \mathcal{K}(\psi)(y)\} = 1\}.$$

Proof. In this proof, let us denote

$$\zeta(u) = \bigvee \{y \in M : \bigvee \{\lambda_{a,b}(u) : [a, b] \subseteq \mathcal{K}(\psi)(y)\} = 1\}.$$

On one hand, for any $y \in M$ and $u \in Fun[W, L]$, $y \leq \psi(u)$ iff $u \in \mathcal{K}(\psi)(y)$, consequently, $\bigvee \{\lambda_{a,b}(u) : [a, b] \subseteq \mathcal{K}(\psi)(y)\} = 1$, since there exists $[a, b] \subseteq \mathcal{K}(\psi)(y)$ such that $u \in [a, b]$. Hence, $y \leq \zeta(u)$. On the other hand, for any $y \in M$ and $u \in Fun[W, L]$, $y \leq \zeta(u)$ implies that there exists $t \geq y$ such that there exists $[a, b] \subseteq \mathcal{K}(\psi)(t)$ such that $\lambda_{a,b}(u) = 1$, that is equivalent to $u \in [a, b]$, which implies that $y \leq t \leq \psi(u)$, by the definition of $\mathcal{K}(\psi)(t)$. ■

A similar result was presented by Banon [2] . In the following, we will present a simplification of this representation for computational functions.

In a collection of intervals, we say that an interval $[a, b]$ is maximal if there does not exist another interval $[a', b']$, distinct of $[a, b]$, such that $[a, b] \subseteq [a', b']$. We denote by $Max(\mathbf{X})$ the set of all maximal intervals of a collection of intervals $\mathbf{X}$.

The *basis* of a computational function ψ at y,denoted $\mathbf{B}(\psi)(y)$, is the collection of all maximal intervals contained in $\mathcal{K}(\psi)(y)$, that is,

$$\mathbf{B}(\psi)(y) = Max(\{[a, b] \subseteq Fun[W, L] : [a, b] \subseteq \mathcal{K}(\psi)(y)\}).$$

Theorem 2 *If ψ is a computational function from $Fun[W, L]$ to M, then it can be represented by, for any $u \in Fun[W, L]$,*

$$\psi(u) = \bigvee \{y \in M : \bigvee \{\lambda_{a,b}(u) : [a, b] \in \mathbf{B}(\psi)(y)\} = 1\}$$

β_ψ					
2	-2	1	2	2	2
1	-2	1	2	2	2
0	-2	1	2	2	2
-1	-2	1	1	1	1
-2	-2	-2	-2	-2	-2
	-2	-1	0	1	2

Fig. 1. The computational function β_ψ from $\{-2,-1,0,1,2\}^2$ to $\{-2,-1,0,1,2\}$.

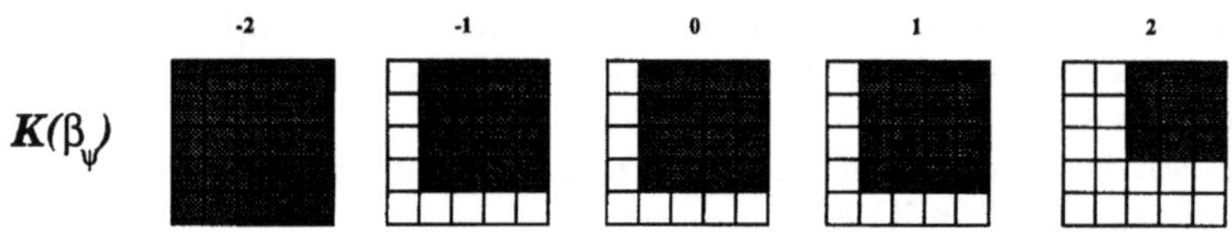

Fig. 2. The kernel of β_ψ.

Proof. For any $[a,b] \subseteq Fun[W,L]$ and $[a',b'] \subseteq Fun[W,L]$, if $[a,b] \subseteq [a',b']$, then, for any $u \in Fun[W,L]$, $\lambda_{a,b}(u) \leq \lambda_{a',b'}(u)$. Besides, for each $[a,b] \subseteq \mathcal{K}(\psi)(y)$ there exists $[a',b'] \in \mathbf{B}(\psi)(y)$ such that $[a,b] \subseteq [a',b']$. These two facts imply that the kernel can be substituted by the basis in the representation of Theorem 1. ■.

The representation of Theorem 2 is simpler than the one of Theorem 1, in the sense that it uses a smaller number of sup-generating functions. By this reason, it is called a minimal representation.

Figures 1 and 2 present, respectively, an example of a computational function and its kernel.

4. K-characteristic functions

In this section, we introduce the family of K-characteristic functions. The fundamental property of the elements of this family is that they are gray-scale locally defined and gray-scale translation invariant.

From now on, to simplify the presentation, we will consider just images with infinite range, that is, $L = M = \mathbb{Z}$.

Let $u \in Fun[W,\mathbb{Z}]$ and $y \in \mathbb{Z}$. The gray-scale *translation* of u by y is given by, for any $z \in W$,

$$(u+y)(z) = u(z) + y.$$

The subset $K = \{-\frac{k-1}{2}, ..., -1, 0, 1, ..., \frac{k-1}{2}\}$, where k is an odd positive integer, will be called a *gray-scale window*. The *windowing* of u at y by the gray-scale window K is the function $u/K_y \in Fun[W,K]$ given by, for any $z \in W$,

$$(u/K_y)(z) = \bigwedge \left\{ \bigvee \left\{ -\tfrac{k-1}{2}, u(z) - y \right\}, \tfrac{k-1}{2} \right\}.$$

A property of the windowing of u by K is that, for any $y, y' \in \mathbb{Z}$,

$$u/K_y = (u + y')/K_{y+y'}.$$

A characteristic function ψ from $Fun[W, \mathbb{Z}]$ to $\mathbb{Z}$ is called *locally defined* in K iff, for any $u \in Fun[W, \mathbb{Z}]$,

$$\psi(u) = u(o) + \beta_{u(o)}(u/K_{u(o)}),$$

where $u(o)$ is the image of u at the origin (i.e., the point $o = (0,0)$) and β_y, for any $y \in \mathbb{Z}$, is a computational function from $Fun[W, K]$ to K.

Note that ψ is characterized by computational functions from $Fun[W, K]$ to K. Hence, (infinite) characteristic functions that are locally defined are characterized by computational functions.

A characteristic function ψ from $Fun[W, \mathbb{Z}]$ to $\mathbb{Z}$ is called *translation invariant* iff, for any $y \in \mathbb{Z}$ and $u \in Fun[W, \mathbb{Z}]$,

$$\psi(u + y) = \psi(u) + y.$$

A characteristic function is called a *K-characteristic function* if it is both locally defined in K and translation invariant.

Theorem 3 *A K-characteristic function ψ is characterized by, for any $u \in Fun[W, \mathbb{Z}]$,*

$$\psi(u) = u(o) + \beta_\psi(u/K_{u(o)}),$$

where β_ψ is a computational function from $Fun[W, K]$ to K

Proof. As ψ is a K-characteristic function it is locally defined in K and translation invariant. By the translation invariance of ψ, for any $u \in Fun[W, \mathbb{Z}]$ and $y \in \mathbb{Z}$, $\psi(u+y) = \psi(u)+y$ and, by its local definition, $(u(o)+y)+\beta_{u(o)+y}((u+y)/K_{u(o)+y}) = u(o) + \beta_{u(o)}(u/K_{u(o)}) + y$. Thus, $\beta_{u(o)+y}((u + y)/K_{u(o)+y}) = \beta_{u(o)}(u/K_{u(o)})$ and, since $(u + y)/K_{u(o)+y} = u/K_{u(o)}$, we have that $\beta_{u(o)+y} = \beta_{u(o)}$. Therefore, for any $y \in \mathbb{Z}$, β_y is the same computational function, denoted β_ψ. ■

Figure 3 presents the K-characteristic function characterized by the computational function of Figure 1.

Furthermore, the mapping, for any $u \in Fun[W, \mathbb{Z}]$, $\psi(u) = u(o) + \beta_\psi(u/K_{u(o)})$, from the set of computational functions, defined from $\{f \in Fun[W, K] : f(o) = 0\}$ to K, to the set of K-characteristic functions constitutes a bijection. The inverse mapping is $\beta_\psi(u/K_{u(o)}) = \psi(u) - u(o)$.

This result holds by the following arguments. On one hand, by construction, $\psi(u) = u(o) + \beta_\psi(u/K_{u(o)})$ is locally defined in K and, by the equalities $\psi(u + y) = u(o) + y + \beta_\psi((u + y)/K_{u(o)+y}) = [u(o) + \beta_\psi((u + y)/K_{u(o)+y})] + y = [u(o) + \beta_\psi(u/K_{u(o)})] + y = \psi(u) + y$, translation invariant, hence, the mapping is surjective. On the other hand, as, for any computational function ϕ from $\{f \in Fun[W, K] : f(o) = 0\}$ to K, $\beta_\psi(u/K_{u(o)}) = [u(o) + \phi(u/K_{u(o)})] - u(o) = \phi(u/K_{u(o)})$, the mapping is injective. As the mapping is injective and surjective, it is a bijection.

Ψ

14	12	13	14	15	16
13	12	13	14	15	15
12	12	13	14	14	12
11	12	13	13	11	12
10	12	12	10	11	12
	10	11	12	13	14

Fig. 3. The characteristic function ψ from $\mathbb{Z}^2$ to $\mathbb{Z}$

5. Representation of WK-operators

In this section, we will introduce and study the representation of a sub-family of the W-operators: the WK-operators. The operators of this family are both windowed and translation invariant in both space and gray-scale domain.

More formally, an image operator is called a *WK-operator* iff it is a W-operator that is characterized by a K-characteristic function.

Theorem 4 *Let Ψ be a WK-operator, characterized by a characteristic function ψ, that is characterized by the computational function β_ψ. Ψ can be represented by , for any $f \in Fun[E, \mathbb{Z}]$ and $x \in E$,*

$$\begin{aligned}\Psi(f)(x) &= (f_{-x}/W)(o) + \\ &\bigvee \left\{y \in K : \bigvee \{\lambda_{a,b}((f_{-x}/W)/K_{(f_{-x}/W)(o)}) : [a,b] \in \mathbf{B}(\beta_\psi)(y)\} = 1\right\}.\end{aligned}$$

Proof. This representation is a consequence of the following results: the characterization of a W-operator by a characteristic function; the characterization of a K-characteristic function by a computational function; the minimal sup-representation of a computational function. ■

It is interesting to note that, for generic images, erosion and dilation by a structuring element b, from W to L, are not WK-operators. However, if the family of images f considered is such that, for any $x \in E$, $\max\{f(y) + b(y) : y \in W_x\} < f(x) + \frac{k-1}{2}$ and $\min\{f(y) - b(y) : y \in W_x\} > f(x) - \frac{k-1}{2}$, then erosion and dilation by b are WK-operators.

The computational function β_ψ that defines the WK-operator Ψ can be learned from observations of input-output pairs of images: $(f, \Psi(f))$. From these pairs of images are extracted the computational function examples: for any $x \in E$,

$$((f/W_x)/K_{f(x)}, \Psi(f)(x) - f(x)).$$

6. Conclusion

In this paper, we have introduced the family of WK-operators. The operators of this family, that is a subfamily of the family of W-operators, are spatially and gray-scale locally defined and translation invariant.

We have also presented a standard morphological representation for WK-operators. This representation has the following structure: *i*-characterizes the operator by a characteristic function that depends on the spatial window; *ii*-characterizes the characteristic function of *i* by a computational function that depends on the gray-scale window; *iii*-decompose the computational function of *ii* in a minimal supremum of sup-generating computational functions.

The family of WK-operators is quite adapted for the design of operators by statistical optimization, since the cardinality of this family is completely controlled by the choice of the spatial and gray-scale windows. In fact, the optimization problem is reduced to the determination of the parameters of the decomposition of *iii*.

The next natural step of this research is to use the WK-operators in the automatic design of gray-scale image operators by statistical optimization.

Acknowledgements

This work has been partially supported by National Science Foundation, grant MIP-9796118, and CNPq, grant ProTeM-CC/AniMoMat-680067/94-9. The figures of this paper were prepared by the student Roberto Hirata Junior.

References

1. G. Banon and J. Barrera. ***Set mapping decomposition by Mathematical Morphology.*** Oxford Series on Optical and Imaging Sciences 10, 1995. Chapter included in the book: 'Mathematical Morphology: Theory and Hardware', edited by R. M. Haralick.
2. G. J. F. Banon. Decomposition of gray-scale windowed image operators by mathematical morphology. In *Brazilian Workshop on Mathematical Morphology II*, IME-USP, Sao Paulo, 1996. Electronic Publication.
3. G. J. F. Banon and J. Barrera. Minimal Representations for Translation-Invariant Set Mappings by Mathematical Morphology. *SIAM J. Appl. Math.*, 51(6):1782–1798, December 1991.
4. G. J. F. Banon and J. Barrera. Decomposition of mappings between complete lattices by mathematical morphology, Part I. General lattices. *Signal Processing*, 30:299–327, 1993.
5. J. Barrera and G. P. Salas. Set Operations on Collections of Closed Intervals and their Applications to the Automatic Programming of Morphological Machines. *Journal of Eletronic Imaging*, 5(3):335–352, July 1996.
6. E. R. Dougherty. Optimal Mean-Square N-Observation Digital Morphological Filters I. Optimal Binary Filters. *CVGIP: Image Understanding*, 55(1):36–54, January 1992.
7. E. R. Dougherty. Optimal Mean-Square N-Observation Digital Morphological Filters II. Optimal Gray-scale Filters. *CVGIP: Image Understanding*, 55(1):55–72, January 1992.
8. E. R. Dougherty and D. Sinha. Computational mathematical morphology. *Journal of Signal Processing*, 38, 1994.
9. E. R. Dougherty and D. Sinha. Computational gray-scale mathematical morphology on lattices (a comparator-based image algebra)-part i: Architecture. *Journal of Real Time Imaging*, 1(1), 1995.
10. P. A. Maragos. *A Unified Theory of Translation-invariant Systems with Applications to Morphological Analysis and Coding of Images.* PhD thesis, School of Elect. Eng. - Georgia Inst. Tech., 1985.
11. G. Matheron. *Random Sets and Integral Geometry.* John Wiley, 1975.

COMPLETE ORDERING AND MULTIVARIATE MATHEMATICAL MORPHOLOGY

Algorithms and Applications

H. TALBOT, C. EVANS and R. JONES
CSIRO, Mathematical and Information Sciences
Locked Bag 17, North Ryde, N.S.W. 2113, AUSTRALIA
Email: {Hugues.Talbot,Ronald.Jones,Carolyn.Evans}@cmis.CSIRO.AU
Fax: (+61) 2 9325-3200

Abstract. The conventional treatment of mathematical morphology applied to multivariate data makes use of the partial ordering relationship between vectors. We find that morphological operators based on this ordering introduce vector values in the output image that are not present in the input image, which can produce counter-inituitive results.

However, artificial complete ordering can be imposed on multivariate data. In this paper we demonstrate that this is the only way to use morphological operators on multivariate images while introducing no new pixel values, and that this approach is quite useful when the imposed ordering is sensible, for example for colour data, when strong correlation between vector values can exist.

We also propose some simple and efficient algorithms based on this idea to extend conventional, univariate mathematical morphology operators to multivariate data, and we show some applications of this approach to real data.

Key words: Colour morphology, Ordering, Transforms, Vector preservation.

1. Introduction

Mathematical morphology (MM) has become popular in the image processing community due mainly to its proven utility and rigorous mathematical description. Initially designed by Matheron [6] to describe sizing procedures, it has since been extended to functionals and to the more general framework of complete lattices [7]. Multivariate MM using the framework of the complete lattices with partial ordering was introduced in [8]. A recent and more thorough treatment of multivariate mathematical morphology can be found in [2], where the use of transformations of the image data to get better ordering properties was first studied.

In the following section, we demonstrate how MM operators on a partially ordered lattice can create new vectors and why this can be unsatisfactory, especially for filtering applications. We show that imposing a total, invertible ordering on the vector space is the only way to use MM operators without generating new vectors. The third section shows some applications of this work, and the last section is concerned with algorithm implementation.

2. Approaches to multivariate mathematical morphology

In this section, we shall briefly discuss the relevant fundamentals of the complete lattice theory, within the framework of multivariate images. A more detailed discussion of the application of complete lattice theory to mathematical morphology can be found in [3], and [7].

2.1. The complete lattice of multivariate images

A set $\mathcal{L}$ with a partial ordering $\preceq$ is called a complete lattice if every subset $\mathcal{S}$ of $\mathcal{L}$ possesses a supremum $\bigvee \mathcal{S}$ and an infimum $\bigwedge \mathcal{S}$.

If $\mathcal{L}$ and $\mathcal{M}$ are two complete lattices, the set of all operators mapping $\mathcal{L}$ into $\mathcal{M}$ itself forms a complete lattice with the partial ordering given by:

$$f, g : \mathcal{L} \to \mathcal{M}, \forall x \in \mathcal{L}, f \preceq g \Longleftrightarrow f(x) \preceq g(x). \tag{1}$$

From this point, we restrict ourselves to the study of multi–dimensional, multi–spectral images. In other words, we only study the space $S_{n,p}$ of functions f from $\mathbf{R}^n$ with values in $\mathbf{R}^p$ ((n,p) positive integers). In this space, we shall only study the behaviour of the supremum operator. The properties and discussions on this operator apply equally well to the infimum operator.

An operator $f : \mathcal{L} \longrightarrow \mathcal{M}$ is *increasing* if $x \preceq y \Rightarrow f(x) \preceq f(y)$. It is called a *dilation* if it commutes with the supremum operator, i.e.: $f(\bigvee_i S_i) = \bigvee_i f(S_i)$ for all collections S_i of subsets of $\mathcal{L}$. It is called an *erosion* if, conversely, it commutes with the infimum operator.

2.2. The generation of new vectors

When dealing with multi–spectral images, the usual order relationship is the vector comparison, i.e.: $x = [x_0, x_1, ..., x_n], y = [y_0, y_1, ..., y_n] \in \mathbf{R}^n, x \preceq y \Longleftrightarrow \forall i, x_i \leq y_i$, where $\leq$ is the usual comparison operator in $\mathbf{R}$.

In this case, the usual supremum operator, here denoted $\bigvee_0$ operates in this way:

$$\forall x \in \mathbf{R}^n, [\bigvee_0 (f,g)](x) = \begin{bmatrix} \vee(f_0, g_0)(x) \\ \vdots \\ \vee(f_p, g_p)(x) \end{bmatrix} \tag{2}$$

Where $\vee$ is the usual supremum operation operating in $\mathbf{R}$. It can be shown (see [8]) that $S_{n,p}$ with any $n, p \in \mathbf{N}^*$, equipped with this supremum operation is a complete lattice, on which morphological operations can be defined as above.

Unfortunately, this supremum operator can generate new vectors not present in the input data, because the $\mathbf{R}$-based supremum operates independently on each component of the input vectors.

2.3. Non vector-preserving supremums

We say that a transformation δ from $S_{n,p}$ into itself does not preserve vectors in the following situation:

$$\exists x \in \mathbf{R}^n, \exists f \in S_{n,p}, \forall y \in \mathbf{R}^n, \delta(f(x)) \neq f(y) \tag{3}$$

Simple examples (see [4] and also section 3) show that, for example, $\bigvee_0$ does not preserve vectors.

2.4. VECTOR-PRESERVING SUPREMUMS

Let us define a family of supremum operators, denoted $\bigvee_\alpha$, which obeys the following conditions:

1. Let S be a compact set of $\mathbf{R}^n, S = \{x_1, x_2, ..., x_i \in \mathbf{R}^n\}$ Then:
$$\forall S, \exists x_i \in S \text{ such that } \bigvee_\alpha (S) = x_i.$$
2.
$$\forall x_1, x_2 \in \mathbf{R}^n, \bigvee_\alpha \{x_1, x_2\} = x_1 \text{ and } \bigvee_\alpha \{x_2, x_1\} = x_2 \Longleftrightarrow x_1 = x_2.$$
3.
$$\forall x_1, x_2, x_3 \in \mathbf{R}^n, \bigvee_\alpha \{x_1, x_2\} = x_1 \text{ and } \bigvee_\alpha \{x_2, x_3\} = x_2 \Longrightarrow \bigvee_\alpha \{x_1, x_3\} = x_1.$$

We have the following:

Property 1 *Operators satisfying conditions 1, 2 and 3 are the supremum operators in $\mathbf{R}^n$ which are vector-preserving.*

Indeed, condition 1 is equivalent to saying that any $\bigvee_\alpha$ is vector preserving. Moreover, conditions 2 and 3 ensure that each such operator define a different total ordering relation on $\mathbf{R}^n$. This can be seen easily by using the operator on pairs of vectors.

$\mathbf{R}^n$ equipped with one of these relations forms a complete lattice; moreover, such supremum operators do exist and are not unique. The simplest example is the *lexicographic order*: when comparing two vectors x_1 and x_2, first compare their first components using the usual ordering of real numbers. The vector with the larger first component is chosen. If the two first components are equal, the second components are compared, and so on, until two component values differ. The vector with the larger value of the latest component is the supremum of the two. If all components of the two vectors are equal, the two vectors are equal. By comparing the vector components in a different order a different supremum operator is defined.

2.5. VECTOR-PRESERVING MULTI–SPECTRAL MORPHOLOGICAL OPERATORS

Let us consider the δ operator on functions belonging to $S_{n,p}$ defined in this way:

Let B be a compact of $\mathbf{R}^n$, f be a function of $S_{n,p}$, x and y be elements of $\mathbf{R}^n$, then
$$\forall x, \delta_B(f(x)) = \bigvee_\alpha (f(x-y), y \in B).$$

Let us further consider a function g also element of $S_{n,p}$. Then we have:

$$\begin{aligned}\delta_B([\bigvee_\alpha(f,g)])(x) &= \bigvee_\alpha(\bigvee_\alpha(f(x-y),g(x-y)),y\in B)\\ &= \bigvee_\alpha(\bigvee_\alpha(f(x-y),y\in B),\bigvee_\alpha(g(x-y),y\in B))\\ &= \bigvee_\alpha(\delta_B(f)(x),\delta_B(g)(x))\end{aligned}$$

Since this operation commutes with $\bigvee_\alpha$, it is a *dilation*. Since the supremum operator is vector-preserving, this dilation is also. If $p = 1$ and we use the ordering of real numbers, this operation correspond to the usual dilation by a flat structuring element. The dual operation using a similar infimum operator is an erosion. We can define in the same way opening and closing which are all extensions of usual operators on the lattice of function with values in $\mathbf{R}$ and are all vector-preserving.

3. Applications

The image shown in Fig. 1a is part of a topographical colour map that has been digitized by a scanner (here only shown in grey-levels due to printing restrictions). Each colour in the map corresponds to a particular ground-cover type, for example water, swampland or pasture. Extra linework complicates this regional information where rivers, walking racks, location names and region boundaries have been marked. We wish to remove this linework from the image so that subsequent segmentation procedures can partition the image into regions by ground-cover type without oversegmenting.

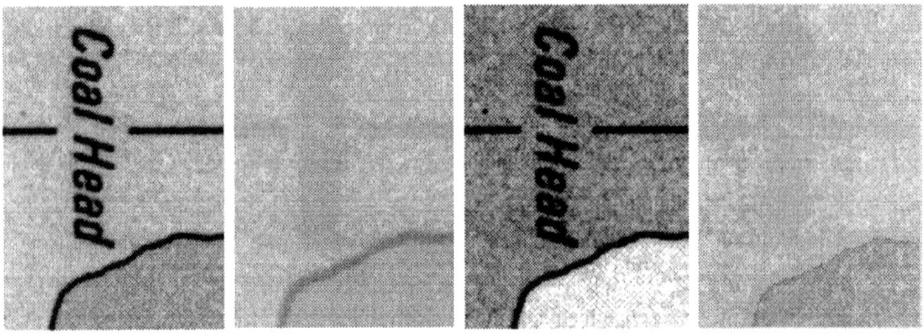

(a) (b) (c) (d)

Fig. 1. Processing a section of digitized colour map. (a) Input image. (b) Result using a closing on each band of the image separately. (c) Reordered image. (d) Colour result using a closing on the reordered image.

Linework in the colour maps is generally dark blue or black so the grey values of the linework in the red, green and blue components of the image are relatively low. The linework is also quite thin, typically less than 15 pixels wide. We will attempt to remove the linework by performing a closing based on the attribute of largest inscribing square [1] with a threshold of 15 pixels.

Fig. 1b shows the result of performing this closing in the traditional componentwise manner on the colour image. The linework has been removed, but a yellow contour has been introduced at the boundary between the blue and red regions. This is because there is a mismatch in the location of the boundary in the blue component of the closed image and the red and green components of the closed image.

Fig. 1c shows the reordered image. The ordering that was imposed was the lexicographic ordering defined in section 2. The red vector components were first compared, then the green and blue. Fig. 1d shows the result of performing the same attribute closing on the reordered image, after this image was converted back to colour. The dark linework has been replaced by the original blue vectors, which were ordered lower than the red vectors, giving a cleaner boundary between the blue and red regions.

While many different total orderings could have been imposed on the colour vectors, the advantage of the chosen imposed ordering is that it commutes with anamorphoses of the separate red, green or blue image values. This is important if a different scanner, with a different response to the red, green or blue illumination levels in the colour image, was used to digitize the colour maps.

4. Efficient Implementation and Algorithms

In this section, we discuss the implementation issues of vector-preserving operators.

4.1. Implementation schemes to avoid

The most straightforward way to implement vector-preserving operator is to hard-code the comparison function (or supremum operator) in the definition of the `max` operator. This should be avoided, however, as it makes experimenting with the comparison function relatively difficult.

Another way of implementing vector-preserving operators is to use an intermediate, invertible transform mapping $\mathbf{R}^n$ into $\mathbf{R}$. The grey-level morphology operations in the mapped space are then performed, and the result mapped back into $\mathbf{R}^n$. This is the approach that was suggested in [2], but in fact not all comparison operator can be associated with an explicit invertible transform in the continuous case. In the discrete case, this is always possible, but can require large look-up tables, because all possible vector comparison must be provided for. The size of such look-up tables would grow exponentially as the number of vector component increases.

4.2. Proposed algorithm

We now present an algorithm to impose a total invertible ordering on multivariate image data with any number of components of any data type. We simply assume that the vector-preserving supremum operator is available as a comparison function.

The basic idea of the proposed algorithm is to build an index of the vectors present in the image under study, and to order these vectors relative to each other. Because no new vector will be generated during morphological operations, we do not need to know the order of any other vectors than those present in the original image.

The difficulty with this approach is to be able to search for the individual vectors (or colours) present in an image, and to later sort them in an efficient way. A solution for this problem is illustrated in Fig. 2 and detailed hereafter.

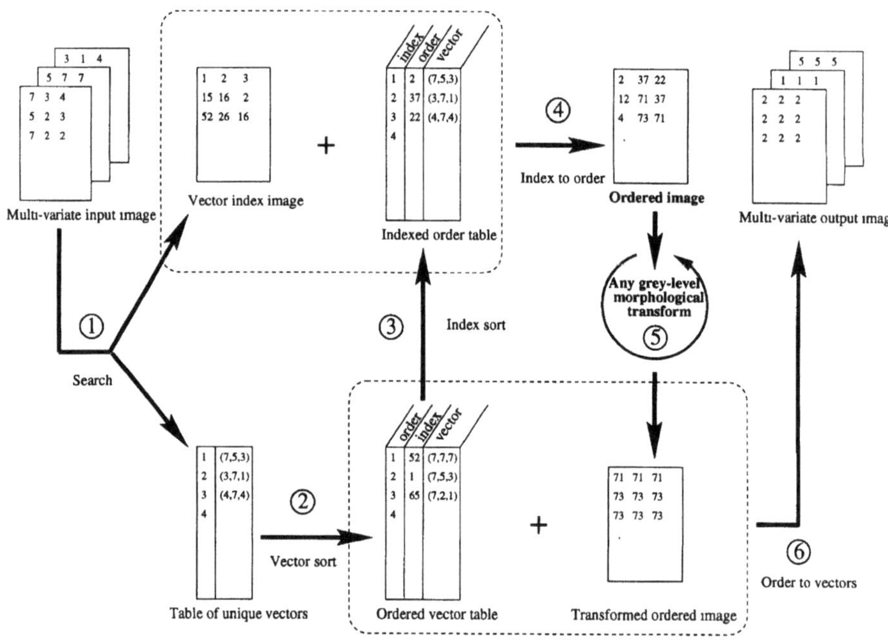

Fig. 2. Proposed algorithm for vector-preserving morphology operators. Steps 1 to 6 are detailed in the text.

Preliminary step: initialization: For searching and sorting vectors, we use *balanced binary trees* (BBTs). BBTs are a very useful family of fundamental data structures which allow searches to be performed in logarithmic time. It is also possible to use BBTs as a sorting mechanism with a complexity equivalent to the classic *quicksort* method. See [5] for a detailed description. The vector comparison function associated with this BBT will define the chosen ordering.
1 Searching and indexing unique vectors: The multivariate input image and the index image are scanned in parallel and in raster order. The BBT is used for searching unique input vectors. Each unique vector has a corresponding integer in the index image.
2 Sorting the vectors: Vectors are sorted using the content of the BBT at the end of the raster scan of the input image. Due to the nature of BBTs, this operation is efficient (it only requires reading the content of the BBT from left to right).
3 Sorting the indexes: Indexes are also sorted (using the quicksort algorithm for example) in order to build a two-way look-up table linking each vector present in the original image and its corresponding unique index with its relative order in the image.
4 Generation of the sorted image: The index image and the sorted vector look-up table constructed above are used to generate an *ordered image*. This produces a single-component, integer image which is an ordered map of the vector image. There are as many grey-levels in this image as there are different vectors in the input image.
5 Morphological transformations: At this point, any combinations of grey-level morphological operations can be performed on the ordered image, which is a normal grey-level image. In fact any conbination of grey-level operations that do not generate new grey levels can be used at this stage, i.e.: not only erosions and dilations (with flat structuring elements), but also rank order filter like medians, as well as area opening, reconstruction operators, etc. This steps generates the *transformed ordered image.*
6 Generation of the output vector image: The output of the morphological processing on the transformed ordered image is converted back to a vector image by simply scanning this image in raster order and replacing each pixel by its corresponding vector using the ordered vector table constructed during step 2 of the algorithm.

Steps 1 to 4 of this algorithm can be implemented in one program that saves the ordered image and the ordered vector table in two different files. Step 6 can be implemented in a second, independent program that generates the output vector image from the transformed ordered image and the saved ordered vector table. Standard implementations of MM operators can be used on the ordered image for step 5.

4.3. Memory requirements and complexity

In addition to the input and output images, the proposed algorithm requires only one intermediate integer image with the same dimensions as the input image. This intermediate image will contain, in succession, the vector indexes and the ordered vector map. In addition, a BBT with as many nodes as there are different vectors in the input image is built during the course of the algorithm. Finally, three arrays with as many entries as there are unique vectors in the input image are also required for the 2-way look-up-table.

The worst-case complexity of the proposed algorithm can be established by the following: if N is the number of pixels in the input image, and C is the number of individual vectors present in the image, ignoring step 5 of the algorithm which is application-dependent, all the other steps only require image scannings (complexity: $O(N)$), sorting on the number of vectors (complexity $O(C.log(C))$) and searching N times for unique vectors (complexity: $O(N.log(C))$).

In the worst case, C cannot be larger than N, and therefore the complexity of the algorithm is $O(N.log(N))$. However, in many real-world images, C is much smaller than N, and in this case the algorithm complexity is really $O(N)$.

5. Conclusion

We have explored the implications of imposing total orderings to multivariate data and we have shown that this allows the application of classical grey-level morphology to such images. This methodology has the benefit that the set of vectors in the output image is a subset of those present in the input image. This property is desirable for some filtering applications. We have presented an application of this work for image filtering. By using the vector-preserving property, we have proposed an efficient algorithm to convert between a multivariate image and an associated grey-level image. This mechanism allows the application of classical grey-level implementations of mathematical morphology operators to process multivariate data.

References

1. E. J. Breen and R. Jones. Attribute openings, thinnings and granulometries. *Computer Vision and Image Understanding*, 64(3):377–389, November 1996.
2. J. Goutsias, H. Heijmans, and K. Sivakumar. Morphological operators for image sequences. 62(3):326–346, 1995.
3. H. Heijmans and C. Ronse. The algebraic basis of mathematical morphology, part I: dilations and erosions. *Computer Vision, Graphics, and Image Processing*, 50:245–295, 1990.
4. R. Jones and H. Talbot. Morphological filtering for colour images with no new colours. In *IVCNZ'96*, pages 149–154, Lower Hutt, 1996. Image and Vision Computing New Zealand.
5. D. Knuth. *The art of computer programming. Volume 3: Sorting and searching.* Addison-Wesley, 1973.
6. G. Matheron. *Random sets and integral geometry.* Wiley, New York, 1975.
7. J. Serra. *Image Analysis and Mathematical Morphology - Volume II : Theoretical Advances.* Academic Press, London, 1988.
8. J. Serra. *Anamorphoses and function lattices (multivalued morphology)*, chapter 13, pages 483–523. Marcel Dekker, New York, 1993.

EXTENSION OF MORPHOLOGICAL OPERATIONS TO COMPLETE SEMILATTICES AND ITS APPLICATIONS TO IMAGE AND VIDEO PROCESSING

RENATO KRESCH
Hewlett-Packard Laboratories Israel
Technion City, Haifa 32000, Israel.
E-mail: renato@hp.technion.ac.il

Abstract. This paper first extends the scope of mathematical morphology from complete lattices to the more general complete semilattices. Then, two particular cases of semilattices, namely, difference and reference semilattices, are defined and considered. Finally, the use of morphological operators in the above semilattices is demonstrated in some video processing applications, like fast-motion detection, innovation extraction, and segmentation-based compression of sequences.

Key words: Complete semilattices, video processing, motion detection, innovation extraction, segmentation-based coding, skeleton, Top-Hat transform, morphological gradient.

1. Introduction

Mathematical morphology is theoretically founded on complete lattices [1], which are partially ordered sets where every subset has a least majorant (supremum) and a greatest minorant (infimum). The supremum and infimum of the complete lattice itself are called respectively greatest element (also universe) and least element (also zero element).

One important example is the complete lattice of grayscale functions, which is traditionally used for morphological image processing, where each image is a function in the lattice [1]. In this complete lattice, a function f is bigger than a function g, if $f(x) \geq g(x)$, for all x in their domain, and the infimum and supremum are the corresponding point-wise operators. The least and greatest elements in the above complete lattice are the all-$(-\infty)$ and the all-$(+\infty)$ functions, respectively.

In many applications, the existence of a least and a greatest elements is intuitive, in which case mathematical morphology on the corresponding complete lattice proves to be very useful. The case where we consider bright (high grayscale valued) objects in an image to form its foreground, and the dark (low grayscale valued) objects to form its background, for example, agrees with the above choice of partial ordering, which leads to the least and greatest grayscale functions defined above.

There are cases, however, that the existence of both a least and a greatest element does not necessarily agrees with one's intuition. For example, consider the difference between two real functions (images), which could represent the error of a prediction operation. In traditional mathematical morphology, this difference function would be processed in the same way as any other, that is, in the framework of the above lattice grayscale of functions, where the all-$(-\infty)$ function is the least element. However, it could be more useful to consider the *null function* to be the least difference-function, because it corresponds to the case when the two original functions are identical. But, then, there does not seem to exist one single greatest function, since both the all-$(-\infty)$ and the all-$(+\infty)$ functions could be considered equally "big".

In such situations, using a complete *semilattice* could be more useful and intuitive than using a complete lattice. A complete *inf semilattice* is a set for which every subset has an infimum (but not necessarily a supremum). It contains a least element (but not necessarily a greatest element). A complete *sup semilattice* is defined dually. Thus, a complete lattice is both a complete inf and sup semilattice.

In this work, we propose the extension of many useful morphological image processing tools for complete inf semilattices (the results are naturally extended to complete sup semilattices too, by duality). These tools include erosion, opening, adjoint dilation, Top-Hat extraction, gradient, and the skeleton.

Also, two types of complete semilattices that are potentially useful in image and video processing are introduced, namely, *difference* semilattices and *reference* semilattices. The former serves as a framework for processing difference images, and the later permits processing of an image *given* a reference one.

In addition, some potential video applications in the above complete semilattices are demonstrated. Specifically: Fast-motion detection, innovation extraction, and contour coding.

2. Morphology on Semilattices

In this section, we extend many of the basic operators of mathematical morphology to complete semilattices. Without loss of generality, we restrict ourselves to complete *inf* semilattices, since all concepts and results are valid in complete sup semilattices as well, by duality. The reader is assumed here to be familiar with the traditional theory of morphology on complete lattices, introduced by Serra in [1] (see [2, chapter 2] for a concise review).

We define *erosions* in complete inf semilattices in practically the same way as they are defined in complete lattices, as an operator that commutes with the infimum[1]. That is, an operator ε in a complete inf semilattice[2] $\mathcal{S}$ is an erosion iff, for all $\{X_i\} \subseteq \mathcal{S}$, $\varepsilon\left(\bigwedge_i X_i\right) = \bigwedge_i \varepsilon\left(X_i\right)$, where $\wedge$ denotes the infimum operator. It is easy to demonstrate that, like in complete lattices, erosions in complete inf semilattices are increasing (i.e., $X \leq Y \Rightarrow \varepsilon(X) \leq \varepsilon(Y)$, $\forall X, Y \in \mathcal{S}$).

The extension of algebraic openings to semilattices is also straightforward. A binary operator γ in an inf semilattice $\mathcal{S}$ is an *algebraic opening* iff it is idempotent ($\gamma\gamma(X) = \gamma(X)$, $\forall X \in \mathcal{S}$), increasing, and anti-extensive ($\gamma(X) \leq X$, $\forall X \in \mathcal{S}$).

Next, we wish to define *morphological* openings in complete inf semilattices. Notice however that, in complete lattices, the morphological opening associated to an erosion is defined using its adjoint dilation, but since there is no general definition of supremum in an inf semilattice, one cannot generally define dilations there[3]. Nevertheless, morphological opening can be completely extended to complete inf

[1] Erosions on complete lattices are required also to preserve the universe. Since there is no universe in a complete inf semilattice, erosions there are not required to preserve any element.

[2] Refer to the concise definition of complete semilattices above in the Introduction, or see [1] for more details.

[3] However, we do define limited versions of supremum and adjoint dilation in the sequel.

semilattices, using the erosion only. Specifically, in a complete inf semilattice $\mathcal{S}$, the morphological opening γ_ε associated to an erosion ε is defined, for any $X \in \mathcal{S}$, by:

$$\gamma_\varepsilon(X) \triangleq \bigwedge\{Y \in \mathcal{S} \mid \varepsilon(X) \leq \varepsilon(Y)\}. \tag{1}$$

Notice that (1) consists of a property of morphological openings in complete lattices. Here, we use this property as the definition for its semilattice counterpart.

It can be easily shown that, for any erosion in a complete inf semilattice, the associated morphological opening always exists, is unique, and is an algebraic opening. Moreover, the following holds: $\varepsilon\gamma_\varepsilon = \varepsilon$. These are fundamental properties in traditional morphology.

We now relate to basic morphological notions that are not directly extendible to inf semilattices, namely, supremum and dilation. Although impossible to define the above operations generally in inf semilattices, we provide limited versions of them.

Unless it is a complete lattice, a complete inf semilattice does not have a well-defined supremum for all its subsets. Nevertheless, a supremum does exist for *some* subsets of the semilattice. Given a complete inf semilattice $\mathcal{S}$, define the set $\mathcal{U}_\mathcal{S}$ of upper-bounded subsets of $\mathcal{S}$ as the set of all the subsets of $\mathcal{S}$ that have a majorant[4]. Supremum is defined over the elements of $\mathcal{U}_\mathcal{S}$, and only over them. Specifically, the least majorant (supremum) $\vee B$ of a set $B \subset \mathcal{S}$ exists iff $B \in \mathcal{U}_\mathcal{S}$, in which case it is equal to $\bigwedge\{Y \in \mathcal{S} \mid Y \geq X, \forall X \in B\}$. In summary, if a subset of the complete inf semilattice has a majorant, then it has a least majorant, and this is the supremum.

Let us consider now dilations in complete inf semilattices. We opt here not to define dilation generically (as an operator that commutes with the supremum), but we do define the *adjoint dilation* of a given erosion, restricted to a sub-domain of the semilattice. This sub-domain is the set of *ε-bounded elements of $\mathcal{S}$*, symbolized by $E_\mathcal{S}(\varepsilon)$, and defined as the set of elements in $\mathcal{S}$ that are smaller or equal to the erosion, by ε, of some element in $\mathcal{S}$. In other words, the adjoint dilation (defined next) can be applied only to those elements in the semilattice that are bounded by the erosion of some other element in the semilattice. The *adjoint dilation* of ε, denoted δ_ε, is defined as:

$$\delta_\varepsilon(X) \triangleq \bigwedge\{Y \in \mathcal{S} \mid X \leq \varepsilon(Y)\}. \tag{2}$$

Notice that (2) is a property of adjoint dilations in complete lattices. Similarly to morphological opening before, we adopt the property as the definition of the semilattice counterpart. It can be easily shown that the adjoint dilation is increasing in $E_\mathcal{S}(\varepsilon)$.

The following interesting property holds *for all* element X in a complete inf semilattice, and for every erosion defined there: $\gamma_\varepsilon(X) = \delta_\varepsilon\varepsilon(X)$. Notice that this closes a circle; the above relation, which is the definition of morphological opening in complete lattices, is obtained here as a general property. Let us stress that the above holds for all elements in the semilattice, since, for any X, $\varepsilon(X)$ is in the domain of δ_ε, $E_\mathcal{S}(\varepsilon)$.

[4] A majorant of a set is an element, not necessarily in the set, that is greater or equal each and every element of the set.

3. Semilattices for Image and Video Processing

In this section, we propose semilattices that are potentially useful for image and video processing, and show examples of basic morphological operations defined in them. In Section 4, we present a few applications of the operators defined in these semilattices.

3.1. Difference Semilattice

We call difference semilattice to a semilattice of real functions with Euclidean domain E, where the least element is the null function. The infimum operator in difference semilattices is defined point-wise, like in grayscale lattices. But here the infimum between two real numbers is set to be the minimum between them, if they are both positive, the maximum between them, if they are both negative, or zero, otherwise.

According to the above definition, the infimum between two functions f and g in a difference semilattice is actually the *center* [1, 3] of $(f, g, 0)$. Generically, the center of (f, g, h) returns for each point x the median of $\{f(x), g(x), h(x)\}$. The center is usually very useful for designing self-dual morphological filters [3].

As in any inf semilattice, an element (function) g in the difference semilattice is smaller than another one f, iff $f \wedge g = g$.

Let us consider the following operator, defined in a difference semilattice:

$$[\varepsilon(f)](x) \triangleq \bigwedge_{z \in B} f(x+z), \tag{3}$$

where B is a pre-defined structuring element in the Euclidean space, i.e., $B \subset E$. It is easy to show that ε is an erosion, and also that it is translation invariant. This erosion is the difference-semilattice counterpart of the erosion by a flat structuring element, defined in grayscale lattices. It is easy to show that the erosions in difference semilattices as defined above are self-dual, in the sense that they satisfy: $\varepsilon(-f) = -\varepsilon(f)$. I.e., negative and positive components are dealt with in the same way. This property is also extended to other operators in a difference semilattice (openings, gradients, skeletons, Top-Hats, etc).

As any morphological opening in a complete inf semilattice, the one associated with the above erosion is well defined by equation (1), and can be calculated as the composition of the erosion with its adjoint dilation. In this case, the adjoint dilation is given by: $[\delta(f)](x) = \bigvee_{z \in B} f(x-z)$. The supremum $\vee$ is also point-wise, and it returns for each point the maximum among the input values, if they are all positive, and the minimum among them, if they are all negative. Otherwise, the supremum is not defined.

Usually, the morphological opening in grayscale lattices is used for filtering out elements that are bright and either smaller or thinner than the structuring element. Similarly, the morphological opening in inf semilattices can be used for filtering small and thin elements of difference images, only that, due to the self-dual nature of the difference semilattice, it removes positive (bright) elements, when the function is positive, and negative (dark) elements, when it is negative.

Gradients and Top-Hat transforms [4] can also be defined in difference semilattices. However, only operators defined in terms of erosions and openings (and

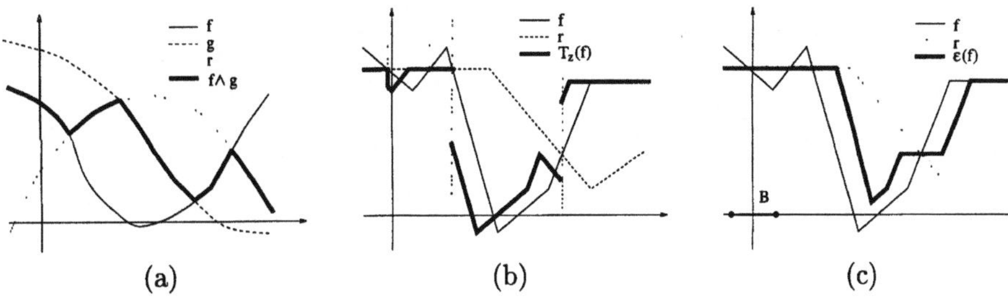

Fig. 1. Operations in a reference semilattice. (a) Infimum, (b) translation by the operator $T_z(\cdot)$, and (c) erosion by a flat structuring element.

not dilations and closings) can be defined there. Therefore, only *internal* gradients ($X - \varepsilon(X)$) and *white* Top-Hat transforms ($X - \gamma_\varepsilon(X)$) can be defined. Another useful morphological tool that can be extended to difference semilattices is the morphological skeleton [5, 6, 7, 2], since it is defined by means of erosions and openings only.

3.2. Reference Semilattice

Another example of complete inf semilattice, which we call *reference semilattice*, consists also of real functions, but its least element is now a given function $r(x)$, called *reference function*. In a reference semilattice $\mathcal{S}_r$, the infimum of two functions f and g are given by the center of (f, g, r). One can see that difference semilattices are particular cases of reference semilattice, where the reference function is null.

Unfortunately, the operator given by eq. (3) is not an erosion in a reference semilattice (unless $r(x)$ is constant). This is because translation does not commute with the infimum operation in $\mathcal{S}_r$. To correct this, we define a translation-like operator, $T_z(f)$ (where f is the input function, and z is the "amount" of translation), that does commute with the infimum. For each point x in the Euclidean domain, $[T_z(f)](x)$ returns $\max[f(x+z), r(x+z)]$, if the latter and $f(x)$ are greater than $r(x)$, $\min[f(x+z), r(x+z)]$, if the latter and $f(x)$ are smaller than $r(x)$, or $r(x)$, otherwise. The above "translation" acts like a regular translation, but with a restriction: Imaginary barriers are constructed at the points x where $f(x) = r(x)$, which the translation of $f(x)$ is not allowed to cross (see example in Fig. 1).

Now, the above "translation" permits us to define the following erosion in $\mathcal{S}_r$: $\varepsilon(f) \triangleq \bigwedge_{z \in B} T_z(f)$. Usually, this erosion has the following interesting property: It causes f to move towards r, since erosions usually cause elements to migrate to the least element. As we see in Section 4, this property can be useful for various image processing tasks, especially if f and r have approximately the same objects, slightly displaced with respect to each other, as is the case of adjacent frames in a video sequence. Erosions in reference semilattices, as defined above, can be seen as a generalization of the well-known *geodesic reconstruction* operation [8].

One can also specify the adjoint dilation and the morphological opening corresponding to the above erosion, and define gradients, Top-Hats, and skeletons. Because of lack of space, we leave the above outside the scope of this article.

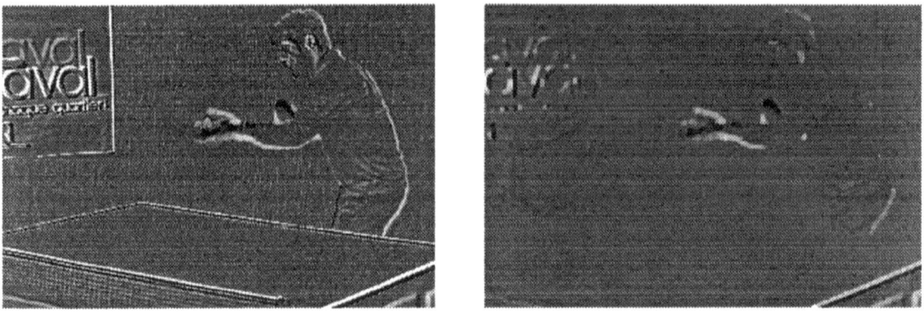

(a) (b)

Fig. 2. Example of fast-motion detection for a real sequence. (a) Difference between frames 65 and 66 of "Table-Tennis", and (b) result of the difference-semilattice scheme: Opening of (a) in a difference semilattice.

4. Applications

4.1. Fast-Motion Detection

Suppose we wish to detect *fast* motion in a video sequence, i.e., given 2 frames from a video sequence, we wish to detect the regions of the second frame that undertook a spatial displacement greater than a fixed number of pixels. This could be required for surveillance purposes, for instance.

We propose here a scheme that provides an approximated solution, using a difference semilattice. It consists of filtering the difference images (given by the pixel-by-pixel difference between adjacent frames) by means of an opening in the corresponding difference semilattice. In difference images, regions with high absolute value usually indicate motion or innovation (objects or regions that appear in only one of the two frames), and, in the former case, the width of the region is usually related to the amount of displacement (i.e., speed of motion). By filtering components with width smaller than a certain threshold out of the difference image, one gets an image which retain mainly innovation and "fast" motion (displacement higher than the filtering threshold). This kind of difference image filtering can be easily performed in a difference semilattice, by means of the opening by a disc-like flat structuring element, with diameter equal to the filtering threshold. As an example, consider frames 65 and 66 of the well-known "Table tennis" sequence (not shown here). The difference image is shown in Figure 2(a). The opening of the difference image by a 3×3 square is given in Figure 2(b). The components of the opening result can be roughly interpreted as those image components with displacement greater than 3 pixels, or innovation.

4.2. Innovation Extraction

Suppose now that we are interested in retaining innovation only. This could be required in applications such as inspection.

Consider, for example, the following specific application: A physician wishes to obtain an image showing the blood vessels in a certain region of the body of a

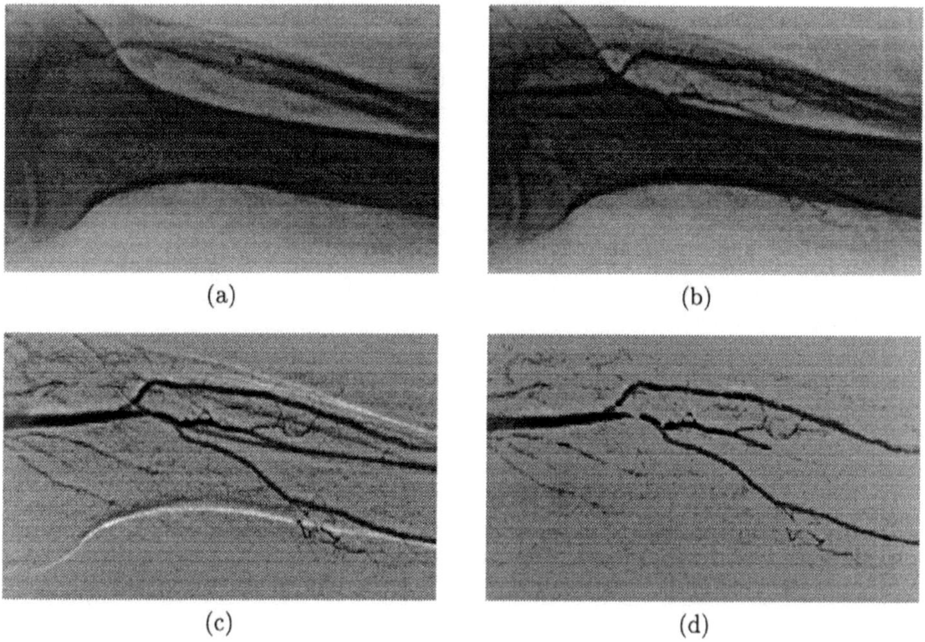

Fig. 3. Innovation extraction. (a),(b) X-ray pictures, taken respectively before and after injection of a contrasting agent, (c) difference image, containing the desired blood vessels, and undesired edge artifacts, (d) final result of the proposed scheme: Difference between (b) and the opening of (a) in the reference semilattice defined by (b).

patient. For this purpose, two X-ray images (f_1 and f_2) of the corresponding region are taken, respectively before and after a contrasting agent is injected and spread in the blood vessels (see Figures 3(a),(b)). The desired image consists of the innovation only. Notice that a simple difference between the images does not provide a good solution, because it contains not only the desired vessels, but also artifacts related to slight movements of the patient (see Figure 3(c)).

We propose now using a reference semilattice framework for solving the problem, with f_2 as the reference image. There, we consider a series of n erosions, ε^n, performed by a unitary disc-like structuring element. This series of erosions causes the edges in f_1 to move up to n pixels towards the edges in f_2. Next, we perform the adjoint series of n dilations, resulting in a morphological opening on the whole. This opening produces a kind of registration between the images, because the edges of f_1 that reach their counterparts in f_2 after the erosions do not return to their original positions after the dilations. Finally, the difference between f_2 and the resulting image provides us with the desired result (see Figure 3(d)).

4.3. Compression of Segmented Sequences

Recently, segmentation-based coding has been emerging as an efficient approach to very high compression of video sequences [9]. It consists of, first, segmenting each frame, and then coding segment contours separately from their interior information.

We propose here a skeleton decomposition in a reference semilattice, aimed as a first step for coding segment borders. This decomposition seems (in synthetic simulations) to efficiently represent the borders, taking into consideration correlation between frames.

We assume that we are given two frames of a sequence of images, f_1 and f_2, containing labeled segments only, without their texture or slow-transition contents.

All one needs to fully characterize a skeleton decomposition is a semilattice, and an erosion defined in it. The erosion determines its associated opening, and the decomposition family, which are used in Lantuéjoul's formula [5, 2], producing the skeleton representation.

Here, we choose as semilattice the reference semilattice defined by f_1 (differently from before, where we choose f_2 as the reference image). In this semilattice, we use the same erosion as in the previous application. The resulting skeleton decomposition is then applied to f_2. As it turns out, most of the skeleton points are attached to edges of f_1. For this reason, and because we assume a coding procedure where both the coder and the decoder already know f_1 when coding f_2, we expect to be able to efficiently code the positions of most of the skeleton points of f_2 (for instance, by an adapted version of the algorithm in [2, chapter 8]).

5. Conclusion

The basic operators of mathematical morphology are extended from complete lattices to complete semilattices. Moreover, two examples of semilattices potentially useful for image processing are presented, namely, difference and reference semilattices, where the former is a special case of the latter. In each of these semilattices, a set of morphological basic operators (erosion, adjoint dilation, and opening) is defined. Finally, three applications in which the above semilattices seem to be useful, according to preliminary simulations, are briefly described.

References

1. J. Serra, editor, *Image Analysis and Mathematical Morphology, Volume 2: Theoretical Advances.* New York: Academic Press, 1988.
2. R. Kresch, *Morphological Image Representation for Coding Applications*, D.Sc. Thesis, Dept. of Electrical Engineering, Technion, Israel, June 1995. Available at `http://www-sipl.technion.ac.il/publications/thesis/renato/renato.html`.
3. H.J.A.M. Heijmans, "Self-dual morphological operators and filters," *Journal of Mathematical Imaging and Vision*, Vol. 6, No. 1, pp.15-36, 1996.
4. E.R. Dougherty, *An Introduction to Morphological Image Processing*, SPIE Volume TT 9, Washington, 1992.
5. J. Serra, *Image Analysis and Mathematical Morphology.* London: Academic Press, 1982.
6. P. Maragos and R.W. Schafer, "Morphological Skeleton Representation and Coding of Binary Images," *IEEE Trans. ASSP*, Vol. 34, No. 5, pp. 1228–1244, Oct. 1986.
7. J. Serra, "Skeleton Decompositions," *SPIE Vol. 1769, Image Algebra and Mathematical Morphology*, San Diego, pp. 65–76, 1992.
8. P. Salembier and J. Serra, "Flat zones filtering, connected operators, and filters by reconstruction," *IEEE Transactions on Image Processing*, Vol. 4, No. 8, pp.1153-1160, 1995.
9. P. Salembier, P. Brigger, J.R. Casas, and M. Pardas, "Morphological operators for image and video compression," *IEEE Transactions on Image Processing*, Vol. 5, No. 6, pp.881-898, 1996.

FUZZY GEODESIC MATHEMATICAL MORPHOLOGY FROM FUZZY GEODESIC DISTANCE

ISABELLE BLOCH
Ecole Nationale Supérieure des Télécommunications, Département Images. CNRS URA 820. 46 rue Barrault. 75013 Paris, France.
Tel: +33 1 45 81 75 85 - Fax: +33 1 45 81 37 94 - Email: bloch@ima.enst.fr

Abstract. We define in this paper fuzzy geodesic mathematical morphology, based on a previous definition of fuzzy geodesic distances. The proposed definition includes the definition of fuzzy balls, based on the comparison of fuzzy numbers, and is valid in any dimension. The main properties of the basic operators are demonstrated. These new operations enhance the set of fuzzy morphological operators, leading to transformations of a fuzzy set conditionally to another fuzzy set.

Key words: Fuzzy geodesic distance, fuzzy balls, fuzzy mathematical morphology, fuzzy geodesic dilation and erosion.

1. Introduction

In mathematical morphology, an important set of operations is constituted by geodesic transformations [14, 9, 13, 10]. They are particularly useful in image processing, where transformations may have to be performed conditionally to a restriction of the spatial domain. Applications can be found for defining operators under reconstruction (e.g. filtering operators), in image segmentation, in pattern recognition, where operations have to be constrained by results of other transformations.

In this paper, we propose to define geodesic transformations on fuzzy sets. To our knowledge, this is the first attempt towards extending geodesic morphology to fuzzy sets, in contrary to Euclidean morphology, that has already motivated several works, e.g [5, 6, 15, 1, 7, 11, 16] and several others since these original works. The aim of this extension is to provide geodesic operators for image processing under imprecision, where image objects are represented as spatial fuzzy sets (i.e. an object in the image is represented as a fuzzy set through a membership function assigning to each point of the image a value in $[0, 1]$ representing its membership degree to the object; with such a representation spatial imprecision, for instance on the limits of the objects, is directly taken into account). We will consider mainly dilation and erosion, which are the two main operators, from which a large set of operators can be built, by iterating and combining these two basic ones.

Let us first introduce some notations and recall some definitions of geodesic morphology on binary sets. In the Euclidean case, the considered space S is equipped with the Euclidean distance d_E, and we denote by $D^r(Y)$ and $E^r(Y)$ the dilation and erosion of a set Y by a ball B_r of size r. In the geodesic case, transformations are defined conditionally to a reference set X. The considered distance is then the geodesic distance in X (i.e. the distance $d_X(x, y)$ from x to y is the length of the shortest path from x to y completely included in X). A geodesic ball of size r and center x is defined as:

$$B_X(x,r) = \{y \in X, d_X(x,y) \leq r\}. \tag{1}$$

Geodesic dilation and erosion of Y conditionally to X of size r are then defined as:

$$D_X^r(Y) = \{x \in \mathcal{S}, B_X(x,r) \cap Y \neq \emptyset\} = \{x \in \mathcal{S}, d_X(x,Y) \leq r\}, \tag{2}$$

$$E_X^r(Y) = \{x \in \mathcal{S}, B_X(x,r) \subset Y\} = X - D_X^r(X - Y). \tag{3}$$

We propose to generalize equations 1, 2 and 3 to fuzzy sets. This calls for extensions of geodesic distance and of geodesic balls to fuzzy sets. We have already proposed definitions for fuzzy geodesic distances in [2, 3]. We recall the definition having the best properties in Section 2. We propose in Section 3 a definition of fuzzy geodesic balls and we give its main properties. In Section 4 we derive definitions of fuzzy geodesic dilation and erosion, and present their algebraic properties.

2. Fuzzy geodesic distance

Let us denote by μ a fuzzy set defined on the space $\mathcal{S}$. Equivalently μ is a function from $\mathcal{S}$ into $[0,1]$ and, for each point x of $\mathcal{S}$, $\mu(x)$ is a value representing the membership degree of x to the fuzzy set μ. The geodesic distance between two points x and y represents the length of the shortest path between x and y that "goes out of μ as least as possible". We have proposed several formalisms for this notion. Here we recall only the one having the best properties. This definition relies on the degree of connectivity, as defined by Rosenfeld [12]. In the case where $\mathcal{S}$ is a discrete bounded space (as is usually the case in image processing), the degree of connectivity in μ between any two points x and y of $\mathcal{S}$ is defined as:

$$c_\mu(x,y) = \max_{L(x,y)} [\min_{t \in L(x,y)} \mu(t)], \tag{4}$$

where $L(x,y)$ denotes a path from x to y, constituted by a sequence of points of $\mathcal{S}$ according to the discrete connectivity defined on $\mathcal{S}$.

We denote by $L^*(x,y)$ a shortest path (in the Euclidean sense) between x and y on which c_μ is reached (this path, not necessarily unique, can be interpreted as a geodesic path descending as least as possible in the membership degrees), and we denote by $l(L^*(x,y))$ its length. Then we define the geodesic distance in μ between x and y as:

$$d_\mu(x,y) = \frac{l(L^*(x,y))}{c_\mu(x,y)}. \tag{5}$$

It corresponds to the weighted geodesic distance (in the classical sense) computed in the α-cut of μ at level $\alpha = c_\mu(x,y)$. In this α-cut, x and y belong to the same connected component (for the considered discrete crisp connectivity).

This definition satisfies the following set of properties (see [3] for the proof):

1. positivity: $\forall (x,y) \in \mathcal{S}^2, d_\mu(x,y) \geq 0$;
2. symmetry: $\forall (x,y) \in \mathcal{S}^2$, $d_\mu(x,y) = d_\mu(y,x)$;
3. separability: $\forall (x,y) \in \mathcal{S}^2$, $d_\mu(x,y) = 0 \Leftrightarrow x = y$;

4. d_μ depends on the shortest path between x and y that "goes out" of μ "as least as possible", and d_μ tends towards infinity if it is not possible to find a path between x and y without going through a point t such that $\mu(t) = 0$;
5. d_μ is decreasing with respect to $\mu(x)$ and $\mu(y)$;
6. d_μ is decreasing with respect to $c_\mu(x,y)$;
7. d_μ is equal to the classical geodesic distance if μ is crisp.

The triangular inequality is not satisfied, but from this definition, it is possible to build a true distance, satisfying triangular inequality, while keeping all other properties. This can be achieved in the following way:

$$d'_\mu(x,y) = \min_{t \in \mathcal{S}} \left[\frac{l(L^*(x,t))}{c_\mu(x,t)} + \frac{l(L^*(t,y))}{c_\mu(t,y)} \right].$$

These properties are in agreement with what can be required from a fuzzy geodesic distance, both mathematically and intuitively.

In this approach, the geodesic distance between two points is defined as a crisp number. It could be also defined as a fuzzy number, taking into account the fact that if the set is imprecisely defined, geodesic distances in this set can be imprecise too. This is left for future work, but all what follows can be applied for any definition of a fuzzy geodesic distance, as a crisp number or as a fuzzy number.

3. Fuzzy geodesic balls

3.1. DEFINITIONS

In this Section, we define fuzzy geodesic balls. Let us denote by $\beta_\mu(x,\rho)$ the fuzzy geodesic ball of center x and radius ρ, conditionally to μ. We define $\beta_\mu(x,\rho)$ as a fuzzy set on $\mathcal{S}$. Intuitively, given that x is in μ, for each point y the value $\beta_\mu(x,\rho)(y)$ should represent the fact that y belongs to μ and that it is at a geodesic distance in μ from x less than ρ. Therefore, $\beta_\mu(x,\rho)(y)$ has to be defined as a conjunction of three terms: the degree to which x belongs to μ, the degree to which y belongs to μ, and the degree $\delta(d_\mu(x,y) \leq \rho)$ to which $d_\mu(x,y) \leq \rho$, i.e.:

$$\forall y \in \mathcal{S},\ \beta_\mu(x,\rho)(y) = t[\mu(x), \mu(y), \delta(d_\mu(x,y) \leq \rho)], \tag{6}$$

where t is a t-norm.

Obviously, $\delta(d_\mu(x,y) \leq \rho)$ should be a decreasing function of $d_\mu(x,y)$. If we consider that ρ is a crisp number, we can choose a simple Heaviside function, such that:

$$\delta(d_\mu(x,y) \leq \rho) = \begin{cases} 1 & if \quad d_\mu(x,y) \leq \rho \\ 0 & else \end{cases} \tag{7}$$

Then we derive:

$$\forall y \in \mathcal{S},\ \beta_\mu(x,\rho)(y) = \begin{cases} t[\mu(x), \mu(y)] & if \quad d_\mu(x,y) \leq \rho \\ 0 & else \end{cases} \tag{8}$$

A fuzzy ball is therefore a subset of μ constituted of points y which are at a geodesic distance from x less than ρ, and whose membership degrees are bounded by $\mu(x)$.

This assumes that the value of interest ρ is precisely defined, which may appear as restrictive in a fuzzy context.

If we consider that some imprecision is attached to ρ, rather than considering it as crisp, then we can choose a smoother function, depending on the amount of imprecision attached to ρ. The problem with this approach is that the chosen decreasing function is somewhat arbitrary, and probably difficult to tune for specific applications.

Therefore, we propose another approach, where the link between this function and the imprecision of ρ is made more explicit. For this aim, we consider ρ as a fuzzy number. Defining $\delta(d_\mu(x,y) \leq \rho)$ calls then for the comparison of fuzzy numbers: $d_\mu(x,y)$ is less than ρ if $d_\mu(x,y)$ is equal to the minimum of $d_\mu(x,y)$ and ρ. The minimum between two fuzzy numbers has been defined in [8] as follows. Let ν and ν' be two fuzzy numbers. From the definition of fuzzy numbers, the α-cuts of ν and ν' are bounded intervals, denoted as $[{\nu_\alpha}^-, {\nu_\alpha}^+]$ and $[{\nu'_\alpha}^-, {\nu'_\alpha}^+]$ respectively. The minimum of ν and ν' is then the fuzzy number the α-cuts of which are:

$$\min(\nu,\nu')_\alpha = [\min({\nu_\alpha}^-, {\nu'_\alpha}^-), \min({\nu_\alpha}^+, {\nu'_\alpha}^+)]. \tag{9}$$

Here, we have to compute the minimum between $d_\mu(x,y)$ and ρ. Let us denote by $[\rho_0, \rho_2]$ the support of ρ and by ρ_1 its modal value. Applying equation 9 in the case where $d_\mu(x,y)$ is a crisp number, we come up with the following result, for all real number z:

– if $d_\mu(x,y) \leq \rho_0$:

$$\min(d_\mu(x,y),\rho)(z) = \begin{cases} 1 \ \ if \ \ z = d_\mu(x,y) \\ 0 \ \ if \ \ z \neq d_\mu(x,y) \end{cases} \tag{10}$$

– if $\rho_0 \leq d_\mu(x,y) \leq \rho_1$:

$$\min(d_\mu(x,y),\rho)(z) = \begin{cases} \rho(z) & if \ \ z < d_\mu(x,y) \\ 1 & if \ \ z = d_\mu(x,y) \\ 0 & if \ \ z > d_\mu(x,y) \end{cases} \tag{11}$$

– if $\rho_1 \leq d_\mu(x,y) \leq \rho_2$:

$$\min(d_\mu(x,y),\rho)(z) = \begin{cases} \rho(z) & if \ \ z \leq d_\mu(x,y) \\ 0 & if \ \ z > d_\mu(x,y) \end{cases} \tag{12}$$

– if $d_\mu(x,y) \geq \rho_2$:

$$\min(d_\mu(x,y),\rho)(z) = \rho(z). \tag{13}$$

To have $d_\mu(x,y) \leq \rho$ is equivalent to $d_\mu(x,y) = \min(d_\mu(x,y),\rho)$, or to $d_\mu(x,y) \subset \min(d_\mu(x,y),\rho)$ and $\min(d_\mu(x,y),\rho) \subset d_\mu(x,y)$. This last form can be easily translated into fuzzy terms, in a way similar that the one used in [6], as:

$$\delta(d_\mu(x,y) \leq \rho) = t[\inf_z T[c(d_\mu(x,y))(z), \min(d_\mu(x,y),\rho)(z)],$$

$$\inf_z T[d_\mu(x,y)(z), c(\min(d_\mu(x,y),\rho))]],$$

where t is a t-norm, c a fuzzy complementation (typically $c(z) = 1 - z$) and T a t-conorm, dual of t with respect to c. This leads to the following result:

$$\delta(d_\mu(x,y) \leq \rho) = \begin{cases} 1 & if\ d_\mu(x,y) \leq \rho_0 \\ \inf_{z \leq d_\mu(x,y)} c(\rho)(z) & if\ \rho_0 \leq d_\mu(x,y) \leq \rho_1 \\ 0 & if\ d_\mu(x,y) \geq \rho_1 \end{cases} \tag{14}$$

Since ρ is increasing on $[\rho_0, \rho_1]$ (since it is a fuzzy number), we obtain:

$$\delta(d_\mu(x,y) \leq \rho) = \begin{cases} 1 & if\ d_\mu(x,y) \leq \rho_0 \\ c(\rho)(d_\mu(x,y)) & if\ \rho_0 \leq d_\mu(x,y) \leq \rho_1 \\ 0 & if\ d_\mu(x,y) \geq \rho_1 \end{cases} \tag{15}$$

Finally, we obtain:

$$\forall y \in \mathcal{S},\ \beta_\mu(x,\rho)(y) = \begin{cases} 0\ if\ d_\mu(x,y) \geq \rho_1 \\ t[\mu(x), \mu(y), c(\rho)(\delta(d_\mu(x,y))]\ else \end{cases} \tag{16}$$

If t is chosen for instance as the product, $\mu(x)$ and $\mu(y)$ appear as weighting factors.

This definition may appear as severe. A more "optimistic" definition can be derived from the relationship "to the left of", as introduced in [4], but applied here in a simpler 1D case, as:

$$\delta(d_\mu(x,y) \leq \rho) = \begin{cases} 1 & if\ d_\mu(x,y) \leq \rho_1 \\ c(\rho)(d_\mu(x,y)) & if\ d_\mu(x,y) \geq \rho_2 \end{cases} \tag{17}$$

These definitions are illustrated in Figure 3.1.

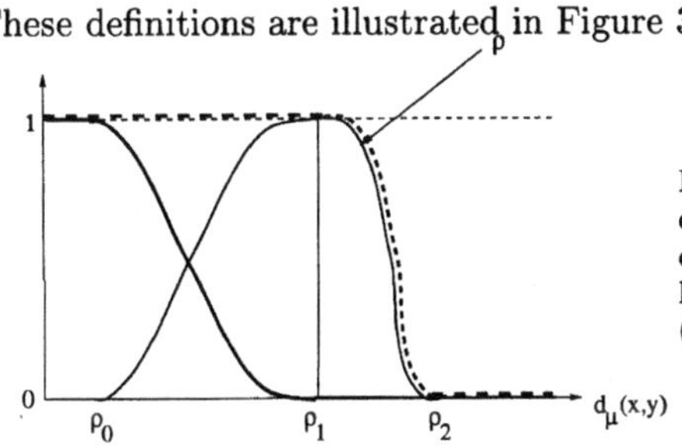

Fig. 1. Illustration of the definition of $\delta(d_\mu(x,y) \leq \rho)$ using the minimum of two fuzzy numbers (continuous dark line) and using the relation "left to" (dashed line).

The proposed definition of a fuzzy geodesic ball applies directly to any other definition of the fuzzy geodesic distance, represented either as a crisp number, or as a fuzzy number. Also the following properties hold, and are not restricted to the particular form of the fuzzy geodesic distance we use.

3.2. Properties

The proposed definitions of fuzzy geodesic balls share the following properties:

1. $\beta_\mu(x,\rho)(x) = \mu(x)$;
2. $\beta_\mu(x,\rho)(y) \leq \mu(x)$;
3. $\beta_\mu(x,\rho)(y) \leq \mu(y)$;
4. if ρ is a crisp number, $\delta(d_\mu(x,y) \leq \rho)$ is binary, and equal to 1 iff $d_\mu(x,y) \leq \rho$;

5. if μ and ρ are crisp, then $\beta_\mu(x,\rho)$ is the crisp geodesic ball, therefore compatibility with the binary case is achieved;
6. spatial invariance: $\beta_\mu(x,\rho)$ is invariant by translation and rotation;
7. monotony with respect to ρ: if ρ and ρ' are such that $\rho_1 \leq \rho'_1$ and $\rho \leq \rho'$ on $[\rho_0, \rho_1]$ (which is typically the case if ρ' is just a translation of ρ), then $\beta_\mu(x,\rho) \leq \beta_\mu(x,\rho')$, expressing that a fuzzy geodesic ball is included in a fuzzy geodesic ball of same center and "larger" radius;
8. a fuzzy geodesic ball is always included in the Euclidean ball of same radius.

These properties are the fuzzy equivalents of the properties of crisp geodesic balls. This shows the consistency of the proposed extension.

4. Fuzzy geodesic mathematical morphology

4.1. Translating binary expressions into fuzzy ones

In order to extend geodesic morphological operations to fuzzy sets, we translate equations 2 and 3 into fuzzy terms. The idea is to replace formally every binary concept by its fuzzy equivalent. The following table summarizes the main definitions of fuzzy equivalents:

crisp concept	equivalent fuzzy concept
set X	fuzzy set / membership function μ
complement of a set	fuzzy complementation c
intersection $\cap$	t-norm t
union $\cup$	t-conorm T
existence $\exists$ / universal symbol $\forall$	supremum / infimum

From these equivalences, more complex relationships can be translated. For instance, the expression $A \subset B$, which is equivalent to $A^C \cup B = \mathcal{S}$ is translated as:

$$\inf_{x \in \mathcal{S}} T[c(\mu_A)(x), \mu_B(x)],$$

which is a number in $[0,1]$ representing the degree to which the fuzzy set μ_A is included in the fuzzy set μ_B.

Such translations have already been used for defining Euclidean morphological operators [6], leading to the following generic expressions for the dilation and erosion of a fuzzy set μ by a fuzzy structuring element ν:

$$\forall x \in \mathcal{S},\ D(\mu,\nu)(x) = \sup_{y \in \mathcal{S}} t[\mu(y), \nu(y-x)], \tag{18}$$

$$\forall x \in \mathcal{S},\ E(\mu,\nu)(x) = \inf_{y \in \mathcal{S}} T[\mu(y), c(\nu(y-x))]. \tag{19}$$

These definitions have good properties in terms of both mathematical morphology and fuzzy sets, as shown in [6]. Therefore we based our work on these definitions. However, the proposed construction of geodesic operators could be applied to any other definition.

4.2. Definitions of basic fuzzy geodesic operators

In the geodesic case, we use similar rules to translate equations 2 and 3 into fuzzy terms. This leads to the following definitions of fuzzy geodesic dilation and erosion of μ' conditionally to μ:

$$\forall x \in \mathcal{S},\ D^{\rho}_{\mu}(\mu')(x) = \sup_{y \in \mathcal{S}} t[\beta_{\mu}(x, \rho)(y), \mu'(y)], \tag{20}$$

$$\forall x \in \mathcal{S},\ E^{\rho}_{\mu}(\mu')(x) = \inf_{y \in \mathcal{S}} T[c(\beta_{\mu}(x, \rho)(y)), \mu'(y)]. \tag{21}$$

From these two basic operators, other can be defined, as is done in classical morphology. For instance fuzzy geodesic opening and closing are simply defined as $O^{\rho}_{\mu}(\mu') = D^{\rho}_{\mu}[E^{\rho}_{\mu}(\mu')]$ and $C^{\rho}_{\mu}(\mu') = E^{\rho}_{\mu}[D^{\rho}_{\mu}(\mu')]$.

4.3. Properties

The proposed definitions of fuzzy geodesic dilation and erosion have the following properties, which are similar to the properties of classical geodesic operators:

1. compatibility with the crisp case: if μ, μ' and ρ are crisp, the definitions are equivalent to the binary geodesic operators;
2. duality with respect to complementation:
$$\forall x \in \mathcal{S},\ D^{\rho}_{\mu}[c(\mu')](x) = c[E^{\rho}_{\mu}(\mu')](x),$$
assuming that the t-norm and the t-conorm used in dilation and erosion respectively are dual with respect to the complementation c;
3. the result of the geodesic dilation of μ' conditionally to μ is included in μ:
$$\forall x \in \mathcal{S},\ D^{\rho}_{\mu}(\mu')(x) \leq \mu(x),$$
expressing that the transformed set stays inside the conditioning set;
4. invariance with respect to geometrical transformations, and local knowledge property;
5. increasingness:
$$\mu' \leq \mu'' \Rightarrow \forall x \in \mathcal{S}, D^{\rho}_{\mu}(\mu')(x) \leq D^{\rho}_{\mu}(\mu'')(x);$$
6. restricted extensivity:
$$\forall x \in \mathcal{S},\ D^{\rho}_{\mu}(\mu')(x) \geq t[\mu(x), \mu'(x)];$$
7. interpretation: rewriting the expression of fuzzy geodesic dilation leads to:
$$D^{\rho}_{\mu}(\mu')(x) = \sup_{y \in \mathcal{S}} t[t[\mu(x), \mu(y), \delta(d_{\mu}(x, y) \leq \rho)], \mu'(y)]$$
and, since a t-norm is commutative, associative and increasing:
$$D^{\rho}_{\mu}(\mu')(x) = t[\mu(x), \sup_{y \in \mathcal{S}} t[\mu(y), \mu'(y), \delta(d_{\mu}(x, y) \leq \rho)]].$$
This represents the intersection of μ with the dilation of μ' performed on a neighborhood containing the points y of μ (the conditioning aspect) such that $d_{\mu}(x, y) \leq \rho$ (the geodesic distance aspect). This interpretation is in complete agreement with what is expected from a geodesic dilation.

5. Conclusion

We presented in this paper an original way to define fuzzy geodesic morphological operators, based on fuzzy geodesic distance. We proposed definitions of these operators and of fuzzy geodesic balls that have nice features: they deal with a direct representation of spatial imprecision in the fuzzy sets, they are consistent with existing binary definitions, they have good formal properties, in agreement with the formal properties of crisp definitions and with intuitive requirements.

Future works aim at investigating further properties of these definitions, at comparing the different possible instantiations of them, and at evaluating their power in image processing problems under imprecision.

References

1. B. De Baets. Idempotent Closing and Opening Operations in Fuzzy Mathematical Morphology. In *ISUMA-NAFIPS'95*, pages 228–233, College Park, MD, September 1995.
2. I. Bloch. Distance floue géodésique en traitement d'images. In *Rencontres francophones sur la Logique Floue et ses Applications*, pages 17–24, Paris, France, November 1995.
3. I. Bloch. Fuzzy Geodesic Distance in Images. In A. Ralescu and T. Martin, editors, *Lecture Notes in Artificial Intelligence: Fuzzy Logic in Artificial Intelligence, towards Intelligent Systems*, pages 153–166. Springer Verlag, 1996.
4. I. Bloch. Fuzzy relative position between objects in images: a morphological approach. In *IEEE Int. Conf. on Image Processing ICIP'96*, volume II, pages 987–990, Lausanne, September 1996.
5. I. Bloch and H. Maître. Constructing a fuzzy mathematical morphology: alternative ways. In *Second IEEE International Conference on Fuzzy Systems, FUZZ IEEE 93*, pages 1303–1308, San Fransisco, California, March 1993.
6. I. Bloch and H. Maître. Fuzzy Mathematical Morphologies: A Comparative Study. *Pattern Recognition*, 28(9):1341–1387, 1995.
7. V. di Gesu, M. C. Maccarone, and M. Tripiciano. Mathematical Morphology based on Fuzzy Operators. In R. Lowen and M. Roubens, editors, *Fuzzy Logic*, pages 477–486. Kluwer Academic, 1993.
8. D. Dubois and H. Prade. *Fuzzy Sets and Systems: Theory and Applications.* Academic Press, New-York, 1980.
9. J. Serra (Ed.). *Image Analysis and Mathematical Morphology, Part II: Theoretical Advances.* Academic Press, London, 1988.
10. C. Lantuejoul and F. Maisonneuve. Geodesic Methods in Image Analysis. *Pattern Recognition*, 17(2):177–187, 1984.
11. A. T. Popov. Morphological Operations on Fuzzy Sets. In *IEE Image Processing and its Applications*, pages 837–840, Edinburgh, UK, July 1995.
12. A. Rosenfeld. The Fuzzy Geometry of Image Subsets. *Pattern Recognition Letters*, 2:311–317, 1984.
13. M. Schmitt and J. Mattioli. *Morphologie mathématique.* Masson, Paris, 1994.
14. J. Serra. *Image Analysis and Mathematical Morphology.* Academic Press, London, 1982.
15. D. Sinha and E. Dougherty. Fuzzy Mathematical Morphology. *Journal of Visual Communication and Image Representation*, 3(3):286–302, 1992.
16. D. Sinha, P. Sinha, E. R. Dougherty, and S. Batman. Design and Analysis of Fuzzy Morphological Algorithms for Image Processing. *IEEE Trans. on Fuzzy Systems*, 5(4):570–584, 1997.

TOTAL ORDERING BASED ON SPACE FILLING CURVES FOR MULTIVALUED MORPHOLOGY

JOCELYN CHANUSSOT AND PATRICK LAMBERT
LAMII/CESALP - BP 806 - 74016 Annecy Cedex - France
(CNRS - G1047 - ISIS - Information Signal ImageS)
tel : (33) 450 666 074 - FAX : (33) 450 666 020
email : Jocelyn.Chanussot@univ-savoie.fr

Abstract. This paper addresses the problem of the extension of morphological operators to the case of multicomponent images. It basically comes to the definition of a well suited vectorial ordering relation. We briefly discuss the possible extensions found in the literature and propose an ordering scheme based on the bit mixing paradigm. This choice is justified using the space filling curve equivalent representation of some classical total ordering relations. Results concerning the extension of alternating sequential filters by reconstruction to the case of colour images are presented.

Key words: vectorial ordering, space filling curves, multivalued morphology.

1. Introduction

Mathematical morphology theory has now become a widely used non linear technique for image processing. Initially designed as a set theory, it was then generalized to the set of gray-level images with the help of umbras, and, more recently, to any complete lattice, which is the appropriate mathematical framework for morphology [6, 9, 12]. This minimum necessary algebraic structure is a set T such that:

• T is induced by a (partial) ordering relation "$\leq$";

• for any family (finite or not) of elements in T, there exists a smallest majorant called the sup (for supremum) and a greatest minorant called the inf (for infimum).

Concurrently, multicomponent image processing has become an important field of research. If the image has N different components, each pixel is now represented by an N-dimensional vector. To extend morphological operators to multicomponent images, it is of course possible to process each component separately and to recombine the results afterwards (Marginal processing). This approach is not fully satisfying for different reasons :

• since the output vector will rarely be one of the input vectors, the appearance of, for instance, false colours (in the case of colour images) is quite inevitable,

• if the image has N different components, N processings are required, leading to time consuming operators,

• finally, this approach does not take into account the inter-component correlation.

For these reasons, it is preferable to use a purely vectorial approach that processes the image all at once. But we then have to define a well suited vectorial lattice structure to work in and, in particular, a vectorial ordering relation is required to define the sup and the inf of any family of N-dimensional vectors. Section 2 presents the unifying approach to multivariate data ordering formalized in [4].

Section 3 presents a specific total vectorial ordering relation proposed in [2] based on the bit mixing paradigm. Section 4 presents the equivalence existing between any total vectorial order and a space filling curve. Before the conclusion, section 5 presents the application of the bit mixing paradigm to the colour extension of ASF by reconstruction.

2. Vectorial ordering relations : a formal unifying approach

The key point to extend morphological operators to the N-dimensional case is the definition of a vectorial ordering relation that induces a lattice structure on the data. Goutsias [4] presents a unifying formalism based on the used of a vector transform h from the working space $\mathbb{R}^N$ into $\mathbb{R}^Q$ (with $Q \leq N$ *a priori*) followed by a marginal ordering on $\mathbb{R}^Q$:

$$h : \mathbb{R}^N \to \mathbb{R}^Q, \qquad X \to h(X)$$

Assuming an ordering relation has previously be defined on $\mathbb{R}^Q$, an order can be defined on $\mathbb{R}^N$ in a straightforward way using the following relation :

$$\forall (X,Y) \in \mathbb{R}^N \times \mathbb{R}^N, X \leq Y \Leftrightarrow h(X) \leq h(Y)$$

Depending on the used transform and on the value of Q, different situations are possible. The two next sections respectively discuss the case $Q > 1$ and $Q = 1$.

2.1. $Q > 1$: PARTIAL ORDERING RELATION

The only existing order on $\mathbb{R}^Q$, with $Q > 1$, is the canonic order (the well known lexicographic order corresponds to the case $Q = 1$ with the appropriate transform). In this structure, two vectors X and Y from the product space $\mathbb{R}^Q$ are compared in the following way:

$$X \leq Y \Leftrightarrow X(i) \leq Y(i), \forall i \in \{1, \ldots, Q\}$$

This order is a partial order. For instance vectors $(1,2,3)$ and $(3,2,1)$ can not be compared. Nevertheless, this canonic lattice structure is often used for multivalued morphology [5, 13]. This ordering relation can be used directly on the input data (case $h = Identity$ application). In this case, basic operations such as the supremum or the infimum, and all their possible compositions, are strictly equivalent to a marginal processing. It is also possible to use other transforms h such as the Karhunen Loeve transform or the principal components analysis [15] and to use the canonic lattice structure on the transformed data after this change of base. But, in every case, a partial order is induced, leading unavoidably to the appearance of false colours : the maximum of a set of vectors is not defined and the supremum is rarely one of the input vectors. Furthermore, using a partial ordering relation poses an algorithmic problem. All existing fast programs have been meant to deal with gray level images, i.e. with totally ordered complete lattices, and most of them cannot be directly adapted to partially ordered lattices.

2.2. $Q = 1$: TOTAL (PRE-)ORDERING RELATION

To avoid the problem of false colour appearance, it is necessary to use a transform h with $Q = 1$. In this case, the maximum of a set $\{X_j\}$ of vectors is always defined.
• If h is not injective, this maximum may not be uniquely defined. The non-antisymmetrical relation induced by such a transform is a **pre-ordering** relation,

leading to what are called "morphological-like" filters (the theoretical background is not strictly respected) [15]. Goutsias [4] explains how to solve this theoretical problem by defining a new equivalence relation between the vectors to restore the antisymmetrical property of the induced relation, but the problem of uniquely defining the supremum (or equivalently here, the maximum) still remains. Nevertheless, some good results may be provided using pre-ordering relations. For instance, at each pixel is associated the value of its ith component, with i fixed *a priori* or adaptively determined :

$$h : \mathbb{R}^N \to \mathbb{R}, \qquad X \to X(i)$$

Other techniques based on **reduced ordering** using different generalized distances (Mahalanobis and s.o.) can also be used [15].

• The only true **total ordering** relation that is commonly presented in the literature is the lexicographic order. It is defined as follows :

$$\forall (X,Y) \in \mathbb{R}^N \times \mathbb{R}^N, X < Y \Leftrightarrow$$
$$\exists k \in \{1,\ldots,N\} : X(i) = Y(i) \ \forall i \in \{1,\ldots,k-1\}, X(k) < Y(k)$$

With the general formalism, if the image has N different components coded each on p bits, this ordering relation corresponds to the following transform h :

$$h : T^N \to \mathrm{Im}_h(T^N), \qquad X \to \sum_{i=1}^{N} X(N+1-i) \cdot 2^{p\cdot(i-1)}$$

where T is the set $\{0, 1, ..., 2^p - 1\}$. This order is not fully satisfying either since it induces an important dissymmetry between the different components of the image.

3. The bit mixing paradigm

Following the same formalism, we propose here a transform h that bijectively maps N-dimensional vectors into $\mathbb{R}$. This transform is based on the binary representation of each component of the considered vector X. If the N components of X are coded with p bits each, the $N \cdot p$ available bits are mixed up together to build the $N \cdot p$ bits long scalar value $h(X)$. Of course, there are many different ways to mix up these $N \cdot p$ bits, but we choose the one that minimizes the unavoidable dissymmetry between the different components. So, we first take the first bit of $X(1)$, then the first bit of $X(2)$, until the first bit of $X(N)$, and we repeat the process with the second bit, and so on until the pth bit. Mathematically, this can be written as follows: $X \in \mathbb{R}^N$ has N components $X(i)$, each one coded on p bits $X(i)_j \in \{0,1\}$ with $j \in \{1, ..., p\}$. The considered mapping h can then be written as follows:

$$h(X) = \sum_{j=1}^{p} \left\{ 2^{N\cdot(p-j)} \cdot \sum_{i=1}^{N} 2^{N-i} \cdot X(i)_j \right\}$$

In the same way, we can define the reciprocal mapping h^{-1}. Let S be an integer scalar coded with $N \cdot p$ bits $S_j \in \{0,1\}$ with $j \in \{1, ..., N \cdot p\}$. Then, each component $X(i)$ of $h^{-1}(S) = X \in \mathbb{R}^N$ is defined by:

$$X(i) = \sum_{j=1}^{p} 2^{p-j} \cdot S_{N\cdot(p-j)+N-i+1}$$

In the following, for the sake of clarity and simplicity, we will restrict this paper, without any loss of generality, to the case of the colour images, represented in the classical 3*8 bits Red Green Blue base. Figure 1 presents the coding and the decoding of such a colour pixel.

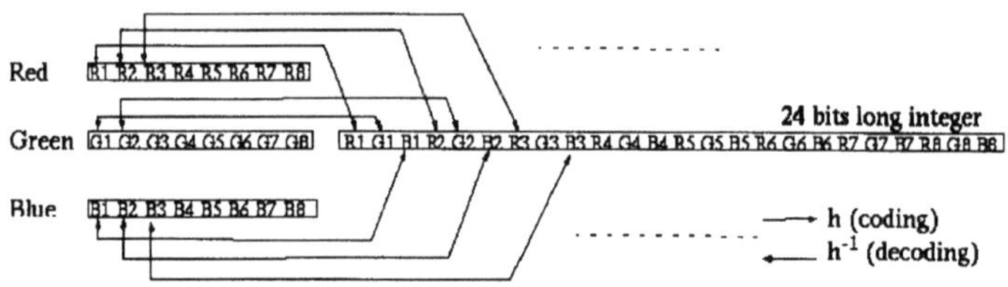

Fig. 1. Coding (h) and decoding (h^{-1}) of one colour pixel.

As previously said, we chose the coding that minimizes the unavoidable dissymmetry inherent to this kind of coding : approximately the same weight is attached to each component of the image (all the components are processed quite symmetrically). Furthermore, no significant changes have been observed in the application presented in section 5 by inverting the order of the components (the remaining small dissymmetry is of no consequence). Nevertheless, if some *a priori* information concerning the relative importance of the different components is available, it can be taken into account by attaching a higher weight to the most significant component (for instance by putting its 2 or 3 first bits at the head of the coding). On the contrary, to increase the symmetry between the relative importance of each component, the following variant can be used : during the coding, the order of the mixed bits is randomly changed at each range, giving for instance the following result, with the same notations as in figure 2 : $R1 \cdot G1 \cdot B1 \cdot G2 \cdot B2 \cdot R2 \cdot B3 \cdot R3 \cdot G3 \ldots$. Of course, the same order is kept for all the pixels of the image. We will see in section 4 that this variant theoretically improves the quality of the coding. The order induced by the canonic lattice structure is probably the most natural and intuitive vectorial order, even if it is just a partial one. A key-point is that the order induced by the bit mixing paradigm or its variant can be regarded as a simple extension of the canonic order (if two vectors are comparable in the canonic structure, they will keep the same order with our relation), but this partial order has been completed into a total one : two vectors are now always comparable.

4. Total vectorial ordering & space filling curves

4.1. Presentation

For a given N-dimensional space, a space filling curve is a curve that goes through each point of the space one single time [10]. It gives a monodimensional representation of a multidimensional space. It thus induces a total order on the space : to compare two vectors, one just has to compare their curvilinear abscissas along the curve. Reciprocally, for each total ordering relation, it is possible to build a corresponding space filling curve by joining the smallest vector of the space to the greatest, passing through all the other points increasingly (according to the chosen ordering relation). Therefore, there is a double equivalence :

(total order on $\mathbb{R}^N$) $\Leftrightarrow$ (bijective application $h : \mathbb{R}^N \to \mathbb{R}$) $\Leftrightarrow$ (space filling curve in $\mathbb{R}^N$)

Bidimensional space filling curves have already been used in image processing as scanning methods. Such curves are presented on figure 2 in the case of a bidimensional space where each component is coded on 4 bits (from 0 to 15). For instance, the Peano scan (figure 2-c), that allows to fully exploit the spatial correlation between adjacent pixels, has been used in [7] to improve the performances of recursive median filter. [14] used a tridimensional Peano curve to improve the compression quality of colour images. The "zig-zag" scan (figure 2-b) has been used in the J-PEG norm for image coding (see for instance [11]). Regazzoni [8] used the curve shown on the figure 2-a to propose a vector median filter with an approach that is quite similar to the one presented in this paper. The layered structure of that curve shows that the corresponding order is essentially based on the euclidean norm of the vectors. Figure 2-h shows the curve corresponding to the lexicographic order in a bidimensional space. The high anisotropy of the curve shows the dissymmetry induced by this order between the different components.

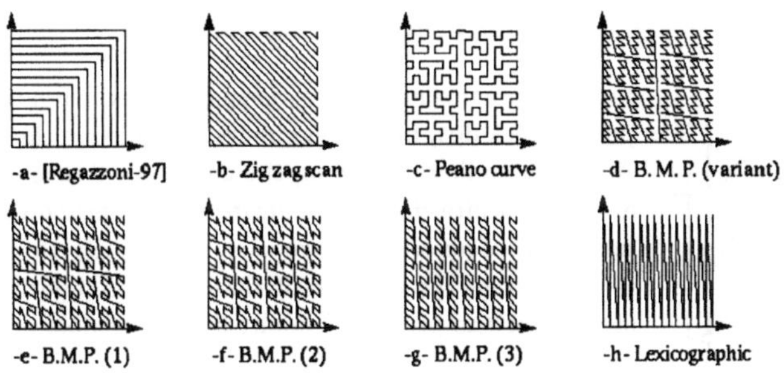

Fig. 2. Examples of bidimensional space filling curves.

4.2. Bit mixing paradigm & space filling curve

The curves corresponding to the order induced by the bit mixing paradigm and its variant are presented respectively on the figures 2-e and -d. We can make some comments that remain true in the general case (N components coded on p bits) :

• The curve is self-similar (fractal structure). That leads to a "hierarchical" comparison of the vectors. On the curve (figure 2-e), this can be seen as a division of the space into four main blocks. If the two vectors we want to compare are in two different blocks (i.e. if the first bits of their coded form are different), it suffices for the comparison. Otherwise, if they are in the same block, this block is divided into four smaller blocks, and so on, taking more and more details of the curve into account, until the comparison is made possible (each scale of the curve corresponds to one bit of the coding).

• Figures 2-f and -g show the curves obtained by mixing the bits respectively two by two and three by three. Mixing them four by four gives, in this case, the lexicographic order. The curve gradually evolves from an almost isotropic curve (figure 2-e) to a highly anisotropic one (lexicographic order), giving more and more importance to one component.

• All the curves, at different degrees, do not respect the space topology : some points that are very close in the original space are moved away on the curve. This can be interpreted in term of noise enhancement in the image. Similarly, the curve can bring closer together points that were far away from each other in the original space. This can mask some transitions in the image. These perturbations are inevitable since no bijective transform from $\mathbb{R}^N$ into $\mathbb{R}$ (the equivalent representation of a space filling curve) can be linear. So, we tried to choose the curve that minimizes this topological distortion. For this, we built a quantitative criterion that measures the perturbation undergone by the neighbourhood notion during the transformation. For each point of the original colour space, we calculated the number of his neighbours that still belong to his neighbourhood along the curve (figure 3) and we averaged the results for the whole space. Table 1 gives the results obtained with some presented curves. Apart from the Peano curve, the topology is preserved at best when the variant of the bit mixing paradigm is used. We nevertheless did not use the Peano curve since the order it induces does not respect the canonic lattice structure. For instance, it decides that vector $(0,8)$ is greater than vector $(15,15)$. As a consequence, we chose the bit mixing paradigm to induce a complete lattice structure on the data set. Next section presents some results obtained with this approach.

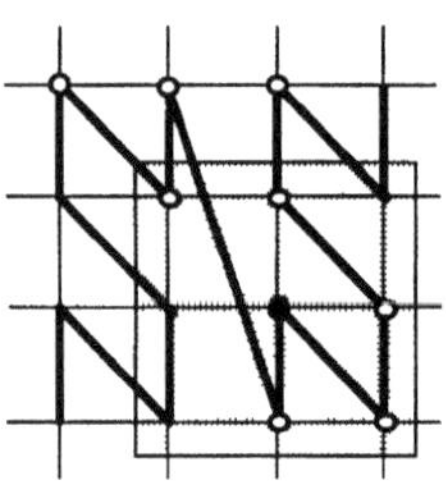

In the original colour space, using 8-connectivity, the considered point (black dot) has 8 neighbours (included in the gray square). After the coding, i.e. along the curve, his 8 nearest neighbours are the 8 white dots. The entire neighbourhood has not been preserved. The criterion we used to quantify this phenomenon is the average intersection of these two neigbourhoods. For instance, for the presented point, five white dots are included in the gray square : 5/8 = 62,5 %.

Fig. 3. Quantitative evaluation of topological distortion.

TABLE I
Topology preservation with different space filling curves

Curve	Neighbourhood preservation ratio
Peano	58 %
Bit mixing paradigm (variant)	50 %
Bit mixing paradigm	47 %
Zig-zag scan	27 %
Regazzoni [8]	26 %
Lexicographic	25 %

5. Application

Using the bit mixing paradigm approach, we tested the colour extension of the alternating sequential filters (ASF) by reconstruction [3]. These operators are connected filters. Connected filters play a major role in mathematical morphology: they are the filters that do not introduce any discontinuity in the image: the edge information is either totally removed or exactly preserved, but the point is that it is never moved or blurred. An ASF by reconstruction is the composition of several morphological openings and closings by reconstruction that are applied alternatively and successively with a growing structuring element size. Those filters have a good ability to reduce noise and to simplify an image and are, as a consequence, very useful in many situations (preprocessing for segmentation, coding ...). The implementation of this vectorial extension of morphological operators to the colour case is very easy and fast. Since only sup- and inf- operations are involved, the processing scheme presented in figure 4 can be used. We can see that the optimal algorithms that are available for the monodimensional case can be used directly on the coded data (in a scalar form).

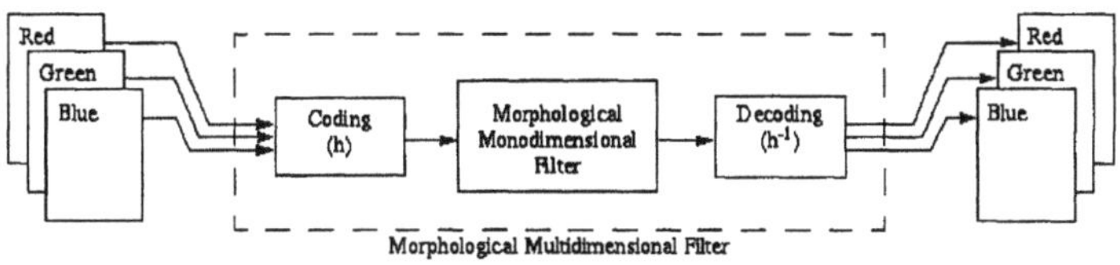

Fig. 4. Processing synopsis to filter a colour image with a "gray-level" filter.

To illustrate the good behaviour of the filter with the proposed structure, we can look at pictures 5(a) & (b). They respectively show the luminance of an original colour image (aerial view from the Canaries islands) and the same image filtered by the following sequence :

$$\text{(filtered image)} = \text{decoding } \circ \overline{\gamma}_5 \circ \overline{\phi}_5 \circ \overline{\gamma}_3 \circ \overline{\phi}_3 \circ \text{ coding (initial image)}$$

where $\overline{\gamma}_n$ and $\overline{\phi}_n$ respectively represent the opening and the closing by reconstruction with a square flat structuring element of size n.

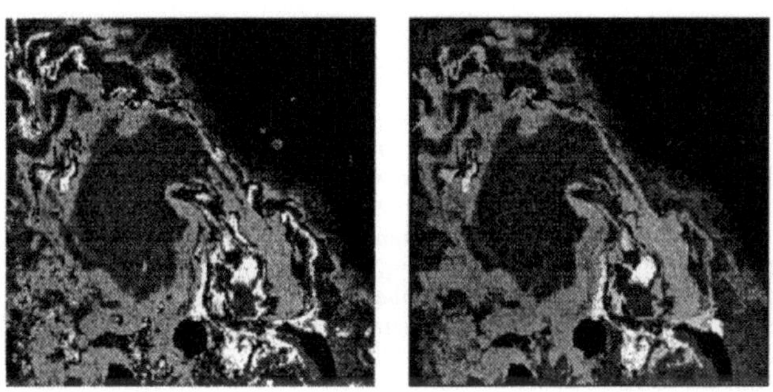

Fig. 5. -a- (left) original image (luminance)(aerial view of the Canarie Islands); -b- (right) filtered image (luminance) ($\overline{\gamma}_5 \circ \overline{\phi}_5 \circ \overline{\gamma}_3 \circ \overline{\phi}_3$).

The filtered image is visibly much simpler than the original one ; some details have been lost, but preserved edges remain absolutely unchanged, which is the typical intended behaviour of the used filter. We also can notice the loss of contrast due to the properties of the openings and the closings (respectively anti-extensivity and extensivity) leading to a kind of "tarnished" result ; the dynamics of the image gradually decreases. This is also typical for this kind of filters.

6. Conclusion

Using the bit mixing paradigm, we proposed a bijective transform that encodes multidimensional data in a scalar form. This transform induces a total ordering relation on the vectors and thus can be used to extend morphological operators to multicomponent images. We justified this choice by studying its equivalent representation as a space filling curve, and we tested the method on the colour extension of alternating sequential filters by reconstruction. Of course, we could have used the very same approach on colour images in the Hue Saturation Intensity base, taking the hue more or less into account adaptively to the saturation for instance [1], or on any other multicomponent images (multispectral, multitemporal and s.o.). The method seems to be very promising. It actually presents several advantages:

- it prevents the appearance of false colours since any output vector is necessarily one of the input vectors,
- since the transform is bijective, the definition of, for instance, the supremum of any set of vectors is unambiguous,
- it is very simple to implement (coding and decoding are very easy and fast),
- it is well suited to all already existing optimized algorithms that deal with the one dimensional case. The existing programs are used directly without any change (except for the coding and decoding steps that have to be added). We just need to check that the program can deal with integers coded on the right number of bits (for instance, 24 and not only 8).
- the properties of the operators we wanted to extend are well preserved.

References

1. T. Carron and P. Lambert. Color edge detection using jointly hue, saturation and intensity. In *Proc. IEEE 1st Intern. Conf. Image Processing (ICIP'94), Austin, USA*, pages 977–981, 1994.
2. J. Chanussot and P. Lambert. Bit mixing paradigm for multivalued morphological filters. In *Proc. IEEE 6th Intern. Conf. Image Processing and its Applications Dublin, Ireland*, pages 804–808, July 1997.
3. J. Crespo, J. Serra, and R. W. Schafer. Theoretical aspects of morphological filters by reconstruction. *Signal Processing*, 47:201–225, 1995.
4. J. Goutsias, H. Heijmans, and K. Sivakumar. Morphological operators for image sequences. *Computer. Vision Image Understanding*, 62(3):326–346, 1995.
5. C. Gu. *Multivalued Morphology and Segmentation based Coding.* PhD thesis, Ecole Polytechnique Federale (EPFL), Lausanne, 1995.
6. H. J. A. M. Heijmans. *Morphological Image Operators.* Acad. Press, New York, 1994.
7. J. Nieweglowski and T. Campbell. Recursive image filters and space filling curves. In *Proc. IEEE Winter Workshop Nonlinear Digital Signal Processing, Tampere, Finland*, pages 4.1/5.1–5.6, Jan. 1990.
8. C. Regazzoni and A. Teschioni. A new approach to vector median filtering based on space filling curves. *IEEE Trans. on Image Processing*, 6(7):1025–1037, July 1997.
9. C. Ronse. Why mathematical morphology needs complete lattices. *Signal Processing*, 21(2):129–154, 1990.
10. H. Sagan, editor. *Space Filling Curves.* Springer Verlag, New-York, 1994.
11. R. Schafer and T. Sikora. Digital video coding standards and their role in video communications. *Proc. of the IEEE*, 83(6):907–924, June 1995.
12. J. Serra, editor. *Image Analysis and Mathematical Morphology. II: Theoretical Advances.* Acad. Press, London, 1988.
13. J. Serra. Anamorphoses and function lattices. In E. R. Dougherty, editor, *Mathematical Morphology in Image Processing*, chapter 13, pages 483–523. Marcel Dekker, New York, 1993.
14. R. Stevens, A. Lehar, and F. Preston. Manipulation and presentation of multidimensional image data using the peano scan. *IEEE Trans. on Pattern Analysis and Machine Intelligence*, PAMI-5(5):520–526, sept. 1983.
15. C. Vertan, V. Popescu, and V. Buzuloiu. Morphological like filters for color images. In *Proc. EUSIPCO, Trieste, Italy*, pages 165–168, September 1996.

MEASURES AND INDICES OF REFLECTION SYMMETRY FOR CONVEX POLYHEDRA

ALEXANDER V. TUZIKOV *
Institute of Engineering Cybernetics
Surganova 6, 220012 Minsk, Belarus

JOS B.T.M. ROERDINK
University of Groningen
P.O. Box 800, 9700 AV Groningen, The Netherlands

HENK J.A.M. HEIJMANS
CWI
P.O. Box 94079, 1090 GB Amsterdam, The Netherlands

and

STANISLAV A. SHEYNIN
Institute of Engineering Cybernetics
Surganova 6, 220012 Minsk, Belarus

Abstract. This paper discusses measures of reflection symmetry for 3D convex sets which are based on Minkowski addition and Brunn-Minkowski inequalities for volume and mixed volume. These measures are invariant to similitude transformations. It is also shown how these measures can be computed efficiently for convex polyhedra.

Key words: Reflection symmetry measure, Minkowski addition, Brunn-Minkowski inequality, Index of reflection symmetry, Convex polyhedra

1. Introduction

For many objects, presence or absence of symmetry is a major feature, and therefore the problem of object symmetry identification is of great interest in image analysis and recognition, computer vision and computational geometry. There exists a vast literature dealing with all kinds of symmetry of shapes and grey-scale images. There exist algorithms for the identification of exact symmetries as well as techniques devoted to the computation of quantitative information about the amount of symmetry in shapes and images. Efficient algorithms for detection of exact symmetries of point sets, polygons and polyhedra can be found e.g. in [9, 19].

Since real images are always disturbed by noise it is useful to have a tool to measure the amount of symmetry in them. Towards this goal Grünbaum [6] introduced the notion of *symmetry measure*. Most of the theoretical results concerning symmetry measures are obtained for convex sets. Some interesting results concerning measures of central symmetry (point reflections) can be found in [6]. Related results for reflection and rotation symmetry of convex sets can be found in [1, 2]. Studies of central symmetry measures for convex sets using morphological transformations

* A. Tuzikov and S. Sheynin were supported by the INTAS grant N 96-785

are reported in [10, 11, 13]. Symmetrization transformations based on Minkowski addition were used in [17] to study rotation and reflection symmetry measures of convex sets.

Most practical algorithms developed for symmetry measurement are applied to the 2D case and only some of them can be extended to 3D (see, for example, the symmetry distance introduced in [20]). Since this paper deals with the 3D case we discuss only some literature concerning the 3D case. The extended Gaussian image representation was used in [15] for finding axes of reflection and rotation symmetries of 3D objects. Octree representations were used in [12] to measure the symmetry degree of 3D objects.

In this paper we extend the approach described in [8] for finding symmetry measures and indices of 2D convex sets to the 3D case. We restrict ourselves to reflection symmetry measures. The measures discussed here are based on Minkowski addition and properties of volume and mixed volume functionals.

As was shown in [18] for the case of convex polygons, the reflection symmetry measure can be computed in polynomial time. Here it is conjectured that finding reflection symmetry indices for convex polyhedra is reduced to the computation of the symmetry measure for a finite number of reflection planes only. This conjecture is based on a more general result obtained in [16] for comparing convex polyhedra.

2. Symmetry measures

Denote by $\mathcal{K}(\mathbb{R}^3)$, or briefly $\mathcal{K}$, the family of all nonempty compact subsets of $\mathbb{R}^3$. The compact convex subsets of $\mathbb{R}^3$ are denoted by $\mathcal{C} = \mathcal{C}(\mathbb{R}^3)$. For two subsets A and B we write $A \equiv B$ if these sets differ only by translation. The group of all linear transformations on $\mathbb{R}^3$ is denoted by G. If $g \in G$ and $A \subseteq \mathbb{R}^3$, then $g(A) = \{g(a) \mid a \in A\}$. Furthermore, we denote by $E \subseteq G$ the reflections in the planes through the origin, and by R the rotations around axes through the origin.

Let u be a vector on the unit sphere S^2 in $\mathbb{R}^3$ and let Π_u be the plane in $\mathbb{R}^3$ orthogonal to the vector u passing through the origin. The reflection w.r.t. the plane Π_u is denoted by e_u.

Definition 1 *A set $A \subset \mathbb{R}^3$ is called* reflection symmetric *if there exists a reflection $e_u \in E$ such that $e_u(A) \equiv A$. We say that e_u is a symmetry of A, and we call Π_u the* plane of reflection symmetry.

To access the symmetry of sets we need a tool to measure the amount of symmetry. Several years ago, Grünbaum [6] introduced the notion of *symmetry measure*. We adapt his definition of central symmetry measure for the case of reflection symmetry in the following way.

Definition 2 *Let $\mathcal{J} \subset \mathcal{K}$. A function $\mu : \mathcal{J} \times E \to [0,1]$ is called a* reflection symmetry measure *if for every $e \in E$ the function $\mu(\cdot, e)$ is continuous on $\mathcal{J}$ with respect to the Hausdorff topology, and if the following conditions hold:*

1. $\mu(A,e) = \mu(A',e)$ *if* $A \equiv A'$;
2. $\mu(A,e) = \mu(e(A),e)$;
3. $\mu(A,e) = 1$ *iff* $e(A) \equiv A$.

*Let $H \subseteq G$ be a set such that heh^{-1} is a reflection if e is a reflection and $h \in H$. We say that a reflection symmetry measure μ is H-*invariant *if*

$$\mu(A, e) = \mu(h(A), heh^{-1}) \quad \textit{for all } h \in H.$$

Below we introduce two reflection symmetry measures based on Minkowski addition and properties of volume and mixed volume functionals. The *Minkowski addition* of two sets $A, B \subseteq \mathbb{R}^3$ is defined by

$$A \oplus B = \{a + b \mid a \in A,\ b \in B\}.$$

It is obvious that for two sets $A, B \subseteq \mathbb{R}^3$ and $g \in G$, we have

$$g(A \oplus B) = g(A) \oplus g(B).$$

Denote by $V(A)$ the volume (Lebesgue measure) of the set $A \subset \mathbb{R}^3$. Given convex sets $A, B \subset \mathbb{R}^3$ and $\alpha, \beta \geq 0$ one gets from the Minkowski theorem on mixed volumes [3, p.353] the following relation:

$$V(\alpha A \oplus \beta B) = \alpha^3 V(A) + 3\alpha^2 \beta V(A, A, B) + 3\alpha\beta^2 V(A, B, B) + \beta^3 V(B). \tag{1}$$

Here $V(A, A, B)$ and $V(A, B, B)$ are called *mixed volumes*. The following inequalities are used below; see Hadwiger [7] or Schneider [14] for a comprehensive discussion.

Brunn-Minkowski inequality. For two arbitrary compact sets $A, B \subset \mathbb{R}^3$ the following inequality holds:

$$V(A \oplus B)^{\frac{1}{3}} \geq V(A)^{\frac{1}{3}} + V(B)^{\frac{1}{3}}, \tag{2}$$

with equality if and only if A and B are convex and homothetic modulo translation, i.e., $B \equiv \alpha A$ for some $\alpha > 0$.

Minkowski inequality. For convex sets $A, B \in \mathcal{C}(\mathbb{R}^3)$

$$V(A, A, B)^3 \geq V(A)^2 V(B), \tag{3}$$

and as before equality holds if and only if $B \equiv \alpha A$ for some $\alpha > 0$.

Given a plane reflection e_u define the transformation $b_u : \mathcal{K} \longrightarrow \mathcal{K}$ by

$$b_u(A) = \frac{1}{2}(A \oplus e_u(A)).$$

It is easy to see that the set $b_u(A)$ is reflection symmetric with respect to the plane Π_u. In the literature the transformation b_u is called *Blaschke symmetrization* [14].

Proposition 1 *Given a set $A \in \mathcal{C}$ and a plane reflection e_u, the inequality $V(b_u(A)) \geq V(A)$ always holds. Furthermore, the following statements are equivalent:*

(i) $e_u(A) \equiv A$, *i.e., e_u is a symmetry of A;*

(ii) $b_u(A) \equiv A$;

(iii) $V(b_u(A)) = V(A)$.

For a proof we refer to [8, Prop.7.2].

Let us introduce the functionals $\mu_1, \mu_2 : \mathcal{C} \times E \to \mathbb{R}_+$ defined for compact convex sets as follows

$$\mu_1(A, e) = \frac{8V(A)}{V(A \oplus e(A))}, \tag{4}$$

$$\mu_2(A, e) = \frac{V(A)}{V(A, A, e(A))}. \tag{5}$$

It is known from properties of mixed volumes that for every linear transformation g, the following identity holds: $V(g(A), g(B), g(C)) = |\det g| V(A, B, C)$. Here $\det g$ denotes the determinant of g. Now, using relation (1) it follows that

$$\mu_2 = \frac{3\mu_1}{4 - \mu_1}. \tag{6}$$

Moreover the following proposition is true.

Proposition 2 *The functionals μ_1 and μ_2 are reflection symmetry measures which are invariant under rotations and scalings.*

Sometimes, one is not so much interested in the symmetry measure for a specific reflection plane Π_u, but rather in the maximum of these values over all planes. We call this maximum the *index of reflection symmetry*. Thus we define:

$$\iota_1(A) = \sup_{e \in E} \frac{8V(A)}{V(A \oplus e(A))}, \tag{7}$$

$$\iota_2(A) = \sup_{e \in E} \frac{V(A)}{V(A, A, e(A))}. \tag{8}$$

Evidently, both indices are related to each other by a formula analogous to (6). It is obvious that both indices ι_1 and ι_2 are invariant under similitude transformations, i.e., $\iota_1(g(A)) = \iota_1(A)$ and $\iota_2(g(A)) = \iota_2(A)$ for every similitude transformation g.

If the supremum in (7) is achieved for $e = e_u$, then we call Π_u the *ι_1-optimal plane of reflection symmetry* (note that because of relation (6), the ι_1-optimal and ι_2-optimal planes coincide).

To find the ι_1-optimal plane of reflection symmetry it is necessary to maximize $\mu_1(A, e_u)$ over all possible positions of reflection planes passing through the coordinate origin. In general this is a time consuming problem. Therefore, one often restricts oneselves to a finite number of reflection planes. Following ideas from classical mechanics [5] one can associate with every body its ellipsoid of inertia. It is known that symmetry planes of reflection symmetrical bodies are orthogonal to the principal axes of this ellipsoid. Therefore, to reduce the computation complexity of the optimization, one can define approximate measures by considering only planes orthogonal to the principal axes.

If one restricts attention to the class of convex polyhedra then it is possible to get additional results. In Section 3 we show that it is possible to reduce the time complexity of the given optimization problem.

3. Convex polyhedra

To compute the indices of reflection symmetry ι_1, ι_2 for a convex polyhedron P it is necessary to find minima of the following functionals

$$V(P \oplus e_u(P)) \quad \text{and} \quad V(P, P, e_u(P)) \tag{9}$$

for $u \in S^2_+$, the hemisphere containing unit vectors with non-negative z-coordinate.

Denote by $r_{u,\alpha}$ the rotation in $\mathbb{R}^3$ about the oriented axis directed along vector u over an angle α (counter-clockwise direction). Given two unit vectors $u_1, u_2 \in S^2_+$, denote by $\alpha(u_1, u_2)$ the angle smaller than 180° between them. The composition of plane reflections can be expressed as a rotation in the following way

$$e_{u_2} e_{u_1} = r_{u_1 \times u_2,\ 2\alpha(u_1,u_2)}, \tag{10}$$

where $u_1 \times u_2$ denotes the outer product of vectors u_1 and u_2.

Given any vector $v \in S^2$ denote by $S_+(v)$ the upper part of the great circle in S^2 which is orthogonal to v. Let us investigate the functional $V(P \oplus e_u(P))$ for $u \in S_+(v)$ (see Fig. 1).

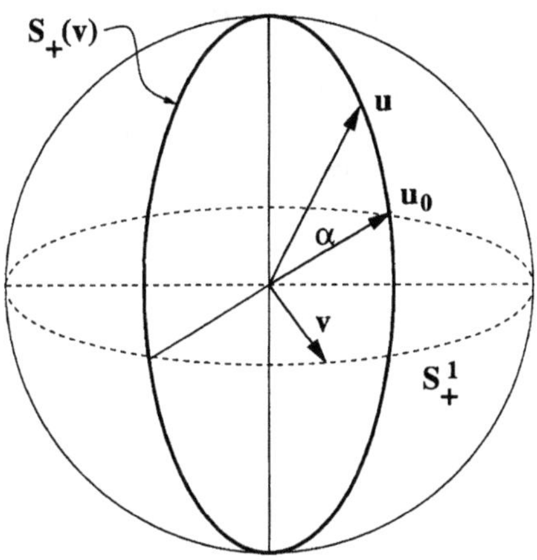

Fig. 1. Vector v is orthogonal to (upper part of) the great circle $S_+(v)$ and $u_0, u \in S_+(v)$.

Proposition 3 *For a vector $v \in S^2$ the functionals $V(P \oplus e_u(P))$ and $V(P, P, e_u(P))$ are piecewise concave in u, when u runs over $S_+(v)$.*

Proof. To prove this result, fix $u_0 \in S_+(v)$ as in Fig. 1. For $u \in S_+(v)$ we have $u \times u_0 = v$, hence

$$e_u e_{u_0} = r_{v,2\alpha},$$

where $\alpha = \alpha(u_0, u)$. Applying e_{u_0} at the right of both expressions and using that $e^2_{u_0} = \mathrm{id}$, we get

$$e_u = r_{v,2\alpha} e_{u_0}.$$

Therefore, putting $Q = e_{u_0}(P)$, we find

$$V(P \oplus e_u(P)) = V(P \oplus r_{v,2\alpha}(Q)).$$

When u runs over $S_+(v)$, the angle α runs over $(0, \pi]$. The concavity of the functional at the right hand-side has been established in [16], and the proof follows. Q.E.D.

Observe that u_0 is the 90°-rotation of v around the z-axis. Thus $Q = e_{u_0}(P)$ depends on v: $Q = Q(v)$. We have shown the following relation:

$$\min_{u \in S_+^2} V(P \oplus e_u(P)) = \min_{v \in S_+^1} \min_{0 < \alpha \leq \pi} V(P \oplus r_{v,2\alpha}(Q(v))).$$

A similar relation holds for the functional $V(P, P, e_u(P))$.

Therefore the optimization problem on the hemisphere S_+^2 is reduced to the optimization problem on the semi-equator S_+^1.

In our report [16] we have dealt with the computation of

$$\min_{u \in S^2,\ 0 \leq \alpha < 2\pi} V(P \oplus r_{u,\alpha}(Q)),$$

$$\min_{u \in S^2,\ 0 \leq \alpha < 2\pi} V(P, P, r_{u,\alpha}(Q)),$$

for arbitrary convex polyhedra P and Q. It was proven that only finitely many vectors $u \in S^2$ have to be checked to compute the minimum. Based on these results we formulate the following conjectures:

Conjecture 1 *There exist a finite number of vectors* $v_1, \ldots, v_k \in S_+^1$ *such that*

$$\min_{u \in S_+^2} V(P \oplus e_u(P)) = \min_{i=1,\ldots,k} \min_{0 \leq \alpha < \pi} V(P \oplus r_{v_i,2\alpha}(Q(v_i))),$$

$$\min_{u \in S_+^2} V(P, P, e_u(P)) = \min_{i=1,\ldots,k} \min_{0 \leq \alpha < \pi} V(P, P, r_{v_i,2\alpha}(Q(v_i))).$$

4. Conclusion

The 3D case is essentially more difficult than the 2D case: for the latter it has been shown [8] that only finitely many cases have to be checked. Basically, the reason for this difference is that in the 2D case, the composition of a reflection and a rotation is a reflection; this no longer true in the 3D case. The conjecture formulated above is still to be proven. Our present work concerns the implementation of algorithms for computing indices of reflection symmetry for convex polyhedra. Such algorithms are based on the slope diagram representation of convex polyhedra [4].

The problem of computation of indices of reflection symmetry is more difficult than computation of central symmetry measures which indicate the amount of central symmetry. Write $\check{A} = \{-x,\ x \in A\}$ and introduce the functionals $\mu_3, \mu_4 : \mathcal{C} \longrightarrow \mathbb{R}_+$ defined for compact convex sets as follows

$$\mu_3(A) = \frac{8V(A)}{V(A \oplus \check{A})},$$

$$\mu_4(A) = \frac{V(A)}{V(A, A, \check{A})}.$$

One can show that the functionals μ_3 and μ_4 define affine invariant central symmetry measures for convex sets in $\mathbb{R}^3$.

References

1. C. K. Chui and M. N. Parnes. Measures of N-fold symmetry for convex sets. *Proceedings of the American Mathematical Society*, 26:480–486, 1970.
2. B. A. deValcourt. Measures of axial symmetry for ovals. *Israel Journal of Mathematics*, 4:65–82, 1966.
3. R. Gardner. *Geometric Tomography*. Cambridge University Press, 1995.
4. P. K. Ghosh and R. M. Haralick. Mathematical morphological operations of boundary-represented geometric objects. *Journal of Mathematical Imaging and Vision*, 6(2/3):199–222, 1996.
5. H. Goldstein. *Classical Mechanics*. Addison-Wesley, Reading, MA, 1950.
6. B. Grünbaum. Measures of symmetry for convex sets. In *Proc. Sympos. Pure Math.*, volume 7, pages 233–270. Providence, USA, 1963.
7. H. Hadwiger. *Vorlesungen über Inhalt, Oberfläche und Isoperimetrie*. Springer, Berlin, 1957.
8. H. J. A. M. Heijmans and A. Tuzikov. Similarity and symmetry measures for convex shapes using Minkowski addition. *IEEE Transactions on Pattern Analysis and Machine Intelligence*, to appear, 1998.
9. X. Y. Jiang and H. Bunke. Detection of rotational and involutional symmetries and congruity of polyhedra. *The Visual Computer*, 12(4):193–201, 1996.
10. M. Jourlin and B. Laget. Convexity and symmetry: Part 1. In J. Serra, editor, *Image Analysis and Mathematical Morphology, vol. 2: Theoretical Advances*, London, 1988. Academic Press.
11. G. Matheron and J. Serra. Convexity and symmetry: Part 2. In J. Serra, editor, *Image Analysis and Mathematical Morphology, vol. 2: Theoretical Advances*, pages 359–375, London, 1988. Academic Press.
12. P. Minovic, S. Ishikawa, and K. Kato. Symmetry identification of a 3d object represented by octree. *IEEE Transactions on Pattern Analysis and Machine Intelligence*, 15(5):507–514, 1993.
13. R. Schneider. On a morphological transformation for convex domains. *Journal of Geometry*, 34:172–180, 1989.
14. R. Schneider. *Convex Bodies. The Brunn-Minkowski Theory*. Cambridge University Press, Cambridge, 1993.
15. C. Sun and J. Sherrah. 3d symmetry detection using the extended Gaussian image. *IEEE Transactions on Pattern Analysis and Machine Intelligence*, 19(2):164–168, 1997.
16. A. Tuzikov, J. B. T. M. Roerdink, and H. J. A. M. Heijmans. Similarity measures for convex polyhedra based on Minkowski addition. Technical Report CS-R9708, University of Groningen, Department of Computer Science, 1997.
17. A. V. Tuzikov, G. L. Margolin, and A. I. Grenov. Convex set symmetry measurement via Minkowski addition. *Journal of Mathematical Imaging and Vision*, 7(1):53–68, 1997.
18. A. V. Tuzikov, G. L. Margolin, and H. J. A. M. Heijmans. Efficient computation of a reflection symmetry measure for convex polygons based Minkowski addition. In *Proceedings of 13th International Conference on Pattern Recognition*, volume B, pages 236–240, 1996.
19. J. D. Wolter, T. C. Woo, and R. A. Volz. Optimal algorithms for symmetry detection in two and three dimentions. *Visual Computer*, 1:37–48, 1985.
20. H. Zabrodsky, S. Peleg, and D. Avnir. Symmetry as a continuous feature. *IEEE Transactions on Pattern Analysis and Machine Intelligence*, 17:1154–1166, 1995.

AN APPROACH TO DISCRETIZATION BASED ON THE HAUSDORFF METRIC

D. WAGNER, M. TAJINE and C. RONSE
LSIIT UPRES-A 7005, Département d'Informatique de l'U.L.P.
7 rue René Descartes, F-67000 Strasbourg, France
Tel. 33-3.88.41.66.38 Fax. 33-3-88.60.26.54
e-mail: {ronse,tajine}@dpt-info.u-strabg.fr
URL: http://gr6.u-strasbg.fr/~ronse

Abstract. In this paper, a new approach to the discretization of n-dimensional Euclidean figures is studied: the discretization of a compact Euclidean set is a discrete set whose Hausdorff distance to that compact set is minimal; in particular the discretization depends on the choice of metric in the Euclidean space, for example the Euclidean or a chamfer distance. The set of possible discretizations is non-void, finite, and closed under union; it is thus convenient to choose *the* discretization as the greatest one among such sets. We give a mathematical description of that set. We have a bound on the Hausdorff distance between a compact set and its Hausdorff discretization, and this discretization converges (for the Hausdorff metric) to the compact set when the grid spacing tends to zero. Of particular interest are what we call *cellular distances*, that is those for which every Euclidean point in a voxel's cell is closer to that voxel than to any other one; for such a distance, the *supercover* discretization made of all voxels whose cell intersects the compact set, minimizes the Hausdorff distance to the compact set. The supercover is the (maximal) Hausdorff discretization only for the chessboard distance. We give also a bound on the Hausdorff distance between a compact set and its morphological discretization by dilation, leading to a similar convergence result.

Key words: Euclidean and cellular distances, compact sets, discretization, Hausdorff metric

1. Introduction

Discretization, in other words the transformation of Euclidean figures into their digital counterparts, is an important problem both in computer graphics and in image processing. In computer graphics, or image synthesis, models of object shapes, generally described by equations in 2D or 3D Euclidean space, must be translated into digital 2D or 3D objects, which can be displayed through 2D digital images.

In image processing, images are not artificially designed, but given by a natural scene and an optical device, and their discretization is generally achieved by an electronic device. Thus the image processing worker has generally no choice on the type of discretization to be applied to a natural image, although it is always possible to study the problem theoretically. However in digital image processing, one is free to choose the *filter* to be applied to the image; in particular one can choose a *convolution mask* in linear filtering, or a *structuring element* in morphological processing. Thus the discretization of a convolution mask or of a structuring element becomes a practical problem.

Several approaches have been studied for the discretization of Euclidean shapes or images, but none is completely satisfactory. What are the desirable qualities of a discretization? We can state some of them here:

- The discretization of an Euclidean object should retain its significant properties, or these properties should lead to digital counterparts.
- The Euclidean object should to a certain extent (dependent upon the resolution of the discrete space) be reconstructible from its discretization; when the resolution of the discrete space tends to zero, the discretization should converge (in some specified mathematical sense) to the original Euclidean object.
- The discretization should be computationally tractable: the discretization of an object given by a set of analytic (in-)equations should be easily computed.

1.1. Previous approaches

The discretization of Euclidean shapes, especially line segments, circles and ellipses, has been studied by many authors in the field of image processing and analysis; several standard schemes are described in [5], and many references on digital straight lines can be found in [13]. In image synthesis, the standard reference for digital straight line segments is Bresenham's algorithm [4], which is equivalent to the grid-intersect discretization in image analysis [5]. All these methods are local, and their authors do generally not address the problem of convergence of the discretized shape to the Euclidean original shape when the grid spacing tends to zero.

There is also an approach to digital geometry where digital shapes are defined *independently* from their Euclidean counterparts [1, 12]. Here the main criterion for a "good" definition of a digital shape (say, a line segment or a circle) is that the digital object should have many geometrical properties which are digital counterparts of the properties of the corresponding Euclidean object; for example: two distinct parallel lines have an empty intersection, while two non-parallel lines intersect at a single location, circles disconnect the plane into two connected components, etc. The main problem with this approach is the multiplicity of alternative definitions of the same object, and the lack of a unified procedure for defining a digital object.

Several authors have considereded the discretization of figures (binary images) from a morphological point of view; we follow here the approach of Heijmans [10, 7, 8]. Let us write $\mathcal{E}$ for the Euclidean space $\mathbf{R}^n$ and $\mathcal{D}$ for the corresponding digital space (for example $\mathbf{Z}^n$). Let A be a Euclidean structuring element ($A \subset \mathcal{E}$). The *discretization by dilation* is the map $\Delta_\oplus^A : \mathcal{P}(\mathcal{E}) \rightarrow \mathcal{P}(\mathcal{D})$ defined by setting:

$$\forall X \subseteq \mathcal{E} \qquad \Delta_\oplus^A(X) = (X \oplus \check{A}) \cap \mathcal{D} = \{p \in \mathcal{D} \mid A_p \cap X \neq \emptyset\}. \qquad (1)$$

Here A_p is the translate of A by p and $\check{A}$ is the symmetric of A ($\check{A} = \{-s \mid s \in A\}$).

As the structuring element A can vary, we obtain in this way a wide variety of discretization schemes. For example taking $A = \{o\}$, where o is the origin, we obtain the classical discretization by intersection which associates to X the discrete set $X \cap \mathcal{D}$. Another example of discretization by dilation is given by the *supercover*. To every discrete point $p \in \mathcal{D}$ we associate the *closed cell* $C(p)$ consisting of all Euclidean points $x \in \mathcal{E}$ which are at least as close to p as to any other point in $\mathcal{D}$:

$$\forall p \in \mathcal{D}, \qquad C(p) = \{x \in \mathcal{E} \mid \forall q \in \mathcal{D} \setminus \{p\},\ d_2(x,p) \leq d_2(x,q)\}. \qquad (2)$$

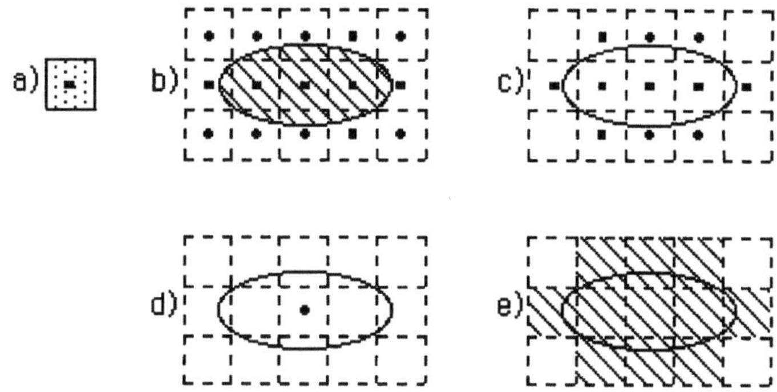

Fig. 1. a) The structuring element A is $C(o)$. b) A Euclidean set X overlayed with the pixels p and their cells $C(p)$. c) $\Delta_{\oplus}^{A}(X) = \Delta_{SC}(X)$. d) $\Delta_{\ominus}^{A}(X)$. e) $E_{\oplus}^{A}(\Delta_{\oplus}^{A}(X))$ (the hatched polygon, not containing its outer border).

Here d_2 denotes the Euclidean distance. For example if $\mathcal{D} = \mathbf{Z}^n$ and $p = (p_1, \ldots, p_n)$, then $C(p)$ is the set of all $(x_1, \ldots, x_n) \in \mathcal{E}$ such that for each $i = 1, \ldots, n$ we have $|x_i - p_i| \leq 1/2$. The *supercover discretization* Δ_{SC} associates to every $X \in \mathcal{P}(\mathcal{E})$ the set of all $p \in \mathcal{D}$ such that $C(p)$ intersects X (see Fig. 1.c):

$$\forall X \subseteq \mathcal{E} \qquad \Delta_{SC}(X) = \{p \in \mathcal{D} \mid C(p) \cap X \neq \emptyset\}. \tag{3}$$

In fact, Δ_{SC} is $\Delta_{\oplus}^{C(o)}$ (compare (3) with (1)). The term "supercover" comes from [6], where some of its properties are stated; [2] gives further properties, and analytical equations for the supercover discretization of 3D line segments and triangles.

One important property of discretization by dilation $\Delta_{\oplus}^{A}$ is that it is a dilation in the algebraic sense, so that there is an *adjoint erosion* $E_{\oplus}^{A} : \mathcal{P}(\mathcal{D}) \to \mathcal{P}(\mathcal{E})$, which is in some way a "Euclideanization" standing dual to the discretization; it associates to every $Y \in \mathcal{P}(\mathcal{D})$ the greatest $X \in \mathcal{P}(\mathcal{E})$ such that $\Delta_{\oplus}^{A}(X) \subseteq Y$ (see Fig. 1.e).

Heijmans studied the properties of $\Delta_{\oplus}^{A}$ under the *covering assumption*, which states that $\mathcal{D} \oplus A = \mathcal{E}$, or equivalently that for every $X \neq \emptyset$ we have $\Delta_{\oplus}^{A}(X) \neq \emptyset$. He showed [7] that (under some precise mathematical conditions) when the grid spacing tends to zero, $E_{\oplus}^{A}(\Delta_{\oplus}^{A}(X))$ converges to X for the hit-or-miss topology.

One can also consider the *dual by complementation* (or *negative*) of $\Delta_{\oplus}^{A}$; this is the *discretization by erosion* $\Delta_{\ominus}^{A} : \mathcal{P}(\mathcal{E}) \to \mathcal{P}(\mathcal{D})$ defined by setting (see Fig. 1.d):

$$\forall X \subseteq \mathcal{E} \qquad \Delta_{\ominus}^{A}(X) = (X \ominus A) \cap \mathcal{D} = \{p \in \mathcal{D} \mid A_p \subseteq X\} = \mathcal{D} \setminus \left(\Delta_{\oplus}^{A}(\mathcal{E} \setminus X)\right).$$

1.2. Our approach

We can raise against previous approaches the following criticism:

- The purely discrete approach of [1, 12] does not allow to obtain automatically the discrete object from its Euclidean counterpart.
- Up to now, discretization methods have relied upon purely local operations, without global criteria.

- In the approach by Heijmans [7], it is not the discretized object $\Delta_{\oplus}^{A}(X)$, but its reconstruction $E_{\oplus}^{A}(\Delta_{\oplus}^{A}(X))$, which converges to X when the grid spacing tends to zero.
- The hit-or-miss topology under which this convergence is shown does not give a measure of "closeness" between two sets. Alternative definitions of convergence based on lattice theory (as in [9]) have the same defect, and they are moreover artificial when applied to geometrical shapes.

Now the "closeness" between a Euclidean and a digital set can be measured if we have a metric on such sets. Given a space of points on which a distance function d is defined, this function d can be extended to a metric on the space of all nonvoid *compact* subsets of that space, namely the ***Hausdorff distance***, written H_d.

We can define a ***Hausdorff discretizing set*** of a nonvoid compact subet K of $\mathcal{E}$ as a nonvoid finite subset S of $\mathcal{D}$ whose Hausdorff distance $H_d(S,K)$ to K is minimal. In fact the set $\mathcal{M}_{H_d}(K)$ of Hausdorff discretizing sets of K is nonvoid and finite, but it can contain sometimes several elements. However this set is closed under the union operation, so that we can define the ***Hausdorff discretization*** of K as the union of all Hausdorff discretizing sets; we write it $\Delta_{H_d}(K)$; NB: $\Delta_{H_d}(K) \in \mathcal{M}_{H_d}(K)$.

We give a mathematical characterization of $\Delta_{H_d}(K)$; the Hausdorff distance between K and $\Delta_{H_d}(K)$ is bounded by the radius of the cells $C(p)$, so that $\Delta_{H_d}(K)$ converges to K (for the Hausdorff metric) when the resolution of $\mathcal{D}$ tends to zero.

We give then the relation between the Hausdorff and supercover discretizations. In general the Hausdorff discretization of K "lies in the neighbourhood" (in a precise mathematical sense) of the supercover discretization of K. Moreover, we have two very interesting results:

- The supercover discretization is always a Hausdorff discretizing set iff the distance d is what we call *cellular*.
- The supercover discretization is always the Hausdorff discretization iff the distance d has its unit ball homothetic to the cell $C(o)$ (for a square grid, that is $\mathcal{D} = \mathbf{Z}^n$, this means that d is a multiple of the chessboard distance).

Our method can also be applied to show that in the discretization by dilation $\Delta_{\oplus}^{A}$, where the covering assumption $\mathcal{D} \oplus A = \mathcal{E}$ is verified, the Hausdorff distance between a compact K and its discretization $\Delta_{\oplus}^{A}(K)$ is bounded by the radius of A; this implies in particular that $\Delta_{\oplus}^{A}(K)$ converges to K (in the Hausdorff metric sense) when the resolution of $\mathcal{D}$ and the size of A tend to zero.

Our results are stated without proofs; most of them can be found in [14]. The next section recalls the definitions of distances, norms, and Hausdorff metric, and introduces some new concepts in that respect. The following one states the definition and properties of Hausdorff discretization. The last one gives our conclusion.

2. Distances, Norms, and Hausdorff Metric

We assume that the reader is acquainted with elementary point-set and metric topology (otherwise, see [11] for an introduction).

Let $\mathcal{E} = \mathbf{R}^n$ be the n-dimensional Euclidean space; write o for the origin in $\mathcal{E}$ (the neutral for vector addition); for $x \in \mathcal{E}$, its coordinates are written $x_1, \ldots, x_n$ (thus $x = (x_1, \ldots, x_n)$). Let $\mathcal{D}$ be an n-dimensional digital space forming a rectangular grid in $\mathcal{E}$, containing the origin o, (that is, $\mathcal{D} = \prod_{i=1}^{n}(\rho_i \mathbf{Z})$, where ρ_i is the resolution or grid spacing along the i-th axis). A *distance* on $\mathcal{E}$ is a function $d : \mathcal{E}^2 \to \mathbf{R}$ satisfying for all $x, y, z \in \mathcal{E}$: (a) $d(x,x) = 0$ and $d(x,y) > 0$ for $x \neq y$, (b) $d(x,y) = d(y,x)$, and (c) $d(x,z) \leq d(x,y) + d(y,z)$. A *norm* on $\mathcal{E}$ is a function $N : \mathcal{E} \to \mathbf{R}$ satisfying for all $\lambda \in \mathbf{R}$ and $x, y \in \mathcal{E}$: (i) $N(o) = 0$ and $N(x) > 0$ for $x \neq o$, (ii) $N(\lambda x) = |\lambda| N(x)$, and (iii) $N(x+y) \leq N(x) + N(y)$. Every norm N leads to a distance d defined by $d(x,y) = N(x-y)$. Conversely, a distance d is induced in such a way by a norm N iff it is both:

- *translation-invariant*, that is, $d(x,y) = d(x+z, y+z)$ for any $x, y, z \in \mathcal{E}$; and
- *compatible with scaling*, that is, $d(\lambda x, \lambda y) = \lambda d(x,y)$ for any $\lambda > 0$ and $x, y \in \mathcal{E}$.

Although several of our results apply to arbitrary distances on $\mathcal{E}$, we assume from now on that $\mathcal{E}$ is endowed with a distance d arising from a norm N. Well-known examples are the L^p norms $\|\cdot\|_p$ for $1 \leq p \leq \infty$, defined by

$$\|x\|_p = \left(|x_1|^p + \cdots + |x_n|^p\right)^{1/p} \quad (1 \leq p < \infty), \qquad \|x\|_\infty = \max\left(|x_1|, \ldots, |x_n|\right).$$

The distance induced by $\|\cdot\|_p$ is written d_p; d_1 is the Manhattan distance, d_2 is the Euclidean distance, and d_∞ is the chessboard distance. Other examples of distances induced by norms include *chamfer distances* [5]. For $r \geq 0$ and $x \in \mathcal{E}$ we define $\mathcal{B}_r^d(x)$, the *closed ball of radius r centered about x*, by

$$\mathcal{B}_r^d(x) = \{y \in \mathcal{E} \mid d(x,y) \leq r\}. \tag{4}$$

We write $\mathcal{B}_r^d$ for $\mathcal{B}_r^d(o)$; by translation-invariance, $\mathcal{B}_r^d(x)$ is the translate of $\mathcal{B}_r^d$ by x.

Let A and B be two non-void *compact* (i.e., bounded and closed) subsets of $\mathcal{E}$. For every $p \in \mathcal{E}$, we define $d(p,B) = \min\{d(p,b) \mid b \in B\}$ (this minimum is attained). Now we define (this maximum is also attained):

$$h_d(A,B) = \max\{d(a,B) \mid a \in A\}. \tag{5}$$

We define then the *Hausdorff distance* between A and B as:

$$H_d(A,B) = \max\bigl(h_d(A,B), h_d(B,A)\bigr). \tag{6}$$

It is well-known [3] that H_d is indeed a distance function (satisfying requirements (a,b,c)) on the space $\mathcal{H}(\mathcal{E})$ of nonvoid compact subsets of $\mathcal{E}$. Note [3] that for every $r \geq 0$ we have $h_d(A,B) \leq r$ iff $A \subseteq B \oplus \mathcal{B}_r^d$, and

$$H_d(A,B) \leq r \iff A \subseteq B \oplus \mathcal{B}_r^d \text{ and } B \subseteq A \oplus \mathcal{B}_r^d. \tag{7}$$

2.1. Covering Radius and Cellular Distances

We introduce here several new concepts that will be used in the next section:

Definition 1 *The* cellular covering radius $r_{cc}(d)$ *and the* covering radius $r_c(d)$ *are:*

$$r_{cc}(d) = \min\{r > 0 \mid C(o) \subseteq \mathcal{B}_r^d\} \qquad \textit{and} \qquad r_c(d) = \min\{r > 0 \mid \mathcal{E} = \mathcal{D} \oplus \mathcal{B}_r^d\}.$$

As d is compatible with scaling, both $r_{cc}(d)$ and $r_c(d)$ are proportional to the grid spacing of $\mathcal{D}$; if this grid spacing is multiplied by a scalar factor $\alpha > 0$ (i.e., if every $p \in \mathcal{D}$ is replaced by αp), then $r_{cc}(d)$ and $r_c(d)$ are multiplied by α.

Definition 2 *The distance d is cellular (w.r.t. $\mathcal{D}$) if for every $p, q \in \mathcal{D}$ and $x \in C(p)$ we have $d(p,x) \leq d(q,x)$. In particular, for $x \in C(p) \cap C(q)$ we have $d(p,x) = d(q,x)$.*

When the norm satisfies $N(x_1, \ldots, x_i, \ldots, x_n) = N(x_1, \ldots, -x_i, \ldots, x_n)$ for $i = 1, \ldots, n$, d is cellular [14]; this is indeed the case for the distances d_p ($1 \leq p \leq \infty$).

Proposition 3 *We have $r_c(d) \leq r_{cc}(d)$. If d is cellular, then $r_c(d) = r_{cc}(d)$.*

3. Hausdorff Discretization

Let $\mathcal{H}(\mathcal{E})$ be the family of all nonvoid compact (i.e., bounded and closed) subsets of $\mathcal{E}$; the Hausdorff distance H_d is defined on $\mathcal{H}(\mathcal{E})$. Let $\mathcal{H}(\mathcal{D}) = \mathcal{H}(\mathcal{E}) \cap \mathcal{P}(\mathcal{D})$; in fact $\mathcal{H}(\mathcal{D})$ consists of all nonvoid finite subsets of $\mathcal{D}$.

Definition 4 *Given $K \in \mathcal{H}(\mathcal{E})$ and $S \in \mathcal{H}(\mathcal{D})$, S is a* Hausdorff discretizing set of *K if S minimizes the Hausdorff distance to K:*

$$\forall T \in \mathcal{H}(\mathcal{D}), \quad H_d(K,S) \leq H_d(K,T).$$

The family of Hausdorff discretizing sets is written $\mathcal{M}_{H_d}(K)$.

In fact there can be several Hausdorff discretizing sets of K:

Proposition 5 *For $K \in \mathcal{H}(\mathcal{E})$, $\mathcal{M}_{H_d}(K)$ is nonvoid, finite, and closed under the union operation: for $K_1, K_2 \in \mathcal{M}_{H_d}(K)$, we have $K_1 \cup K_2 \in \mathcal{M}_{H_d}(K)$.*

We can thus choose the greatest element of $\mathcal{M}_{H_d}(K)$ as the discretization of K:

Definition 6 *Given $K \in \mathcal{H}(\mathcal{E})$, the* Hausdorff discretization of K *is the greatest element $\Delta_{H_d}(K)$ of $\mathcal{M}_{H_d}(K)$:*

$$\Delta_{H_d}(K) = \bigcup \mathcal{M}_{H_d}(K) = \bigcup_{S \in \mathcal{M}_{H_d}(K)} S.$$

Let us insist that this choice of $\Delta_{H_d}(K)$ as *the* discretization is just a matter of convenience, it does not mean that $\Delta_{H_d}(K)$ is in any way a "better" discretizing set than any other element of $\mathcal{M}_{H_d}(K)$.

3.1. Characterization and Convergence Properties of the Hausdorff Discretization

We will show that the Hausdorff discretization of a compact set K is in fact the discretization by dilation with a ball $\mathcal{B}_r^d$ whose radius r can be mathematically characterized and depends on K:

Definition 7 *For $r > 0$ and $X \in \mathcal{P}(\mathcal{E})$, the* discretization of radius r of X *is the set Δ_r^d defined by*

$$\Delta_r^d(X) = \Delta_{\oplus}^{\mathcal{B}_r^d}(X) = (X \oplus \mathcal{B}_r^d) \cap \mathcal{D} = \{p \in \mathcal{D} \mid \mathcal{B}_r^d(p) \cap X \neq \emptyset\}. \qquad (8)$$

Fig. 2. Left: A compact set K overlayed with pixels p and their cells $C(p)$. Right: $\Delta_{H_d}(K)$ for $d = d_2$, the Euclidean distance; we show the circles of radius $r_H(K,d)$ centered about its pixels.

Definition 8 *Given $K \in \mathcal{H}(\mathcal{E})$, let*

$$\mathcal{R}(K,d) = \{r \geq 0 \mid K \subseteq \Delta_r^d(K) \oplus \mathcal{B}_r^d\} \qquad \text{and} \qquad r_H(K,d) = \inf \mathcal{R}(K,d). \tag{9}$$

$r_H(K,d)$ is called the Hausdorff radius *of K for the distance d.*

Theorem 9 *$\Delta_{H_d}(K) = \Delta^d_{r_H(K,d)}(K)$ and $H_d\big(\Delta_{H_d}(K), K\big) = r_H(K,d) \leq r_c(d)$.*

This result is illustrated in Fig. 2. An immediate consequence is that $r_H(K,d) \leq r_{cc}(d)$, which is proportional to the resolution, i.e., grid spacing; as this resolution tends to zero, so does $H_d\big(S,K\big)$ for $S \in \mathcal{M}_{H_d}$: thus the Hausdorff discretization converges to the compact set in the Hausdorff metric sense.

3.2. Relation with the Supercover and Morphological Discretizations

Although we have a mathematical characterization of the Hausdorff discretization, we still need a practical way to construct it. On the other hand, the supercover discretization $\Delta_{SC}(K)$ defined in (3) is generally easy to construct [2, 6]. We see below that $\Delta_{H_d}(K)$ is in fact restricted to a "neighbourhood" of $\Delta_{SC}(K)$:

Proposition 10 *For $K \in \mathcal{H}(\mathcal{E})$,*

$$\Delta_{H_d}(K) \subseteq \Delta_{SC}(K) \oplus \Delta_{SC}\big(\mathcal{B}^d_{r_{cc}(d)}\big) \qquad \text{and} \qquad H_d(K, \Delta_{SC}(K)) \leq r_{cc}(d).$$

For cellular distances we can say more:

Theorem 11 *The supercover discretization is always a Hausdorff discretizing set, that is:*

$$\forall K \in \mathcal{H}(\mathcal{E}), \quad \Delta_{SC}(K) \in \mathcal{M}_{H_d}(K),$$

iff the distance d is cellular.

Theorem 12 *The supercover discretization is always the Hausdorff discretization, that is:*

$$\forall K \in \mathcal{H}(\mathcal{E}), \quad \Delta_{SC}(K) = \Delta_{H_d}(K),$$

iff $C(o) = \mathcal{B}^d_{r_{cc}(d)}$. For $\mathcal{D} = (\rho\mathbf{Z})^n$, this means that $d = \lambda \cdot d_\infty$ for some $\lambda > 0$.

It would be interesting to apply our analysis to the discretization by dilation studied by Heijmans. We obtain the following result:

Proposition 13 *Let A be a structuring element satisfying the* covering assumption $\mathcal{D} \oplus A = \mathcal{E}$. *Let $r_d(A) = \min\{r > 0 \mid A \subseteq \mathcal{B}_r^d\} = \sup\{N(x) \mid x \in A\}$. Then $r_d(A) \geq r_c(d)$, and for every $K \in \mathcal{H}(\mathcal{E})$, we have $H_d(K, \Delta_\oplus^A(K)) \leq r_d(A)$.*

It follows that if the grid spacing and the size of A are both multiplied by a scalar α, when $\alpha \to 0$, $\Delta_\oplus^A(K)$ converges to K in the Hausdorff metric sense.

4. Conclusion

We have introduced a new theory of discretization, based on the Hausdorff distance. It is mathematically well-founded, and provides the convergence (in the Hausdorff metric sense) of the discretization to the original object when the resolution tends to zero. This discretization is based on a global criterion, and it adapts itself to the object to be discretized. It seems that the (greatest) Hausdorff discretization $\Delta_{H_d}(K)$ is not always easy to compute. However for *cellular distances* (such as the usual d_p or chamfer distances), the *supercover discretization* $\Delta_{SC}(K)$ is a Hausdorff discretizing set, and so it can be chosen as a suitable discretization; moreover, it is generally easy to compute, and it preserves some topological properties of Euclidean objects [2, 6]. This gives thus additional support for this type of discretization.

Further work will be needed on the topological properties of Hausdorff discretization, on its extension to grey-level images, and on the discretization of geometrical and morphological operations.

References

1. E. Andrès. Cercles discrets et rotations discrètes. Thèse de Doctorat, Université Louis Pasteur, Strasbourg (France), Dec. 1992.
2. E. Andrès, P. Nehlig, and J. Françon. Tunnel-free supercover 3D polygons and polyhedra. In D. Fellner and L. Szirmay-Kalos, editors, *Proc. Eurographics'97*, 16(3), pages C3–C13, Budapest, Sept. 1997.
3. M. F Barnsley. *Fractals Everywhere*. Academic Press, second edition, 1993.
4. J. E. Bresenham. Algorithm for Computer Control of a Digital Plotter. *IBM Systems Journal*, 4(1):25–30, 1965.
5. J.M. Chassery and A. Montanvert, *Géométrie Discrète en Analyse d'Images* Hermès, Paris, 1991.
6. D. Cohen-Or and A. Kaufman. Fundamentals of surface voxelization. *Graphical Models and Image Processing*, 57(6):453–461, Nov. 1995.
7. H. J. A. M. Heijmans. Morphological discretization. In U. Eckhardt et al., editors, *Geometrical Problems in Image Processing*, pages 99–106. Akademie Verlag, Berlin, 1991.
8. H. J. A. M. Heijmans. Discretization of morphological operators. *Journal of Visual Communication and Image Representation*, 3(2):182–193, June 1992.
9. H. J. A. M. Heijmans and J. Serra. Convergence, continuity and iteration in mathematical morphology. *J. Visual Communication and Image Representation*, 3(1):84–102, Mar. 1992.
10. H. J. A. M. Heijmans and A. Toet. Morphological sampling. *Computer Vision, Graphics and Image Processing: Image Understanding*, 54(3):384–400, 1991.
11. J.G. Hocking and G.S. Young *Topology*. Dover Publications Inc., New York, 1988.
12. J.-P. Reveillès. Géométrie discrète, calcul en nombres entiers et algorithmique. Thèse de Doctorat d'Etat, Université Louis Pasteur, Strasbourg (France), Dec. 1991.
13. C. Ronse A bibliography on digital and computational convexity. *IEEE Transactions on Pattern Analysis and Machine Intelligence*, 11(2):181–190, Feb. 1989.
14. D. Wagner. Distance de Hausdorff et problème discret-continu. Mémoire de D.E.A. (M.Sc. Dissertation), Université Louis Pasteur, Strasbourg (France), June 1997. URL: http://gr6.u-strasbg.fr/Reports/DEA_97_Wagner.ps.gz

MATHEMATICAL MORPHOLOGY BASED ON METRIC SPACES

J. L. DíAZ DE LEÓN S. and J. H. SOSSA-AZUELA.*
Centro de Investigación en Computación-IPN
Av. Juan de Dios Bátiz esquina con M. Othón de Mendizábal
México D.F. 07738, México.
Fax: (52 5) 729 6000 ext. 56607
e-mail:jdiaz,hsossa@pollux.cic.ipn.mx

Abstract. A new framework and a theoretical basis toward the construction of very fast morphological operators (of zero complexity) for an infinite (countable) family of regular metric spaces is presented. The framework is completly defined by the three axioms of metric. The theoretical basis here developed points out properties of some metric spaces and relationships between metric spaces in the same family, just in terms of the properties of the four basic metrics stated in this work. The results her developed, are general in the sense that they cover the most commonly used metrics such as the chamfer, the city block and the chess board metrics. Generalizations and new results related with distances and distance transforms, which in turn are used to develop the morphologic operations in constant time, in contrast with the polynomial time algorithms are also given.

Key words: Mathematical Morphology, Fast Distance Transform, Zero Order Complexity, Constant Time Algorithms, Metric Spaces

1. Introduction

Any morphological operator depends on an ordered class of sets called structural elements [2], [3] and [4]. Usually, these sets are symmetric. The most useful, however, are those based on a metric space. In many applications (for example squeleton extraction and reconstruction, filtering, connexity testing, region filling and proximity testing) the use and manipulation of big discs is necessary. However, due to the nature of the operations of erosion and dilation, the use of bigger discs as structural elements results in slower algorithms for which the time increases in a quadratic way with respect to the radii of the disc (linearly with respect to the cardinality of the structural element).

The purpose of this work is to show that using a new approach based on the metric space structure, it is possible to reduce the complexity of the morphological operators to zero complexity for any regular discrete metric space and to eliminate the use of the structural element. In particular, this is done for an infinite family of metric spaces further defined. The use of the distance transformation is proposed for it comprises the information concerning all the discs included in a region. A useful framework to solve some computer vision problems is also provided. The philosophy to achieve these objectives is as follows: 1) Prove some important relations between metrics and some computer vision concepts, 2) Prove that for any linear combined metric space based on three metrics (in fact any regular discrete metric space), the

* H. Sossa is also with the Automatic Control Section, CINVESTAV-IPN, Mexico. E-mail:hsossa@ctrl.cinvestav.mx

distance transform can be computed in constant time (zero complexity), and 3) Prove that the basic morphological operations (erosion, dilation, opening and closing) are obtained by combinations of thresholdings over distance transformations.

2. Main Results

The following results can be extended to X^n without too much effort. However, in most cases, they will be given for Z^2. Hereafter, on this work, α, β, γ and χ are assumed positive real numbers unless something different is stated. The demonstrations and a more exhaustive explanation of the given results can be found in [1]. In the body of this work it is specified which results give in the direct or in the indirect way an algorithm as a by-product. These algorithms are also given in [1].

2.1. Results concerning Metric and Pseudometric Spaces on Z^2

Whenever X is a subset of the real numbers ($\Re$), the symbol X^+ means $\{a \in X \,|a \geq 0\}$.
Definition. Let d be a (pseudo) metric on X. The function $d_s = \alpha \cdot d$, is called equivalent image of d.

The more commonly used discrete metrics on Z^2, which will be the basis of this work, are defined as follows: Let $p_1 = (x, y)$ and $p_2 = (u, v)$ real valued vectors. Then we have:

$$\begin{aligned} d_4(p_1, p_2) &= |u - x| + |v - y| \, . \\ d_8(p_1, p_2) &= \max\{|u - x|, |v - y|\} \, . \\ d_{6R}(p_1, p_2) &= \max\{|u - x|, |v - y|, |u - x + v - y|\} \, . \\ d_{6L}(p_1, p_2) &= \max\{|u - x|, |v - y|, |u - x - v + y|\} \, . \end{aligned}$$

these metrics will form the basis to construct any metric for a particular family of metrics, as will be shown later.
Definition. Let $N_8(c) = \{(c_x + a, c_y + b) \,|a, b \in \{1, 0, -1\}, c = (c_x, c_y) \in Z^2\}$ a 3×3 neighborhood centered on c. The set

$$K(c) = \{(p - c, d(p, c)) \,|p \in N_8(c) \, \text{and} d \text{is metric on} Z^2\}$$

is called the distance neighborhood of c induced by d.

Notice that if the metric in question is invariant to translations, also it is the induced distance neighborhood.
Lemma 2.1. Let the metrics d_4, d_8, d_{6R}, d_{6L} defined as above, then it holds that:

$$d_4 + d_8 = d_{6R} + d_{6L}.$$

From the above discussion we can see that either set $\{d_4, d_8, d_{6R}\}$ or $\{d_4, d_8, d_{6L}\}$ of metrics (in fact any triplet from the 4 basic metrics in discussion) can be used instead of the complete set of metrics $\{d_4, d_8, d_{6R}, d_{6L}\}$ because any of these metrics can be found as a linear combination of the other three.
Theorem 2.1. There exist $\alpha, \beta \in Z^+$ such that $d = \alpha \cdot d_4 + \beta \cdot d_8$ has a distance neighborhood which is an equivalent image of:

$$\begin{bmatrix} q & p & q \\ p & 0 & p \\ q & p & q \end{bmatrix}$$

with $p, q \in Z^+; 0 < p \leq q \leq 2 \cdot p$.

Theorem 2.2. There exist $\alpha, \beta, \gamma \in Z^+$ such that $d = \alpha \cdot d_4 + \beta \cdot d_8 + \gamma \cdot d_{6R}$ has a distance neighborhood which is an equivalent image of:

$$\begin{bmatrix} s & p & q \\ p & 0 & p \\ q & p & s \end{bmatrix}$$

with $0 < p \leq q \leq s \leq 2 \cdot p$.

Theorem 2.3. There exist $\alpha, \beta, \gamma \in Z^+$ such that $d = \alpha \cdot d_4 + \beta \cdot d_8 + \gamma \cdot d_{6L}$ has a distance neighborhood which is an equivalent image of:

$$\begin{bmatrix} q & p & s \\ p & 0 & p \\ s & p & q \end{bmatrix}$$

with $0 < p \leq q \leq s \leq 2 \cdot p$.

Theorems 2.1 to 2.3 mean that a wide family of the so called "chamfer metrics" are represented in a member of the family of metric spaces (X, d_l). In particular, an important chamfer metric, the chamfer $(5, 7)$ metric is fully described by the following linear combined metric:

$$d = 2 \cdot d_4 + 3 \cdot d_8.$$

2.2. Results concerning Distance Transforms, Contours, Neighborhoods and R-borders

Definition. A mapping DT:

$$\begin{aligned} DT &: F \to \Delta = \{(x, \delta(x))\}; x \in X. \\ F &= \{(x, f(x))\}; f : X \to \{0, 1\}. \\ f(x) &= 1 \Leftrightarrow x \in R \subset X \text{ and } f(x) = 0 \Leftrightarrow x \notin R. \\ \delta &: X \subseteq \Re^n \to \Re^+. \\ \delta(x) &= \min\{d(x, y) \,|y \notin R\}. \end{aligned}$$

where d is a metric, is called distance transformation, F is an image and δ is called distance transform function.

Theorem 2.4. Let $\delta_4, \delta_8, \delta_{6R}, \delta_{6L}$ be distance transform functions for the metrics d_4, d_8, d_{6R}, d_{6L} respectively. Then in general:

$$\alpha\delta_4 + \beta\delta_8 + \gamma\delta_{\{6L,6R\}} \leq \delta_{\alpha \cdot d_4 + \beta \cdot d_8 + \gamma \cdot d_{\{6R,6L\}}}; \alpha, \beta, \gamma \in Z^+;$$

From this Theorem we can see that the linear combination of distance transform functions is not the distance transform function for the associated metric although it is bounded by this.

Definition. Let $R \subset X \subseteq \Re^n$. Then the contour of R is defined as:

$$B_k(R) = \{x \in R \,|\inf\{d_k(x, y)\} \Rightarrow \inf(D - \{0\}), y \notin R\}.$$

where D is the range of the metric d. Notice that if $X \subset Z^n$ and $D \subset Z^+$;

$$\inf\{d_k(x,y)\} = \min\{d_k(x,y)\}; \inf(D - \{0\}) = \min(D - \{0\})$$

Definition. The set $N_k(x) = \{y \in Z^2 \,|d_k(x,y) \Rightarrow \inf(D - \{0\})\}$, is called the $k-$Neighborhood of x in the specified metric. D is the range of the metric.
Lemma 2.2. Let $A_k = \{x\,|\exists y \notin R \text{ and } y \in N_k(x)\}$; then $A_k = B_k(R)$.
Theorem 2.5. For any linear combined metric it holds that:

$$B_l(R) \in \{B_4(R), B_8(R), B_{\{6R,6L\}}(R)\}.$$

Theorem 2.6. Let $B(x) \in \{B_l(R), B_8(R), B_{\{6R,6L\}}(R)\}$. Then it holds that $B_4(R) \subseteq B(R)$ and consequently $card(B_4(R)) \leq card(B(R))$. Here $card(X)$ means cardinality of X.
Definition. Given a metric space (Z^2, d) and a set $R \subset Z^2$ then the set defined by:

$${}_rB(R) = \{x \in R\,|\min\{d(x,y)\} = r; y \notin R, r \in D\}$$

where D is the range of the metric; is called r-bound of R.
Theorem 2.7. Let $R_0 = \emptyset; R_1 = R; R_2 = R_1 - B(R_1); R_k = R_{k-1} - B(R_{k-1})$ where: $B \in \{B_4, B_8, B_{\{6R,6L\}}\}$. Then it holds that:

$${}_rB(R) = B(R_r).$$

Theorems 2.5 to 2.7 are useful to verify a) the validity of the classical methods for the norms 0 and ∞, and b) their usefulness for the linear combined metrics.
Theorem 2.8. Let a metric $d_k \in \{d_4, d_8, d_{6R}, d_{6L}\}$. For any $p, q \in Z^2$ it holds:
i. – $d(p,q) = \min\{d_k(c,q)\,|c \in N_k(p)\} + 1$.
ii. – $d_k(p,q) = \max\{d_k(c,q)\,|c \in N_k(p)\} - 1$.
iii. – $\{c \in N_4(p)\,|d_4(p,q) = d_4(c,q)\} = \emptyset$.
Corollary 1. It is clear that: $\min\{d(c,q)\,|c \in N(p)\} \geq \min\{\delta(c)\,|c \in N(p)\}$ and by Theorem 8 it holds that:

$$\begin{aligned}\delta(p) &= \min\{\min\{d(c,q)\,|c \in N(p)\}\,|y \notin R\} + 1 \\ &\Rightarrow \delta(p) = \min\{\delta(c)\,|c \in N(p)\} + 1.\end{aligned}$$

With this result, we can construct Algorithm 1 (see [1]), to obtain the true distance transform function $\delta(x)$ trough its initial value $\delta_0(x)$ using Theorem 2.8.
Theorem 2.9. Let $d_l = \alpha \cdot d_4 + \beta \cdot d_8 + \gamma \cdot d_{\{6R,6L\}}; \alpha, \beta, \gamma \in Z^+$. Then d_l is regular and given $p, q \in Z^2; q \notin N_8(p)$ it holds that:

$$\exists z \in N_8(p) \text{ such that } d_l(z,q) + d_l(z,p) = d_l(p,q).$$

Theorem 2.10. Let $d_l = \alpha \cdot d_4 + \beta \cdot d_8 + \gamma \cdot d_{\{6R,6L\}}; \alpha, \beta, \gamma \in Z^+$. Then given $p, q \in Z^2$ it holds:

$$d_l(p,q) = \min\{d_l(c,q) + d_l(c,p)\,|c \in N_8(p)\}.$$

Corollary 2. It is clear that: $\min\{d(c,q)\,|c \in N(p)\} \geq \min\{\delta(c)\,|c \in N(p)\}$ and by Theorem 11 it holds that:

$$\begin{aligned}\delta(p) &= \min\{\min\{d(c,q) + d(c,p)\,|c \in N(p)\}\,|y \notin R\} \\ &\Rightarrow \delta(p) = \min\{\delta(c) + d(c,p)\,|c \in N(p)\}\end{aligned}$$

due to the properties of the minimum.

Corollary 3. In general Theorem 2.10 and Corollary 2 hold only for any metric on which for all $p, q \in Z^2; q \notin N(p)$

$$\exists z \in N(p) \text{ such that } d(z,q) + d(z,p) = d(p,q).$$

Not only for $d = d_l$, and $N(p)$ can be any neighborhood containing p and z.

Now we can construct Algorithm 2 (see [1]), to calculate the distance transform $\delta(x)$ of an input image using linear combined metrics trough its initial value $\delta_0(x)$.

Notice that Algorithm 2 includes Algorithm 1 as a particular case. This is because Theorem 2.8 is a particular case of Theorem 2.10. Consequently, Algorithm 2 is slower that Algorithm 1.

2.3. Results concerning Erosions and Dilations

Definition. Let (Z^2, d_k) be a metric space with $d_k \in \{d_4, d_8, d_{\{6R,6L\}}, d_l\}$. The set defined by:

$$E(R) = \{x \in R \,|\min\{d_k(x,y)\} > \min(D - \{0\}); y \notin R\}$$

where D is the range of d_k; is called *eroded set of* R, and the operator E_k is called *simple erosion* for metric k.

Definition. Let (Z^2, d_k) a metric space with $d_k \in \{d_4, d_8, d_{\{6R,6L\}}, d_l\}$. The set defined by:

$${}_rE(R) = \{x \in R \,|\min\{d_k(x,y)\} > r \in D; y \notin R\}$$

where D is the range of d_k; is called *r-simultaneous eroded set of* R, and the operator ${}_rE_k$ is called *simultaneous erosion* for metric k.

Theorem 2.11. Let (Z^2, d_k) be a metric space with $d_k \in \{d_4, d_8, d_{\{6R,6L\}}\}$. Then it holds that

$${}_nE(R) = E^n(R).$$

In general this is not true for the linear combined metrics. On the other hand, a direct consequence of theorem 11 for basic metrics is the following:

Theorem 2.12. Let (Z^2, d_k) be a metric space with $d_k \in \{d_4, d_8, d_{6R}, d_{6L}\}$. Then it holds that:

$${}_nE(R) = \{x \in R \,|\delta(x) > n\}.$$

With this theorem we are able to construct Algorithm 3 (see [1]), to obtain the n-simultaneous erosion of R.

Theorem 2.13. Let (Z^2, d_l) be a metric space, then it holds that:

$$E_l^n \in \{E_4^n, E_8^n, E_{6R}^n, E_{6L}^n\}.$$

From Theorem 2.13 we can see that the consecutive (classical) erosion has no meaning for linear combined metrics; the result would be the same erosion.

Definition. Let (Z^2, d_k) be a metric space with $d_k \in \{d_4, d_8, d_{\{6R,6L\}}, d_l\}$. The set defined by:

$$T(R) = R \cup \{y \notin R \,|\min\{d(x,y)\} = \min(D - \{0\}); x \in R\}.$$

where D is the range of d_k; is called *dilated set of* R, and the operator T_k is called *simple dilation* for metric k.
Definition. Let (Z^2, d_k) be a metric space with $d_k \in \{d_4, d_8, d_{\{6R,6L\}}, d_l\}$. The set defined by:

$$_rT(R) = R \cup \{y \notin R | \min\{d(x,y)\} \leq r, r \in D; x \in R\}.$$

where D is the range of d_k; is called *simultaneous dilated set of* R, and the operator T_k is called *simultaneous dilation* for metric k.
Theorem 2.14. Let (Z^2, d_k) be a metric space with $d_k \in \{d_4, d_8, d_{\{6R,6L\}}, d_l\}$. then it holds:

$$T(R) = \overline{E(\bar{R})}.$$

Theorem 2.15. Let (Z^2, d_k) be a metric space with $d_k \in \{d_4, d_8, d_{\{6R,6L\}}, d_l\}$. then it holds:

$$_rT(R) = \overline{_rE(\bar{R})}.$$

Theorem 2.11*. Let (Z^2, d_k) be a metric space with $d_k \in \{d_4, d_8, d_{\{6R,6L\}}\}$. Then it holds that

$$_nT(R) = T^n(R).$$

Theorem 2.12*. Let (Z^2, d_k) be a metric space with $d_k \in \{d_4, d_8, d_{6R}, d_{6L}\}$. Then it holds that:

$$_nT(R) = \{x | \overline{\delta}(x) \leq n\}.$$

where $\overline{\delta}(x)$ is the distance transform of $\overline{R}$.

Using Theorem 2.12* we acan construct Algorithm 4 (see [1]), to obtain the n-simultaneous dilation of R.
Theorem 2.13*. Let (Z^2, d_l) be a metric space, then it holds that:

$$T_l^n \in \{T_4^n, T_8^n, T_{6R}^n, T_{6L}^n\}.$$

Once again, we can see that the consecutive (classical) dilation has no meaning for linear combined metrics since the result is exactly the same that for a basic one.

3. Conclusions and Present Research

A set of results concerning metric spaces, distance transforms and mathematical morphology were presented. These results point out the strong connection between these three topics and are valid for the following family of metric spaces: $(X \subseteq R^2, d_l)$ where $d_l = \alpha \cdot d_4 + \beta \cdot d_8 + \gamma \cdot d_{6R} + \chi \cdot d_{6L}$ with $\alpha, \beta, \gamma, \chi$ natural numbers. They include:

1. A theory for some aspects of mathematical morphology and distance transforms based on metric spaces.
2. The practical elimination of the structural element, when these are discs on a metric space.
3. All the results are general for a countable infinite family of metrics spaces, comprising those used in Computer Vision.

4. The algorithms constructed based on these results are efficient with the lowest complexity (i.e. constant time regardless the size of the desired operation).
5. Two topics apparently not related (Distance Transform and Mathematical Morphology) resulted to be strongly connected with the same axiomatic.

Table I resumes the performance for generic classical morphological and distance transforms operations against the proposed methods, based on algorithm complexity and arithmetic density.

TABLE I
Performance of the proposed algorithms to compute distance transforms, erosions and dilations against the classical methods.

Characteristic	Metrics 4 and 8	Hexagonal metrics	Euclidean Approximation	Proposed approach
Complexity	$\vartheta(1)$	$\vartheta(1)$	$\vartheta(2)$	$\vartheta(0)$
Arithmetic density	low	low	high	low
Irrational fucntions	no	no	yes	no
Extensibility	none	none	none	infinite family
Euclidean approx.	<85%	<85%	>96%	94-96%

TABLE II
Processing time in seconds for the erosion and the dilation.

n	clas 4	Diaz 4	clas 8	Diaz 8	clas 6R	Diaz 6R	clas 6L	Diaz 6L
1	0.0549	0.0549	0.0549	0.0549	0.0549	0.0549	0.0549	0.0549
2	0.1098	0.0549	0.1098	0.0549	0.1098	0.0549	0.1098	0.0549
3	0.1648	0.0549	0.1648	0.0549	0.1648	0.0549	0.1648	0.0549
4	0.2197	0.0549	0.2197	0.0549	0.2197	0.0549	0.2197	0.0549
6	0.3296	0.0549	0.3296	0.0549	0.3296	0.0549	0.3296	0.0549

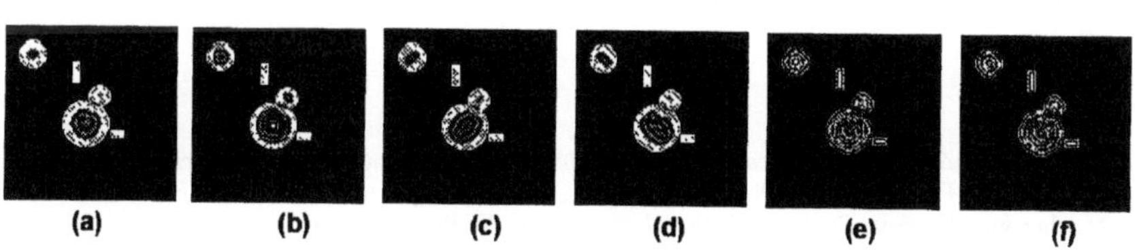

Fig. 1. Distance tranform. a) $d = d_4$, b) $d = d_8$, c) $d = d_{6R}$, d) $d = d_{6L}$, e) $d = 2d_4 + 3d_8$, f) $d = d_4 + d_8$.

Table II, in the other side, compares the processing times required by the proposed algorithms to compute dilations and erosions against the classical ones, and this for different sizes of the operator. In all cases, the times were independent of the image and of the metric space in all experiments. The experiments were done in 486 processor running at 25Mhz. In the case of the opening and clossing operations, the processing time is exactly the sum of the times for erosion and dilation.

Figures 1, 2 and 3 show respectively, the distance transform and the r-borders, the erosion and the dilation of a binary image.

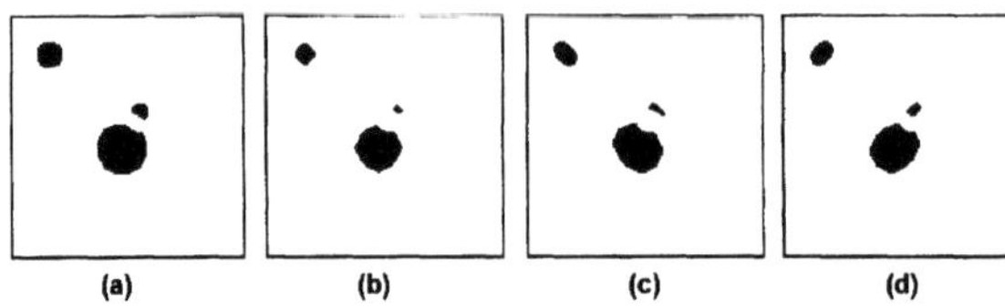

Fig. 2. Erosion. a) $d = d_4$, b) $d = d_8$, c) $d = d_{6R}$, d) $d = d_{6L}$.

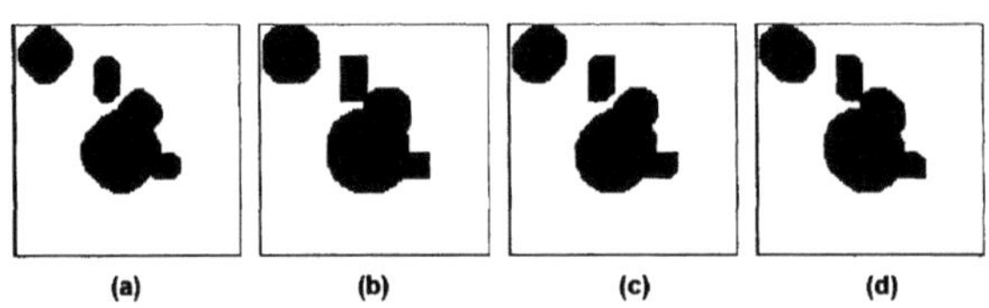

Fig. 3. Dilation. a) $d = d_4$, b) $d = d_8$, c) $d = d_{6R}$, d) $d = d_{6L}$.

At the moment, we are preparing a complete parallel version of these results. We are also extending these results to cover a wider family of metric spaces. Another interesting extension of the proposed framework is also the inclusion of "even discs". As an even disc we mean discs whose center is not an element of the metric space. This extension can be useful to solve some parallelization problems in several kind of problems in Computer Vision such as squeletonization.

References

1. J. L. Diaz-de-León. *Mathematical Morphology based on Linear Combined Metric Spaces in* Z^2. PhD thesis, CINVESTAV - IPN, Electrical Engineering Deparment, 1996. In Spanish.
2. G. Matheron. *Random Sets and Integral Geometry.* Willey, New York, 1975.
3. J. Serra. *Image Analysis and Mathematical Morphology, Vol. 1.* Academic Press, San Diego, 1982.
4. J. Serra. *Image Analysis and Mathematical Morphology, Vol. 2.* Academic Press, San Diego, 1988.

GREY SCALE CONVEX HULLS: DEFINITION, IMPLEMENTATION AND APPLICATION

PIERRE SOILLE
Ecole des Mines d'Alès & EERIE
Laboratoire de Génie Informatique et d'Ingénierie de la Production
Parc scientifique G. Besse, F-30000 Nîmes Cedex, France

Abstract. The convex hull of a set is the smallest convex set containing this set. An equivalent definition consists in intersecting all half-planes containing the set. We show that this latter definition is nothing but an algebraic closing that can be applied to grey scale images. An implementation of this definition leading to a decreasing family of convex hulls with increasing precision is proposed. Finally, an application of grey scale convex hulls to the direct grey scale extraction of features for character recognition is presented.

Key words: Convex hull, pattern recognition, shape description, image analysis, discrete geometry, mathematical morphology, algorithms, character recognition.

1. Introduction

A Euclidean set is *convex* if and only if the line segments connecting any two pair of points of the set are included in the set. The convexity property is an important shape descriptor. In the discrete case, there are problems with the above mentioned definition because there may be more than one connected digital line segment linking two points. Hence, several definitions have been proposed for determining whether a discrete set is convex or not:

1. A set of vertices S of a graph $\mathcal{G}$ is convex if for all $u, v \in S$, every vertex on *all* digital shortest paths linking u to v is also in S.
2. A set of vertices S of a graph $\mathcal{G}$ is convex if for all $u, v \in S$, there exists *at least one* shortest path or geodesic linking u to v while remaining in S.
3. A set of points S within a grid is convex if it is equivalent to points falling within the intersection of all half-planes containing S.

These definitions and further discussions about digital convexity issues can be found in [17, 8, 10, 11, 12, 7]. In [16], a set is said to be *weakly* or *strongly* convex depending on whether it satisfies the first or the second definition.

Since the intersection of any family of convex sets is also a convex set, there exists a smallest convex set containing any given set. The smallest convex set containing a set is called the *convex hull* of this set. A non-convex pattern can be analyzed by computing the set difference between its convex hull and the pattern itself. The *concavity regions* of a set are defined as the connected components of this difference image. The use of concavity trees for shape recognition is presented in [3].

Computational geometry algorithms for computing the convex hull of a set of points of the Euclidean plane can be found in [15, chaps. 4-5], [20] and [14]. Algorithms for extracting the convex hull or concavity regions of binary images are proposed in many papers (e. g. [21, 17, 2, 18, 1, 4]).

In this paper, we show that the convex hull of a set can be defined in terms of morphological transformations suited to the processing of grey scale images (Sec. 2). We then present a simple and fast algorithm implementing the proposed definition (Sec. 3). Before concluding, an application to pattern recognition is detailed (Sec. 4).

2. Definitions

2.1. Morphological closing

Morphological image operators are grounded on the intuitive notion of probing an image with a shape of a given size and modifying the image values depending on whether the shape fits, or does not fit, within the image components. Denoting by ϕ_B the closing operator with a structuring element B and ^{c} the set complementation, we have the following relationship: $\phi_B(X) = [\bigcup\{B \mid B \subseteq X^c\}]^c$. We propose here an equivalent formulation by considering the intersection of all translations of the complement of the structuring element B such that it contains X:

$$\phi_B(X) = \bigcap\{B^c \mid X \subseteq B^c\}. \tag{1}$$

Closings are increasing, extensive, idempotent operators. Due to the increasing property, closings as defined in Eq. 1 can be extended to grey scale images using the threshold decomposition principle [13]. We consider a grey scale image $f : F \to I$ as a function which maps a discrete grid of pixel coordinates F to a finite chain of nonnegative integers I whose upper bound is denoted by $t_{\max}$. We denote by $CS_t(f)$ the cross-section of the image f at level t. This operator maps a grey tone image into a binary image as follows: $CS_t(f) = T_{[t,t_{\max}]}(f)$, where $T_{[t_i,t_j]}$ is the threshold operator which sets all pixels x of the input image f whose values lie in the range $[t_i, t_j]$ to 1 and the other ones to 0. We therefore obtain the following threshold decomposition for the closing of a grey scale image f: $\phi_B(f) = \sum_{t=1}^{t_{\max}} \phi_B[CS_t(f)]$.

2.2. Binary convex hull

The third definition of a convex set (see introduction) provides us with an interesting algorithm for building the convex hull of a set, i.e., by intersecting all half-planes containing the set. Rather than considering all half-planes, let us first restrict our attention to those half-planes which have a given orientation θ. There are two such half-planes the second one being the complement of the first. We denote these two half-planes by π_θ^+ and π_θ^-, i.e., $(\pi_\theta^+)^c = \pi_\theta^-$. We now consider the intersection of all translations of these half-planes which contain the set X under study:

$$\bigcap\{\pi_\theta^+ \mid X \subseteq \pi_\theta^+\} \quad \text{and} \quad \bigcap\{\pi_\theta^- \mid X \subseteq \pi_\theta^-\}. \tag{2}$$

Note that the intersection of a series of half-planes of a given orientation and sideness is itself a half-plane of identical orientation and sideness. Moreover, from Eq. 1, it can be seen that the intersection of the half-planes given in Eqs. 2 are nothing but closings with the complement of the corresponding half-plane:

$$\bigcap\{\pi_\theta^+ \mid X \subseteq \pi_\theta^+\} = \phi_{\pi_\theta^-}(X) \qquad \text{and} \qquad \bigcap\{\pi_\theta^- \mid X \subseteq \pi_\theta^-\} = \phi_{\pi_\theta^+}(X).$$

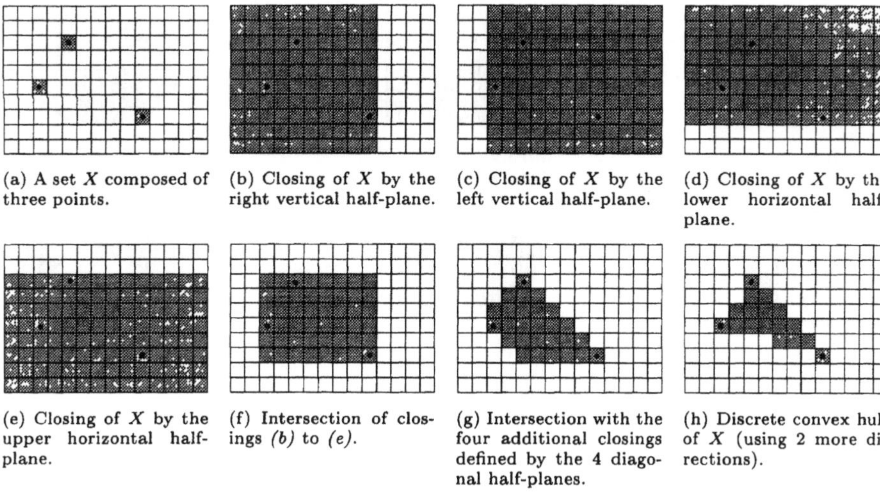

(a) A set X composed of three points. (b) Closing of X by the right vertical half-plane. (c) Closing of X by the left vertical half-plane. (d) Closing of X by the lower horizontal half-plane.

(e) Closing of X by the upper horizontal half-plane. (f) Intersection of closings *(b)* to *(e)*. (g) Intersection with the four additional closings defined by the 4 diagonal half-planes. (h) Discrete convex hull of X (using 2 more directions).

Fig. 1. From closings to convex hulls: a didactical example illustrating the steps leading to the convex hull of a binary pattern composed of three pixels.

Closings of a set of three points with both horizontal and vertical half-planes are shown in Figs. 1b to 1e. In order to compute the convex hull CH of a set X, we need to intersect all half-planes containing the set. This comes down to intersect $\phi_{\pi_\theta^+}(X)$ and $\phi_{\pi_\theta^-}(X)$ for all possible orientations:

$$CH(X) = \bigcap_\theta [\phi_{\pi_\theta^+}(X) \cap \phi_{\pi_\theta^-}(X)]. \tag{3}$$

This definition is illustrated in Fig. 1. The convex hull using only vertical and horizontal directions is shown in Fig. 1f. It corresponds to the smallest rectangle containing the input set. Better approximations of the Euclidean convex hull are obtained by increasing the number of directions. In Fig. 1g, the four diagonal half-planes have also been considered. We define the *discrete convex hull* of a set of pixels of a discrete image as the set of pixels whose *centres* belong to the Euclidean convex hull of the input pixels. The discrete convex hull of the original set of three pixels (Fig. 1a) is shown in Fig. 1h. It uses only 2 more directions (slopes $3/2$ and $-2/7$). In general, by increasing the number of directions we always get a series of discrete convex sets which monotonously decrease and tend to the discrete convex hull of the original set. The convex hull of a real binary pattern is shown in Fig. 2 with increasing precision. The convex hull being a closing, we define the convex hull top-hat transform as the difference between the convex hull and the original set. It follows that the convex hull top-hat outputs the concavity regions of the processed pattern. An example is provided in Fig. 2f.

2.3. Grey scale convex hull

In the previous section, a definition of the convex hull operator in terms of an intersection of morphological closings has been proposed. It follows that the convex

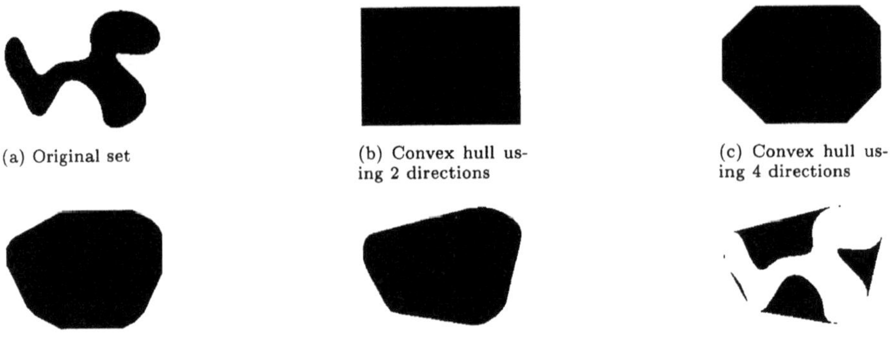

(a) Original set

(b) Convex hull using 2 directions

(c) Convex hull using 4 directions

(d) Convex hull using 8 directions

(e) Convex hull using 48 directions

(f) Arith. diff. between (e) and (a)

Fig. 2. Convex hull of binary shape using an increasing number of directions for the half-planes used in the closing operations.

hull operator is an algebraic closing that can be applied to grey scale images using the threshold superposition principle. In fact, rather than computing the convex hull of each cross-section, we can directly perform the point-wise minimum $\wedge$ between the closings with half-planes in all possible orientations:

$$CH(f) = \bigwedge_{\theta} [\phi_{\pi_\theta^+}(f) \wedge \phi_{\pi_\theta^-}(f)]. \tag{4}$$

The grey scale convex hull of a grey scale image is illustrated in Fig. 3. Again, the output images monotonously decrease and tend to the discrete convex hull of the input image when the number of directions increases.

3. Implementation

3.1. Principle

In order to compute the convex hull of an image using Eq. 4, we need to implement the closing operator with a half-plane of arbitrary orientation. Our implementation is illustrated in Fig. 4 using a 7×5 grey scale image and a half-plane of orientation $\pi/4$. A detailed explanation of the algorithm follows. Given a half-plane of arbitrary direction, we position it next to an image corner so as to have an empty intersection between the half-plane and the image. Assuming that the values outside the image are set to 0, we then progressively translate the half-plane so as to process all image pixels. The maximum value along the new image pixels falling within the half-plane after each translation of the half-plane is computed. It is then compared to the value obtained at the previous translation (at the very first step, the previous value is initialized to 0). The current value is then set to the maximum between the previous value and the computed maximum value among the new pixels. These pixels are then set to the current value. The half-plane is then further translated and the procedure is repeated until all pixels have been processed. The resulting image is nothing but the closing of the original image by a half-plane.

(a) Original grey scale image
(b) Convex hull using 2 directions
(c) Convex hull using 4 directions
(d) Convex hull using 8 directions
(e) Convex hull using 48 directions
(f) Arith. diff. between (e) and (a)

Fig. 3. Convex hull of a grey scale image using an increasing number of directions for the half-planes used in the closing operations.

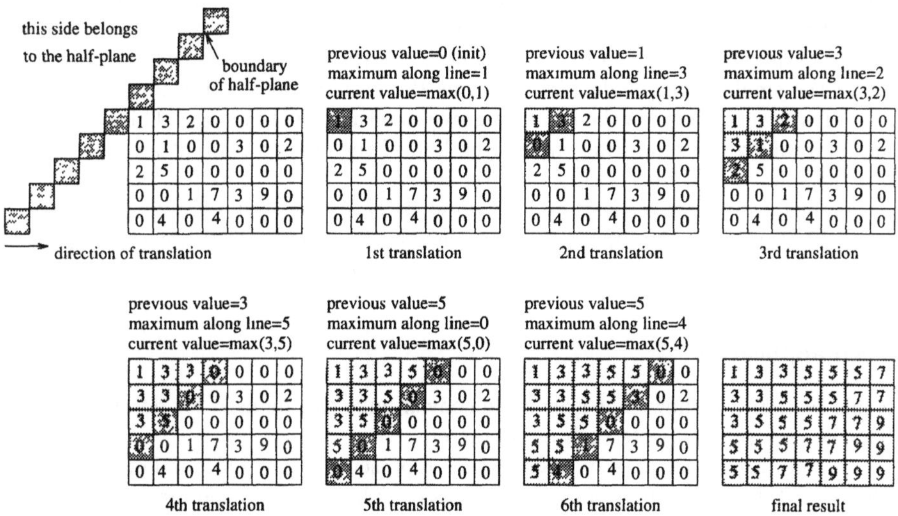

Fig. 4. Closing of a small grey scale image by a half-plane of orientation $\pi/4$. The light grey regions have already been closed. The dark grey pixels represent the new pixels falling within the image frame after a translation of one pixel of the half-plane along the horizontal direction (see text for a detailed explanation).

In practice, we store 2-D images in a 1-D array; that is, one line after another. By doing so, the new pixels falling within the image frame after each translation are directly accessed thanks to a unique array of indices called the p array. For instance, $p(x) = x + n * ncol$ sets the index array for processing the nth line of f (i.e., when closing with an horizontal half-plane) and $p(x) = x * ncol + n$ for processing the nth column (i.e., when closing with a vertical half-plane). A line at arbitrary orientation within the 2-D array is constructed using Bresenham's scan-line conversion algorithm [5, 6]. The pixels of the resulting line are addressed by simply loading the p array with the appropriate indices. This approach has already been detailed in [19] for the computation of erosions and dilations along discrete lines at arbitrary angles. Here, the maximum value is computed along each indexed line and compared to the previous value.

3.2. Performance

Complexity is linear with respect to the number of image pixels and the number of orientations. The algorithm is highly parallel since the closings by each pair of dual half-planes can be computed in parallel before considering the point-wise minimum between all of them. As concerns memory requirements, the algorithm uses three images of the same size and depth: the first holds the original image, the second contains the current intersection of all already computed closings, and the third is used for computing the closing by the current dual half-planes.

Note that for an arbitrary image, the precision of the computed convex hull increases with the number of directions and converge to the actual discrete convex hull of the image. The required number of directions for converging depends on the shape, the size and the orientation of the image objects. The upper bound of the number of directions is given by $2n - 1$ where n is the length in pixels of the longest side of the smallest rectangle with contains the cross-section of the image at level $t_{min} + 1$ where t_{min} equals the smallest grey scale value of the image. This rectangle can be obtained by computing the convex hull of the cross-section at level $t_{min} + 1$ with the sole vertical and horizontal directions.

The proposed implementation can be enhanced in the particular case of binary images. Indeed, once the value 1 has been encountered along a line, it can be directly propagated to the image region that has not yet been processed.

4. An application

In this section, we show that grey scale convex hulls can be used for describing the shape of grey scale objects such as characters.

Starting from a set of grey scale images representing decimal numbers (see first column of Table I), we first compute their convex hull (second column). The difference between the convex hull and the original image provides us with an image of the concavity regions of each number (third column). Concavity regions which are visible from the left (resp. right) side are then computed and shown in the fourth (resp. fifth) columns. Inner regions are defined as those concavity regions that are not visible form the left and right sides nor from the top and bottom sides of the image. These regions are shown in the right most column of Table I. All regions are then used for featuring the numbers. For example, a zero is the sole character whose inner region is almost identical to its concavity region. This approach is currently investigated for the automatic recognition of car plates. Note that our approach does not require to threshold the characters beforehand. Hence, it fits well the direct grey scale extraction of features for character recognition recommended in [22].

TABLE I
Decimal numbers extracted from car plates together with their convex hulls and the corresponding concavity regions.

Decimal number	Convex hull	Concavity regions	Left regions	Right regions	Inner regions

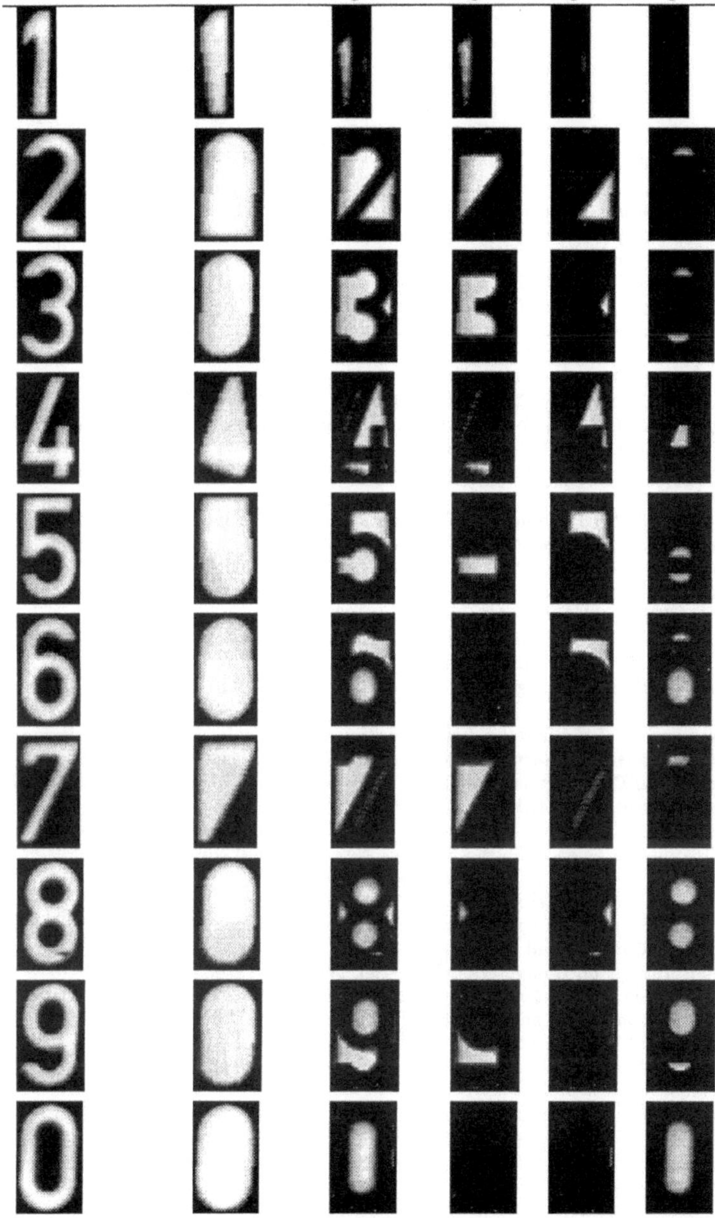

5. Concluding remarks

The definition of the convex hull of a set in terms of an intersection of morphological closings with a series of half-planes allows us to extend the convex hull transform to grey scale images. It is also at the basis of a very efficient implementation. The extension of this implementation to 3-D grey scale data is possible using a similar approach to that developed in [9] for line segments.

References

1. S. Akl and G. Toussaint. Fast convex hull algorithm. *Information Processing Letters*, 7(5):219–222, August 1978.
2. C. Arcelli and L. Levialdi. Concavity extraction by parallel processing. *IEEE Transactions on Systems, Man and Cybernetics*, 1:394–396, September 1971.
3. B. Batchelor. Using concavity trees for shape description. *Computers and Digital Techniques*, 2(4):157–167, August 1978.
4. G. Borgefors and G. Sanniti di Baja. Analysing nonconvex 2D and 3D patterns. *Computer Vision and Image Understanding*, 63(1):145–157, January 1996.
5. J. Bresenham. Algorithm for computer control of digital plotter. *IBM System Journal*, 4:25–30, 1965.
6. J. Foley, A. van Dam, S. Feiner, and J. Hughes. *Computer graphics —Principles and Practice—*. Addison-Wesley, 2nd edition, 1990.
7. F. Harary, R. Melter, and I. Tomescu. Digital metrics: A graph-theoretical approach. *Pattern Recognition Letters*, 2:159–163, March 1984.
8. F. Harary and J. Nieminen. Convexity in graphs. *Journal of Differential Geometry*, 16:185–190, 1981.
9. R. Jones and P. Soille. Periodic lines and their applications to granulometries. In P. Maragos, W. Schafer, and M. Butt, editors, *Mathematical Morphology and its Applications to Image and Signal Processing*, pages 264–272. Kluwer Academic Publishers, 1996.
10. C. Kim and A. Rosenfeld. Convex digital solids. *IEEE Transactions on Pattern Analysis and Machine Intelligence*, 4(6):612–618, November 1982.
11. C. Kim and A. Rosenfeld. Digital straight lines and convexity of digital regions. *IEEE Transactions on Pattern Analysis and Machine Intelligence*, 14(2):149–153, March 1982.
12. C. Kim and J. Sklansky. Digital and cellular convexity. *Pattern Recognition*, 15:359–367, 1982.
13. P. Maragos and R. Ziff. Threshold superposition in morphological image analysis systems. *IEEE Transactions on Pattern Analysis and Machine Intelligence*, 12(5):498–504, May 1990.
14. M. McQueen and G. Toussaint. On the ultimate convex hull algorithm in practice. *Pattern Recognition Letters*, 3:29–34, 1985.
15. F. Preparata and I. Shamos. *Computational geometry —An introduction—*. Texts and Monographs in Computer Science. Springer-Verlag, New York, 2nd edition, 1985.
16. M. Schmitt and J. Mattioli. Strong and weak convex hulls in non-Euclidean metric: theory and application. *Pattern Recognition Letters*, 15:943–947, September 1994.
17. J. Sklansky. Recognition of convex blobs. *Pattern Recognition*, 2:3–10, 1970.
18. J. Sklansky, L. Cordella, and S. Levialdi. Parallel detection of concavities in cellular blobs. *IEEE Transacations on Computers*, 25:187–196, February 1976.
19. P. Soille, E. Breen, and R. Jones. Recursive implementation of erosions and dilations along discrete lines at arbitrary angles. *IEEE Transactions on Pattern Analysis and Machine Intelligence*, 18(5):562–567, May 1996.
20. G. Toussaint. A historical note on convex hull finding algorithms. *Pattern Recognition Letters*, 3:21–28, 1985.
21. S. Unger. Pattern detection and recognition. In *Proc. of IEEE*, volume 47, pages 1737–1752, 1959.
22. L. Wang and Y. Pavlidis. Direct gray-scale extraction of features for character recognition. *IEEE Transactions on Pattern Analysis and Machine Intelligence*, 15(10):1053–1066, October 1993.

DIGITIZATION AND CONNECTIVITY

MICHEL SCHMITT
École Nationale Supérieure des Mines de Paris
Centre de Géostatistique
35, rue Saint Honoré
77305 Fontainebleau-Cedex, FRANCE

Abstract. Digitizations of a set can preserve its connectivity and its homotopy. We show that the best two notions are the topological connectivity and the geodesic-connectivity with bounded diameter. The measurability of these notions is also investigated in the context of random closed sets.

Key words: Connectivity, Mathematical Morphology, Geodesic Distance

Although connectivity is a common life word – objects consisting of one piece – its mathematical translation can be expressed in many ways, giving rise to different and non equivalent notions. In this paper, some of them are investigated: topological connectivity, arc-connectivity and geodesic-connectivity. All the notions are *models* of the real world. The underlying mathematics ensure their coherence and in our sense, the best model is the one for which the most numerous notions my be derived. Here, the notions under study are digitization on the one hand and random sets on the other.

The paper is organized as follows:

- First, we recall the different notions of connectivity.
- Then, we examine digitization procedures which preserve the connectivity and especially their convergence.
- Finally, the question of measurability of connected sets is investigated and we conclude by some open questions.

Only some of the proofs are included here.

1. Reminder on Connectivity

For simplicity reasons concerning mainly notations and illustrations, we suppose that the objects under study are embedded in the Euclidean plane R^2. However, the results presented here can be easily extended to dimension n.

1.1. Topological Connectivity

This notion is perhaps the least intuitive and explicitely makes use of the topology. For a complete presentation of these notions, see any text book on general topology, like [1] or [7].

Definition 1.1 *The set $X \subseteq R^2$ is connected if it is not the union of two non-empty disjoint closed sets for the induced topology on X.*

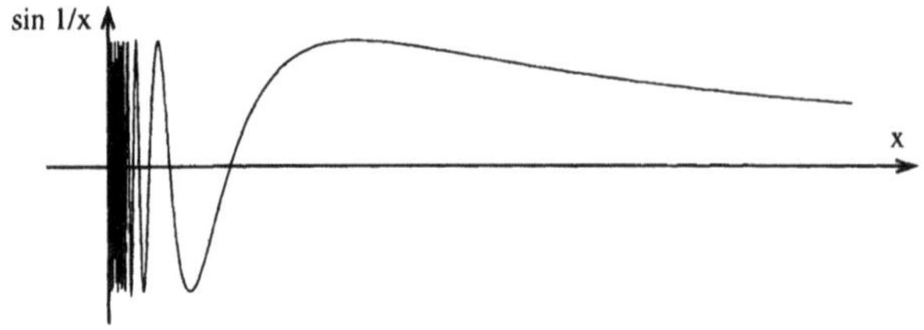

Fig. 1. Graph of function $\sin\frac{1}{x}$, $x \in]0, 6\pi[$ with the line segment from $(0,-1)$ to $(0,1)$. This set is compact, connected, but not arc-connected.

Let us explain this definition. A subset $Y \subseteq X$ is closed for the induced topology (or simply closed in X) if there exists a closed subset $Z \subseteq \mathsf{R}^2$ verifying $Y = Z \cap X$. Then the connectivity criterion is written as: if $X = X_1 \cup X_2$, with X_1 and X_2 closed in X and disjoint, then either X_1 or X_2 is empty.

1.2. Arc-Connectivity

This notion is different and uses the idea that an object is composed of one piece if one can travel from any point to any point in the object. In mathematical words:

Definition 1.2 *A path from x to y in $X \subseteq R^2$ is any continuous function γ from a compact interval $[a,b] \subset R$ into X, such that $\gamma(a) = x$ and $\gamma(b) = y$.*

Definition 1.3 *The set $X \subseteq R^2$ is arc-connected if there exists a path between any pair of points in X.*

The relationships between these two first notions are the following:

Theorem 1.4 *Arc-connectivity implies connectivity. Moreover, if $X \subseteq R^2$ is an open set, then arc-connectivity is equivalent to connectivity.*

In the case of compact sets, the two notions are not equivalent as presented in fig. 1. However, a very similar notion may be derived for compact sets. The principle is that two points do not need to be connected by a (continuous) path, but only by a succession of small jumps.

Definition 1.5 *An ε-path from x to y in $X \subseteq R^2$ is a sequence of points of X, $(x_i)_{i=0}^{i=p}$, such that the distances $d(x_{i-1}, x_i) \leq \varepsilon$, $i = 1 \ldots p$, $x_0 = x$ and $x_p = y$.*

Theorem 1.6 *A compact subset $X \subset R^2$ is connected if and only if any pair of points in X can be connected by an ε-path in X for any $\varepsilon > 0$.*

1.3. Geodesic-Connectivity

If we carefully examine the notion of path, we see that sometimes the path does not correspond to the intuition: especially, it may have an infinite length. An example is depicted in fig. 2. If we want all the paths to have a finite length, the simplest way is to use the geodesic distance on the object [3] (see also the chapter devoted to geodesy in [5]). Although the geodesic distance may be defined on any subset in R^2, it is better to consider now that X is compact.

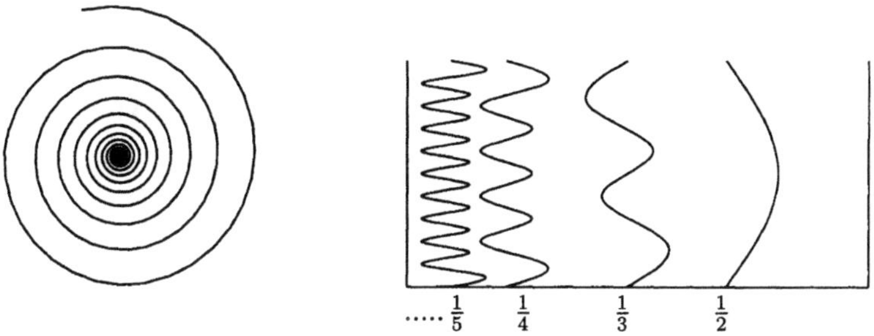

Fig. 2. Two strange objects: (left) a logarithmic spiral with its center is a compact, arc-connected set. Each point of the spiral is at an infinite distance of the center. (right) compact geodesic-connected set with an infinite geodesic diameter. At the left part, there are infinitely many curves with a length going towards infinity.

Definition 1.7 *The* geodesic distance $d_X(x,y)$ *in* $X \subseteq R^2$ *is the lower bound of the lengths of the paths from* x *to* y *into* X. $d_X(x,y) = \infty$ *if there exists no path or if any path is of infinite length.*

This geodesic distance defines on X a topology which is equal or finer than the topology induced by that of R^2.

Definition 1.8 *Let* $X \subseteq R^2$ *be compact.* X *is* geodesic-connected *if it is connected for the geodesic topology induced by* d_X.

Theorem 1.9 *Let* $X \subseteq R^2$ *be compact. If* X *is geodesic-connected, then* X *is arc-connected and connected. But arc-connectivity does not imply geodesic-connectivity, as presented in fig. 2 (left).*

Note however that the **geodesic diameter** (upper bound of the geodesic distance between any two points in X) of a geodesic-connected set may be infinite as depicted in fig. 2 (right).

1.4. Connected Components and Homotopy

The three notions of connectivity described in the previous section can be used to define connected components and homotopy. Here X is assumed to be compact.

Definition 1.10 *For any connectivity notion, a connected component of* X *is a maximal (with respect to inclusion) connected set included in* X.

The set of connected (arc-connected, geodesic-connected) components of $X \subseteq R^2$ build a partition of X, *i.e.* they cover X and are pairwise disjoint. But recall that the connected components depend on the connectivity type: for instance, the spiral is connected and arc-connected, but has two geodesic-connected components (the spiral itself and its center).

The aim of **homotopy** is to describe how the connected components are organized in the plane. We restrict ouselves here to the case where X is composed of a finite number of connected components as well as X^c. The homotopy is described by a tree [6] (see fig. 3):

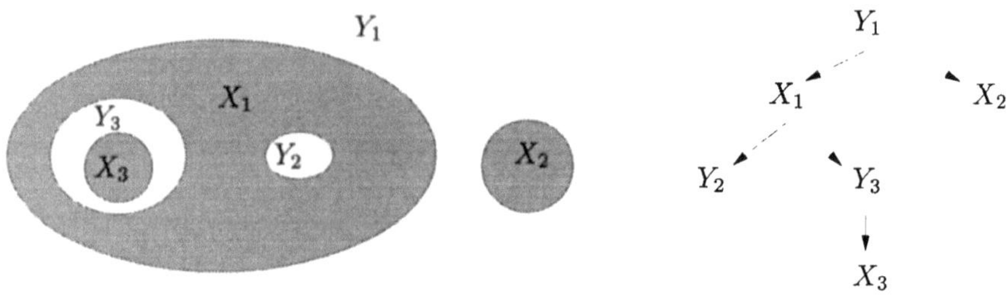

Fig. 3. The homotopy of $X = \cup X_i$ is described by a tree.

- Its root is the unbounded connected component of X^c, here Y_1.
- Its sons at level 1 are the connected components of X which are adjacent to its father, here X_1 and X_2.
- The sons at level 2 are the connected components of X^c which are adjacent to its father. Here Y_2 and Y_3 are sons of X_1.
- and so on.

The even level are composed of connected components of X^c, the odd ones of connected components of X.

Definition 1.11 *A compact set is simply connected if it is connected and its complementary set is connected as well.*

2. Digitization of Sets

In this section, we investigate how sets may be digitized in order to preserve either connectivity or homotopy. The simplest digitization procedure, namely taking all the points of the lattice falling into X does not preserve either connectivity or homotopy (see fig. 4, right).

2.1. Set Hitting

We use a sequence of digitization grid $\mathcal{G}_n$ defined as follows: the space R^2 is decomposed into closed squares $C^n_{kl} = [k/2^n, (k+1)/2^n] \times [l/2^n, (l+1)/2^n]$, k, l integers. These squares have an edge length of $1/2^n$. So, two squares of the tesseletion have in common an edge, a vertex or are disjoint.

We consider the following digitization procedure, called here hitting digitization. Let $X \subseteq \mathsf{R}^2$ be a compact set and construct its digitized version at level n by:

$$X_n = \bigcup \{C^n_{kl} \mid C^n_{kl} \in \mathcal{G}_n, \quad C^n_{kl} \cap X \neq \emptyset\} . \tag{1}$$

So, at level n of description, X_n is the union of the squares (pixels) hitting set X. Note that for X_n, all the notions of connectivity are equivalent.

The hitting procedure preserves the (topological) connectivity:

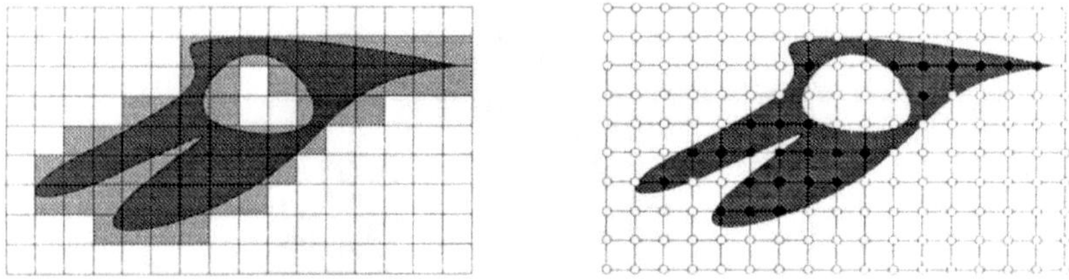

Fig. 4. Two digitization procedures: by hitting at left and by covering at right. As the continuous set is connected, note that the digitized set by hitting is also connected, but not the covering one.

Theorem 2.1 *Let $X \subseteq R^2$ be a compact set. Then X is connected iff X_n is connected for all n.*

The sequence X_n is a decreasing sequence and $X = \cap_n X_n$. However, the monotony is not required to ensure the connectivity of the limit:

Theorem 2.2 *Let $(X_n)_{n>0}$ be a sequence such that X_n is a union of squares in $\mathcal{G}_n$. If X_n is connected for n sufficiently large and if X_n converges to a compact set (for the Hausdorff metric), then $\lim X_n$ is connected.*

Proof: (of theorem 2.2) Let x and y be two points of X. As $X_n \to X$, there exists x_n and y_n in X_n verifying $x_n \to x$ and $y_n \to y$ (see [4]). Let $\varepsilon > 0$ and take n such that $\varepsilon > \sqrt{2}/2^{n-2}$ and such that the Hausdorff distance between X_n and X is less than $\varepsilon/4$. As X_n is connected at least for n sufficiently large, there exists a chained list of squares $(C_1, \ldots, C_p)$ included in X_n such that $x_n \in C_1$ and $y_n \in C_p$. For each of these squares, there exists a point in X at distance less than $\varepsilon/4$. So we build a sequence of points $(a_1, \ldots, a_p)$ in X which is an ε-path. Using theorem 1.6, X is connected. ■

Proof: (of theorem 2.1) Suppose X is connected. Let us show that X_n is connected for any n. Take $\varepsilon < \sqrt{2}/2^{n-1}$ and two squares C_1 and C_2 of X_n. Let $x \in X \cap C_1$, $y \in X \cap C_2$. Then let $(a_1, \ldots, a_p)$ be an ε-path in X from x to y. Each of these points a_i are included in a square C_i of X_n. Because the (a_i) build an ε-path, the (C_i) are adjacent or equal. In other words, X_n is connected.
Conversely, the X_n build a sequence of connected sets converging towards X. Using theorem 2.2, X is connected. ■

Unfortunately, the connectivity of X_n does not tell anything about the arc-connectivity and geodesic-connectivity of X (see the examples of the spiral or of $\sin \frac{1}{x}$). The problem comes from the fact that a sequence of adjacent squares (in 4- or 8-connectivity), that builds a digital path, does not necessarily converge to a path: discontinuities or an infinite length may occur, due to oscillations. However, if the sequence of digital paths is restricted to be uniformly bounded, then, when it has a limit, the limit has nice properties. More precisely:

Theorem 2.3 *Let $(X_n)_{n>0}$ be a sequence such that X_n is a union of squares in $\mathcal{G}_n$. If X_n is connected for n sufficiently large, if the geodesic diameter of X_n is bounded by M (independent of n) and if X_n converges (for the Hausdorff metrics), then $\lim X_n$ is geodesic-connected and of bounded diameter.*

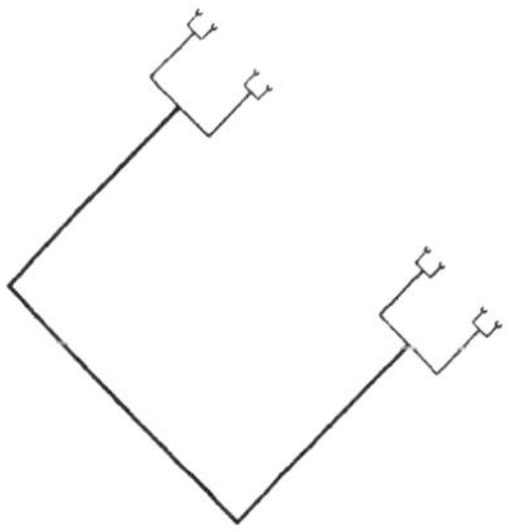
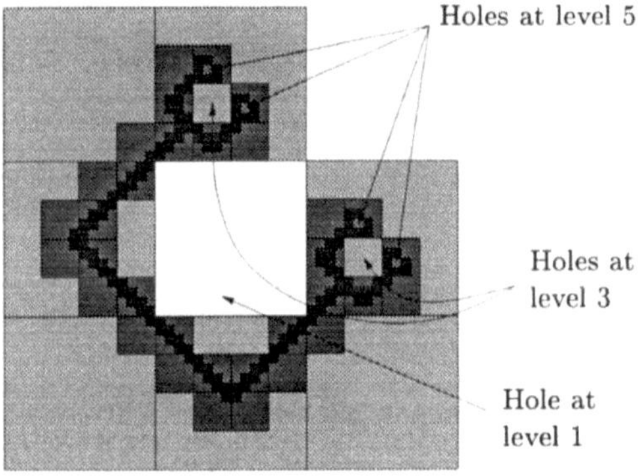

Fig. 5. Simply connected set whose covering digitization has holes at any scale. Only three scales are represented at right.

Proof: we use the fact that the geodesic distance "drops" at the limit (due to F. Maisonneuve and quoted in [5]): if $x_n \to x$, $y_n \to y$ and $X_n \to X$ (for the Hausdorff topology), then $\liminf(d_{X_n}(x_n, y_n)) \geq d_X(x,y)$. Now, take two points x and y of X. As $X_n \to X$ for the Hausdorff topology, there exists x_n and y_n in X_n such that $x_n \to x$ and $y_n \to y$. Applying the geodesic inequality and because $d_{X_n}(x_n, y_n) \leq M$, we obtain that $d_X(x,y) \leq M$. So X is geodesic-connected and of bounded diameter. ■

So, the two nice cases for digitizing are the topological connectivity and the geodesic-connectivity with bounded diameter.

2.2. Preserving Simple Connectivity

The previous section seems to indicate that with the hitting procedure as digitization process, all the good properties of connectivity are preserved. However, concerning the simple connectivity, fig. 5 shows that the limit of digitized sets by hitting may be simply connected, although all the digitized sets are not!

Is it possible to design a digitization procedure which preserves the simple connectivity? The answer is yes and the procedure is the simplest one can imagine: it is sufficient to fill the holes!

Theorem 2.4 *Let $X \subseteq R^2$ be a compact simply connected set and X_n the digitized sets by hitting. Let $\hat{X}_n$ be the sequence of sets obtained by filling the holes of X_n. By construction, $\hat{X}_n$ is simply connected. Then $\hat{X}_n$ converges towards X.*

The proof is omitted here.

2.3. Preserving Homotopy

The procedure for constructing simply connected digitized sets can be extended to preserve the homotopy of X in the case where X has a finite number of connected components as well as its complementary set and is not presented here.

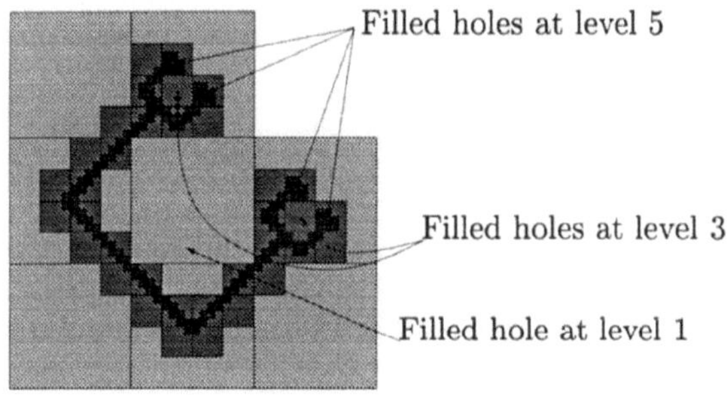

Fig. 6. Digitization by hitting of the compact set presented in the previous figure. The holes have been filled and now, the simple connectedness is preserved.

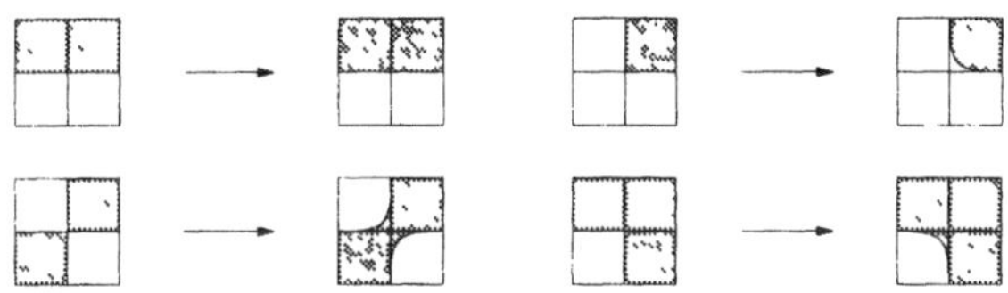

Fig. 7. Corner smoothing procedure for obtaining open-closed sets with respect to a ball.

2.4. Morphological Open-Closed Sets

We have seen so far how a set X may be digitized by X_n and at the same time respect the homotopy. In [6], J. Serra shows that for any closed set which is open and closed with respect to a ball of radius ε, its digitization by covering respects the homotopy as soon as the digitization lattice is sufficiently fine. Our procedure may be completed in order to yield open-closed sets: some circular arcs have to be added in order to round off convex and concave corners. The configurations are depicted in fig. 7.

For our procedure with squares, a similar result exists. Note that X_n is morphologically open with respect to a square of edge length $1/2^n$ and also morphologically closed with respect to a square of edge length strictly less than $1/2^n$. Then, for any digitization by covering of X_n with a square grid of spacing less or equal to $1/2^n$, the homotopy is preserved.

3. Measurability Properties

This last section is devoted to the measurability of different classes of closed sets. This measurability is taken according to the hit-or-miss σ-algebra, denoted by σ_f [4]. These measurability properties are important, because they give a meaning to some assertions, like "is the random compact set X almost surely (a.s.) connected"?

Theorem 3.1 *The set of connected compact sets and the set of bounded geodesic-connected compact sets are both closed for the Hausdorff topology, hence measurable for σ_f.*

In other words, *the probability that a random compact set is connected* or *is geodesic-connected and bounded* makes sense.

Proof: (only for connected sets. The case of geodesic-connected and bounded sets is very similar) Let $\mathcal{X}_n$ denote the set of all sets which are a connected finite union of squares in $\mathcal{G}_n$. Theorem 2.1 shows that any connected set is the decreasing limit of sets in $\mathcal{X}_n$, that is a limit in the sense of the Hausdorff topology. Conversely, any compact limit of sets in $\mathcal{X}_n$ is connected (theorem 2.2). So the set of connected compact sets is $\overline{\cup_n \mathcal{X}_n}$, where $\overline{\mathcal{U}}$ stands for the topological closure of $\mathcal{U}$ in the Hausdorff sense. ■

A last notion has been proposed by Ch. Lantuéjoul [2], working on the class of closed sets X equal to the closure of their interior. For him, X is connected if the interior of X is connected (also arc-connected because the interior is an open set). This notion gives rise to very good measurability properties.

4. Conclusion

We have seen that digitizations of a set can preserve connectivity and homotopy of a set. We have shown that the two best notions are the topological connectivity and the geodesic-connectivity with bounded diameter. The measurability of these notions have been also investigated in the context of random closed sets.

We have not been able to derive a proof of measurability or non-measurability of the following classes of sets (at least they are not closed in the set of compact sets):

- the class of arc-connected sets,
- the class of geodesic-connected sets.

References

1. G. Choquet. *Cours d'analyse, Tome 2 : Topologie, 2ème édition.* Masson, Paris, France, 1973.
2. Ch. Lantuéjoul. Conditional simulation of object-based models. In D. Jeulin, editor, *Proceedings of Advances in Theory and Applications of Random Sets, Fontainebleau, October 9-11, 1996*, pages 271–288. World Scientific, 1997.
3. Ch. Lantuéjoul and F. Maisonneuve. Geodesic methods in quantitative image analysis. *Pattern Recognition*, 17(2):177–187, 1984.
4. G. Matheron. *Random Sets and Integral Geometry.* John Wiley and Sons, New York, 1975.
5. M. Schmitt and J. Mattioli. *Morphologie Mathématique.* Logique - Mathématiques - Informatique. Masson, Décembre 1993.
6. J. Serra. *Image Analysis and Mathematical Morphology.* Academic Press, London, 1982.
7. K.T. Smith. *Primer of Modern Analysis.* Springer-Verlag, 1983.

TOUCHING DISTANCES AND DISTANCE TRANSFORM

V. STAROVOITOV
Institute of Engineering Cybernetics, National Academy of Sciences
Surganov Str. 6, Minsk, 220012, Belarus
E-mail: ValeryS@newman.basnet.minsk.by

Abstract. We present a new metric and call it as a touching metric. Disks, circles and digital paths based on this metric have interesting properties. The distance transform based on the metric is introduced.

Key words: Digital geometry, Metric, Distance transform, Short path

1. Introduction

The geometrical properties of quantized spaces have led to a non-Euclidean geometry known as digital geometry. Distance functions form an important part of the geometry. They are used in digital image processing and pattern recognition. Very often metrics form a basis for morphological transforms and via the transforms they are applied for image analysis [2, 6, 9, 12, 17].

Usually the Euclidean, the Hausdorff, the city-block, the chessboard, the chamfer metrics [1] and their generalization called m-neighbor distances [3] are used. Interesting knight-like metrics [4] have sense in digital geometry only. More complicated are the so-called "path-generated" and the N- sequences metrics [3, 7, 14, 23]. A path in a digital space consists of several adjacent points and their neighborhood is considered with respect to a particular criterion. The octagonal metric is an example.

Metrics may be defined on the whole space or on a part of it. In the later case they are called geodesic or constrained metrics [6, 11]. Several metrics are constructed from distances on graphs whose vertices are the points of the space lattice [15]. Different applied problems led to other metric generalizations, to the usage of more composite functions, to the calculation of distances by several iterations [12, 17] and to the usage of non-metric functions for distance and distance-like (cost-function) transforms.

In the paper we present a touching metric as a generalization of the "flower-shop distance" (**FS**) introduced by Klein for the Euclidean normed space [5]. They may be based on any of the metrics mentioned. Well-known geometrical figures and transforms have interesting properties with respect to these metrics. Several of them are shown.

The paper is organized as follows. Touching functions and metrics are described in the next section. Section 3 describes touching paths, circles and disks. Distance transforms based on the metrics are developed in Section 4. Section 5 concludes our paper.

2. Touching distances

Let (E, d) be a metric space with a basic metric $d(x, y)$. Let $t \in E$ be a fixed point in the space. The "flower-shop distance" was introduced by Klein for the Euclidean normed space in the following way:

$$\mathbf{FS}_t(x,y) = \begin{cases} 0, & \text{if } x = y, \\ d(x,t) + d(t,y), & \textit{otherwise}. \end{cases}$$

Different generalizations of the $\mathbf{FS}_t$-path give rise to other functions. For example, to reach a point "y" we first have to touch a fixed non-empty set of *contacting points* T.

The distance between the two points x and y is equal to the sum of the two minimum distances with respect to the chosen basic metric $d(.,.)$. One can calculate the length of the minimal touching path between one point x (or a set X) and another point y (or a set Y) with respect to the basic metric $d(.,.)$.

Proposition 1 *Let (E, d) be a metric space, d be an arbitrary basic metric, a nonempty set $T \subset E$ and $x, y \in E$, then the touching function*

$$\mathbf{TM}_T(x,y) = \inf_{t \in T}\{\mathbf{FS}_t(x,y)\} = \begin{cases} 0, & \textit{if } x = y, \\ \inf_{t \in T}\{d(x,t) + d(t,y)\}, & \textit{otherwise} \end{cases}$$

be a metric on E.

Proof. By definition we have

$$\mathbf{TM}_T(x,y) = 0 \iff y = x; \qquad \mathbf{TM}_T(x,y) = \mathbf{TM}_T(y,x);$$

Function $d(.,.)$ is a metric and we use its properties below. For every $\epsilon > 0$, there is a point $t_0 \in T$ such that

$$\mathbf{TM}_T(x,y) = \inf_{t \in T}[d(x,t) + d(t,y)] \geq d(x,t_0) + d(t_0,y) - \epsilon.$$

Let x, y, z be arbitrary points of E, then

$$\mathbf{TM}_T(x,y) \stackrel{def}{=} \inf_{t \in T}\{d(x,t) + d(t,y)\} \stackrel{\forall t_2}{\leq} d(x,t_2) + d(t_2,y) \stackrel{\forall z, t_3}{\leq}$$

$$\leq d(x,t_2) + d(t_2,z) + d(z,t_3) + d(t_3,y);$$

$$\mathbf{TM}_T(x,z) \stackrel{def}{=} \inf_{t \in T}\{d(x,t) + d(t,z)\} \stackrel{\forall \epsilon\, \exists t_2}{\geq} d(x,t_2) + d(t_2,z) - \epsilon;$$

$$\mathbf{TM}_T(z,y) \stackrel{def}{=} \inf_{t \in T}\{d(z,t) + d(t,y)\} \stackrel{\forall \epsilon\, \exists t_3}{\geq} d(z,t_3) + d(t_3,y) - \epsilon.$$

As this holds for every $\epsilon \to 0$, the triangular inequality follows

$$\mathbf{TM}_T(x,y) \leq \mathbf{TM}_T(x,z) + \mathbf{TM}_T(z,y) \qquad \square.$$

Proposition 1 is correct if the basic metric $d(.,.)$ and the function $\mathbf{TM}_T(x, y)$ are geodesic, i.e. they are defined on a subspace of E.

If $T = E$, then $TM_T(x, y) = d(x, y)$. Using the metric $\mathbf{TM}_t$ one can calculate **TM**-distances between a point and a set of points. Let us denote distance from a point to a set based on the $\mathbf{TM}_T$-metric by **TD**:

$$\mathbf{TD}_T(x, Y) = \inf_{y \in Y}\{\mathbf{TM}_T(x, y)\} = \begin{cases} 0, & \text{if } \exists y_0 = x \in Y, \\ \inf_{t,y}\{d(x,t) + d(t,y)\}, & \text{otherwise} \end{cases}$$

$$\mathbf{TD}_T(X, y) = \inf_{x \in X}\{\mathbf{TM}_T(x, y)\} = \begin{cases} 0, & \text{if } \exists x_0 = y \in X, \\ \inf_{x,t}\{d(x,t) + d(t,y)\}, & \text{otherwise} \end{cases}$$

3. Touching circles and disks

Definition 1 *A touching circle (***T-circle***) of radius R_T centered at a point x is a set of points $Y = \{y\}$ situated at the distance $R_T = \mathbf{TM}_T(x, y)$ from x.*

Definition 2 *A touching disk (***T-disk***) of radius R_T centered at a point x is a set of points $Y = \{y\}$ situated at distances $r \leq R_T = \mathbf{TM}_T(x, y)$ from x.*

There are three types of circles and disks with respect to the $\mathbf{TM}_T$-metric. Examples of touching circles and disks are given in Fig.1 and 2. Properties of different types of touching circles and disks depend of the radius value:

- if the radius is less than distance between the center x and a fixed set T, i.e. $R_T < \inf_{t \in T}\{d(x,t)\}$, then a **T-disk** $= \{x\}$ and a **T-circle** cannot exist.
- if the radius is equal to the distance $R_T = \inf_{t_k \in T}\{d(x, t_k)\}$ and there are $k \geq 1$ such t-points, then a **T-disk** $= \{x, t_1, t_2, \ldots, t_k\}$ and a **T-circle** $= \{t_1, t_2, \ldots, t_k\}$.
- if radius R_T is greater than $\inf_{t_k \in T}\{d(x, t_k)\}$, then a **T-disk** (a **T-circle**) is union of "ordinary" disks with radii $r = R_T - \inf_{t_k \in T}\{d(x, t_k)\}$ centered at the points t_k. The **T-disk** also includes the point x, i.e.

$$\textbf{T-disk} = \bigcup_{t_k \in T} \{y \mid d(y, t_k) \leq R_t - d(x, t_k)\} \bigcup \{x\};$$

$$\textbf{T-circle} = \bigcup_{t_k \in T} \{y \mid d(y, t_k) = R_t - d(x, t_k)\}.$$

These properties are fulfilled both in continuous and digital spaces. The figures discussed above are constructed with respect to the chosen basic metric. One can note the following unusual properties of such figures:

- they may have their centers outside the "ordinary" circles and disks;
- **T-circles** may have no points for the given T-radius;
- they may be disconnected even in the Euclidean geometry;
- they are not invariants with respect to any isometric transform (rotation, shift, symmetry) if a set T is fixed, but they are invariants if the set T is also transformed.

4. Calculation aspects of T-metrics

The length of the shortest path with respect to the touching metric can be calculated in a brute force way. Let $X = \{x_k\}$, $Y = \{y_l\}$ and $T = \{t_j\}$. We can calculate two sets of distances $\{d(x_k, t_j)\}$ and $\{d(t_j, y_l)\}$ for all k, l and j. Then

$$\mathbf{TD}_T(X,Y) = \inf_{k \in X} \{\mathbf{TM}_T(x_k, Y)\} = \inf_{x_k, t_j, y_l} \{d(x_k, t_j) + d(t_j, y_l)\}.$$

To calculate $\mathbf{TD}_T$ in this way for 10-points sets X, Y and T we have to calculate of 100 distances and to compare of 1000 sums of the distances! To decrease the number of calculation one can use the non-interior points of Y and T only, i.e. δY and δT.

Proposition 2 *Let (E, d) be a metric space, then $\mathbf{TD}_T(x, Y) = \mathbf{TD}_{\delta T}(x, \delta Y)$ for arbitrary X, Y and $T \subset E$ with respect to any basic metric d.*

Proof. Let δT and δY denote the borders of T and Y, respectively.

I). Let $x \in X \cap T$, i.e. $x = t$, then $\mathbf{TD}_T(x, Y) = \inf_{y \in Y}\{d(x, y)\}$. The shortest path between points $x = t$ and y must intersect δT at a point t_0 and we get $d(x, y) = d(x, t_0) + d(t_0, y)$, i.e. $\mathbf{TD}_T(x, Y) = \mathbf{TD}_{\delta T}(x, Y)$.

II). Let $x \in X \setminus X \cap T$, $t \in T \setminus X \cap \delta T$ and $y \in Y$. For every border point $t_b \in \delta T$

$$d(x, t_b) + d(t_b, y) \leq d(x, t_b) + d(t_b, t) + d(t, y)$$

thanks to the triangular inequality. The shortest path between x and t must intersect δT at a point t_0 and $d(x, t) = d(x, t_0) + d(t_0, t)$, then

$$d(x, t_0) + d(t_0, y) \leq d(x, t) + d(t, y).$$

By definitions we have

$$\inf_{t_b \in \delta T}\{d(x, t_b) + d(t_b, y)\} \leq d(x, t_0) + d(t_0, y) \leq \inf_{t \in T \setminus \delta T}\{d(x, t) + d(t, y)\}$$

and

$$\mathbf{TD}_{\delta T}(x, Y) = \mathbf{TD}_T(x, Y) \leq \mathbf{TD}_{T \setminus \delta T}(x, Y).$$

Similarly, $d(t_0, y) \geq d(t_0, y_0)$ for $\forall y \in Y$ and $y_0 \in \delta Y$, this proves the proposition. □

In digital image processing every set of contacting points is finite, but contrary to the Euclidean geometry the shortest path between two points may not be unique. To apply Proposition 2 in digital spaces we have to use the appropriate definitions of digital path and border. When the set boundaries contain a lot of points and the space E is geodesic or contains a number of obstacle sets, the brute force way of the shortest path calculation takes a lot of time.

If one need to calculate distances for *every* point of a given set X to a set Y, the *Distance transform* (**DT**) is applied to this problem. The main idea of the transform is to use a regular digital metric for a local propagation of distances from one point to the neighbor point in the path sequence connecting two given sets with respect to the metric [1, 16]. Different algorithms and architectures can be found in the literature (see ref). The result of the transform is called a *distance map*.

An algorithm for the distance transform may be adapted for the touching metric. Let us define the digital distance transform based on an arbitrary touching metric and the touching metric may be based on an arbitrary digital metric.

Let a basic metric d be defined on a subspace $E = \{e_i\} \subseteq \mathbf{Z}^n$. Let these be three point sets in the subspace E: a set of object points X, a set of reached points Y, and a set of contacting points T. Accordantly to our **DT**-description [19] we denote the touching distance transform as $\mathbf{DT}_{TM_T}(X|Y)$. That means it calculates distances from *every point* x_k of the set $X \subset E$ to *the nearest point* of the set $Y \subset E$ touching the fixed set $T \subset E$ and uses $\mathbf{TM}_T$-metric. The distances should be the lengths of the shortest paths inside the subspace E and they should be equal

$$\mathbf{DT}_{TM_T}(X|Y) = \{\mathbf{TM}_T(x_k, Y) = \min_{t \in T, y \in Y} \{d(x_k, t) + d(t, y)\} \mid \forall x_k \in X\}.$$

All three types of "usual" **DT** algorithms (sequential, parallel, ordered queue) may be adapted to the touching **DT**. To realize this transform a uniform way is the use of a queue technique by bucket sort (efficient versions of similar algorithms are published in [13, 21, 22]). With respect to above **TM** definition we should calculate two sets of distances $\{d(x_k, t_j)\}$ and $\{d(t_j, y_l)\}$. Then we have to choose $\mathbf{TM}_T(x_k, Y) = \min_{j,l}\{d(x_k, t_j) + d(t_j, y_l)\}$ for every point $x_k \in X$. The following variant of the touching **DT** based on a queue technique is suggested.

Scheme of the touching *DT* algorithm

1: Select all non-interior points $\delta T = \{t_j\}$ and $\delta Y = \{y_l\}$.

2: Insert $\{y_l\}$ in a queue. Starting from queued points y_l, perform the first "ordinary" $\mathbf{DT}_d(E|\delta Y)$ for points E while the distances for all points t_j are calculated.

3: Sort all points $\{t_j\}$ with respect to their distance $d(t_j, \delta Y)$ and insert them in another queue keeping the distance values.

4: Starting from the queued points $\{t_j\}$ and their intermediate distances $d(t_j, \delta Y)$, perform a propagation of these distances by an application of the second "ordinary" $\mathbf{DT}_d(E|\delta T)$ for points E while the distances for all points x_k are calculated.

We can calculate distances not for all points of the given subspace E, but for those which belong to the shortest paths connecting point pairs of sets δY and δT, δT and X, respectively.

After the last step of the procedure, all points $x_k \in X$ have their required touching distance values. As we use a metric function $\mathbf{TM}_T$ we obtain touching distance values from *every* point x_k to the nearest point $y_l \in Y$ contacting the given point set T. The set of object points X have transformed in a touching distance map. To calculate the shortest touching distance between sets X and Y, one can stop the above procedure when the first point x_k is reached.

A graphical explanation of a geodesic variant of the touching distance transformation and two intermediate distance maps are shown in Fig.3. The set E has an arbitrary shape indicated by grey background. The sets X, Y, T are also shown by different grey brightness. Distance values are shown by letters from 1 to $D = 13$. At the first stage we perform an ordinary distance transformation by propagating of local chessboard distances from the border of the set Y up to the last point of the set T to get a distance value. Then we omit all middle distance values except those distances that are assigned to points of the contacting set T (see Fig.3b). After that we propagate these distances inside E while all points of the given set X obtain the distance values. The result of the second distance transform gives the required touching distance transform.

5. Conclusion

We present a new notion – *touching distances* calculated by the *touching metric*. The metric may be based on any other kind of metric. The main idea of such functions is to calculate distances between points or a point and a set of points via a given set of contacting points.

Several interesting facts follow from the metric application in the continuous and discrete spaces. There are three types of disks and circles based of these metrics. They may consist of a finite number of points even in continuous spaces.

The distance transform based on the touching metric is discussed and a general scheme of a transform algorithm is given.

These distances and distance transform can be used in different applications to geometrically solve optimization problems. For example, several vehicles or robots are situated at different points in town (building,...), what will be the first to reach a given point or one point of a given set of points if each is required first to visit some other fixed points (petrol stations, gates, depositories and so on).

References

1. G. Borgefors. Distance transformations in arbitrary dimensions. *Computer Vision Graphics and Image Processing*, 27:321-345, 1984.
2. T. Boult, R.A. Melter, F. Skorina and I. Stojmenovic. Application of G-neighbors to image processing and morphology. *Mashine Graphics & Vision*, 4(1/2):39-58, 1995.
3. P.P. Das, P.P. Chakrabarti and B.N. Chattereji. Distance functions in digital geometry. *Inform. Sci.*, 42:113-136, 1987.
4. P.P. Das and J. Mukherjee. Metricity of super-kniht's distance in digital geomrtry. *Pattern Recognition Letters*, 11:601-604, 1990.
5. H.J.A.M. Heijmans. *Morphological Image Operators*, Academic Press, Boston, 1994.
6. C. Lantuejoul and S. Beucher. On the use of geodesic metric in image analysis. *J. Microsc.*, 121:39-49, 1981.
7. R. Melter and I. Tomescu. Path-generated digital metrics. *Pattern Recognition Letters*, 1:151-154, 1983.
8. R. Melter. A survey of Digital Metrics. *Contemporary Mathematics*, 119:95-106, 1991.
9. U. Montanari. A method for obtaining skeletons using a quasi-Euclidean distance. *J. Assoc. Comput. Mach.*, 15:600-624, 1968.
10. J.C. Mullikin. The vector distance transform in two and three dimensions. *CVGIP*, 54:526-535, 1992.
11. J. Piper and E. Granum. Computing distance transformations in convex and non-convex domains. *Pattern Recognition*, 20:599-615, 1987.
12. F. Preteux. Watershed and skeleton by influence zones: a distance-based approach. *Journal of Mathematical Imaging and Vision*, 1:239-255, 1992.

13. I. Ragnemalm. The Euclidean distance transform. PhD dissertation No.304, Dep. of Electrical Engineering, Linköping University, Sweden, 1993.
14. F. Rhodes. Digital metrics generated by families of paths. *Discrete Mathematics*, 135:265-277, 1994.
15. F. Rhodes. Metric subgraphs of the chamfer metrics and the Melter-Tomescu path generated metrics. *Discrete Mathematics*, 147:197-209, 1995.
16. A. Rosenfeld and J.L. Pfaltz. Sequential operations in digital image processing. *J. ACM.*, 13:471-494, 1966.
17. P. Soille. Generalized geodesic distances applied to interpolation and shape description. *Proc. Seconds Workshop Mathematical Morphology and its Application to Image Processing*, pages 193-200, France, Sep. 1994.
18. V. Starovoitov. Morphological analysis based on distance transforms. *in Proc. of SPIE Conf.*, USA, 1994, Number 2300, pages 343-354, 1994.
19. V. Starovoitov. Toward a distance transform generalization. *in Proc. of 8th Scandinavian Conf. on Image Analysis*, Vol.1, pages 499-506, Sweden, June 1995.
20. V. Starovoitov. Basic morphological operations by local propagation transforms. *Proc. of 2nd Int. Conf. on Applications of Computer Systems*, pages 191-195, Poland, Nov. 1995.
21. B.J.H. Verwer. Distance Transform: Metrics, Algorithms and Applications. Ph D. thesis, Delft University of Technology, 1991.
22. L. Vincent. Exact Euclidean distance function by chain propagations. *in Proc. Computer Vision and Pattern Rec. Conf.*, pages 520-525, Hawaii, 1991.
23. M. Yamashita and T. Ibaraki. Distances defined by neighborhood sequences. *Pattern Recongnition*, 19:237-246, 1986.

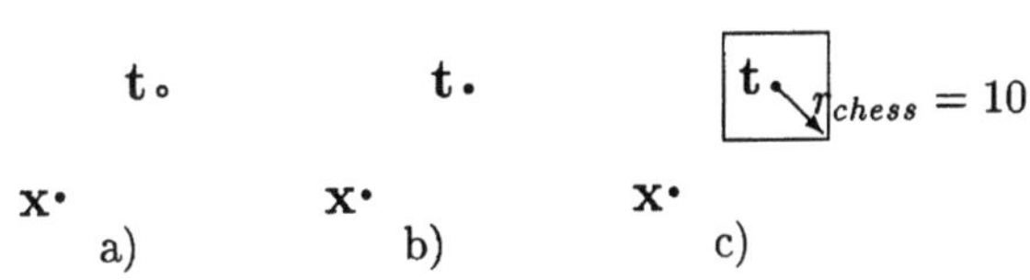

Fig. 1. Three types of touching circles and disks with different radii centered at the point x. The chess-board metric is the basic one. (a) $R_1 < d_{chess}(x,t)$: circle has no points, disk has only one point x; (b) $R_2 = d_{chess}(x,t)$: circle has only one point t, disk has two points x and t; (c) $R_3 = d_{chess}(x,t) + 10$: circle contains only the points on the border of the square, disk contains the point x and the square around the point t.

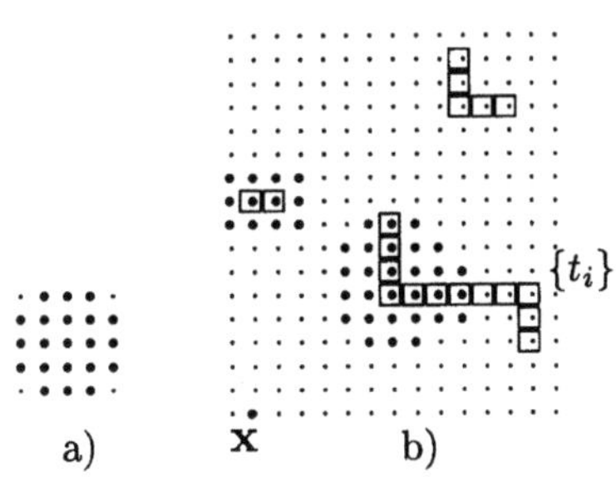

Fig. 2. (a) The octagonal disk of radius 2. (b) The touching digital disk based on the octagonal metric and centered at x, $R_T = 10$. The digital contacting segments $\{t_i\}$ are shown by squares. T-disk is shown by black blobs.

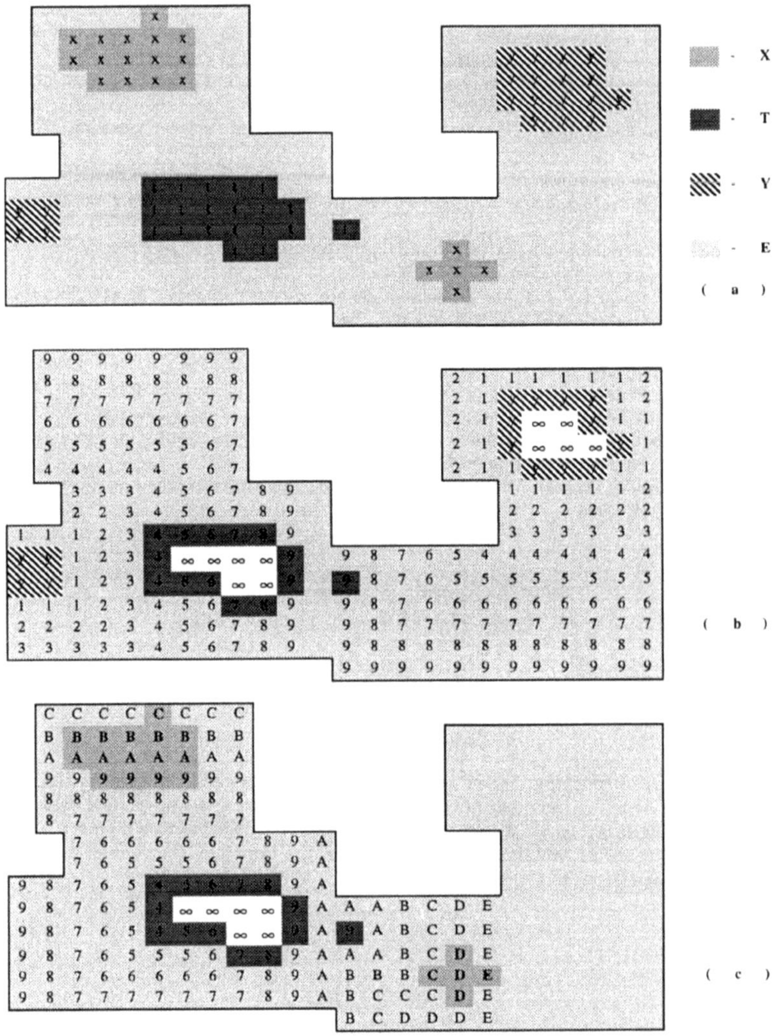

Fig. 3. Idea of the touching $DT_{TM}(X|Y)$ realization. The example is given for the chessboard basic metric d. (a): Initial sets X,Y,T and the subspace E are shown by different textures. (b): The distance map of the first ordinary geodesic $DT_d(\delta T|\delta Y)$. (c): The final distance map after the second $DT_d(X|\delta T)$.

HAUSDORFF DISTANCES AND INTERPOLATIONS

JEAN SERRA
Centre de Morphologie Mathématique - Ecole des Mines de Paris
35, rue Saint-Honoré - 77305 Fontainebleau (FRANCE)

Abstract. The space $\mathcal{K}'(E)$ of the non empty compact sets of a metric space E is itself metric for the so called Hausdorff distance. It is shown in this paper that Hausdorff metric admits at least two different geodesic, with provide interpolators between pairs of sets of $\mathcal{K}'(E)$. When E is affine, the space quotient of $\mathcal{K}'(E)$ under translation admits in turns reduced geodesics, that generate better set interpolators. Finally, by introducing a Hausdorff distance by erosion, one can provide a symmetric form to interpolators, but are no longer geodesics.

1. Introduction

This paper belongs to a series of three texts (ref. [1][2][3]) by S. Beucher, F. Meyer, and J. Serra respectively, originally in French and written in 1994. They have resulted in a patent [4], consequently we could not publish them for a certain time. The paper by Meyer was the subject of a communication in ISMM'96 in Atlanta [5], and the one by Beucher is presented in ISMM'98 [6]. These three related works, which develop three facets of a same idea, transcribe in a written form a number of fruitful discussions at the CMM. This paper is more upstream than the two other ones, and, unlike [5], uses the term "geodesic" in its mathematical sense, i.e. "segment in a metric space".

Before getting to the heart of the matter, we will remind briefly the classical Hausdorff distance, and introduce a variant of it (by erosions).

Let E be a metric space, of distance d, and let $\mathcal{K}'$ be the class of the non empty compact sets of E. Put

$$d(x,Y) = \inf\{d(x,y), y \in Y\} \qquad x \in E\,;\, Y \in \mathcal{K}'$$

and introduce the mapping $\mathcal{K}' \times \mathcal{K}' \to \mathbb{R}_+$

$$\rho(X,Y) = \max\left\{\sup_{x\in X} d(x,Y)\ ;\ \sup_{y\in Y} d(y,X)\right\} \tag{1}$$

Mapping ρ turns out to be a distance, called "Hausdorff distance", which holds on $\mathcal{K}'$ and no longer on E. Rel. (1) can be equivalently written by means of the dilations by the balls of space E. If $B_\lambda(x)$ stands for the compact ball of centre x and of radius λ, the dilation of $X \in \mathcal{K}'$ by λ has an expression

$$\delta_\lambda(X) = \cup\{B_\lambda(x)\ ,\ x \in X\} = \cup\{y : B_\lambda(y) \cap X \neq \emptyset\}. \tag{2}$$

Then, rel.(1) takes the following form, for all $X, Y \in \mathcal{K}'$:

$$\rho(X,Y) = \inf\{\lambda : X \subseteq \delta_\lambda(Y)\ ;\ Y \subseteq \delta_\lambda(X)\}. \tag{3}$$

This last expression, more geometrical than rel. (1) shows in particular a semi-sensitivity to noise: a small lacuna in set X_i or Y_i slightly changes ρ, whereas an isolated point far away from X and added to it modifies ρ drastically. Hausdorff distance satisfies a few algebraic properties, associated (of course) with union and dilation. Moreover it extends easily to grey tone and colour functions by means of their subgraphs [7][8]. Therefore all derived geodesics and interpolators apply also to functions (see S. Beucher's paper [6] in the present issue).

Consider now the subclass $\mathcal{A}$ of $\mathcal{K}'(E)$, of the regular compact sets, i.e.

$$X \in \mathcal{K}' \quad \text{and} \quad \overline{\mathring{X}} = X \;.$$

Then the non negative number

$$\sigma(X,Y) = \inf\{\lambda : \varepsilon_\lambda(X) \subseteq Y \quad ; \quad \varepsilon_\lambda(Y) \subseteq X\}$$

where ε_λ is the erosion adjoint to dilation δ_λ, is a distance on $\mathcal{A} \times \mathcal{A}$. Indeed, if $\sigma(X,Y) = 0$, then

$$Y \supseteq \bigcup_{\lambda>0} \varepsilon_\lambda(X) = \mathring{X} \quad \Rightarrow \quad Y \supseteq \overline{\mathring{X}} \supseteq X \qquad X,Y \in \mathcal{A}$$

and similarly $X \supseteq Y$, hence they are equal (the other axioms are proved as for distance ρ). We shall call σ the *Hausdorff distance by erosions.*

2. First Hausdorff geodesic

If it exists, a geodesic between X and Y, for the Hausdorff distance by dilations, will be a shortest segment from X to Y in space $\mathcal{K}'$, i.e. a family $\{Z_\alpha, 0 \le \alpha \le 1\}$ of non empty compact sets that interpolate X and Y. For $\alpha = 0$ we should obtain $Z_\alpha = X$, for $\alpha = 1, Z_\alpha = Y$, and as α increases, Z_α should progressively leave X and go to Y.

But does such a geodesic exist? In many metric spaces, there are no geodesics, in some other ones, several ones, or even an infinity. What about our current case? To answer the question, we need a small lemma.

Lemma 1 *Given S and X in $\mathcal{K}'$, if $\delta_\rho(S) \supseteq X$ for a $\rho \ge 0$, then $\delta_\rho(\delta_\rho(X) \cap S) \supseteq X$*

Proof: By hypothesis, every point $x \in X$ belongs to $\delta_\rho(S)$, i.e. there exists a point $s \in S$ such that $x \in \delta_\rho(s)$ or, equivalently, such that $s \in \delta_\rho(x)$, hence $s \in \cup\{\delta_\rho(x), x \in X\} \cap S$. Therefore every point $x \in X$ is covered by the ρ-dilate of an element of $\delta_\rho(X) \cap S$. Q.E.D.

The lemma allows to construct a first class of geodesics, as follows:

Theorem 1 *First geodesics: let E be a metric space of distance d, of compact balls δ_λ, and whose $\mathcal{K}'(E)$ denote the class of the non empty compacts sets. Then every pair X,Y in $\mathcal{K}'(E)$ from ρ apart, (for Hausdorff distance by dilations) admits the following geodesic:*

$$\{Z_\alpha = \delta_{\alpha\rho}(X) \cap \delta_{(1-\alpha)\rho}(Y) \quad , \quad \alpha \in [0,1] \;\} \tag{4}$$

Proof : Fix α and put $S = \delta_{(1-\alpha)\rho}(Y)$. Since $\delta_{\alpha\rho}\left[\delta_{(1-\alpha)\rho}\right] = \delta_{\rho}$, we draw from rel. (3) that $\delta_{\alpha\rho}(S) = \delta_{\rho}(Y) \supseteq X$. By applying lemma 1 we have

$$\delta_{\alpha\rho}(Z_{\alpha}) = \delta_{\alpha\rho}\left(\delta_{\alpha\rho}(X) \cap \delta_{(1-\alpha)\rho}(Y)\right) \supseteq X \quad .$$

On the other hand, the definition of Z_{α} implies $\delta_{\alpha\rho}(X) \supseteq Z_{\alpha}$, hence

$$\rho(X, Z_{\alpha}) \leq \alpha\rho$$

One proves similarly that $\rho(Y, Z_{\alpha}) \leq (1-\alpha)\rho$. By combining these two results with the triangular inequality, we finally obtain

$$\rho \geq \rho(X, Z_{\alpha}) + \rho(Y, Z_{\alpha}) \geq \rho(X, Y) = \rho$$

which achieves the proof. Q.E.D.

Discussion: What is the practical value of algorithm (4) ? When we apply it to two disjoint sets X and Y, it often swells the intermediate Z_{α}'s. Typically, the midway set between the two *fine* and *horizontal* segments of the x-axis is a *thick* and *vertical* lens, which inflates more and more as the segments depart from one another.

At least, this first geodesic will have highlighted where the problems do lie. Indeed, the discussion suggests we focus

i/ on other possible geodesics, with finer interpolators,

ii/ on an approach which should treat separately the differences between the relative *positions* of X and Y, and their differences in *shape.*

iii/ on an approach which should not be restricted to extensive dilations, which always "swell", for example by involving dilations and erosions in a symmetrical manner.

The problems are set, it remains to examine how they can be solved.

3. Second Hausdorff geodesic

3.1. Convex case

The geodesic Z_{α} of rel. (4) does not involve dilations of X and Y by each other, but of both of them by the δ_{λ}'s. However, if we focus on the sub-class $\mathcal{C}' \subseteq \mathcal{K}'$ of those elements of $\mathcal{K}'$ that are convex, we can easily exhibit another geodesic, based on cross dilations. In the following, αX stands for the transform of set X under the similitude of ratio α.

Proposition 1 *Let X and Y be two non empty compact convex sets in $\mathbb{R}^n$, at Hausdorff distance ρ from each other. Then the interpolator*

$$Z'_{\alpha} = (1-\alpha)X \oplus \alpha Y \tag{5}$$

is at distance $\alpha\rho$ from X and $(1-\alpha)\rho$ from Y, i.e. the family $\{Z'_{\alpha}\ ,\ \alpha \in [0,1]\}$ is a geodesic in space $\mathcal{C}'$.

Proof : For every set $X \in \mathcal{K}'$ and for any $\alpha \in [0,1]$ we have

$$(1-\alpha)X \oplus \alpha X \supseteq X \quad , \tag{6}$$

the equality being obtained iff X is convex [9]. Therefore, for all $X, Y \in \mathcal{K}'$ of distance ρ apart, we can write

$$Z'_\alpha \oplus \alpha\rho B = (1-\alpha)\, X \oplus \alpha\,(Y \oplus \rho B) \supseteq (1-\alpha)\, X \oplus \alpha X \supseteq X\,. \tag{7}$$

where B is the closed unit ball centered at the origin.

If, in addition, sets X and Y are convex, we have also

$$X \oplus \alpha\rho B = (1-\alpha)\, X \oplus \alpha\,(X \oplus \rho B) \supseteq (1-\alpha)\, X \oplus \alpha Y \supseteq Z'_\alpha\,. \tag{8}$$

Hence, $\rho\,(Z'_\alpha, X) \leq \alpha\rho$, and similarly $\rho\,(Z'_\alpha, Y) \leq (1-\alpha)\rho$. thus we have

$$\rho\,(Z'_\alpha, X) + \rho\,(Z'_\alpha, Y) \leq \rho = \rho\,(X, Y)$$

as well as the inverse inequality (triangular inequality for distances), which achieves the proof. Q.E.D.

In general, the second geodesic $\{Z'_\alpha, \alpha \in [0,1]\}$ does not reduce to the first one, since rel. (8), and its homolog version for Y, imply, when compared with rel. (4) that

$$Z_\alpha\,(X,Y) \supseteq Z'_\alpha\,(X,Y) \qquad \alpha \in [0,1] \quad , \quad X, Y \in \mathcal{C}' \tag{9}$$

For example, in the case of the two above segments, interpolator $Z'_{0,5}$ is itself a segment of same length as X and Y, and placed between them. We are far away from the thick lens $Z'_{0,5}$!

Another very positive feature lies in the fact that Z'_α commutes under translation on X or on Y. Indeed, the translate of X by vector h is nothing but the dilate $X \oplus \{h\}$ of X by $\{h\}$ (i.e. h considered as an element of $\mathcal{P}\,(\mathbb{R}^n)$). Therefore, the equality

$$Z'_\alpha\,(X \oplus \{h\}\,, Y) = (1-\alpha)\,(X \oplus \{h\}) \oplus \alpha Y = Z'_\alpha\,(X,Y) \oplus (1-\alpha)\,\{h\} \tag{10}$$

shows that in Z'_α the effect of the *shapes* of X and Y (which determines Z'_α up to a translation) is separated from the effect of their relative *positions* (which induce only translations on Z'_α).

Geodesics $\left\{Z'_\alpha\right\}$ admits an extension to the whole space $\mathcal{K}'$, given by the following theorem

Theorem 2 *Second geodesics: every pair X, Y in $\mathcal{K}'(E)$ from ρ apart for Hausdorff distance for dilations admits the following geodesic*

$$\{Z''_\alpha = \delta_{\alpha\rho}\,(X) \cap \delta_{(1-\alpha)\rho}\,(Y) \cap (1-\alpha)\, X \oplus \alpha Y \quad , \quad \alpha \in [0,1]\}$$

Proof : We will prove first that

$$x \in X \quad \Rightarrow \quad \delta_{\alpha\rho}\,(x) \cap \delta_{(1-\alpha)\rho}(Y) \cap Z'_\alpha \neq \emptyset$$

Given $x \in X$, there exists a $y \in \delta_\rho\,(Y)$ such that $x \in \delta_\rho\,(Y)$, i.e. that $U\,(\rho) = \delta_{\alpha\rho}\,(x) \cap \delta_{(1-\alpha)\rho}(Y) \neq \emptyset$. Therefore, one can find a value $0 < \rho_0 \leq \rho$ with $U\,(\rho_0)$

reduced to one point, s say. Point s belongs to $U(\rho_0)$ and to the vector $\overrightarrow{xy}$ of unit $\overrightarrow{u}$. More precisely, we have

$$\overrightarrow{s} = \overrightarrow{x} + \alpha\rho b\overrightarrow{u} = \overrightarrow{y} - (1-\alpha)\,\rho b\overrightarrow{u}$$

hence $\overrightarrow{s} = (1-\alpha)\,\overrightarrow{x} + \alpha\,\overrightarrow{y}$, i.e. $s \in Z'_\alpha$ and finally $s \in \delta_{\alpha\rho}(x) \cap \delta_{(1-\alpha)\rho}(Y) \cap Z'_\alpha \subseteq Z''_\alpha$. Since $x \in \delta_{\alpha\rho}(s)$,then $\delta_{\alpha\rho}(Z''_\alpha)$ covers point x, hence $\delta_{\alpha\rho}(Z''_\alpha) \supseteq X$. The brother inclusion $\delta_{\alpha\rho}(X) \supseteq Z''_\alpha$ derives from the definition of Z''_α, which achieves the proof. Q.E.D.

Fig.1 An example of interplator $Z''_{0.5}$ between two sets X and Y.

Fig. 1 shows an example of such a geodesic. The parasite swelling effect is partly mastered. However independence under translation is lost (it was true only for convex sets) and the implementations of the Z'_α require dilations of two arbitrary sets by each other. Perhaps could we try and follow a new lead?

4. Reduced Hausdorff distances

Since the shape of the geodesic $\{Z_\alpha\}$ depends on the respective locations of sets X and Y, we could try and place them in the "most favourable" locations, by shifting and rotating Y (for example). The idea of studying the variation of $\rho(X,Y)$ as a function of the translates of Y was recently applied to template matching problems [9]. Here we will develop it to exhibit new metrics, and, from them, new geodesics. We restrict the displacements to translations (the rotations admit the same theoretical approach, but their implementation is less easy).

In the following, E stands for a compact (but possibly large) region in $\mathbb{R}^n$ or $\mathbb{Z}^n$, equipped with a unit ball B. Denote by X_a (resp. X_b) the translate of set X (resp. Y) by vector a (resp. b), and put

$$\rho_1(X,Y) = \inf\{\rho(X_a, X_b)\ ,\ a, b \in E\} \tag{11}$$

Introduce the quotient space $\mathcal{K}_1$ of $\mathcal{K}'$ for the equivalence under translation, i.e. $X \equiv X_a$, $a \in E$. Operator ρ_1 maps $\mathcal{K}_1 \times \mathcal{K}_1$ into $\mathbb{R}_+$ since it does not change as X and Y are translated, and it is a symmetrical function of its two arguments. Clearly, given $X, Y \in \mathcal{K}'$, the mapping $(a,b) \to \rho\,(X_a, Y_b)$ from $E \times E$ into $\mathbb{R}_+$ is continuous, therefore the compacity of E implies that there exists a pair $(a,b) \in E \times E$ with

$$\rho_1(X,Y) = \rho(X_a, Y_b) \tag{12}$$

In particular, if $\rho_1(X,Y)=0$ then $X_a = Y_b$, i.e. $X = Y$ and, of course, $X = Y$ implies $\rho_1(X,Y)=0$.

Finally, operator ρ_1 satisfies the triangular inequality. To prove it, consider the quantity $\rho_1(X,Z)+\rho_1(Z,Y)$, $X,Y,Z \in \mathcal{K}'$. Since ρ_1 is translation invariant, we can always replace Z by its translate Z_0 centred at the origin. If X_a and X_b denote the locations that minimize $\rho(X,Z_0)$ and $\rho(Y,Z_0)$ respectively, we can write

$$\rho_1(X,Z)+\rho_1(Z,Y)=\rho(X_a,Z_0)+\rho(Z_0,Y_b)\geq\rho(X_a,X_b)\geq\rho_1(X,Y)$$

Hence, ρ_1 is a distance. Moreover, its metrics admits geodesics of the first type. To show it, take a pair (X,Y_b) that minimizes ρ. Given an $\alpha \in [0,1]$, consider the first type interpolator Z_α between X and Y_b (rel. (4)). Z_α turns out to be in optimal position not only w.r. to ρ, but also w.r. to metric ρ_1, i.e. $\rho(X,Z_\alpha)=\rho_1(X,Z_\alpha)$. Indeed, we have

$$\rho=\rho_1(X,Y)=\rho(X,Y_b)\leq\rho(X,Z_\alpha)+\rho(Z_\alpha,Y_b)$$

but $\rho_1(X,Z_\alpha)\leq\alpha\rho$ and also $\rho_1(Z_\alpha,Y_b)\leq(1-\alpha)\rho$, hence

$$\rho\leq\rho_1(X,Z_\alpha)+\rho_1(Z_\alpha,Y_b)\leq\alpha\rho+(1-\alpha)\rho=\rho\ .$$

The equality results, which shows that Z_α is in optimal position for metric ρ_1. In conclusion, we can state

Theorem 3 *Let $\mathcal{K}_1$ denote the quotient space, under translation, of $\mathcal{K}'(\mathbb{R}^n)$ or $\mathcal{K}'(\mathbb{Z}^n)$. Then, the mapping $\mathcal{K}_1 \times \mathcal{K}_1 \to \mathbb{R}_+$ by rel. (11) defines a distance. Given X and Y in $\mathcal{K}'$ there exists at least one geodesic in the sense of ρ_1, which coincides with the geodesic $\{Z_\alpha\}$ in the sense of ρ, whose extremities are one of the pairs, (X_a, X_b) say, that minimizes $\rho(X,Y)$.*

From the reduction $\rho \to \rho_1$, it is always possible to go back easily. Having fixed set X and determined a translation b of Y such that $\rho(X,Y_b)=\rho_1(X,Y)$, it suffices then to take interpolator

$$Z_1(\alpha)=[(X\oplus\alpha\rho B)\cap(Y_b\oplus(1-\alpha)\rho B)]_{-\alpha b}$$

which generates in $\mathcal{K}'$ a segment from X to Y, as spans [0,1]. The proposed approach is valid in $\mathbb{R}^n$ as well as in $\mathbb{Z}^n$. In the latter case, the balls are digital polyhedra. An exact computation of the optimal vector b might be time consuming. In practice, one may content oneself with vector b^*, of which each coordinate $b_i^*, i \in [1,n]$ minimizes the 1-D Hausdorff distance between the projections of X_i of X and Y_i of Y on the corresponding axis No i. If x_i and x_i' (resp. y_i and y_i') stand for the extreme points of X_i (resp. Y_i), component b_i is

$$b_i^*=\left[\left(x_i+x_i'\right)-\left(y_i+y_i'\right)\right]/2$$

5. Interpolations for nested sets

This section is closer to the two related works [5] and [6], and establishes a common basis. Nice developments based on distance function have also been proposed by J.R. Casas [11] and also by P. Moreau and Ch. Ronse [12]. From now on, we particularize the pair (X, Y) under study, to be *ordered*, by taking $X \subseteq Y$. This dissymmetry suggests to play with both Hausdorff distances ρ (for dilations from X) and σ (for erosions from Y).

A point m at a distance $\leq \lambda$ from X and $\geq \lambda$ from Y^c belongs to set $(X \oplus \lambda B) \cap (Y \ominus \lambda B)$, hence to set

$$M = \cup \{(X \oplus \lambda B) \cap (Y \ominus \lambda B) \ , \quad \lambda \geq 0\} \ . \tag{13}$$

Conversely, every point $m \in M$ belongs to one of the terms of the union, so there exists a $\lambda \geq 0$ with $d(m, X) \leq \lambda$ and $d(m, Y^c) \geq \lambda$. In this sense interpolator M is midway between X and Y, and we shall call it the *median* of X and Y. It is easy to notice that the boundary of M is nothing but the skeleton by zone of influence, or skiz, between X and Y^c. This implies the immediate following consequences:

Proposition 2 *The median set $M(X, Y)$ is compact and comprised between X and Y. Its boundary is the locus of those points of $\mathbb{R}^n$ which are equidistant from X and Y^c, and these distances are smaller or equal to*

$$\mu = \inf \{\lambda : \lambda \geq 0 \ , \ X \oplus \lambda B \supseteq Y \ominus \lambda B\} \tag{14}$$

the equality being reached for at least one point of ∂M.

(Direct consequence of the SKIZ properties).Here is now an instructive property which shows how both Hausdorff distances are involved in the median $M(X, Y)$.

Theorem 4 *Given $X, Y \in \mathcal{K}'(\mathbb{R}^n)$, the median element $M(X, Y)$ is at Hausdorff dilation distance μ from X and $X \bullet \mu B$, and at Hausdorff erosion distance μ from Y and $Y \circ \mu B$, where μ is defined by rel. (14).*

Proof : Let ρ be the Hausdorff distance by dilation between sets M and X. We draw from prop. 2 that $X \subseteq M \subseteq X \oplus \mu B$ hence $\mu \leq \rho$. Now for all ρ strictly smaller than μ, the set $X \oplus \mu B$ cannot contain the non empty set $(X \oplus \mu B) \cap (Y \ominus \mu B)$, which is a subset of the boundary $\partial (X \oplus \mu B)$ (prop. 2 again). Hence $\mu = \rho$.

Consider now the Hausdorff distance by dilation ρ' between the closing $X \bullet \mu B$ and M. We have $X \bullet \mu B \oplus \mu B = X \oplus \mu B \supseteq M$, and also $M \oplus \mu B \supseteq X \oplus \mu B = X \bullet \mu B \oplus \mu B$, hence $\rho' \leq \mu$. Just as previously if ρ' is strictly smaller than μ, then $X \bullet \mu B \oplus \rho' B = X \oplus \rho' B$ cannot contain set M, hence $\rho' = \mu$.

The corresponding results for Hausdorff distance by erosions derive by duality. Q.E.D.

Note that in all these results, the distances between sets X and Y (by dilation or by erosion) do not intervene. Indeed they are not associated with μ, but rather with the sum of the successive median sets between X and Y. For example, if M_1 is the median between M and Y, of parameter μ, M_2 the median between M_1 and

Y, etc., then $\mu + \sum \mu_i = \rho$, where ρ is the Hausdorff dilation distance between X and Y (similar result for the erosions by going from Y to X). This allows a series of progressive interpolations from X to Y [6], which are distinct from those, [5][10], which are obtained by replacing M by

$$M_\alpha = \bigcup_\lambda \{(X \oplus \alpha\lambda B) \cap (Y \ominus (1-\alpha)\lambda B)\ ,\ \alpha \in [0,1]\}\ .$$

6. Conclusion

There are various ways to define interpolators between two sets X and Y of a metric space E. If X and Y are supposed to be compact and non empty, Hausdorff distance provides one possible approach, because, as it has been proved in this paper, this distance admits geodesics, i.e. series of "best" interpolators going from X to Y.In fact, several geodesics co-exist, which are not equivalent. Therefore, some additional constraints allow more specific, hence better, interpolators. It is in particular the case when i) X, Y are convex ; ii) E is equipped with a translation, and interpolations are introduced up to a translation ; iii) when $W \subseteq Y$. Up to now, the last case has been the most studied [1] [2] [11].

References

1. Beucher S, Interpolation d'ensembles, de partitions et de fonctions, Tech. rep. N-18/94/MM, Ecole des Mines de Paris, May 1994.
2. Meyer F, Interpolations, Tech. rep. N-16/94/MM, Ecole des Mines de Paris, May 1994.
3. Serra J, Interpolations et distance de Hausdorff, Tech. rep. N-15/94/MM, Ecole des Mines de Paris, May 1994.
4. Patent No 94-14162, first application in France, Nov.1994.
5. Meyer F., A morphological interpolation method for mosaic images, in *Mathematical Morphology and its applications to image and signal processing,* Maragos P. et al. eds. Kluwer, 1996.
6. Beucher S, Interpolation of sets, of partitions and of functions, (to be published for *ISMM'98*).
7. Dougherty E., Application of the Hausdorff metric in gray scale morphology via truncated umbrae, *JVCIR* **2**(2), 1991, pp. 177-187.
8. Serra J. Equicontinuous functions: a model for mathematical morphology, SPIE San Diego Conf., Vol. 1769, pp. 252-263, July 1992.
9. Matheron G., *Random Sets and Integral Geometry*, Wiley, 1975.
10. Huttenlocher D.P., Klunderman G.A., Rucklidge W.J., Comparing images using the Hausdorff distance, *IEEE PAMI*, **15**(9), Sept. 1995.
11. Casas J.R., *Image compression based on perceptual coding techniques*, PhD thesis, UPC, Barcelona, March 1996.
12. Moreau P., and Ronse Ch., Generation of shading-off on images by extrapolations of Lipschitz functions, *Graph. Models and Image Processing*, **58**(6), July 1996, pp. 314-333.

CURVE EVOLUTION, WAVE PROPAGATION, AND MATHEMATICAL MORPHOLOGY

HÜSEYIN TEK and BENJAMIN B. KIMIA
LEMS, Brown University, USA

Abstract. Set theoretic, algebraic, mathematical morphology operations can be viewed as geometric deformations described by curve evolutions governed by a partial differential equation. While typical curve evolution implementations rely on embedding the curve as the zero level set of an evolving surface, this approach cannot handle open curves and junctions, and is inefficient due to the additional dimension. This paper presents a novel approach relying on an analytic wave propagation framework, but one which is implemented on a discrete grid and using discrete directions, by relying on subpixel geometric models. The resulting implementation is exact for the class of piecewise circular curves and its usefulness is demonstrated for extracting skeletons and for smoothing open curves and shapes.

1. Introduction

Classically, mathematical morphology treats binary images as sets and grayscale images as functions, and operates on them in the spatial domain via *morphological transformations,* using *structuring elements* [19, 27, 11, 28]. While there is a growing interest in applying mathematical morphology to abstract spaces such as lattices [28, 12], graphs [35] or manifolds [25] there is also an interest in considering *geometric interpretations* of morphological transformations, where basic morphological operators are formulated as partial differential equations governing the geometric evolution of shapes. To illustrate, consider the dilation transformation of a shape with a disk structuring element. The transformed shape is the union of all disks centered on points of the original shape. The boundary of the transformed shape is a curve parallel to the boundary of the original shape with a distance equal to the radius of the disk. This is precisely Huygens' principle for wavefront propagation [13], relating operations on algebraic constructs, *i.e.*, sets of points, on the one hand, and operations on geometric entities, *i.e.*, curves representing the boundary on the other. Can this deformation of a curve onto another as a result of the dilation can be described as a sequence of infinitesimal deformations each of which is only locally dependent on the original boundary's shape, and thus represented by partial differential equations?

Mathematical Morphology and Curve Evolution: The answer in certain circumstances is positive and founded on three observations. First, observe that for a class of structuring elements, $\hat{\mathcal{B}}$, n repeated dilations (or erosions) with a structuring element $\mathcal{B} \in \hat{\mathcal{B}}$, is equivalent to a single dilation with a structuring element of the same shape, but which is "scaled up" n times, $\mathcal{B}(\lambda)$, *e.g.*, 10 dilations with a circle of radius of 1, $\mathcal{B}(1)$, is exactly a single dilation with a circle of radius 10, $\mathcal{B}(10)$.

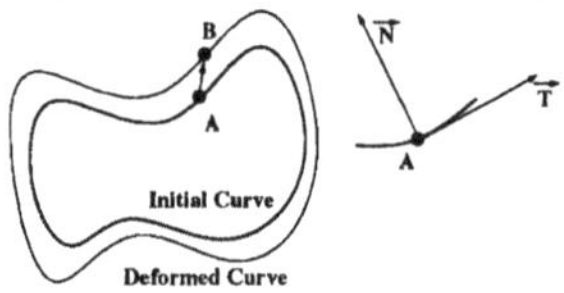

Fig. 1. The points on the initial curve A move to B to generate a new curve. The direction and magnitude of this motion is arbitrary in order to capture general deformations [15].

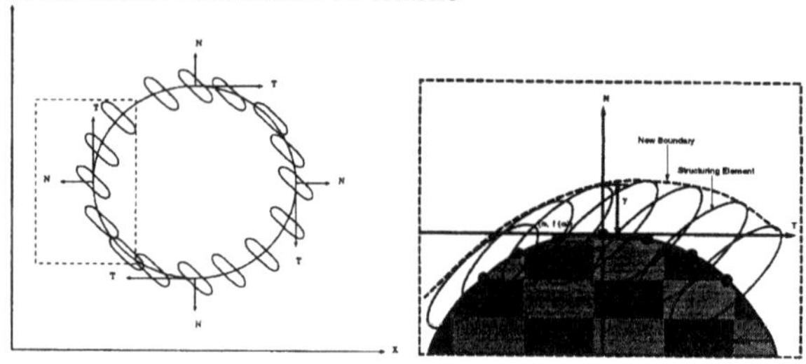

Fig. 2. Morphological operations in the extrinsic and intrinsic coordinates for an arbitrary convex structuring element.

The class of structuring elements $\hat{\mathcal{B}}$ for which this property holds is exactly the set of *convex* structuring elements [19, Theorem 1.5.1, pp. 22–23]. Now, conversely, given a single operation with a convex structuring element, the operation can be decomposed as a sequence of n operations with the same structuring element but of "size" $1/n^{\text{th}}$ the original. In the limit, as $n \to \infty$, this notion of size constitutes the "time" axis for the differential deformation. Second, observe that the outcome of mathematical morphology operations with a shape $\mathcal{S}$, can be determined solely from its boundary, $\partial\mathcal{S}$ [35]. Thus, the mapping from the shape $\mathcal{S}$ to $\mathcal{S}'$ by morphological operations can be reduced to operations taking $\partial\mathcal{S}$ to $\partial\mathcal{S}'$. Third, for a sufficiently small but finite structuring elements, mathematical morphology operations along the smooth portions of the boundary can be viewed as a deformation along the normal by a certain amount, Figure 1. Formally, let some shape $\mathcal{C}(s,t)$ evolve by a mathematical morphology operation, say dilation, by some structuring element $\mathcal{B}(\lambda)$ of size λ, to a new shape $\mathcal{C}(s,t+\lambda)$. This evolution can be represented as $\mathcal{C}(s,t+\lambda) - \mathcal{C}(s,t) = \Gamma(s,t,\lambda)\vec{N}(s,t)$, where Γ is the distance along $\vec{N}$ that a point on the boundary moves in a dilation operation with a structuring element of size λ. Since $\Gamma(s,t,0) = 0$,

$$\frac{\partial \mathcal{C}}{\partial t} = \lim_{\lambda \to 0} \frac{\Gamma(s,t,\lambda) - \Gamma(s,t,0)}{\lambda}\vec{N} = \frac{\partial \Gamma(s,t,\lambda)}{\partial \lambda}|_{\lambda=0}\vec{N} = \beta(s,t)\vec{N}. \tag{1}$$

It has been shown that the amount of differential deformation, β, of a shape $\mathcal{S}$ at a point P due to dilation (erosion) with a convex structuring element $\mathcal{B}$, is the maximal (minimal) projection of $\mathcal{B}$ onto the normal $\vec{N}$ of the boundary at P [2, 3], so that β can only depend on the orientation of the target to the shape $\vec{T}(s,t)$, leading to

$$\begin{cases} \frac{\partial \mathcal{C}}{\partial t} & = \beta(\vec{T}(s,t))\vec{N} \\ \mathcal{C}(s,0) & = \mathcal{C}_0(s), \end{cases} \tag{2}$$

or $\frac{\partial \mathcal{C}}{\partial t} = \beta(\theta)\vec{N}$ where $\theta = \angle(\vec{T}, x\text{-axis})$. The erosion transformation is similarly modeled.

For example, for an ellipse structuring element $\mathcal{B}$, with major and minor axes $(2A, 2B)$, and angle between major axis and the x-axis $\theta_{\mathcal{B}}$, then dilations with $\mathcal{B}$ is achieved using [3] $\beta(\theta) = \sqrt{A^2 \sin^2(\theta - \theta_{\mathcal{B}}) + B^2 \cos^2(\theta - \theta_{\mathcal{B}})}$. For a circular structuring element $A = B$ and $\beta(0) = 1$, *i.e.*, $\frac{\partial \mathcal{C}}{\partial t} = \pm\vec{N}$. Similarly, for a rectangular structuring element the maximum projection is always given by one of its four corner points which after rotation by an arbitrary amount $\theta_{\mathcal{B}}$ (this gives a square for $\theta_{\mathcal{B}} = 0$ and a diamond for $\theta_{\mathcal{B}} = 90^o$) [2],

$$\beta = \begin{cases} asin\theta_\xi + bcos\theta_\xi & 0 \leq \theta_\xi \leq \frac{\pi}{2} \\ asin\theta_\xi - bcos\theta_\xi & \frac{\pi}{2} \leq \theta_\xi \leq \pi \\ -asin\theta_\xi + bcos\theta_\xi & \pi \leq \theta_\xi \leq \frac{3\pi}{2} \\ -asin\theta_\xi - bcos\theta_\xi & \frac{3\pi}{2} \leq \theta_\xi \leq 2\pi \end{cases} \tag{3}$$

where $\theta_\xi = \theta_{\mathcal{B}} - \theta$, and $(2a, 2b)$ are the length and width of rectangle, respectively. See a different formulation of these results later in this section from [5, 26] . Similarly, it is not difficult to obtain β for other convex structuring elements, since $\beta(\theta)$ is the maximal projection of the structuring element $\mathcal{B}$ onto a rotating axis.

Unfortunately, direct simulations of equation (2) are not numerically stable. Observe that shapes undergoing morphological operations change their topology in that they merge and split, leading to problems in administrating newly formed boundaries and annihilated ones, some boundary points bunch-up and merge to form singularities, leading to numerical errors, *etc.* Typically, a reparametrization can be employed, but this is at the expense of smoothing singularities, the salient features of shape. To effectively deal with these problems, Osher and Sethian [21] proposed a numerical method based on embedding the original curve evolution equation in a higher dimension, by representing the evolving curve $C(s,t)$ as the zero level set of an evolving surface, $\phi(x,y,t) = 0$

$$\begin{cases} \phi_t + \beta(\phi_x^2 + \phi_y^2)^{1/2} = 0 \\ \phi(x,y,0) = \phi_0(x,y), \end{cases} \tag{4}$$

where ϕ_0 is the distance transform of $C_0(s)$. Using $\vec{T} = (\cos\theta, \sin\theta) = \frac{(-\phi_y, \phi_x)}{\sqrt{\phi_x^2+\phi_y^2}}$, this can be rewritten as

$$\begin{cases} \phi_t + \mathcal{H}(\phi_x, \phi_y) = 0 \\ \phi(x,y,0) = \phi_0(x,y), \end{cases} \tag{5}$$

where $\mathcal{H}(\phi_x, \phi_y) = \beta(\phi_x, \phi_y)(\phi_x^2 + \phi_y^2)^{1/2}$ is the Hamiltonian associated with this Hamilton-Jacobi equation. For the elliptical structuring elements we get

$$\mathcal{H}(\phi_x, \phi_y) = \sqrt{\frac{[A^2\chi^2 + B^2\zeta^2]\phi_x^2 + 2(A^2 - B^2)\chi\zeta\phi_x\phi_y + [A^2\zeta^2 + B^2\chi^2]\phi_y^2}{\phi_x^2 + \phi_y^2}} \tag{6}$$

where $\chi = cos\theta_{\mathcal{B}}$ and $\zeta = sin\theta_{\mathcal{B}}$. Figure 4 shows the results of dilation of an ellipse by another.

Related Work: Similar differential equations were derived by Brockett and Maragos [5] for the evolution of signals $f(x)$ which dilate (or erode) by functions $g(x)$ or structuring element $\mathcal{B}$ whose domain and range is scaled by t. Specifically, for 1D sets $\frac{\partial\alpha(x,t)}{\partial t} = \pm|\frac{\partial\alpha(x,t)}{\partial x}|$ and for 2D sets $\frac{\partial\alpha(x,y,\tau)}{\partial t}$ is equal to $(|\alpha_x|^2 + |\alpha_y|^2)$ for a disk, $= (|\alpha_x| + |\alpha_y|)$ for a square, and $= max(|\alpha_x|, |\alpha_y|)$ for a diamond. They also described evolution equations for dilations/erosions by functions. Alvarez, Guichard, Lions, and Morel [1] proved that under certain initial conditions required for a multi-scale analysis $\mathcal{T}_t$ transforming an intuitive $u_o(x)$ into $u(x,t) = (\mathcal{T}_t u_0)(x)$, the smoothed signal must satisfy a partial differential equation $\frac{\partial u}{\partial t} = F(D^2u, \nabla u, t)$ where D^2 is a second-order differential operator and ∇ is the gradient. For example, $\frac{\partial u}{\partial t} = \pm|\nabla u|$ represents dilation and erosion. As another example, Catte, Dibos and Koepfler [6] cast Euclidean curve shortening flow $\frac{\partial C}{\partial t} = \kappa\vec{N}$, also known as mean curvature motion, in the mathematical morphology framework by considering a line segment structuring element of length $2\sqrt{2t}$, orientation $\theta\epsilon[0,\pi]$ and three operators $I_t u(x) = \inf_{\theta\in[0,\pi]} \sup_{y\in X+S(\theta,t)} u(x)$, $S_t u(x) = \sup_{\theta\in[0,\pi]} \inf_{y\in X+S(\theta,t)} u(x)$, $C_t u(x) = \frac{1}{2}[(S_{2t}u)(x) + (I_{2t}u)(x)]$ and show that these operators correspond to $\frac{\partial C}{\partial t} = -\kappa^{-}\vec{N}$, and $\frac{\partial C}{\partial t} = \kappa^{+}\vec{N}$, and $\frac{\partial C}{\partial t} = \kappa\vec{N}$, respectively. Boomgard's work is another example relating morphology and PDE approaches [9].

Thus, key applications of mathematical morphology, such as, (*i*) smoothing of shapes and images, (*ii*) skeletal representation of shape, and (*iii*) segmentation by watersheds computation can all be formulated in the curve evolution framework. For the connection to smoothing case, see [15, 16, 17, 14] where Gaussian and curvature-driven smoothing of curves and images are represented together with morphological operations in a unified reaction-diffusion framework driven by $\frac{\partial C}{\partial t} = (\beta_0 - \beta_1\kappa)\vec{N}$. The extraction of shocks, whose geometric locus is the crest lines of the distance transform, has been implemented using wave propagation scheme [29, 33]. Finally, watershed segmentation can be cast in the form of the bubble framework where fourth-order shocks (seeds) are initialized at local minima of the gradient and whose growth is inversely proportional to the gradient [32]. These applications highlight the need for a more general and more efficient curve evolution framework presented next.

2. Analytic Propagation of Geometric Descriptions On a Discrete Grid

While the implementation of curve evolution as zero level sets of an embedding evolving surface is robust and even has subpixel ability [30], it cannot handle *junctions* and *open curves*, and is also *computationally inefficient* due to the additional embedding dimension. We now propose an approach relying on a view of curve evolution as propagation of waves from boundary segments, inspired by Blum's "grass fire" [4]. In this approach, each boundary initiates waves carrying geometrical information, labels, *etc.* which if traveling with constant speed yields the *distance transform*, among other constructs. Recall that the distance transform maps a binary image onto an image where the value at each point is the Euclidean distance from the object, and thus distance represents "time" and a constant distance set represents a wavefront. We shall show that this wave propagation approach based on distance transform can be robustly and reliably applied to subpixel curves, but implemented on a discrete grid with discrete directions. Unlike curve evolution, this approach does not require that shapes be segmented from images but rather is easily applied to open contour segments, possibly with junctions, as available in edge maps. In addition, this approach is computationally efficient since it does not require the higher-dimensional embedding surface. In the following, we first describe a subpixel boundary model, review the current discrete distance transforms, introduce an analytic distance transform (ADT) and show how the explicit representation of a discrete front leads to an exact wave propagation scheme.

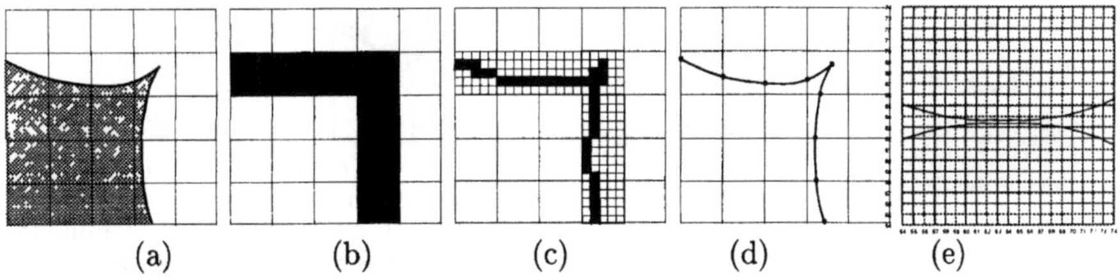

Fig. 3. This figure illustrate that a geometric intra-pixel model is needed for accurate and efficient subpixel boundary representation. The binary representation of a curve (a) on a discrete grid (b) discards much of the geometric information. Higher resolution representation around the boundary (c) is also jagged and does not represent geometric information well. An alternative method to fine sampling is the use of a piecewise circular geometric model (GENO) in each pixel [30], (d). (e) Simulation results from [30] showing that GENO is able to track multiple curves evolving within a pixel extracted from the distance transform map, essential in many morphological operations.

Subpixel Boundary Representation: Traditionally, distance transforms are designed for objects represented by discrete boundary locations [7, 39, 23, 24, 36]. However, observe that image structure is rarely in binary form, and in fact the human visual system has developed the ability to detect features at resolutions an order of magnitude better than retinal resolution (hyperacuity). Many low-level vision algorithms have also developed similar capabilities [8, 31], which immediately raises challenging issues: how to represent each boundary with subpixel resolution *without* recruiting tremendous memory and computational resources? How to represent multiple curves within a pixel? How to represent singularities of curves? The intuitive approach of increasing the underlying grid resolution only at boundary points leads to a jagged representation, does not explicitly represent singularities, and imposes unnecessary memory and computational demands. The use of a piecewise circular geometric model of a curve within each pixel, on the other hand, allows a model of curvature as well as a model of singularities, if needed. Such a model can be effectively represented by the location of its intersections with horizontal and vertical grid lines, a value for curvature(s), and the location of singularity, if required. Chains of piecewise circular approximations *within each pixel* are extremely powerful not only in modeling arbitrary image curves but also in other applications [10].

The usefulness of this subpixel curve model for wave propagation is dependent on the reliable recovery of a such a model, *e.g.*, from a distance transform surface, especially in the presence of corners, junctions, and nearby curves. Siddiqi, Shu, and Kimia developed a three stage approach for the recovery of piecewise-circular subpixel models from their distance transform, relying on a class of non-linear interpolations technique [22]. In this approach, interpolation of a 1D signal in each interval between two samples relies on either the left or on the right neighborhood, whichever is maximally consistent with the interval, thus avoiding interpolating across singularities, without explicitly detecting them. In addition, the interpolation places subpixel singularities wherever necessary [30]. The extension of this technique to geometric curves relies on (*i*) the application of ENO interpolation to horizontal and vertical grid lines to detect zero-crossing points reliably, even in the presence of multiple curves, where all pixels may contain positive numbers Figure 3e, (*ii*)

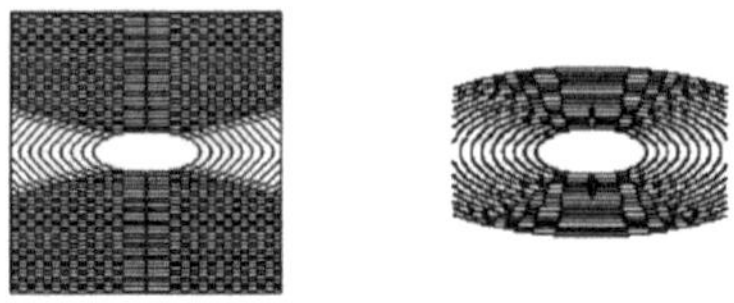

(a) (b)

Fig. 4. This figure illustrates the dilation of an ellipse by a structuring element of the same shape, implemented using the algorithm presented in [37]. (a) While the results when dilation is implemented in a single step are highly accurate, the iterations of a small structuring element in this case do not produce desirable results, as the ellipse is *not* digitally-scalable. (b) The geometric evolution produce accurate results [3].

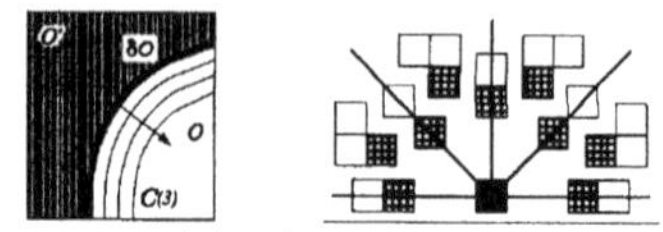

Fig. 5. (a) Three iterations of the propagation of a boundary δO at one pixel per iteration. (b) The minimal set of masks (only a mirror image constructs the bottom half of the full picture): The black pixel represents the source while the grey and white pixels represent subsequent propagation of waves. Black lines represent directly supported directions of propagation. for the intermediate directions of propagation, larger masks are required. (Adapted from [23])

spatially ordering these points, and (*iii*) geometric ENO interpolation (GENO) of these points into a subpixel piecewise circular model [30].

Our goal in this section is to propagate such subpixel geometric descriptions on a discrete grid and construct a computationally efficient and error-free distance transform which propagate not only distance but also geometric and other information. Traditional pixel to pixel descriptions of binary shape is then a special case.

Discrete Distance Transforms: Distance transform methods may be classified into those based on raster scans or those that are contour-based. Danielsson showed that the Euclidean Distance Transform (EDT) can be computed by comparing neighborhood pixels with vector-valued masks used to evaluate distance steps along the axis of the supporting grid. A raster scan version for EDT (REDT) was described in [7] and later extended to signed EDT [39] and then to a contour-based EDT (CEDT) [23, 24]. Previously, REDT and CEDT have been considered of similar usefulness when used for wave propagation and symmetry computation mainly because REDT is relatively simpler to implement, due to the independence of embedded shapes. In addition, CEDT implementations for symmetry set elicitation have often relied upon dilation/erosion of chain-coded representations of contour [38, 36], requiring a pre-processing of contour features. Thus, CEDT has received relatively little attention. However, for simulating wave propagation, we propose that CEDT has a key advantage over REDT in that it provides an explicit representation of geometry, wave labels, *etc.*

The basic design of CEDT follows ideas originated in Montanari [20], brought to the foreground by more recent work, *e.g.*, [34, 36] and, Ragnemalm [23, 24]. While Montanari had the key insight that wave propagation from boundary feature was potentially more efficient than raster-scan sequential DT, Ragnemalm imported the idea of using vectored values for DT [7] and studied different masks and their properties for propagating various metrics from contours. Figure 5 illustrates the minimal complete set of masks required for Euclidean metric propagation on a 2D rectangular distance grid from a point source. Each masks maintains a distance vector (L_x, L_y) from its origin, thus *explicitly* representing the direction and distance of the propagating wavefront. The metric $L^2 = (L_x^2 + L_y^2)$ is also carried to optimize operations.

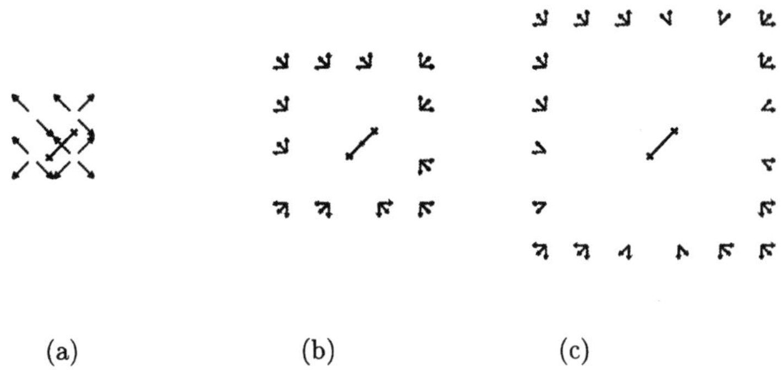

Fig. 6. The subpixel CEDT algorithm: (a) Geometric description of a subpixel boundary (bold line segment) is propagated to adjacent pixels first. (b) These in turn propagate in discrete directions to their neighbors but resort to analytic computations. (c) This procedure is repeated until all waves have been extinguished or reached the boundary.

A further optimization uses ***buckets*** to store wavefront distance values in order of metric L^2, thus leading to constant speed propagation. Other advantages of CEDT include: CEDT is nearly optimal in terms of numerical complexity when compared to raster-scan based DT, and is extended to 3D by defining additional masks [18]. However, like REDT, CEDT is not a totally error-free algorithm. Though exact distance transforms are proposed, in general they are not very intuitive and computationally expensive. In the following two key ideas of (i) analytic propagation of geometric description in conjunction with the distance information and (ii) an explicit representation of the discrete front leads to an exact, efficient, and intuitive framework for curve evolution via wave propagation.

Analytic Distance Transform (ADT): The CEDT algorithm which typically uses discrete wave directions from discrete boundary locations is now generalized to boundaries represented by subpixel piecewise circular arc segments, yet maintaining propagation using discrete grid locations and directions. The propagation is based on Huygen's principle which states that the wavefront at time $t + \Delta t$ is the envelope of propagation at time Δt of waves initiated at the wavefront at time t. In our approach, instead of propagating only distance information (time), we also propagate geometric models of the subpixel boundary in its piecewise circular form. Analytic distance values from source boundary segments are then computed at each step of propagation process, leading to the proposed Analytic Distance Transform (ADT) algorithm. Specifically, each piecewise circular boundary segment within a pixel initiates discrete waves in every discrete directions (8) on the discrete grid locations surrounding the boundary segment. Some wave directions are chosen to overlap with each other to cover the whole image domain. These waves create a discrete wavefront satisfying a minimum distance value where propagation should continue. The first version of the ADT algorithm is described as follows:
(i) **Discrete Wavefront Initialization:** Discrete grid points which enclose the subpixel boundaries are marked for propagation of discrete waves, Figure 6a. These grid points are marked as the ***discrete wavefront.***

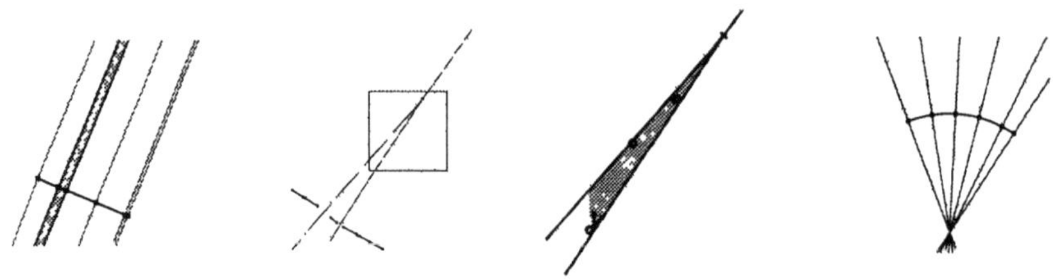

(a) (b) (c) (d)

Fig. 7. This figure illustrates that the ADT algorithm presented in the previous section is almost, but not totally error-free. (a) A line consisting of five subpixel boundary intervals (darker lines). The smallest line segments may not have access to grid locations, *e.g.*, shaded area, thus disappearing entirely. (b) this is not only limited to the small boundary intervals but also may happen when the image contains three or more non-connected boundary intervals. The equi-distance lines, the *shocks* of these non-connected boundary segments, marks the end of wave zone for each boundary segment. (c) The problem area which is marked with rectangular window in (b) is shown in detail. Note that waves from the middle line segment will not be able to propagate into the shaded region, but are rather absorbed by the waves from the left and right boundary segment. This problem has much more severe effects for boundary segments represented by circular arcs where all the waves become narrower towards the center of an arc, (d).

(*ii*) **Initialization of Discrete Wave Directions:** Each grid location on the *discrete front* can now be considered as a source in analogy to the discrete CEDT. Thus, each source attempts to propagate waves to its immediate neighbors (8 neighbors), Figure 6a. When a wave minimizes analytic distance at a point from its boundary source, the *discrete front* will be updated to this grid location. Discrete propagation directions, Vertical (V), Horizontal (H), Diagonal (D) are assigned to each wavefront location based on its location and the orientation of the initial boundary. The two constraints of (*a*) discrete waves should cover the entire propagation space and (*b*) discrete direction overlap should be minimal for efficiency lead to H, V, D and combined HD, VD discrete directions, Figure 6b. An additional concern is to prevent duplication of propagation by simultaneously propagating the wavefront. However, since the wavefronts at discrete points arrive at slightly different times, the propagation takes place from the "oldest" wavefront, by ordering distance values at discrete points.

(*iii*) **Propagation From Discrete Wavefront:** The discrete points on the *discrete wavefront* with minimum values are considered as sources for propagation. Among waves arriving at a point the first one (minimal time/distance) is added to the discrete front. This stage is repeated for every point on the *discrete wavefront* which is updated by propagating each point in the direction along which the point had propagated from, Figure 6c.

The above ADT algorithm is almost, but not totally error-free, mainly due to the use of discrete wave directions. For example, Figure 7a illustrates that for small boundary intervals the discrete grid may not sustain the propagation of waves. While merging these intervals with their neighboring intervals may solve this particular situation, the problem is fundamental to the nature of propagation, *e.g.*, consider

Figure 7b with three or more non-connected segments, where middle line's wave zone becomes increasingly narrower and eventually is unable to propagate into the shaded region, due to the waves from left and right boundary intervals, Figure 7b. Note that the two grid locations in the shaded area, Figure 7c, will receive incorrect waves. This problem has much more severe effects for boundary segments represented by circular arcs where all the waves become narrower towards the center of an arc Figure 7d.

The main difficulty is that the notion of a discrete wavefront is implicitly represented : as Figure 7 illustrates, the error in the distance transform stems from the fact that the discrete front is represented via discrete waves. The implication of such an approximation is that the notion of a front separating interior and exteriors of a propagating wavefront can be compromised. Our proposed remedy is an explicit representation of a discrete wavefront, by including propagation of *rays* from end points of open curves. The propagation takes these ray into account using Huygen's principle which is now applied to sources located both on discrete locations *and* on end interval ray intersections with horizontal and vertical grid lines. Specifically, in addition to previously described wave propagation, two rays (each moving in opposite directions) are initiated in the normal direction from the end points of each piecewise circular boundary segments within a pixel. These rays and discrete waves *together* create an explicitly represented discrete wavefront propagating in a a synchronous fashion. Specifically, the new ADT algorithm modifies step (*ii*) by adding:

(**modified** *ii*) In addition to (*ii*) described previously, each subpixel boundary segment initiates two rays from each subpixel end points. Each ray carries geometric descriptions of boundary segments and in conjunction with other ray from the other end point labels waves into regular and rarefaction waves. At each iteration, each ray will propagate until it intersects with the next coordinate axes. In order to optimize efficiency, discrete waves and rays must be propagated synchronously in each time step starting from the "oldest" wavefront. These updated discrete waves and ray points are marked as *discrete wavefront*, Figure 8.

This algorithm is order of magnitude more efficient than standard implementations of curve evolution via an embedding surface. In addition it is exact in the class of subpixel piecewise circular curves. Figures 9 and 10 illustrate opening and closing operations respectively. Figure 11 illustrates the extraction of skeletons from shapes and grayscale images.

Acknowledgments We gratefully acknowledge the support of the National Science Foundation grant IRI-9700497.

References

1. L. Alvarez, F. Guichard, P.-L. Lions, and J.-M. Morel. Axioms and fundamental equations of image processing. *Arch. Rational Mech. Anal.*, 123:199–257, 1993.
2. A. Arehart, L. Vincent, and B. B. Kimia. Mathematical morphology: The Hamilton-Jacobi connection. Technical Report LEMS 108, Brown University, June 1992.
3. A. Arehart, L. Vincent, and B. B. Kimia. Mathematical morphology: The Hamilton-Jacobi connection. In *ICCV*, pages 215–219, 1993.
4. H. Blum. Biological shape and visual science. *J. Theor. Biol.*, 38:205–287, 1973.

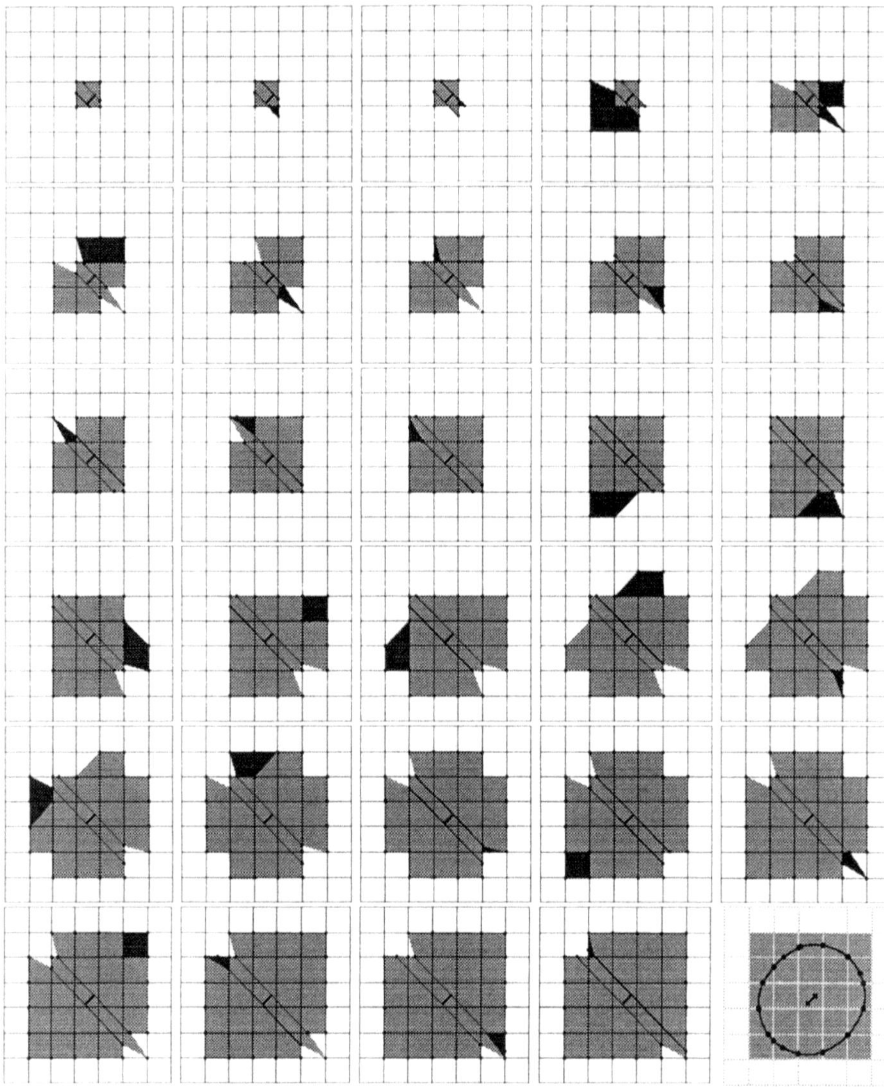

Fig. 8. This figure illustrates the explicit representation of the *discrete front* for a small subpixel line segment (dark line). In each step, the front is updated either by propagating from discrete grid point sources or from end interval ray sources, thus always maintaining a correct separation between the interior and the exterior of the wavefront. Observe that while the wavefront looks jagged, this is only visual effect due to the fact that each point on the wavefront has arrived at a slightly different, but analytically correct, time of arrival. A subpixel front at any time can be correctly computed as shown in the last box for $t = 2.0$.

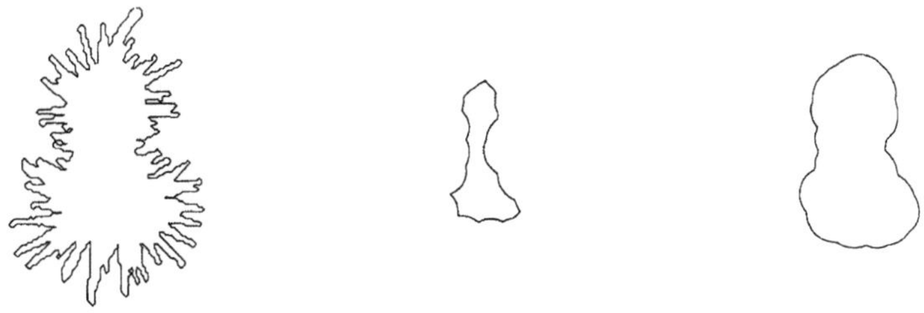

Fig. 9. An example of the opening operation using subpixel evolution.

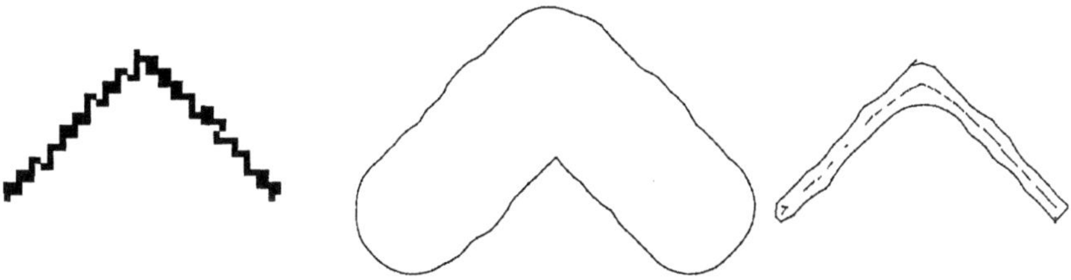

Fig. 10. Smoothing open curves by subpixel closing. Observe how the location and orientation of the line is maintained to subpixel accuracy.

5. R. Brockett and P. Maragos. Evolution equations for continuous-sacle morphology. In *In IEEE Conf. on Acoustics, Speech and Signal Processing*, March 1992.
6. F. Catte, F. Dibos, and G. Koepfler. A morphological scheme for mean curvature motion and applications to anisotropic diffusion and motion of level sets. *SIAM Journal of Numerical Analysis*, 32-6:1895–1909, 1995.
7. P. Danielsson. Euclidean distance mapping. *Computer Graphics and Image Processing*, 14:227–248, 1980.
8. E. DeMicheli, B. Caprile, P. Ottonello, and V. Torre. Localization and noise in edge detection. *IEEE Trans. Pattern Analysis and Machine Intelligence*, 11:1106–1117, 1989.
9. R. V. den Boomgaard. Mathematical morphology: extensions towards computer vision. Ph.D. dissertation, University of Amsterdam, March 1992.
10. P. Giblin and T. Banchoff. Symmetry sets of piecewise circular curves. *Proceedings of the Royal Society of Edinburgh*, 123A:1135–1149, 1993.
11. R. M. Haralick, S. R. Sternberg, and X. Zhuang. Image analysis using mathematical morphology. *IEEE Trans. Pattern Analysis and Machine Intelligence*, 9(4):532–550, July 1987.
12. H. Heijmans. Theoretical aspects of gray-level morphology. *PAMI*, 13:568–582, 1991.
13. F. John. *Partial Differential Equations.* Springer-Verlag, 1971.
14. B. B. Kimia and K. Siddiqi. Geometric heat equation and nonlinear diffusion of shapes and images. *CVIU*, 64(3):305–322, November 1996.
15. B. B. Kimia, A. R. Tannenbaum, and S. W. Zucker. Toward a computational theory of shape: An overview. In *ECCV*, pages 402–407, 1990.
16. B. B. Kimia, A. R. Tannenbaum, and S. W. Zucker. Nonlinear shape approximation via the entropy scale space. In *Proceedings of the SPIE's Geometric Methods in Computer Vision II*, volume 2031, pages 218–233, San Diego, California, July 1993.
17. B. B. Kimia, A. R. Tannenbaum, and S. W. Zucker. Shapes, shocks, and deformations, I: The components of shape and the reaction-diffusion space. *IJCV*, 15:189–224, 1995.
18. F. Leymarie and B. B. Kimia. Surface based euclidean distance transform (SEDT). Technical report in preparation, LEMS, Brown University, April 1998.
19. G. Matheron. *Random Sets and Integral Geometry.* Wiley, 1975.
20. U. Montanari. A method for obtaining skeletons using a quasi–euclidean distance. *JACM*,

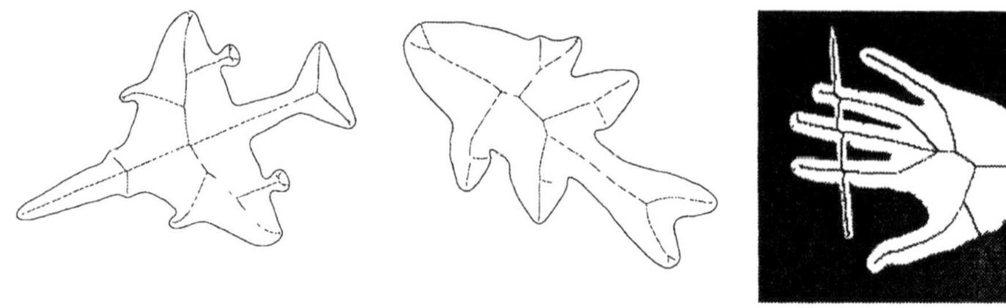

(a) (b) (c)

Fig. 11. This figure shows extraction of shocks from binary shapes (a) and (b), and from grayscale images (c).

15(4):600–624, October 1968.
21. S. Osher and J. Sethian. Fronts propagating with curvature dependent speed: Algorithms based on Hamilton-Jacobi formulations. *Journal of Computational Physics*, 79:12–49, 1988.
22. S. Osher and C.-W. Shu. High-order Essentially Non-Oscillatory schemes for Hamilton-Jacobi equations. *SIAM Journal of Numerical Analysis*, 28:907–922, 1991.
23. I. Ragnemalm. Generation of Euclidean distance maps. Licenciate Thesis No 206, Linkoping University, Sweden, 1990.
24. I. Ragnemalm. Neighborhoods for distance transformations using ordered propagation. *CVGIP: Image Understanding*, 56(3):399–409, 1992.
25. J. Roerdink. Manifold shape: from differential geometry to mathematical morphology. In *Proceedings NATO Shape in Picture Workshop*, Driebergen, NL, September 1992.
26. G. Sapiro, B. B. Kimia, R. Kimmel, D. Shaked, and A. Bruckstein. Implementing continuous-scale morphology. *Pattern Recognition*, 26(9):1363–1372, 1993.
27. J. Serra, editor. *Image Analysis and Mathematical Morpholpogy.* Academic Press, 1982.
28. J. Serra, editor. *Image Analysis and Mathematical Morpholpogy, Part II: Theoretical Advances.* Academic Press, 1988.
29. K. Siddiqi and B. B. Kimia. A shock grammar for recognition. In *CVPR*, San Francisco, California, June 1996.
30. K. Siddiqi, B. B. Kimia, and C. Shu. Geometric shock-capturing ENO schemes for subpixel interpolation, computation and curve evolution. *Graph. Models and Im. Processing*, To Appear.
31. S. Smith and J. Brady. SUSAN - a new approach to low level image processing. In *Int. Journal of Computer Vision*, 1997. in publication.
32. H. Tek and B. B. Kimia. Volumetric segmentation of medical images by three-dimensional bubbles. *Computer Vision and Image Understanding*, 64(2):246–258, February 1997.
33. H. Tek, P. Stoll, and B. B. Kimia. Shocks from images: Propagation of orientation elements. In *CVPR*,June 15-16 1997.
34. B. Verwer. Improved metrics in image processing applied to the hilditch skeleton. In *CVPR*, pages 137–142, 1988.
35. L. Vincent. Graphs and mathematical morphology. *Signal Processing*, 16(4):365–388, 1989.
36. L. Vincent. Exact euclidean distance function by chain propagations. In *Proceedings of the IEEE Computer Vision and Pattern Recognition '91*, pages 520–525, Maui HI, 1991.
37. L. Vincent. Morphological transformations of binary images with arbitrary structuring elements. *Signal Processing*, 22(1):3–23, 1991.
38. Y. Xia. Skeletonization via the realization of the fire front's propagation and extinction in digital binary shapes. *PAMI*, 11(10):1076–1086, 1989.
39. Q.-Z. Ye. The signed Euclidean distance transform and its applications. In *ICPR*, pages 495–499, 1988.

GEOMETRIC PARTIAL DIFFERENTIAL EQUATIONS AND ITERATIVE FILTERING

F. GUICHARD
Dart, Inrets,
2, Av. Gen. Malleret-Joinville, 94114 Arcueil cedex. France.
fguichar@ceremade.dauphine.fr

and

J-M. MOREL
CMLA, ENS Cachan,
61, Av. Président Wilson, 94235 Cachan cedex. France.
morel@ceremade.dauphine.fr

Abstract.
It is well known that a conveniently rescaled iterated convolution of a linear positive kernel converges to a Gaussian. As a consequence, all iterative linear smoothing methods of a signal or an image boil down to the application to the signal of the heat equation. This survey explains how a similar analysis can be performed for image iterative smoothing by contrast invariant monotone operators. Among them, the median operator which computes a local "mean value" independent of contrast changes. We prove that all monotone and contrast invariant operators are asymptotically equivalent (when they become more and more local) to a motion of the image by its curvature. The iteration of these filters is shown to be equivalent to the application to the image of a nonlinear heat equation.

1. The bottom-up approach : from local filters to Partial Differential Operators

Let $u_0(\mathbf{x})$ be a real function defined on the plane, which we interpret throughout this paper as a real image, $\mathbf{x} = (x, y)$ being a point of the image plane and $u_0(\mathbf{x})$ the "grey level" at $\mathbf{x}$. In signal and image processing, "filtering" means an operation, applied to the image, which tends to smooth out asperities, yields a more reliable value at each point and permits the computations of some local features of the image. Linear filtering methods are often defined by a modification of the Fourier spectrum of the image, which attenuates high frequencies. It is therefore a global operation, whose local result depends on global events such as the size and location of the observation window of the image. There has been a growing tendency in image processing to define filtering methods which are as independent as possible of the observation window, which in practice means that they are *local.* A local operator is, intuitively, an operator T which transforms the initial image u_0 into a new image $(Tu_0)(\mathbf{x})$ and such that the value $(Tu_0)(\mathbf{x})$ only depends on the values $u_0(\mathbf{y})$ where $\mathbf{y}$ lies in an *a priori* fixed neighborhood of $\mathbf{x}$. This neighborhood should be as small as possible.

1.1. The linear case

As for an obvious (but useful) example, let us mention the case where $Tu_0(\mathbf{x}) = (M_h u_0)(\mathbf{x})$ is obtained as the mean value of u_0 in a neighborhood of $\mathbf{x}$,

$$M_h u_0(\mathbf{x}) = \frac{1}{\pi h^2} \int_{D(\mathbf{x},h)} u_0(\mathbf{y}) d\mathbf{y}, \tag{1}$$

In this case, the locality means that only values of u_0 inside a disk $D(\mathbf{x}, h)$ around $\mathbf{x}$ matter. When h tends to zero, we get a more and more local estimate of the value of u_0 at $\mathbf{x}$. The parameter h characterizes the "locality" of the considered operator, that is, the size of the neighborhood involved in the filtering operation. The question arises of whether such an operation can be made independent of h. A first answer is to analyze the asymptotic development of $M_h u_0(\mathbf{x})$ as h tends to zero. Clearly, if u_0 is continuous at $\mathbf{x}$, we have $M_h u_0(\mathbf{x}) \to u_0(\mathbf{x})$. If we assume that u_0 is twice continuously differentiable (C^2) at $\mathbf{x}$, then it is easily seen that in fact [1]

$$\frac{(M_h u_0)(\mathbf{x}) - u_0(\mathbf{x})}{h^2} = \frac{1}{8} \Delta u_0(\mathbf{x}) + \varepsilon(h) \tag{2}$$

Where $\Delta u = \frac{\partial^2 u}{\partial x^2} + \frac{\partial^2 u}{\partial y^2}$ is the laplacian.

This formula suggests the following result : consider for any t a sequence of real numbers h and integers n such that $nh^2 \to t$. Set $u^n = T_h^n u_0$. Then it is to be expected that $u^n(\mathbf{x}) \to u(t, \mathbf{x})$ where $u(t, \mathbf{x})$ satisfies the heat equation associated with (2),

$$\frac{\partial u}{\partial t}(t, \mathbf{x}) = \frac{1}{8} \Delta u(t, \mathbf{x}), \quad u(0) = u_0. \tag{3}$$

It is important to notice that the result is fairly independent of the kind of linear smoothing effectuated. More precisely, setting $\mathbf{x} = (x_1, ..., x_N) \in I\!R^N$, and denoting by $(g(\mathbf{x}))^{n*}$ the convolution of a function $g(\mathbf{x})$ with itself n times, one can prove the following theorem.

Theorem 1 *Let $g(\mathbf{x}) = g(|\mathbf{x}|)$, $\mathbf{x} \in I\!R^N$, be a radial function, nonnegative, compactly supported and piecewise C^1, satisfying the normalization conditions $\int_{I\!R^N} g(\mathbf{x}) d\mathbf{x} = 1$ and $\int_{I\!R^N} x_1^2 g(\mathbf{x}) d\mathbf{x} = 2$. Then for every $t > 0$ and every sequence h of real numbers and of integers n such that $t = nh^2$ and $n \to \infty$, one has*

$$(g_h(\mathbf{x}))^{n*} \to \frac{1}{(4\pi t)^{\frac{N}{2}}} e^{-\frac{|\mathbf{x}|^2}{4t}}, \tag{4}$$

the convergence being verified pointwise, uniformly and in $L^2(I\!R^N)$. As a consequence, defining, for every initial bounded image $u_0(\mathbf{x})$, $(L_h u_0) = g_h * u_0$, we can assert that

$$(L_h)^n u_0 \to T_t u_0 \quad \textit{uniformly},$$

[1] We always denote in this text by $o(h)$ a function of h which tends to zero faster than h, by $O(h)$ a function such that, for some constant C, $|O(h)| \leq C|h|$, and by $\varepsilon(h)$ a function tending to zero as h tends to zero. Thus an $o(h)$ function can also be written : $h\varepsilon(h)$.

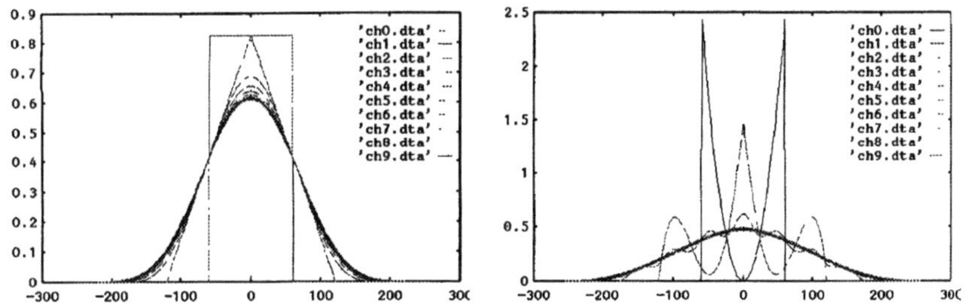

Fig. 1. In this experiment on one dimensional functions, it can be appreciated how fast an iterated convolution of a positive kernel converges to a gaussian. On the left, we display nine iterations of the convolution of a characteristic function with itself, with the adequate rescaling. On the right, the same experiment is repeated with a by far more irregular kernel. Convergence to the gaussian is almost equally fast. (See Theorem 1.)

where $(T_t u_0)(\mathbf{x}) = u(t, \mathbf{x})$ *and* $u(t, \mathbf{x})$ *is solution of the heat equation*

$$\frac{\partial u}{\partial t} = \Delta u, \quad u(0, \mathbf{x}) = u_0(\mathbf{x}). \tag{5}$$

In other terms, the preceding theorem means that from the asymptotic, scale independent viewpoint, all local linear "low pass" iterated filters are equivalent to the heat equation (and to the convolution with gaussians with increasing width).

1.2. Contrast Invariant filters

Our aim here is to generalize the above linear results to a wide class of nonlinear filters, the so called contrast invariant filters. A filter T is called contrast invariant if it satisfies the following properties. (We denote by u, v, ... real functions on $I\!R^N$ representing N-dimensional images. T transforms a function u into a new function Tu.)

a- Monotonicity. If $u \geq v$, then $Tu \geq Tv$.

b- Contrast invariance. If $g(s)$ is a nondecreasing continuous real function, which we call a "contrast change", then $T(g(u)) = g(T(u))$. This means that the action of T does not depend upon the contrast of the image ; an assumption which is of course false for linear filters.

c- Locality. Basically, this assumption means that $(Tu)(\mathbf{x})$ only depends upon the values of u in a fixed neighborhood of $\mathbf{x}$, for instance a disk $D(\mathbf{x}, h)$.

The contrast invariance assumption was proposed by Matheron [26]. From the contrast invariant viewpoint, an image $u(\mathbf{x})$ defined on $I\!R^N$ with values in $I\!R$ is not a reliable information since it has been submitted to an unknown (and arbitrary) contrast change g, where g is an increasing continuous function. When we apply an arbitrary contrast change g to u, what is left of u ? Essentially the family of level sets,

$$\mathcal{X}_\lambda(u) = \{\mathbf{x}, u(\mathbf{x}) \geq \lambda\}.$$

Indeed, it is obviously true that if g is an increasing contrast change, then

$$\mathcal{X}_{g(\lambda)}(gou) = \{\mathbf{x}, g(u(\mathbf{x})) \geq g(\lambda)\} = \mathcal{X}_\lambda(u). \tag{6}$$

Conversely, a function u can be recovered from its level sets $\mathcal{X}_\lambda(u)$ by

$$u(\mathbf{x}) = \sup\{\lambda, \mathbf{x} \in \mathcal{X}_\lambda(u)\} \tag{7}$$

and it is easily seen that if two functions u and v have the same level sets, then they can be deduced from each other by a contrast change. Matheron defined contrast invariant operators T on binary images (or, equivalently, on subsets of $I\!R^N$, which we denote by X, Y,...) by the fundamental property

Set Monotonicity. If $X \subset Y$ then $TX \subset TY$

This property permits a formal extension to a grey level image u by simply applying T to each level set $\mathcal{X}_\lambda(u)$ of u and then constructing Tu from the $T(\mathcal{X}_\lambda u)$ by the formula deduced from (7),

$$Tu(\mathbf{x}) = \sup\{\lambda, \mathbf{x} \in T(\mathcal{X}_\lambda(u))\}, \tag{8}$$

so that the level sets of Tu are the images by T of the level sets of u. (Without risk of ambiguity, we take the same notation for T acting on sets or on functions). Notice that this reconstruction is only possible because of the Monotonicity, which preserves the obvious inclusion property of level sets,

$$\text{If} \quad \lambda \geq \mu, \quad \text{then} \quad \mathcal{X}_\lambda(u) \subset \mathcal{X}_\mu(u). \tag{9}$$

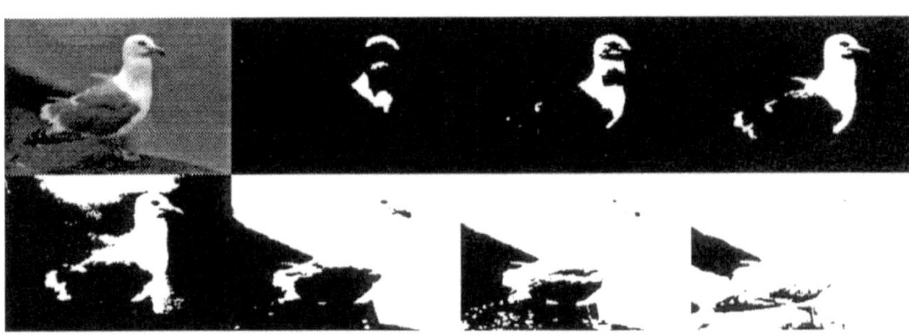

Fig. 2. In this figure, we display the image seabird2, which has a range of grey levels from 0 to 255, and then all level sets whose level is a multiple of 32, in decreasing order from 224 to 32. Notice how essential features of the shapes are contained in the boundaries of level sets, the level lines. Each level set (in white) is contained in the next, (formula 9).

Conversely, one can extend T to compact sets when it is defined on continuous functions in the following way. Let X be a compact subset of $I\!R^N$ and let $u(\mathbf{x})$ be a continuous function such that

$$X = \{\mathbf{x},\ u(\mathbf{x}) \leq 0\} \text{ and } u(\mathbf{x}) \geq C > 0 \text{ for } \mathbf{x} \text{ large enough.} \tag{10}$$

(X is the zero level set of $-u$). It is easily seen that the distance to X, that is $d(\mathbf{x}, X) = \inf\{|\mathbf{x}-\mathbf{y}|, \mathbf{y} \in X\}$ defines such a function. We can define $T(X)$, by using (10) in the following way.

Theorem and Definition 1.1 *(Evans-Spruck). Let T be a monotone and contrast invariant operator (that is, commuting with continuous nondecreasing contrast changes) which is defined on a set $\mathcal{F}$ containing all continuous functions. Then we can extend T to all closed and bounded subsets of $\mathbb{R}^N$ by setting*

$$T(X) = \{\mathbf{x}, Tu(\mathbf{x}) \leq 0\}. \tag{11}$$

The set $T(X)$ so defined does not depend upon the particular choice of u satisfying (10).

Thanks to the preceding theorems, we can canonically associate with a contrast invariant monotone operator defined on continuous functions a monotone operator defined on sets and conversely.

There is a theorem, due to Matheron, giving a characterization of all contrast invariant filters.

Theorem 2 *Let T be a translation invariant monotone operator acting on a set of subsets of $\mathbb{R}^N$. Then there exists a family of sets $\mathbb{B} \subset \mathcal{P}(\mathbb{R}^N)$, which can be defined as $\mathbb{B} = \{X, 0 \in T(X)\}$, such that*

$$T(X) = \bigcup_{B \in \mathbb{B}} \bigcap_{\mathbf{y} \in B} (X - \mathbf{y}) = \{\mathbf{x}, \exists B \in \mathbb{B}, \mathbf{x} + B \subset X\} \tag{12}$$

Let T be a translation invariant monotone and contrast invariant operator defined on a space of continuous and bounded functions. Then there exists a family of closed sets $\mathbb{B} \subset \mathcal{P}(\mathbb{R}^N)$, which can be defined as $\mathbb{B} = \{X, 0 \in T(X)\}$, such that

$$Tu(\mathbf{x}) = \sup_{B \in \mathbb{B}} \inf_{\mathbf{y} \in \mathbf{x}+B} u(\mathbf{y}). \tag{13}$$

Conversely, if $\mathbb{B}$ is a set of subsets of $\mathbb{R}^N$, Relation (12) defines a monotone and translation invariant operator and (13) defines a contrast invariant monotone translation invariant operator.

$\mathbb{B}$ is called the family of structuring elements of T. Every statement about T can be adapted to the "dual" operators of the kind

$$T'u(\mathbf{x}) = \inf_{B \in \mathbb{B}} \sup_{\mathbf{y} \in \mathbf{x}+B} u(\mathbf{y}). \tag{14}$$

Indeed, $T'(u) = -T(-u)$. When $\mathbb{B}$ is reduced to a single element, B, we call T "erosion, (resp. dilation) by B". A particular, important, case is when $B = D(0,t)$ is a disk centered at 0 with radius t. In that case, we define

• Dilation with radius t : $TX = X^t$, the t-neighborhood of X, or, in an equivalent way,

$$T_t u(\mathbf{x}) = \sup_{\mathbf{y} \in D(\mathbf{x},t)} u(\mathbf{y});$$

• Erosion with radius t : $TX = X^{-t}$, the t-interior neighborhood of X, or, in an equivalent way,

$$T_t u(\mathbf{x}) = \inf_{\mathbf{y} \in D(\mathbf{x},t)} u(\mathbf{y}).$$

Let us mention that these schemes are associated a PDE :

Proposition 1 *(Lax formula). Set $u(t,\mathbf{x}) = D_t u_0(\mathbf{x})$ (resp. $E_t u_0(\mathbf{x})$). Then $u(t,\mathbf{x})$ is a viscosity solution of*

$$\partial u/\partial t = ||Du|| = (\frac{\partial u}{\partial x}^2 + \frac{\partial u}{\partial y}^2)^{\frac{1}{2}}.$$

(resp. $\partial u/\partial t = -||Du||$.)

Another very important and useful class of contrast invariant monotone filters are the so called median filters.

Median filters.

The median value of a function u on a bounded set B is the result of an attempt to define a "mean value" which does not depend upon contrast changes. We define for any function u and every set B with bounded measure, the *median filter on B* by

$$\underset{B}{med}\, u = \inf_{meas\, B' \geq \frac{meas B}{2}, B' \subset B} \sup_{\mathbf{x} \in B'} u(\mathbf{x}). \tag{15}$$

1.3. Relevant differential operators in Mathematical Morphology

Before arriving to statements for the contrast invariant operators which will play the same role for contrast invariant filters as Theorem 1 plays for linear filters, we need to define some differential operators which will be canonically associated with those filters, in the same way as the Laplacian $\Delta u = u_{xx} + u_{yy}$ is associated with all linear local low pass filters. From now on, we assume that the dimension N is 2, because of the particular attention we pay to image processing operators. Let us assume that u is twice differentiable (C^2) in a neighborhood of $\mathbf{x}$. We denote by $Du(\mathbf{x}) = (u_x, u_y)(\mathbf{x})$ the gradient of u at $\mathbf{x}$, by $|Du(\mathbf{x})| = (u_x^2 + u_y^2)^{\frac{1}{2}}(\mathbf{x})$ its Euclidean norm. We call "orientation of the gradient" the normalized vector

$$\vec{n}(\mathbf{x}) = -\frac{Du(\mathbf{x})}{|Du(\mathbf{x})|}. \tag{16}$$

We consider the "level" line of u passing by $\mathbf{x}$, defined as the set $\{\mathbf{y}, u(\mathbf{y}) = u(\mathbf{x})\}$. If we assume that $Du(\mathbf{x}) \neq 0$, by the Implicit Function Theorem, we know that the "level line" indeed is a curve, which inherits the smoothness of u in a neighborhood of $\mathbf{x}$. We set $Du^{\perp} = (-u_y, u_x)$, a vector orthogonal to Du and denote by $curv(u)(\mathbf{x})$ the curvature of the level line of u passing by $\mathbf{x}$. Then

$$curv(u)(\mathbf{x}) = \frac{1}{|Du|^3} D^2 u(Du^{\perp}, Du^{\perp}) = \frac{u_{xx}u_y^2 - 2u_{xy}u_x u_y + u_{yy}u_x^2}{(u_x^2 + u_y^2)^{\frac{3}{2}}}. \tag{17}$$

Easy calculations show that this last formula can be written in a compact way as

$$curv(u)(\mathbf{x}) = div(\frac{Du}{|Du|})(\mathbf{x}), \tag{18}$$

where we define, as usual, $div = \frac{\partial}{\partial x} + \frac{\partial}{\partial y}$. In the remainder of this text, we shall be essentially concerned with the above mentioned differential operators for a very simple reason : they are invariant with respect to contrast changes. Indeed, level lines do not depend upon the contrast for, if $u(\mathbf{x}) = u(\mathbf{y})$, then $g(u(\mathbf{x})) = g(u(\mathbf{y}))$. The tangent vector to the level line and the normal vector are therefore contrast-insensitive and so is the curvature of the level line. This can be of course directly checked by substituting $g(u)$ to u in the formula (17) defining the curvature. **As will be clear from the next section, the curvature plays the same role in mathematical morphology as the Laplacian in the linear theory.**

1.4. Asymptotic behavior of contrast invariant filters

The next theorem fixes what happens asymptotically with any contrast invariant filter which is isotropic and associated with a uniformly bounded family of structuring elements. We know from the results mentioned above that any contrast invariant operator can be given a "sup inf" or the "inf sup" form. Thus, in the following, we consider operators directly defined in this form.

Theorem 3 *Let $\mathbb{B}$ be a family of structuring elements in $\mathbb{R}^2$ which is bounded ($\exists M$, $\forall B \in \mathbb{B}$, $B \subset D(0, M)$) and isotropic (if $B \in \mathbb{B}$, then $RB \in \mathbb{B}$ for every rotation R of $\mathbb{R}^2$).*

Let $Tu(\mathbf{x}) = \sup_{B\in\mathbb{B}} \inf_{\mathbf{y}\in\mathbf{x}+B} u(\mathbf{y})$ or resp. $\inf_{B\in\mathbb{B}} \sup_{\mathbf{y}\in\mathbf{x}+B} u(\mathbf{y})$ and define the corresponding rescaled operator (where $h > 0$ tends to 0) T_h by SI_h (or resp. IS_h), where

$$SI_h u(\mathbf{x}) = \sup_{B\in h\mathbb{B}} \inf_{\mathbf{y}\in\mathbf{x}+B} u(\mathbf{y})$$

$$IS_h u(\mathbf{x}) = \inf_{B\in h\mathbb{B}} \sup_{\mathbf{y}\in\mathbf{x}+B} u(\mathbf{y}).$$

Then, considering the real function $H(h) = T(x + hy^2)(0)$ and assuming that u is a C^3 function on $\mathbb{R}^2$,

(i) if $H(0) \neq 0$, then

$$(T_h u)(\mathbf{x}) = u(\mathbf{x}) + hH(0)|Du|(\mathbf{x}) + O(h^2)$$

and we obtain an operator behaving as an erosion or a dilation depending on the sign of $H(0)$;

(ii) if $H(0) = 0$, and $Du(\mathbf{x}) \neq 0$, then

$$T_h u(\mathbf{x}) = u(\mathbf{x}) + h|Du(\mathbf{x})|H(\frac{1}{2}hcurv(u)) + O(h^3);$$

and we obtain a "curvature motion";

(iii) if $H(0) = 0$ and $Du(\mathbf{x}) = 0$, then the above formulas are not valid, but

$$T_h u(\mathbf{x}) = u(\mathbf{x}) + h^2 T(\lambda_1 x^2 + \lambda_2 y^2) + O(h^3),$$

where λ_1 and λ_2 are the eigenvalues of $D^2 u(\mathbf{x})$.

The application of the preceding theorem to the above mentioned filters (median, weighted median, etc.) is straightforward : one only needs to compute in each case the associated function H.

Fig. 3. The median filter does a smoothing linked to the curvature of the level set. Left : image of a simple shape. Right : difference of this image and the same smoothed by one iteration of the median filter. We see in black, the points which have changed. The width of the difference is proportional to the curvature, (Theorem 5).

As a direct application of Theorem 3, we consider the case where the structuring elements all are segments with the same length and centered at zero. Surprisingly enough, this operator has the same effect as the median filter. More precisely, we deduce the following result, [11].

Theorem 4 *Let $\mathbb{B}$ be the set of all segments of the plane with length 2 and centered at zero. Let u be a C^3 function in $\mathbb{R}^2$. Then, if $Du(\mathbf{x}_0) \neq 0$*

$$\frac{1}{2}(IS_h + SI_h)u(\mathbf{x}_0) = u(\mathbf{x}_0) + h^2 \frac{1}{4} curv(u)|Du|(\mathbf{x}_0) + O(h^{2+\frac{1}{3}})$$

and if $Du(\mathbf{x}_0) = 0$,

$$|\frac{1}{2}(IS_h + SI_h)u(\mathbf{x}_0) - u(\mathbf{x}_0)| \leq |\Delta u(\mathbf{x}_0)|h^2 + O(h^3)$$

We now examine the convergence of iterated median filters.

Theorem 5 *[7, 8, 15] Let u_0 be a lipschitz function on $\mathbb{R}^2$ and set*

$$u_h(\mathbf{x}, t) = \underset{D(\mathbf{x}, h^{\frac{1}{2}})}{med} u_h(\mathbf{x}, (n-1)h) \qquad \text{for} \quad (n-1)h \leq t \leq nh$$

$$\textit{and} \quad u_h(\mathbf{x}, 0) = u_0(\mathbf{x}).$$

Then $u_h(\mathbf{x}, t)$ tends uniformly on every compact set of $\mathbb{R}^2 \times \mathbb{R}^+$ to a function $u(\mathbf{x}, t)$ when $n \to +\infty$, where $u(\mathbf{x}, t)$ is the unique viscosity solution of

$$\frac{\partial u}{\partial t} = \frac{1}{6}|Du| curv(u) \tag{19}$$

This equation has recently received a lot of attention because of its geometrical interpretation : indeed, (see S. Osher and J. Sethian [29], L. C. Evans and J. Spruck [15]...) the level lines of the solution move in the normal direction with a speed proportional to their mean curvature (curvature in two dimensions). (This "mean curvature motion" effect will be showed in the experimental results presented below). As we have seen, this "mean curvature motion" property implies that the proposed scale space satisfies the main "contrast invariant" axiom, that is, the invariance of the scale space under any rearrangement of the grey level scale. Thus, the mean curvature motion is a contrast invariant operator. It is quite close to the basic operations of mathematical morphology. Indeed, dilation and erosion can be interpreted as a (forward or backward) motion of level curves at constant speed in the direction of the normal. A theory of weak solutions based upon the so-called viscosity solutions theory has been proposed by Y. G. Chen, Y. Giga and S. Goto [12], L.C. Evans and J. Spruck [15], Y. Giga, S. Goto, H. Ishii and M.H. Sato [16].

Theorem 5 gives a differential interpretation of the median filter and somehow explains why it works so well. Indeed, equation (19) means that all level curves of u are smoothed at a speed depending upon their local radius curvature, so that all small level sets are quickly removed and larger ones have their asperities smoothed out.

Fig. 4. On the first column, seabird1 and its level lines for all levels multiple of 12. On the second column, a median filter on a disk with radius 2 has been iterated twice on seabird2. On the third column, an infsup and then a supinf filter based on segments have been applied. On the bottom : the corresponding level lines of the results, which, according to the theoretical results, move at a normal speed proportional to their curvature, (Theorems 5 and 6).

Let us now consider the case of unbounded families of structuring elements, which was excluded in Theorem 3. One could think it spurious to consider such families for defining local operations on the image. This is, however, not the case when families of structuring elements are "naturally" unbounded ; our main example in view here is the case of affine invariant families. The interest of such families is to define filtering operators on an image which are -in a restrictive sense- projective invariant : indeed, projection invariance reduces to affine invariance when photographed plane objects are far enough with respect to the focal distance.

Asymptotic behavior of affine and contrast invariant filters.
Let $\mathbb{B}$ be a family of plane closed nonempty bounded sets which is invariant by the special linear group $SL(\mathbb{R}^2)$, that is, $\mathrm{A}B \in \mathbb{B}$ whenever $\det(\mathrm{A}) = 1$ and $B \in \mathbb{B}$. An obvious example of such a family is the family of ellipses with area 1, and, more generally, we can consider the set $\{\mathrm{A}B, \mathrm{A} \in SL(\mathbb{R}^2)\}$ where B is an arbitrary bounded set of the plane.

In what follows, we consider two scaling parameters s and h such that $s = h^{\frac{3}{2}}$, and we set $\mathbb{B}_s = \{s^{\frac{1}{2}}B, B \in \mathbb{B}\}$. The meaning of this scaling is that s behaves like

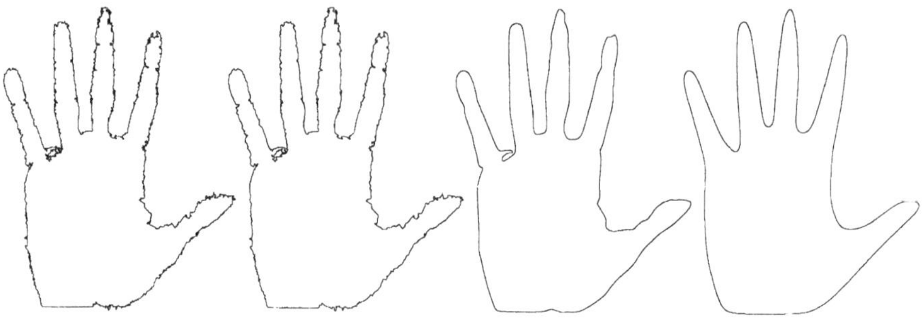

Fig. 5. Affine scale space of a "hand" curve. (scales 0, 1, 20, 400). Evolution obtained by an iterated Matheron affine invariant operator implemented as a curve evolution scheme. (L.Moisan,[28]).

the area of elements of $\mathbb{B}_s$, while the use of h makes statements on the asymptotic behavior simpler.

$SI_h u$ is understood as an "affine erosion" of u and $IS_h u$ as an "affine dilation". Our main concern is the behavior of $IS_h u(\mathbf{x})$ and $SI_h u(\mathbf{x})$ when $u(\mathbf{x})$ is a smooth (C^3) function and $h \to 0$. Under fairly general assumptions on $\mathbb{B}$, there is a theorem proving that all affine invariant contrast invariant operators have one and the same behavior.

We consider a family $\mathbb{B}$ of convex sets, symmetric with respect to zero and invariant under the special linear group (all linear maps with determinant 1). All sets in $\mathbb{B}$ have area s.

Denoting by T_h one of the schemes IS_h, SI_h or $SI_h IS_h$, we define the result of their iteration n-times by

$$u_h(\mathbf{x},(n+1)h) = T_h u_h(\mathbf{x}, nh),$$
$$\text{and} \quad u_h(\mathbf{x},0) = u_0(\mathbf{x}).$$

As the last result to be stated in this text, on can prove the convergence of every iterated affine filter to the Affine Contrast Invariant Scale Space (AMSS) :

Theorem 6 *There exists a constant $c_{\mathbb{B}} > 0$ such that if $u_0(\mathbf{x})$ is any Lipschitz function on $\mathbb{R}^2$, then $u_h(\mathbf{x}, nh)$ tends uniformly on every compact set to $u(\mathbf{x},t)$, where $u(\mathbf{x},t)$ is the unique viscosity solution of*

$$\frac{\partial u}{\partial t} = |Du| g(curv(u)), \tag{20}$$

where :

$$\lim_{s\to 0} \frac{(T_s u)(\mathbf{x}_0) - u(\mathbf{x}_0)}{s^{2/3}} = c_{\mathbb{B}} |Du| g(curv(u))(\mathbf{x}_0)$$

$$\begin{aligned} \textit{where } g(r) &= c_{\mathbb{B}}(r^-)^{\frac{1}{3}} \quad \textit{if} \ T_h = SI_h \\ &= c_{\mathbb{B}}(r^+)^{\frac{1}{3}} \quad \textit{if} \ T_h = SI_h \\ &= c_{\mathbb{B}}(r)^{\frac{1}{3}} \quad \textit{if} \ T_h = SI_h\, IS_h. \end{aligned}$$

As a consequence, there is a single possibility to smooth an image in a contrast invariant and affine invariant way. All scale space equations considered here have a unique viscosity solution. We have no space to give a generic proof, but the use of iterated filters also is a mathematical tool to prove existence of a solution to any PDE model associated with an iterated contrast invariant filtering. It applies, in particular, to the mean curvature motion (the Osher-Sethian equation).

Conclusion. Mathematical analysis shows that there are only two ways of smoothing iteratively an image with contrast invariant local filters : both are described by a PDE. The first way is motion by mean curvature and is well simulated by an iterated median filter. (PDE scheme are, however a better subpixelian implementation). The second way is affine shortening, which although definable as an iterated inf-sup operator is better characterized by a PDE.

Acknowledgments : We gratefully acknowledge partial support by EC project ERBCHRXCT930095 (Human capital and mobility). A preliminary version of this work has been published for the tutorial ICIP95, we thank Michael Unser for his support and encouragement. Part of this work has been done by using the MegaWave image processing environment (authors : Jacques Froment and Sylvain Parrino), and Cognitech, Inc. image processing facilities (http://www.cognitech.com). We also thank Luis Alvarez for much interaction on the subject of this work and Jean-Michel Lasry for a key idea which he confided to us and never published [22].

References

1. L. Alvarez, F. Guichard, P.L. Lions and J.M. Morel (1992), "Axioms and fundamental equations of image processing" *Report 9216, CEREMADE. Université Paris Dauphine.* Arch. for Rat. Mech. 123, 3, 199-257 (1993).
2. S. Angenent, G. Sapiro, A. Tannenbaum (1994), "On the affine heat equation for nonconvex curves" Technical report.
3. A. Arehart, L. Vincent and B.B. Kimia. (1993) "Mathematical Morphology : the Hamilton-Jacobi connection." *Int. J. of Comp. Vis.*, 215-219.
4. H. Asada and M.Brady (1986), "The curvature primal sketch", *IEEE Transactions on Pattern analysis and machine intelligence* **8** No. 1.
5. J. Babaud, A. Witkin, A. Baudin and R. Duda. (1986) " Uniqueness of the gaussian kernel for scale-space filtering." *IEEE Trans. PAMI 8,*.
6. G. Barles (1985), "Remarks on a flame propagation model", *Technical Report No 464 INRIA Rapports de Recherche.*
7. G. Barles and C. Georgelin (1992), "A simple proof of convergence for an approximation scheme for computing motions by mean curvature", SIAM Journal of Numerical Analysis, 32(2): 484-500, 1995.
8. G. Barles and P.E. Souganidis (1989), "Convergence of approximation schemes for fully nonlinear second order equation", *Asymp. Anal.*, 4, 271-283.
9. Bart M. ter Haar Romey (Ed.) (1994) *Geometry-Driven Diffusion in Computer Vision* Kluwer Academic Publishers.
10. W. Brockett and P. Maragos (1992), "Evolution Equations for continuous-scale morphology", *ICASSP, San Francisco* 23–26.

11. F.Catté, F.Dibos and G.Koepfler (1993) "A morphological approach of mean curvature motion" *SIAM journal of Numerical Analysis, SINUM 32.6, 1995*
12. Y-G. Chen, Y. Giga and S. Goto (1991), "Uniqueness and existence of viscosity solutions of generalized mean curvature flow equations", J. Diff. Geometry, 33, 749-786.
13. M.G. Crandall, H. Ishii and P.L. Lions (1992), *User's guide to viscosity solutions of second order partial differential equation*, Bull. Amer. Math. Soc., 27, 1-67, 1992.
14. L.C. Evans (1993) "Convergence of an algorithm for mean curvature motion", *Indiana Univ. Math. J., 42, 533-557*
15. L.C Evans. and J. Spruck (1991) "Motion of level sets by mean curvature 1", *J. Diff. Geometry* 33, 635-681.
16. Y. Giga, S. Goto, H. Ishii and M.M. Sato (1991) "Comparison principle and convexity preserving properties for singular degenerate parabolic equations on unbounded domains" *Indiana Univ. Math. J.*, 40, 443-470.
17. F. Guichard, J-M. Morel, (1995) "Partial Differential Equation and Image Iterative Filtering", Tutorial of ICIP 95, Washington D.C.
18. M. Grayson (1987), "The heat equation shrinks embedded plane curves to round points", *J. Differential Geometry* **26**, 285–314.
19. H. Ishii (1995), "A generalization of the Bence, Merriman and Osher algorithm for motion by mean curvature", Curvature flows and related topics (Levico, 1994), 111–127, GAKUTO Internat. Ser. Math. Sci. Appl., 5, Gakkotosho, Tokyo.
20. B.B. Kimia, A. Tannenbaum, and S.W. Zucker (1992), "On the evolution of curves via a function of curvature, 1: the classical case", *J. of Math. Analysis and Applications* **163**, No 2.
21. J.J. Koenderink and A.J. van Doorn (1986), "Dynamic shape", *Biol. Cyber.* **53**, 383–396.
22. J.M. Lasry (1992) Unpublished notes, December 1992.
23. Y. Lamdan, J.T.Schwartz and H.J.Wolfson (1988) "Object recognition by affine invariant matching" *In Proc. CVPR 88*
24. C. Lopez, J.M. Morel (1992) "Axiomatisation of shape analysis and application to texture hyperdiscrimination" *Proceedings of the Trento Conference on Surface Tension and Movement by Mean Curvature* De Gruyter Publishers, Berlin.
25. P. Mascarenhas (1992), "Diffusion generated motion by mean curvature", CAM Report 92-33, University of California, Los Angeles.
26. G. Matheron (1975), *Random Sets and Integral Geometry*, John Wiley, N.Y.
27. B. Merriman, J. Bence and S.Osher (1992), "Diffusion Generated motion by Mean Curvature", *CAM Report 92-18. Depart. of Mathematics. University of California, Los Angeles CA* 90024.1555.
28. L. Moisan, (1998) "Affine Plane Curve Evolution : A Fully Consistent Scheme", *IEEE Transactions on Image Processing, vol. 7, no. 3.*
29. S. Osher and J. Sethian(1988), "Fronts propagating with curvature dependent speed : algorithms based on the Hamilton-Jacobi formulation", *J. Comp. Physics* **79**, 12–49.
30. D. Pasquignon, (1997) "Approximation of viscosity solution by morphological filters." *Preprint. Submited in Tohoku Journal of Mathematics Japan.*
31. G. Sapiro and A. Tannenbaum (1994) "On affine plane curve evolution" *Journal of Functional Analysis*, 119, 1, 79-120.
32. A.P. Witkin (1983), "Scale-space filtering", *Proc. of IJCAI, Karlsruhe* 1019–1021.

2D & 3D FIGURAL MODELS OF ANATOMIC OBJECTS FROM MEDICAL IMAGES

STEPHEN M. PIZER, DANIEL S. FRITSCH, KAH-CHAN LOW and JACOB D. FURST
Medical Image Display & Analysis Group
University of North Carolina
Dept. of Computer Science, Univ. of NC, Chapel Hill, NC 27599-3175, USA
www.cs.unc.edu/Research/Image

Abstract. Figural object models represent object shape in terms of a hierarchy of simple shapes, protrusions, and indentations that we call "figures". Each figure has an unbranching, coarsely sampled net of medial primitives which carry width information. The figural hierarchy and its medial primitives thus define the boundary with a tolerance proportional to the corresponding medial widths, and associate with each boundary location a normal vector representing the figural boundary at a scale proportional to the medial width. Such a model is efficiently and effectively deformable to match image information according to, and producing, probabilistic measures of object shape and measures of image match. This paper specifies the makeup of figural object models and describes methods for building models of anatomic objects.

1. Figural Models

We have developed a shape-based segmentation, registration, rendering, and shape-analysis technique that performs unusually effectively, as tested in a number of medical image analysis applications in 2D and 3D. The method is based on a figural model of shape, which we believe has merit for a variety of purposes of image analysis, graphics, and scientific analysis of shape.

What would one wish of a shape model for objects or groups of objects? First, it must handle both geometric and statistical aspects of shape:

1) Reflection of the geometric aspects of shape, certainly the invariance to translation, rotation, and zoom, and perhaps the differential properties that underlie notions such as curvature. Included in these geometric aspects are matters of spatial scale that allow a wrinkled face and a smooth face both to be said to have the shape of a face while allowing a prune and a wrinkled face both to be said to have a wrinkled shape.

2) Dealing with the statistical properties of shape related to the fact that a shape is a distribution of possibilities, with varying probability, which will be reflected in a sample of encountered shapes that might be provided to the modeler. The approach thus must provide a means to quantitate the probability of a particular instance of a shape.

Second, it should reflect the principle that form follows function and handle the fact that shapes occur in a variety of dimensions:

3) Relation to the functional properties of an object, typically made up of named components that can articulate or separately vary in shape.

4) Providing intuitive shape properties such as curving and widening.

5) Generality for handling shape in 2 spatial dimensions, in 3 spatial dimensions, and in space-time.

Finally, it should be widely and efficiently applicable:

6) Allowing the extraction of shape information from images of objects with that shape and application of the shape information in extracting such objects or designing such objects.

7) Leading to efficiency both in extracting the model from the training data and in using the model for the applications of image analysis; graphics design, simulation, and presentation; and scientific analysis of shape.

We suggest that these concerns are particularly well handled by a type of model which focuses on those figures that can separately articulate and deform and which separates the concerns for small-scale detail that are typically characterized as boundary texture from the larger scale concerns as to figural and interfigural shape that should deal not only with positional relations but also with spatial tolerance.

In this paper we first describe the components of our model, first focusing on the part of the model representing figural aspects and then briefly describing the part of the model that characterizes boundary detail. Within the figural part we first present what a medial primitive is and then how medial primitives are linked first to represent a figure and then to form the hierarchy of figures that represents an object. Second, we address how this allows a probability of a shape to be defined and how it allows a measure of match to image data to be defined.

The second half of the paper focuses on means for building such models, and the paper closes with a discussion of the relation of this modeling approach to other approaches and of anticipated directions for development of our ideas and methods.

1.1. Medial Primitives

As illustrated in Figure 1, it is natural in both 2D and 3D to divide objects into a collection of a main figure and protrusions, indentations, neighboring objects, and included objects, with the defining property of a figure being that a mesh of points ordered along the boundary are specially related as being opposite from a corresponding mesh of points ordered in opposing directions along the boundary. These related point pairs have been called medial involutes. The medial relation was characterized by Blum [2] to include a medial position $\bar{x}$ and a medial width r; to this we added typically two, but possibly more, boundary normal directions $\vec{n}_1$ and $\vec{n}_2$. We call such a tuple $(\bar{x}, r, \{\vec{n}_i\})$ a medial primitive (Figure 2). A medial primitive implies two or more corresponding boundary positions $\bar{x} \pm r\vec{n}_i$.

In the figural model open circles indicate medial points, and the attached arrowed dark lines indicate the combination of two associated normal directions and a common width. Grey lines indicate inter-primitive links between adjacent medial

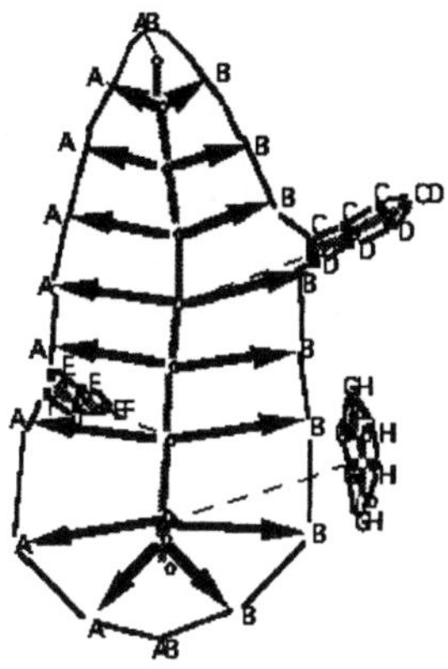

Fig. 1. Figural definition and a 2D figural model for a test object. Illustrated are 4 figures in an object with a noncontiguous boundary and the series of medial involute pairs (A:B, C:D, E:F, G:H) which form the natural sequence of opposite points along each figure. To one figure (A:B) are associated a protrusion figure (C:D), an indentation figure (E:F), and a neighboring figure (G:H).

primitives in a figure, and dashed lines indicate interfigural links between a medial primitive of the child and medial primitive of the parent.

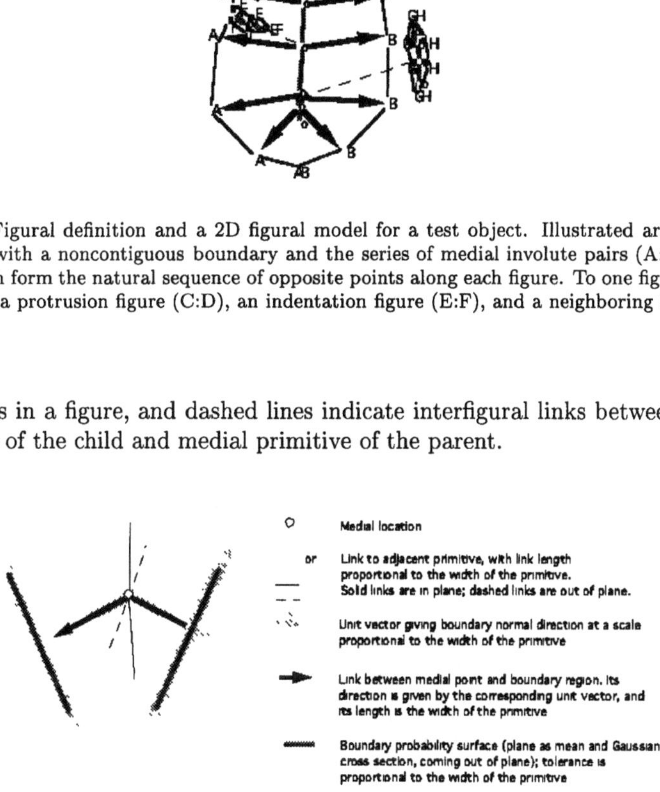

Fig. 2. A 3D medial primitive, made from a medial location, a width, and 2 boundary-normal directions, and implying a boundary probability. 3D medial primitives have the same form as 2D primitives, but they are generally linked into a 2D mesh, except for tubular figures, which have a circular cross-section and thus a curvilinear interconnection of their primitives as in 2D.

The requirements of including spatial scale as a geometric variable and providing zoom invariance lead to the properties that

1) The boundary implied by a figural model is smooth and located by a medial primitive only with a tolerance proportional to r. That is, the medial primitives

imply a cloud of probability of boundary locations.

2) It is the object boundary probability cloud to which the vectors that are elements of a medial primitive are normal. For example, these vectors are in its direction of maximum second directional derivative of the probability function. As such, the normals reflect information at a scale proportional to r. Alternatively, if the normals are derived from images of the object as image intensity gradients, these gradients should be derived from the image using apertures of width proportional to r.

3) Relationships of relative position, relative orientation, relative width, and relative narrowing rate along the direction bisecting the medial primitive's normals change at a rate limited by a width-proportional aperture in proportion to $1/r$. Therefore, spatial sampling of medial primitives (and the implied boundary positions) is most appropriately done with spacing proportional to r. Similarly, relationships between a medial primitive of a parent figure and the proximal medial primitive of its subfigure should be expressed as multiples of r (e.g., the geometric mean of the r values of the two adjacent primitives).

Figure 3 illustrates the rich shape information of change in position, orientation, width, and narrowing rate in the link between two adjacent medial primitives.

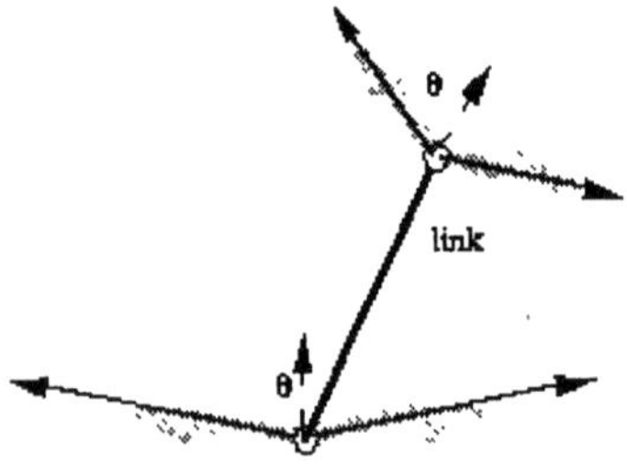

Fig. 3. Shape information between two adjacent (linked) medial primitives. In each primitive the medial position is indicated by a circle, the boundary normals are indicated by the grey arrows, and the vectors connecting the medial position to the implied boundary positions are indicated by the black arrows. The bisector between the normals is indicated by the dashed arrow. The translation (position change) for the link corresponds to the vector difference between the circle centers at the two ends of the link. The rotation (orientation change) for the link corresponds to the difference between the two dashed vectors at the two ends of the link. The width difference (scale change) for the link corresponds to the difference between the lengths of the grey arrows at the two ends of the link. The narrowing rate change for the link corresponds to the difference between the cosines of the angles θ at the two ends of the link.

1.2. Figures as Nets of Medial Primitives and Interprimitive Links

By the definition of a figure, the medial primitives of a figure form a nonbranching simple medial manifold. This manifold, sampled according to the spacing defined by just-mentioned property 3 to produce a mesh of medial primitives, produces an efficient representation of the shape of a figure that satisfies all of the requirements listed earlier. While in principle a full representation of shape can involve the relationship among all primitives [3, 6], efficiency is achieved by only representing adjacent interprimitive relations, and it appears that a considerable part of relevant shape information can be derived from just these links between adjacent medial primitives. From a Bayesian point of view, we are representing the probability of a shape by a first-order Markov Random Field [1], and in particular by a Gibbs distribution [23].

1.3. Figural Hierarchies and Interfigural Links

An object can be represented by a hierarchy of figures, or in the most general case, by a directed acyclic graph of figures. Typically this hierarchy has smaller scale objects as subfigures of larger width objects (although typically a figure's primitives cover a range of widths), but it is also possible to have parents be more stable across the populations of instances of a shape than their subfigures.

Each parent figure-to-subfigure relationship can be represented by a interprimitive link of the type just discussed. Thus (see Figure 1) the link indicates the position, orientation, width, and narrowing rate of the proximal end of the subfigure relative to the parent figure.

1.4. The Probability of an Instance of a Shape

The probability of a particular shape can be derived from the Gibbs distribution referred to above. According to the Gibbs distribution, the log of the probability of a particular network of medial primitives that is homologous to (has the same link structure as) a particular shape model is, except for an additive constant and a constant multiplier, equal to the sum of logs of the probabilities of the individual interprimitive links in the instance. The probability associated with a particular link is provided by a metric between the link in the instance and the homologous link in the model, combined over the various aspects of link difference described above. More than one such combination has been suggested [23, 18].

The log probability of an instance of a shape provides a measure that can be used to calculate shape difference or to discriminate shapes among classes according to the methods of statistical pattern recognition. The measure of shape difference can be used in an objective function that is maximized to segment an object from an image or provide a registration transformation between a target image and a reference image. In a Bayesian view of these image analysis objectives, the shape log probability term forms the log prior summand in a log posterior, which then is combined with the boundary displacement probability described shortly.

1.5. Medialness

A medial primitive with its particular tuple values, when placed on an image, measures agreement with the image information by summing the boundariness at each of the boundary positions derived from the tuple, in the direction along the corresponding normal in the tuple, and at a scale proportional to the radius in the tuple. We thus define the medialness of a primitive in an image as the sum over the normals in the primitive of these boundariness values [21]. This medialness, summed over the medial primitives, can be treated as the figural component of a log likelihood function contributing to a log posterior that is maximized for segmentation or registration [10]. The behavior of this medialness at a primitive across a set of training images can be analyzed statistically to set a normalizing weight related to the expected medialness value at that primitive.

1.6. Representation of Boundary Displacements

The view of boundary detail as a displacement from the medial-primitive implied boundary is different from the more common view of the boundary position as the only primitive [6], but our view should not be taken to imply that these displacements should be neglected. Our image analysis techniques have shown that optimizing an objective function related only to the figural components first, followed by a further optimization of the boundary related to boundary detail [10, 4], is a particularly stable way of operating.

The displacement function (along the medial primitive-implied normals) should be thought to be applied not simply at a boundary point sampling corresponding to the medial primitive sampling. Rather the medially-implied boundary and the corresponding normals should be interpolated into a sampling corresponding to the finer scale of the boundary detail. At these samples, it is reasonable to model that detail by the autocovariance of the scalar displacement function, a function which can be extracted from a set of training images. The probability of any particular displacement function would then be measured by a Gaussian random process with the specified autocovariance. To form the probability of an object, this probability could then be multiplied by the figural probability discussed above.

2. Building a 2D or 3D Model of a Figure

In this section we discuss the extraction of model from a single training image containing an object or from the characteristic function obtained by segmenting the object from an image, by whatever means, e.g., manually.

2.1. Via Cores

A core [20] is a height ridge [13] of medialness, computed from image data, either an original image or a characteristic image of the object(s) to be modeled. Each core approximates the locus of a potential medial primitive of a figure. For a 2D image the 1D ridge has this property, except for an image of a disk, for which a 0D ridge (the center, radius, and all directions as normals) is needed. For a 3D image a 2D ridge has this property, except for an image of objects with a circular cross section (tubes), for a which a 1D ridge is needed, or of balls, for which a 0D ridge is needed.

The choice of which of the many cores available are to be used to represent an object requires comprehension of the object, probably by a user. Each core needed can be specified by indicating interactively a medial primitive value near the core. Means of optimization and ridge following [20, 11, 12] can then be applied to extract a fragment of the core. We have implemented and used two approaches. One has the form of a predictor-corrector, for which each recurrent step involves offsetting from the present core position in a direction approximating the core tangent and then iteratively optimizing medialness in the directions for which the core is specified to be optimal. This approach, described in [8, 9, 20] is effective for 2D images, but it does not seem to be feasible on images of higher dimension.

For 3D (and also 2D) images, an algorithm called marching ridges that generalizes the marching cubes algorithm [17] has been developed [11, 12]. In the marching ridges algorithm an m-dimensional core from an n-dimensional image is computed by finding in a voxel (or pixel) simultaneous zero crossings of directional derivatives of medialness in $n - m$ directions specified by the local optima of medialness with respect to orientation and scale, with the additional constraint that the second derivative in each specified direction is negative.

Because one core may not form the full locus of medial primitives for a figure, as a result of image noise, blurring, or interfering figures, the fragment corresponding to one core may need to be connected to another interactively or via the mathematical loci called ridge connectors [7]. It is also necessary to locate the end curves closing off the figural meshes via maxima in an endness measure along the core [5].

2.2. Via Cylinder Axes

If for a 3D figure a space curve roughly forming a generalized cylinder axis for the figure is provided, either by user drawing or from one or more attached 1D core fragments, a mesh of medial primitives representing the figure can frequently be derived with little further interaction. We have seen that for most figures occurring in natural or manufactured objects, a reslicing of the image data orthogonal to the axis and at a spatial sampling proportional to the width parameter of the cylinder axis, provides a set of 2D images that allows the propagation of a single sampled 1D core from one of these 2D images to form the desired 2D mesh. The 1D core is produced by the interactive specification of a single medial primitive, as described in the previous subsection. This sampled core is then propagated into adjacent resliced images via the 2D figural model deformation technique described in [10]. The resulting deformed model forms the model that is further deformed into the next slice, etc. Finally, to limit the dependence on the original cylinder axis, one can deform the resulting 3D mesh into the 3D image data via our figural deformation method, which should be fast because little deformation will be needed.

2.3. Forming a Hierarchy of Figural Representations

Pairs of figural models thus created can be linked as a parent and child, either by interactive pointing to one medial primitive in each mesh or by using properties on a parent primitive that signal a subfigure [15, 7]. The resulting hierarchy forms the component of the object model that describes figural information. The boundary implied by this figural model, using interpolation among the medial primitives, can be used as the basis of either an active contour method [19] or optimization of displacement along the normals, using the prediction from medial primitives as a prior [4]. An analysis using the assumption of ergodicity can yield the autocovariance function that models the displacement function.

3. Results

We have generated figural models for a variety of objects in 2D and 3D medical images using the approaches outlined in the previous section. Here we present a few of those results.

Figure 4 shows a five-figure linked model representing the shape of the ventricle in an axial slice from an MR brain scan. To generate this model, the medial track for each figure was first extracted automatically using core extraction algorithms. Following sampling of the medial track, each medial mesh point was associated with its medial involutes to generate the linked boundary mesh. Under user interaction, the figural models were linked by connecting adjacent boundary points. This model has been successfully used to demonstrate automatic segmentation of similar shapes in data sets acquired from other patients.

Fig. 4. A linked five-figure model for a ventricle in an MR slice through the brain. The medial mesh was produced automatically via core extraction.

We have also constructed figural models for objects in 3D images. An early result using the approach of generalized cylinder axes is described below for a gourd in a CT scan containing a collection of similar shaped objects and a star fruit (see Figure 5).

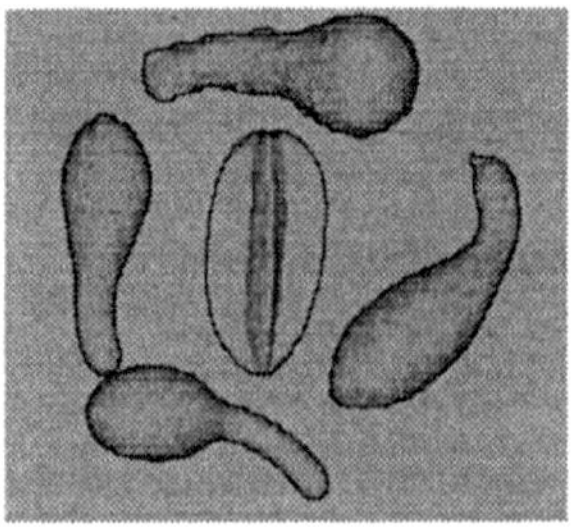

Fig. 5. Volume rendering of a CT scan containing various fruits and vegetables. The modeled object is the gourd at the far left.

To generate the figural model, we first segmented the object of interest using manual contouring and display of the tiled surface in a program designed for interactive specification of the generalized cylinder axis. The user drew a curve through the surface such that the intersections of the planes normal to the curve and the object surface produce shapes that are consistent. Once a generalized cylinder axis was defined, a 3D figural model was constructed on a set of normal planes extracted from samples along the cylinder axis. A representative slice is used to first construct a 2D figural model and the model is then propagated to adjacent slices using the 2D deformation program. Figure 6 illustrates this approach.

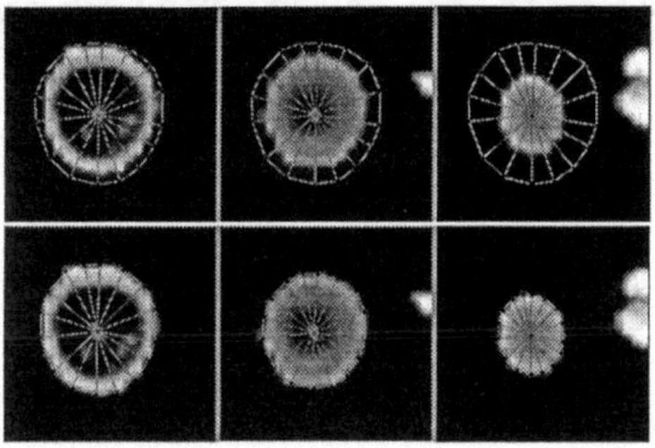

Fig. 6. Slice-to-slice propagation of a 2D figural model for purposes of creating a 3D figural model. Top: Characteristic model superimposed on slices normal to the generalized cylinder axis. Bottom: Result of running the 2D deformation on each of the slices.

This 2D model is propagated between slices, and the corresponding medial and boundary sample points are linked. Figure 7 illustrates the resulting figural mesh.

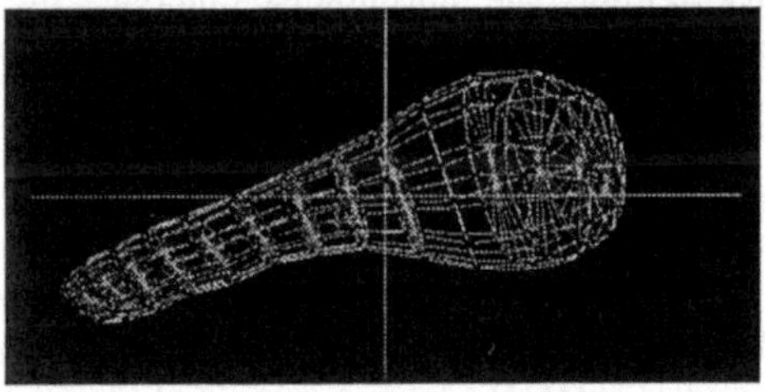

Fig. 7. A 3D figural model of a gourd.

Such an approach has also been used to build models of anatomical structures in 3D MR images. Shown in Figure 8 is a six-figure model of the brain ventricle shown along with orthogonal slices through the volumetric data. This model has been used successfully to segment ventricles from data sets acquired from other patients.

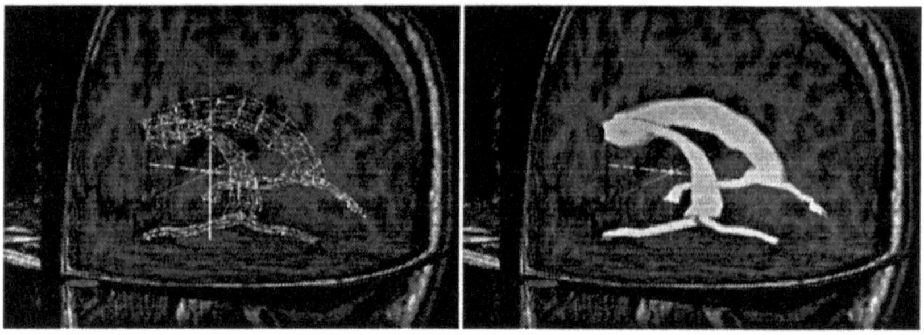

Fig. 8. A six-figure model of the lateral ventricles displayed along with orthogonal planes from the model MR data set. Left: Linked medial and boundary meshes. Right: Surface rendering of the boundary mesh.

4. Discussion

4.1. Relation with Mathematical Morphology

While in principle the theory of figural models described above is based on Gaussian linear weighting functions (apertures) because they fall out from requirements of invariance and a few other desirable properties [16, 20], the operations used to form a figural model and the associated boundary displacement model can efficiently be approximated by operations from mathematical morphology. Selection and application of medial scales correspond to opening with spherical structuring operators. When combined with parabolic dilation [22], this approach can be applied to the intensity graphs forming an image (greyscale morphology). Care needs to be taken to separate the analysis of different figures, but this seems easier with approaches of mathematical morphology than with Gaussian linear weighting. Given the figural analysis and the ability to apply the morphologic difference operator, morphologic analysis at smaller scales can characterize the boundary displacement function.

4.2. Towards Models Based on Many Training Images

Models need to reflect the variabilities in a population of objects taken to have a common shape (as well as the variabilities in the way the intensity distributions in images reflect these shapes). The theory of Gibbs distributions on which the figural model is based provides methods for estimating the associated hyperparameters [23], given m homologous models extracted from a sample of training images.

Such a set of models can be obtained by bootstrapping from a model extracted on a single training image. That model can then be deformed into the remaining training images, and the resulting set of models can be used to estimate the needed hyperparameters. The resulting probabilistic model yields the deformation penalty term used in the objective function of our object extraction method. Applying this method with this objective function to the full sample of training images yields a new set of models that can be used to estimate the final hyperparameters of the Gibbs probability distribution which, together with the model topology, describes the shape.

4.3. Design-Based Models

Besides the image-based models upon which this paper has focused, it is possible to design objects by interactively specifying hierarchically linked meshes of medial primitives. Such a design process appears to have a variety of advantages over the presently used spline representations of object boundaries or constructive solid geometric approaches: The design process has the simplicity of constructive solid geometry while providing a rich set of primitive shapes. Matching of the resolution needs of a rendering becomes efficient. Physically based modeling can be done at a succession of decreasing scales, providing gains in efficiency. Morphing becomes particularly natural and efficient.

4.4. Spatiotemporal Models

Finally, these ideas can be extended beyond models of spatial objects to objects that change over time, such as the beating heart or an anatomic structure varying in shape with the progression of a disease or with drug treatment. We have barely begun work in this direction.

4.5. Alternative Models of Shape

Many models of shape, like the figurally based models described here, involve the spatial conformation of primitives. Alternatives to the medial primitives we use include landmarks [3], voxels [14], and boundary sample points possibly equipped with normals [6]. As proposed to date, these typically do not carry the scale, i.e, tolerance and aperture, information that is an inherent part of figurally based models. This property, which gives stability to figurally based models, is compensated for in the other models by using all the $\mathcal{O}(N^2)$ relations among the N primitives (e.g., [3]) or by redefining primitives into linear combinations that require only $\mathcal{O}(N)$ relations [6]. While we believe that our figural modeling approach meets particularly well the requirements of a shape model listed earlier in this paper, it remains for further research to determine the relative strengths of these approaches.

Acknowledgements

The work reported here was accomplished partially under the support of NIH grants P01 CA47982, R01 CA67812, and R01 LM05508. The UNC Departments of Computer Science, Radiation Oncology, and Radiology also provided important support. We gratefully acknowledge the ideas of Stephen Aylward, Gregory Clary, James Damon, Valen Johnson, Robert Keller, and Jason Miller, which have contributed to this work.

References

1. J. Besag. Spatial interaction and the statistical analysis of lattice systems. Read before the Royal Statistical Society, March 1984.
2. H. Blum and R. N. Nagel. Shape descriptions using weighted symmetric axis features. *Pattern Recognition*, 10:167–180, 1978.

3. F. L. Bookstein. *Morphometric Tools for Landmark Data.* Cambridge University Press, 1991.
4. D. Chen. *Volume Rendering Guided by Multiscale Medial Models.* PhD thesis, University of North Carolina at Chapel Hill, 1997.
5. G. J. Clary, S. M. Pizer, D. S. Fritsch, and J. R. Perry. Left ventricular wall motion tracking via deformable shape loci. In H. U. Lemke, M. W. Vannier, and K. Inamura, editors, *Computer Assisted Radiology and Surgery*, pages 271–276, 1997.
6. T. F. Cootes, A. Hill, C. J. Taylor, and J. Haslam. The use of active shape models for location structures in medical images. In H. H. Barrett and A. F. Gmitro, editors, *Information Processing in Medical Imaging*, volume 687 of *Lecture Notes in Computer Science*, pages 33–47. Springer-Verlag, 1993.
7. J. N. Damon. Properties of ridges and cores for two dimensional images. *Journal of Mathematical Imaging and Vision*, 1998. to appear.
8. D. Eberly. *Ridges in Image and Data Analysis.* Computational Imaging and Vision Series. Kluwer Academic Publishers, Dordrecht, NL, 1996.
9. D. S. Fritsch, D. Eberly, S. M. Pizer, and M. J. McAuliffe. Simulated cores and their application in medical imaging. In *Proceedings of the International Conference on Image Processing in Medical Imaging*, pages 365–368, 1995.
10. D. S. Fritsch, S. M. Pizer, L. Yu, V. Johnson, and E. L. Chaney. Localization and segmentation of medical image objects using deformable shape loci. In H. H. Barrett and A. F. Gmitro, editors, *Information Processing in Medical Imaging*, Lecture Notes in Computer Science, pages 127–140. Springer-Verlag, 1997.
11. J. Furst, S. Pizer, and D. Eberly. Marching cores: a method for extracting cores from 3D medical images. In *Proceedings of the workshop on Mathematical Methods in Biomedical Image Analysis*, pages 124–130. IEEE Computer Society Technical Committee on Pattern Analysis and Machine Intelligence, June 1996.
12. J. D. Furst. *Height Ridges of Oriented Medialness.* PhD thesis, University of North Carolina at Chapel Hill, 1998. In preparation.
13. J. D. Furst, R. S. Keller, J. E. Miller, and S. M. Pizer. Image loci are ridges in geometric spaces. In B. ter Haar Romeny, L. Florack, J. Koenderink, and M. Viergever, editors, *Scale Space Theory in Computer Vision*, Lecture Notes in Computer Science, pages 176–187. Springer-Verlag, 1997.
14. U. Grenander. *Elements of Pattern Theory.* Johns Hopkins University Press, 1996.
15. K. P. Jones and D. S. Fritsch. Finding subfigure cores, 1995. Internal Report; UNC-CH Department of Computer Science.
16. T. Lindeberg and B. M. ter Haar Romeny. Linear scale-space I: Basic theory. In B. M. ter Haar Romeny, editor, *Geometry-Driven Diffusion in Computer Vision*, Computational Imaging and Vision, pages 1–38. Kluwer Academic Publishers, 1994.
17. W. E. Lorensen and H. E. Cline. Marching cubes: a high resolution 3D surface construction algorithm. *Computer Graphics*, 21(4):163–169, July 1987.
18. K. C. Low. A figural deformation approach to segmentation, 1997. Internal Report; UNC-CH Department of Computer Science.
19. M. McAuliffe, D. Eberly, D. Fritsch, E. Chaney, and S. Pizer. Scale-space boundary evolution initialized by cores. In *Visualization in Biomedical Computing*, 19996.
20. S. M. Pizer, D. Eberly, B. S. Morse, and D. S. Fritsch. Zoom-invariant visioin of figural shape: The mathematics of cores. *Computer Vision and Image Understanding*, 1998. to appear.
21. S. M. Pizer, D. S. Fritsch, V. Johnson, and E. Chaney. Segmentation, registration and measurement of shape variation via image object shape, 1996. tutorial notes *Visualization in Biomedical Computing.*
22. R. van den Boomgaard and A. W. M. Smeulders. The morphological structure of images, the differential equations of morphological scale space. *IEEE Trans. PAMI*, pages 1101–1113, November 1994.
23. A. Wilson and V. Johnson. Towards a framework for automated image analysis. In K. V. Mardia, C. A. Gill, and I. L. Dryden, editors, *Proceedings in Image Fusion and Shape Variability Techniques*, pages 13–20, 1996.

GENERATION OF THE EUCLIDEAN SKELETON BY A BISECTOR DECISION RULE ON A TWO-SHORTEST-VECTOR DISTANCE MAP

H. LI*
Engineering Physics Department, TsingHua University, Beijing, 100084, China.

and

A. M. VOSSEPOEL
Pattern Recognition Group, Faculty of Applied Sciences, Delft University of Technology, Lorentzweg 1, 2628 CJ Delft, The Netherlands.
E-mail:{lihong, albert}@ph.tn.tudelft.nl

Abstract. The Euclidean skeleton is essential for general shape representation. The common methods based on the local maxima of the distance transform can only find parts of the exact Euclidean skeleton. This paper provides a novel method to extract the whole exact Euclidean skeleton by a bisector decision (BD) rule on a two-shortest-vector distance (TSVD) map. The method succeeds in generating a well-connected Euclidean skeleton without any linking algorithm, and this is supported by a theoretical analysis and many experiments. The average complexity of the whole algorithm is linear in the number of pixels. This has also been verified by experiments.

Key words: Euclidean, skeleton, medial axis, distance transform, shape representation

1. Introduction

The Euclidean skeleton is ideal among all skeleton techniques, because it is mathematically well-defined and well-satisfying common requirements, such as isotropy and reconstructibility. In the literature, it is also known as 'medial axis'(MA), and 'axis of symmetry', which were first introduced by Blum [4, 5].

Numerous skeletonization algorithms have been developed. In recent literature, the Euclidean distance transform (DT) seems indispensable to realize Blum's original idea [12, 15]. The algorithms based on DT try to detect the peaks and 'ridge' points [1, 2, 9, 13], also known as local maxima, or centers of maximal disks (CMDs), from a distance map. But some essential problems with the discrete maximal disks [9, 17] determine that the detected results are not connected. In the following sections, we will concentrate on the methods to extract well-connected exact Euclidean skeleton based on Euclidean distance maps for binary images.

---

* Dr. Hong Li is a visiting scholar in the Pattern Recognition Group, Delft University of Technology, where this work was finished and supported partly by the Fellowship Foundation of Delft University of Technology and the Netherlands Organization for Scientific Research (NWO), project nr. 612-21-001.

1.1. THE PROBLEMS FOR DISCRETE MAXIMAL DISK

The idea to find maximal disks is based on continuous inscribed circles [6, 9]. Unfortunately, pixels and boundaries in discrete images have thickness, this prohibits the discrete disks either from representing all possible continuous circles in the object, or from checking the inclusion relation between the circles.

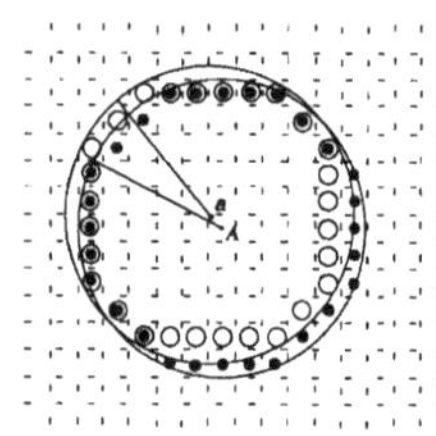

Fig. 1. Circles centered in one square, with different associated disks

First, the concept of one tangent point between two continuous circles does not hold in a discrete image, and the different errors of discrete disks with different radii and orientation may cause extra uncertainty when comparing them. A discrete disk can contain another discrete disk while their associated continuous circles do not have the same relation [9]. Secondly, the centers of continuous maximal circles are usually not exactly at the centers of digital grids. In Fig. 1, equal-radius circles centered at square center 'A' and at another point 'a' in the square, have different associated disks marked by '•' and '∘', respectively[1]. This problem has been analyzed for small disks by enumerating all possibilities [19], but so far it defies generalization. It is obvious that with only one discrete disk, neither can we present all continuous circles centered in a pixel square, nor can we check the tangent relation among such circles.

Ge [9] also showed that a discrete disk not contained by all its 8-neighboring disks may be contained by another disk centered some small distance away. This means that when you want to find local maximal disk with the disks of its neighbors, you do not know whether the neighbor area is sufficiently large. So, it is very difficult to find an algorithm with linear complexity.

1.2. OUR NEW ALGORITHM

We start from Blum's idea of 'equidistant points' [4]. The first and the second shortest vectors pointing to the boundary are used to check 'equidistancy' by a continuous geometric concept: bisector. Our scheme is simple: a) getting a TSVD map which stores the two shortest vectors (TSVs). b) extracting skeleton by a bisector decision rule.

S. Ho and C. R. Dyer [10] used the idea of TSVs to measure the performance of medial axis pixels for shape smoothing. Their work is useful to establish a hierarchical Euclidean skeletonization method.

[1] The discrete disk includes all pixels with centers inside its associated continuous circle; and its boundary is defined as the pixels in the disk that have at least one 4-connected background neighbor.

2. Notation

We consider a binary image, discrete on a square grid. The foreground (objects) F has value 1, with 8-connection; the background has value 0, with 4-connection. The neighborhood of a pixel p consists of all its eight nearest pixels or neighbors, the four neighbors in vertical and horizontal directions are called direct (or D-) neighbors; the other four (diagonal) neighbors are called indirect (or I-) neighbors.

2.1. Discrete skeleton with continuous image model

The 4-connected outside contour is defined as a discrete boundary $B(F)$. A continuous boundary $CB(F)$ is defined by the straight line segments connected by the centers of D-neighbors in $B(F)$.

The continuous object surrounded by the $CB(F)$ has a continuous skeleton. If a pixel covers a segment of the continuous skeleton, the pixel is put into the skeletal set. It is not necessary to prove the connectivity of the continuous skeleton of the continuous image. So, the discrete Euclidean skeleton defined as above is (8-) connected.

If the continuous skeleton crosses only a corner of a pixel square (like 'a' and 'b' in Fig. 2), the pixel is not a skeletal one. But if a square edge is on the continuous skeleton, the two pixels (like 'c' and 'd') sharing the edge belong to the skeleton.

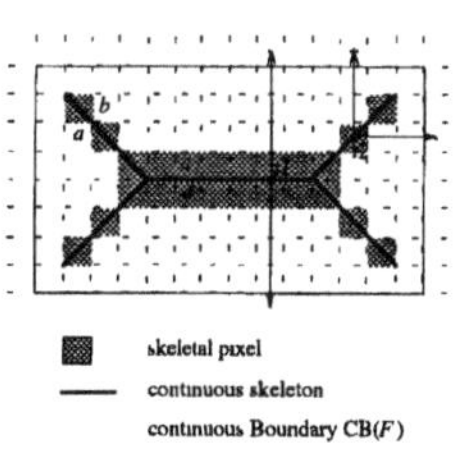

Fig. 2. Continuous and discrete skeleton

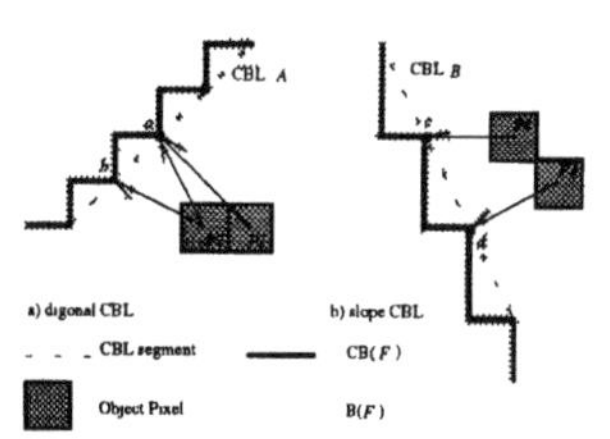

Fig. 3. Slope continuous boundary lines (CBL)

2.2. The Two-Shortest-Vector Distances(TSVD) map

The algorithm TSV_8SED is a modified version of Danielsson's 8SED algorithm [7] to scan and store the TSVs of each pixel into the TSVD map.

It should be emphasized that the TSVs should not point to one continuous straight line. Smeulders *et al.* [16] discussed the effect of the digitization to straight lines. We extract three kinds of continuous boundary lines (CBL) from the $CB(F)$: vertical, horizontal and slope. The slope lines can only be found from repeating horizontal or vertical segments with direct connection. In Fig. 3, CBL B is formed by 3-pixel vertical segments. For diagonal CBL A, pixel $P2$ has equal long vectors pointing to a and b, we just store the first one met.

3. Extraction of skeleton from the TSVD map

3.1. Bisector condition

For a pixel on TSVD map, the two endpoints of its TSVs (on the $CB(F)$) have a bisector. If the bisector crosses[2] the pixel square, the bisector condition is true.

3.2. Performance of the bisector condition

The endpoints of the TSVs on the $CB(F)$ must be on an orthogonal (vertical or horizontal) segment (like that of pixel $t2$ in Fig. 2), or at a pixel center locating at a $CB(F)$ corner (like the center of pixel $T1$ in Fig. 4). For the case that the boundary includes a slope CBL, it is similar. In Fig. 3, the nearest $CB(F)$ point of every point in pixel $P1$ and a half of pixel $P2$ is at point a; and the nearest $CB(F)$ point of every point in pixel $P3$ is at d. But the nearest $CB(F)$ point of every point in the lower half of pixel $P4$ is on a vertical $CB(F)$ segment. Therefore, we just discuss three cases of the endpoints of the TSVs: both on orthogonal segments; one at a corner, the other on a segment; and both at corners.

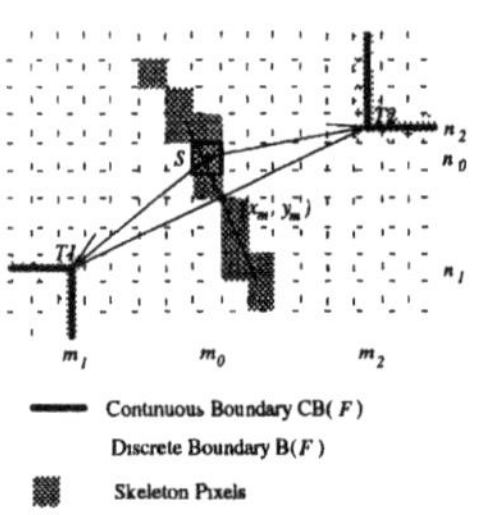

Fig. 4. Bisector condition

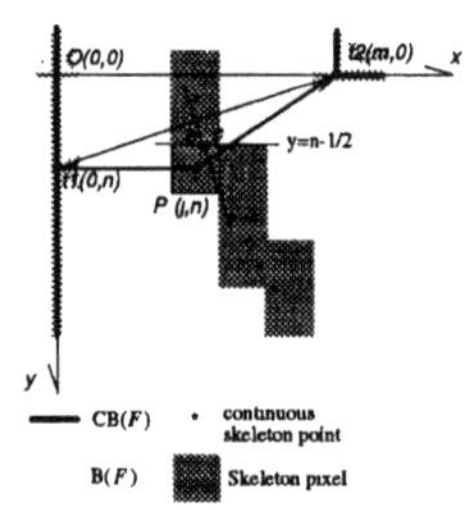

Fig. 5. One at corner, the other on a vertical segment

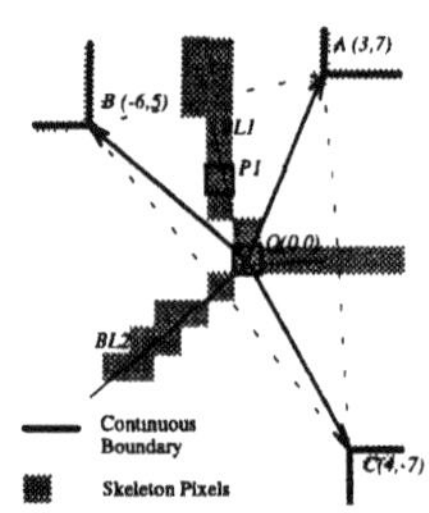

Fig. 6. All endpoints at corners of an acute triangle

3.2.1. *Both on orthogonal segments*

In this case, the bisector is the continuous skeleton itself (like that in Fig. 2), the bisector condition is sufficient and necessary.

3.2.2. *One at corner, the other on a orthogonal segment*

In the area ($x, y \in (0, m)$, $x, y \in \mathbb{R}$, $m \in \mathbb{N}$) in Fig. 5, the continuous skeleton is a parabola. For a pixel $p(j,n)$ ($n, j \in (0, m)$, $n, j \in \mathbb{N}$), the lines on its edges $y = n \pm 1/2$ have crossing points with its bisector and the parabola, which locate at b and c, with x_b and x_c as the position on the x coordinate, respectively, we have:

$$x_b = (m^2 + n^2 \pm n)/(2m)$$

$$x_c - x_b = 1/(8m).$$

If point b is on the pixel square, because $m, n \in \mathbb{N}$,

$$x_b = j \pm k/(2m),\ k \in [0, m-1], k \in \mathbb{N},$$

$$j - 1/2 + 5/(8m) \leq x_c \leq j + 1/2 - 3/(8m).$$

[2] In this paper, when we say a line crosses a pixel square, it means the line and the square have more in common than just one (*corner*) point.

So, it is clear that points b and c are always both in or both outside a distinct pixel square, and the bisector condition is both sufficient and necessary when one of the TSVs endpoints is on an orthogonal segment and the other at a $CB(F)$ corner.

3.2.3. *Both at CB(F) corners*

When the endpoints of TSVs are at two $CB(F)$ corners, as shown in Fig. 6, the bisector $BL1$ is part of the true continuous skeleton until it meets another bisector. For pixels near $P1$, their TSVs endpoints have the same bisector $BL1$, so the bisector condition here is both sufficient and necessary.

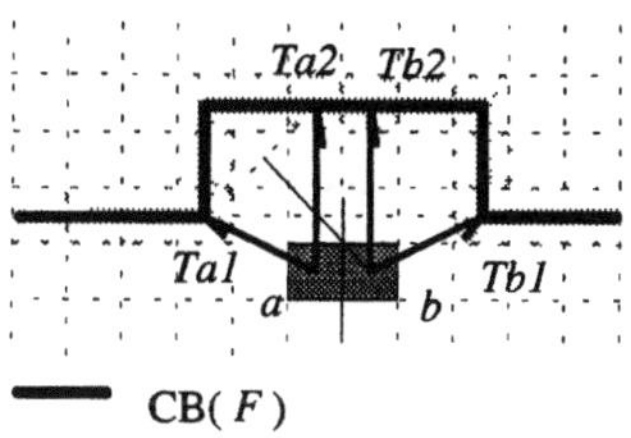

Fig. 7. Interactive neighbors

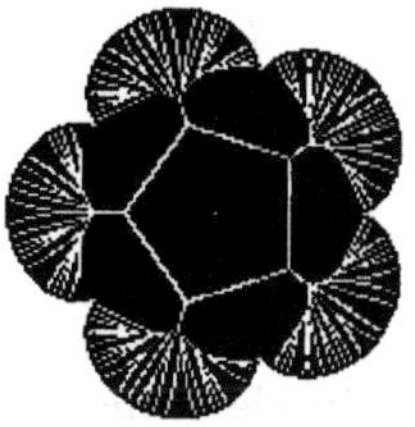

Fig. 8. Missed pixels by the bisector condition

It is more complex for the pixels near an intersection of continuous skeleton branches. For pixel Q in Fig. 6, the bisector condition works; but in Fig. 7, the skeleton pixels a and b are missed, because their bisectors just cross a corner of the respective square. We call such neighbors the interactive neighbors. Fig 8 shows these missed skeleton pixels

Such things only happen in the neighborhood of an intersection of continuous skeleton branches. We therefore add a `supplemental condition`:

For one object pixel a, if its bisector condition is false, and it has a interactive neighbor b, the shortest vector of a may form new bisectors with each vector of TSVs of pixel b. If a new bisector crosses the square of pixel a, a belongs to the skeleton.

3.3. The bisector decision (BD) rule

So, for any object pixel, if the bisector condition or the supplemental condition is true, the pixel belongs to the skeleton. We call this the BD rule.

Although it is very difficult to give a full proof of the sufficiency of the BD rule, we fail in finding evidence to the contrary. With the above analysis, the BD rule seems justified. Essentially, it is an effective rule for extracting the Euclidean skeleton from a discrete image, because the bisector is a continuous geometry concept.

4. Experiments and results

Our algorithms are programmed in C, and are run on a Sun Ultra 2 station. Example 1 is named 'Flower' with five petals and a hole at the center. The petals are not exactly round. The Euclidean skeletons in Fig. 8 and Fig. 9 show good, though not complete, isotropy. Only very few skeleton pixels are missed in Fig. 8, without the supplemental condition. Fig. 9 shows well-connected Euclidean skeletons. Because

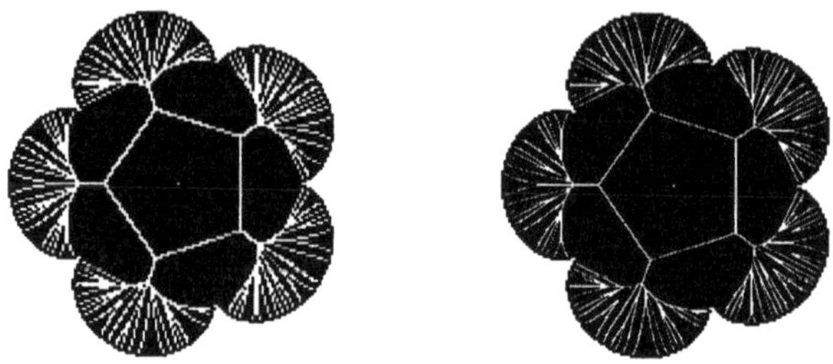

(a) 256×256 (b) 512×512

Fig. 9. The Euclidean skeleton of the 'Flower' by the BD rule

the smallest step in Fig. 9-b is 2 pixel-long, more branches exist than in Fig. 9-a. The concise inside connectedness and the trivial details near the boundary are interesting. It is obvious that the performance of the bisector condition is already good, whereas the performance of the BD rule is even better.

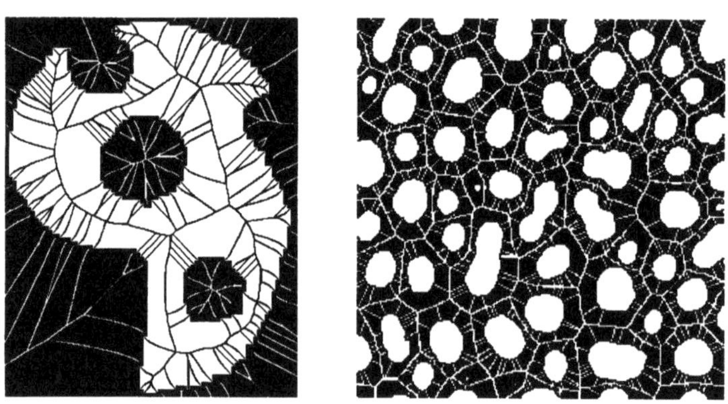

(a) Example 2,(320×400) (b) Example 3,(256×256)

Fig. 10. The Euclidean skeleton extracted by the BD rule.

Example 2 is a figure duplicated from reference [9]. It also can be found in earlier literature [3, 8, 14]. The symmetries inside and outside the objects were discussed in [4, 20]. Fig. 10-a combines the Euclidean skeletons of the foreground and the background of Example 2. The smallest edge step here is 4 pixel-long due to the enlargement. The well-connected skeleton of the foreground is as good as the best result in Ge's paper [9]. We note that Ge has to use a linking algorithm while we just use the BD rule. The figure named 'Cermet' in Example 3 in Fig. 10-b is used in some works with quasi-Euclidean distance [18, 11].

TABLE I
Timing(Second) of our algorithm with Example 3

	256×256	512×512	1024×1024
TSV_8SED	0.37	1.23	4.07
BD rule	0.20	0.73	2.67

TABLE II
Timing (Second) on Example 2. Ge's algorithms ran on a Sun SPARC 2, Dorst's and ours on Sun Ultra 2

Methods	256×256	400×400	1024×1024
Ge's algorithm 3+2 [9]	-	19	-
Ge's exhaustive [9]	-	370	-
Dorst's [8]	0.05	-	0.63
TSV_8SED + BD rule	0.16	0.58	2.14

The average complexity of our skeleton algorithm with TSV_8SED and the BD rule is linear in the number of pixels. Table I shows this. Table II list the timing comparison with Ge's Euclidean algorithms and Dorst's pseudo-Euclidean skeleton algorithm. It should be noted that the values of Ge's algorithm are measured on other computers with enormously different performances. The average complexity of Ge's method [17] is not linear, so it may be very slow for large images. Table II shows the speed of our algorithm is comparable with Dorst's pseudo-Euclidean skeleton algorithm.

5. Conclusions

Well-connected exact Euclidean skeletons can be extracted without any linking algorithm. Experiments shows that the method using the BD rule on the TSVD map works well for images different in size and complexity. This is also supported by a theoretical analysis. We fail to find evidence to the contrary. Trying to use more geometry may be a way to develop digital image processing and recognition.

Although the average complexity of our method is linear in the number of pixels, and the speed is comparable to a quasi-Euclidean skeleton algorithm, more work can be done to improve the speed and to reduce the memory demand.

It is desirable to establish a hierarchical, or multi-scale Euclidean skeleton to deal with the fragment branches and the significant influence of noise. The TSVs may be very useful for labeling the skeleton pixels [10].

Acknowledgements

The authors wish to thank Prof. I. T. Young for his thoughtful advice, Dr. Lucas van Vliet for his help, and the reviewers for their comments and advice.

References

1. C. Arcelli and G. Sanniti di Baja. Euclidean skeleton via centre-of-maximal disc extraction. *Image and Vision Computing*, 11(3), 1993.
2. C. Arcelli and G. Sanniti di Baja. Ridge points in Euclidean distance maps. *Pattern Recognition Letters*, 13(4):237-243, 1992.
3. C. Arcelli and G. Sanniti di Baja. A width independent fast thinning algorithm. *IEEE Trans. on PAMI*, 7(4):463-474, 1985.
4. H. Blum. A transformation for Extracting New Descriptors of Shape. In W. Wathen-Dunn, editor, *Symposium on Models for Perception of Speech and Visual Form*, MIT Press, Cambridge, Mass.,1967.
5. H. Blum. Biological Shape and Visual Science. *J. Theoret. Biol.*, 38, 205-287,1973.
6. J. W. Brandt and V.R. Algazi. Continuous Skeleton Computation by Voronoi Diagram. *CVGIP - Image Understanding*, 55(3):329-338, 1992.
7. P. E. Danielsson. Euclidean distance mapping. *Computer Graphics and Image processing*, 14, 227-248, 1980.
8. L. Dorst. Pseudo-Euclidean skeletons. In *Proc, 8th Int. Conf. on Pattern Recognition*, page 286-288, Paris, October 27-31, 1986.
9. Y. R. Ge and J. M. Fitzpatrick. On the generation of skeletons from discrete euclidean distance maps. *IEEE Trans. on PAMI*, 18(11):1055-1066, 1996.
10. S. Ho and C. R. Dyer. Shape smoothing using Medial Axis Properties. *IEEE Trans. on PAMI*, 8, 512-520, 1986.
11. L. Ji and J. Piper. Fast homotopy-preserving skeletons using mathematical morphology. *IEEE Trans. on PAMI*, 14(6):653-664, 1992.
12. F. Leymarie and M. D. Levine. Simulating the grassfire transform using an active contour model. *IEEE Trans. on PAMI*, 14(1):56-75, 1992.
13. F. Meyer. Digital Euclidean Skeleton. In *Proc. SPIE Visiual Comm. and Image Processing*, 1306, 251-262, 1992.
14. C. W. Niblack, D. W. Capson, and P. B. Gibbons. Generating skeletons and centerlines from the medial axis transform. In *Proc. 10th Int. Conf. on Pattern Recognition*, subconference B, page 881-885, Atlantic City, NJ, June 16-21, 1990.
15. R.L. Ogniewicz and O. Kubler. Hierarchic Voronoi skeletons. *Pattern Recognition*, 28(3):343-359, 1995.
16. A. W. M. Smeulders et al. Some Shape Parameters for Cell Recognition. In E. S. Gelsema and L. N. Kanal, editors, *Pattern Recognition in Practice*, North-Holland Publishing Company, 1980.
17. H. Talbot and L. Vincent. Euclidean Skeletons and Conditional Bisector. In *Proc. SPIE Visual Comm. and Image Processing '92*, vol. 1818, 862-876, 1992.
18. B. J. H. Verwer. Improved metrics in image processing applied to the Hilditch skeleton. In *Proc. 9th Int. Conf. on Pattern Recognition*, page 137-142, Rome, Nov. 1988.
19. A. M. Vossepoel. Estimating the size of circular pre-images from coarsely digitized representations. In *Proc. 11th Int. Conf. on Pattern Recognition*, Volume III, 365-368, IEEE Computer Society Press, 1992.
20. M. W. Wright, R. Cipolla, and P. J. Giblin. Skeletonization using an extended Euclidean distance transform. *Image and Vision Computing*, 13(5):367-375, 1995.

A FIGURE SPLITTING METHOD BASED ON THE MULTI-RESOLUTION SKELETON

H.HONTANI, M.NAKAO and K.DEGUCHI
Department of Mathematical Engineering and Information Physics, University of Tokyo, 7-3-1, Hongo, Bunkyo-ku, Tokyo, Japan

Abstract. We propose a contour figure splitting method. The purpose is to split the given figure into shape components, and to represent the split components hierarchically. Most of conventional methods fail to split off some protrude components which have not two curvature negative minimal points on the contour. In this paper, we propose to split the given figure at the ligature and the semi-ligature parts of skeletons. Our method can split off such the protrude components.

In order to represent the split components hierarchically, a technique of the multi-resolution analysis is introduced. The given figure is observed at various resolutions. At every resolution, the given figure is split by lines into components. As the resolution becomes lower, the number of the components decreases. The splitting lines which exist at the lower resolutions are considered stronger. We make a hierarchical representation of split components based on the strength of the splitting lines. In order to measure the strength, we must make correspondences between splitting lines obtained at different resolutions. We developed a method to make these correspondences. Some experimental results show that the proposed method is applicable for many types of figure analyses.

Key words: shape segmentation, multi-scale analysis, skeleton, shape description

1. Introduction

For figure shape analysis, it is essential to split the figure into components, and at the same time, to obtain hierarchical structure of the split components in the original figure. Many methods have been proposed for splitting the given figure into some components[1][2][7]. However, most of the methods only cut off protrude components by splitting the figure at the concave points of the contour. Such methods, first, search for curvature minimal points on the contour, then connect pairs of those to form splitting lines. These methods fail to extract some type of protrude components. As shown in Fig.1, for example, if the protrude component has only one curvature minimal point on the contour, such the component is not extracted. Other than this weakness, most of the conventional methods were not for analysis of hierarchical structure of the figure.

In this paper, we propose a figure splitting method based on its skeletons. The method splits the given figure at the *ligature* and the *semi-ligature* parts of the skeletons. This enables to split off the protrude components which have only one curvature minimal point on the corresponding contour. Our method also generates a hierarchical representation of the split components. In this representation, the larger size shape components are in higher layers. In order to measure the size of the shape component, we employ the *multi-resolution skeleton*[6].

In this technique, a given figure is observed at various resolutions. As the resolu-

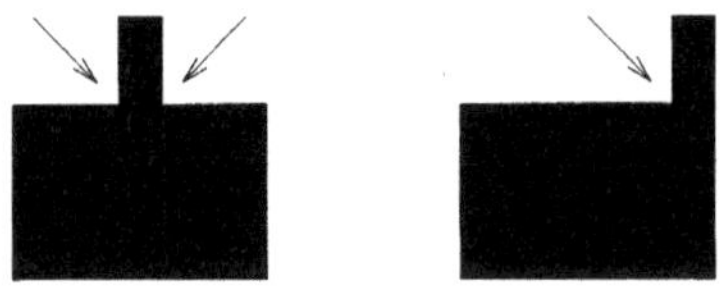

Fig. 1. Examples of protrude components. Right one has only one curvature minimal point on the contour(indicated by arrows).

tion becomes lower, small shape components are smoothed out. We split the figure into shape components at every resolution, and assign the resolution at which the component disappears to the size of the shape component.

For this assignment, we must make correspondences between the shape components split at different resolutions. First, we make correspondences between skeleton branches extracted at different resolutions. Then, we obtain the shape component correspondences.

In the next section, the method for skeleton extraction is explained. Section 3 introduces the multi-resolution skeleton. In Sec.4, methods for the figure segmentation and for a hierarchical representation are described. Section 5 shows some experimental results, and finally, conclusion is described in Sec.6.

2. Skeleton Extraction

We employ a skeleton extraction method proposed by Kimmel et al.[4]. The skeleton of a contour figure can be defined as follows by using a notion of *maximal disk.*

Definition 1 *A skeleton is a set of center points of maximal disks.*

Definition 2 *A maximal disk is a disk contained in the figure, and not contained in other disks which are contained in the figure.*

The skeleton generating point is defined with the maximal disk.

Definition 3 *A maximal disk touches the contour of the figure at least at two points. These touching points are called the skeleton generating points.*

Every point on the contour of the figure becomes a skeleton generating point for at least one skeleton point. The algorithm to generate the skeletons is as follows.

1. Find the positively maximal curvature points $s_1, s_2, \cdots, s_N$ on the contour, which are called *contour splitting points,* and split the contour into segments $C_1, C_2, \cdots, C_N$ at those points. That is, the end points of the segment C_i $(i = 1, 2, \cdots, N)$ are s_i and s_{i+1} $(s_{N+1} = s_1)$.
2. At every point p inside of the contour figure, calculate the minimal Euclidean distance maps $D_i(p)$ $(i = 1, 2, \cdots, N)$ to every segment C_i.
3. Compose a tentative skeleton as the zero-level sets of differences of smallest two distance maps. That is; $\hat{\phi} = \{(x, y) : \forall i, j \ ; \ i \neq j, (D_i = D_j) \cap (\cap_{k \neq i,j} D_i < D_k)\}$, where $i, j, k \in \{1, 2, \cdots N\}$.

Fig. 2. An example of the extracted skeleton. Left:the original figure; Middle: the tentative skeleton; Right: the skeleton obtained finally.

4. Eliminate all points which lie on branches of the tentative skeleton starting at points s_i and their distances from s_i are less than $\frac{1}{\kappa(s_i)}$. Here, the $\kappa(s_i)$ is the curvature of the contour at the point s_i. As the result, some of tentative skeleton branches shorten or disappear, and finally the skeleton of the contour figure is obtained.

An example of this process is shown in Fig.2.

3. Multi-resolution skeleton

As will be described in next section, the given figure is split into its shape components based on the skeleton branches. In order to split the figure into hierarchical components, we need the strength of the splitting line.

To evaluate the strength of the split, first, we account the importance of every skeleton branch by constructing the multi-resolution skeletons proposed by Pizer et al.[6].

To obtain the figures at various resolutions, we blur the given figure with various scales. As the scale changes, we obtain different skeletons.

3.1. Blurring

The blurring operation for the multi-resolution skeleton should have the property that it creates no new curvature maximal point. One of blurring methods which has this property is to develop the contour according to the following diffusion equations[5].

$$\begin{cases} \frac{\partial R(s;t)}{\partial t} &= -\frac{1}{2}\frac{\partial^2}{\partial s^2}R(s;t)n(s;t) \\ R(s;0) &= R_0(s), \end{cases}$$

where s is the arc length, $R_0(s)$ is the coordinates of the contour point of the given original figure, n is a unit normal vector which directs outward, and t (≥ 0) is the scale parameter which corresponds to the resolution. Figure 3 shows examples of blurred figures. The structures smaller than t are smoothed out.

3.2. Change of skeletons

Figure 4(middle) shows the change of the curvature maximal points. The x-axis shows the position on the contour normalized by the total contour length. The y-axis shows $\log t$. At every scale, the curvature positive maximal points are dotted.

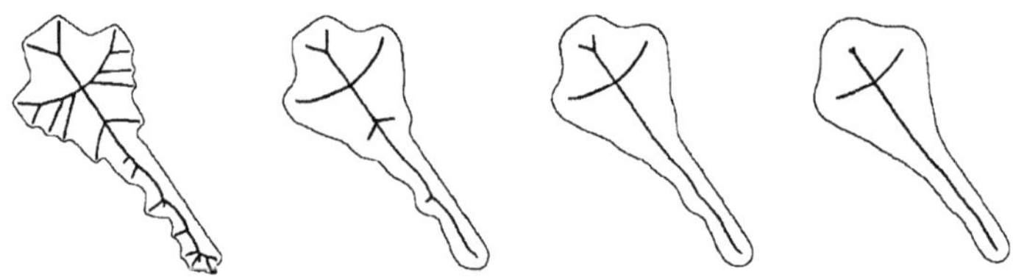

Fig. 3. The change of the figure and skeleton. The contour length of the original figure is 188, and the scale t = 0, 100, 200, and 400, from left to right.

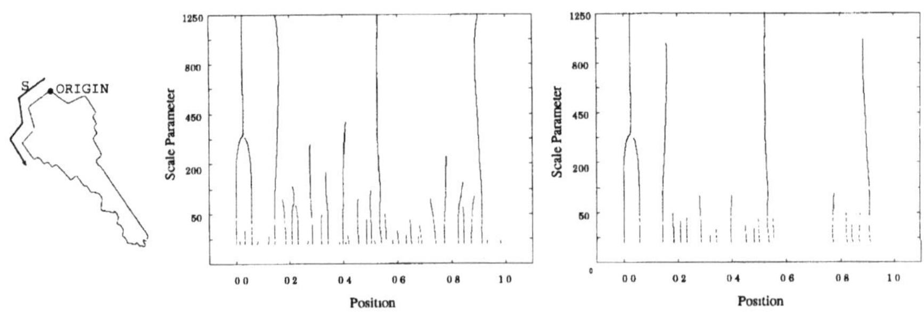

Fig. 4. Scale-space. Left: the position of the origin; Middle: the curvature positive maximal points plotted in the scale-space; Right: contour splitting points.

This x-(log)t space is called a *scale-space.* The scale-space representation helps to identify skeletons extracted at different scales.

3.3. Correspondences of Skeletons

The skeleton branches extracted from the original figure must be identified with ones extracted at every scale.

Here, the curvature maximal points at which skeleton branchs start are called *contour splitting points*, which split the contour of the figure into segments. We call these segments the *skeleton generating segments.*

In the scale-space, the contour splitting points extracted at every scale also generate a set of plottings. Figure 4(right) shows an example. As the scale increases, the number of the contour splitting points decreases.

A skeleton branch has corresponding two skeleton generating segments. A skeleton generating segment at scale $t = t_0$ is represented in the scale-space as one interval on a line $t = t_0$ bounded by two contour splitting points.

In the scale space, if an interval on a line $t = t_0$ can be connected with an interval on a line $t = t_1 (\neq t_0)$ without crossing any plottings of the contour splitting points, then these two skeleton generating segments are identified.

The plottings of contour splitting points disappear as the scale increases in two types. One is that a contour splitting point disappears alone, and the other is that two or more splitting points merge into one point.

When the contour splitting point disappears alone, two neighboring skeleton

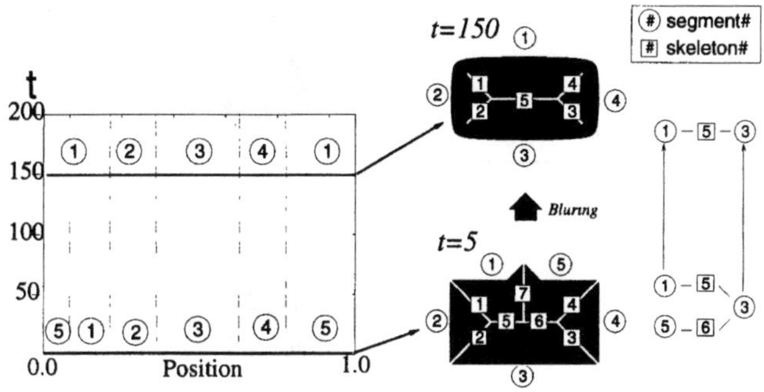

Fig. 5. Matching of skeletons extracted at different resolutions

generating segments merge into one segment. Then, a skeleton generated by these two segments disappears. Figure 5 shows an example. Two skeleton generating segments #1 and #5 at scale $t = 0$ merge into one segment #1 at scale $t = 150$, and the skeleton #7 which is generated by these two segments at $t = 0$ disappears.

When one skeleton disappears, other two neighbor skeletons which are divided by the skeleton may merge into one skeleton. Figure 5 also shows an example. The skeleton branches #5 and #6 at $t = 0$ merge into #5 at $t = 150$.

When two or more contour splitting points merge into one at some scale, one or more skeleton generating segments disappear(see Fig.6). Then corresponding skeleton branches also disappear. Figure 6 shows an example for this case. When the skeleton generating segment #4 at $t = 0$ disappears, the corresponding skeleton branches #3 and #4 disappear.

4. Figure segmentation

We propose a figure segmentation method which refers to the *ligature* and the *semi-ligature* parts of skeletons (see Fig.7).

Usually, a part of the contour corresponds to a part of the skeleton branch. But, sometimes, a contour point corresponds to a part of the skeleton branch. Such a part of the skeleton branch is called ligature or semi-ligature.

Definition 4 *A ligature is a set of skeleton points which share just only two points as their skeleton generating points.*

Definition 5 *A semi-ligature is a set of skeleton points which share one point as one of their skeleton generating points.*

Actually, the ligature and the semi-ligature are detected as following. If we move a skeleton point along a skeleton branch, then the corresponding two generating points also move on the contour. If the ratios of lengths L_1 and L_2 of the two generating points correspond to the length Δ along the skeleton branch we move, which are

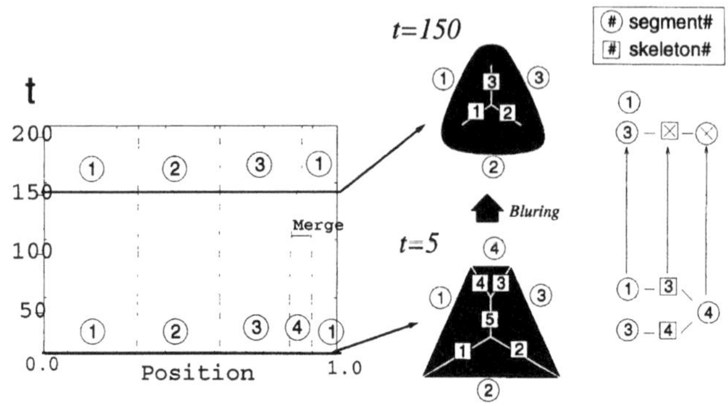

Fig. 6. Matching of skeletons extracted at different resolutions

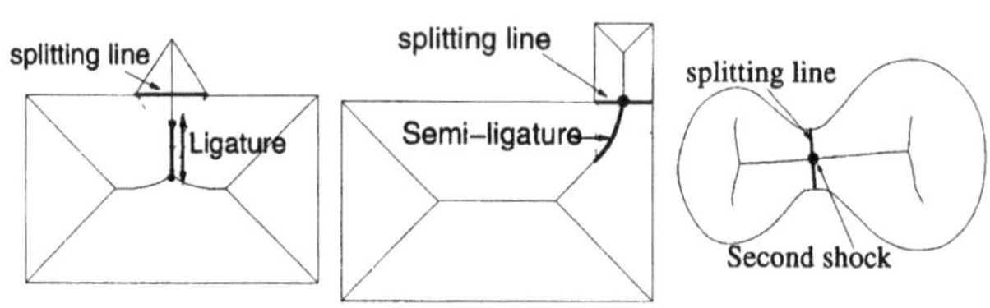

Fig. 7. Splitting a figure. Left: By a ligature, Middle: By a semi-ligature, and Right: By a second shock

L_1/Δ and L_2/Δ, are both smaller than a threshold M, then the skeleton point is detected as a point of the ligature. If only one of the values is smaller than M, then the skeleton point is detected as a semi-ligature. By using the latter concept, the method can split off a protrude part which does not have two corresponding concave parts on the contour.

Next, on every ligature and semi-ligature, we find skeleton points whose radius of the maximal disk is minimum. Then, two skeleton generating points of this skeleton point are connected by a straight line. We call this line a *splitting line.* By this splitting line, the figure is split into components.

In addition to this, we split the figure at a narrow part. For this splitting, we employ the concept of *second shock* proposed by Kimia et al.[3]

Definition 6 *A second shock is a skeleton point at which the radius of the maximal disk is minimal along the skeleton branch (see Fig.7(right)).*

Two skeleton generating points of the second shock are connected by a straight line, and the figure is split into two parts. Figure 8 shows some examples of figures split by the proposed method.

To construct a hierarchy of split components of a given figure, we evaluate the *strength* of each splitting line. In order to obtain this strength, the blurring operation

Fig. 8. The split figures. Protrude parts and narrow parts were split.

Fig. 9. Left: The extracted skeleton and the skeleton points correspond to the splitting lines; Right: the change of the splitting lines. The scale $t = 10$, 60, 120, and 400 respectively.

explained in Sec.3.1 is used.

When the given figure is blurred, some splitting lines disappear because small protrude parts or narrow parts are smoothed out. As the scale of the blurring operator increases, the total number of splitting lines decreases.

We assign the scale at which the splitting line disappears to its strength. Because every splitting line corresponds to one skeleton branch, the multi-resolution skeleton enables to make correspondences between splitting lines obtained at different scales. By using these correspondences, the algorithm for making a hierarchical representation of the split figure is given as follows.

1. Blur the given figure with various scales, and extract splitting lines at every scale.
2. Find correspondences of splitting lines extracted at every scale with ones of the original figure.
3. Evaluate the strength of every splitting lines.
4. Split the figure into components by the splitting lines in order of their strength. Every component to be split is represented as a parent, and the split components are represented as children.

5. Experimental Results

Figure 9(left) shows an example of the extracted skeleton. The dots are skeleton points correspond to the splitting lines. The contour figure and its splitting lines changed as the scale increased as shown in Fig.9(right). Figure 10 shows the hierarchical representations of the split components obtained by proposed method.

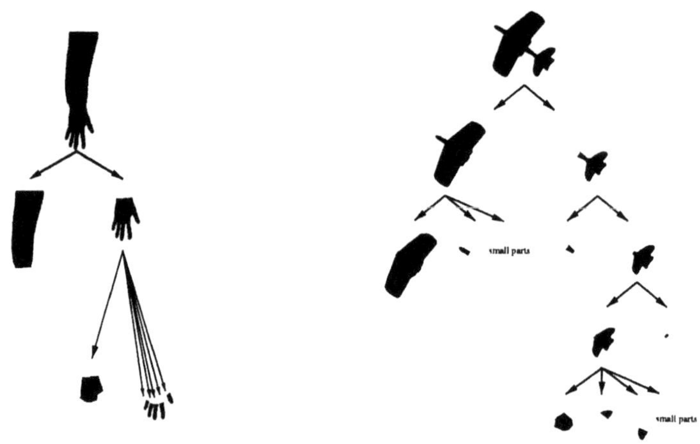

Fig. 10. The hierarchical representation of split components of figures

6. Conclusion

We proposed a contour figure splitting method to represent the figure hierarchically with the split components. In order to make this representation, we employed the idea of multi-resolution skeleton, and developed a method to make correspondences between skeletons extracted at different resolutions.

The proposed method split the figure at the ligature and the semi-ligature parts of the skeleton. Therefore, the method splits off the protrude parts which have only one corresponding concave part of the contour. Some experimental results show that the proposed method is applicable for wide variety of figure analyses.

References

1. D. D. Hoffman. Codon constraints on closed 2d shapes. *CVGIP*, 31:156–177, 1985.
2. D. D. Hoffman and W. A. Richards. Parts of recognition. *Cognition*, 18:65–96, 1985.
3. B. B. Kimia, A. R. Tannenbaum, and S. W. Zucker. Shapes, shocks, and deformations i: The components of two-dimensional shape and the reaction-diffusion space. *International Journal of Computer Vision*, 15:189–224, 1995.
4. R. Kimmel, D. Shaked, and N. Kiryati. Skeltonization via distance maps and level sets. *Compuiter Vision and Image Understanding*, 62(3):382–391, 1995.
5. F. Mokhtarian and A. Mackworth. A theory of multiscale, curvature-based shape representation for planner curves. *IEEE Transactions on Pattern Analysis and Machine Intelligence*, 14(8):789–805, 1992.
6. S. Pizer, W. R. Oliver, and S. H. Bloomberg. Hierarchical shape description via the multiresolution symmetric axis transform. *IEEE Transactions on Pattern Analysis and Machine Intelligence*, 9(4):505–511, 1987.
7. K. Siddiqi and B. B. Kimia. Parts of visual form: Computational aspects. *IEEE Transactions on Pattern Analysis and Machine Intelligence*, 17(3):239–251, 1995.

ADVANCES IN DIFFERENTIAL MORPHOLOGY: IMAGE SEGMENTATION VIA EIKONAL PDE & CURVE EVOLUTION AND RECONSTRUCTION VIA CONSTRAINED DILATION FLOW

PETROS MARAGOS and MUHAMMAD AKMAL BUTT
School of E.C.E., Georgia Institute of Technology, Atlanta, GA 30332, USA.
P.M. is also with the Institute for Language & Speech Processing, Athens, Greece.

Abstract.
In differential morphology multiscale dilations and erosions are modeled via nonlinear PDEs in scale-space. Curve evolution offers stable and efficient algorithms to numerically solve such PDEs on grids. In this paper, we approach two important tasks in morphological image analysis, image segmentation by watersheds and opening-by-reconstruction filtering, by using two nonlinear PDEs related to the dilation PDE. In particular, for segmentation we use the eikonal PDE of geometrical optics driven by a refractive index field equal to the image gradient magnitude. The opening by reconstruction we model via an edge-constrained dilation flow. Further, we use the numerical algorithms of curve evolution to solve these PDEs. Our results show that the eikonal PDE-based segmentation yields regions whose boundaries are much closer to the true Euclidean geometry than the results of the currently available digital watershed algorithms. For the reconstruction problem, the PDE approach offers an interesting new interpretation.

Key words: Differential Morphology, PDEs, Curve Evolution, Segmentation, Reconstruction.

1. Introduction

In 1992 the continuous multiscale morphological operators of dilation, erosion [1, 3, 18] and opening, closing [3] were modeled via nonlinear partial differential equations (PDEs) acting in scale-space. These advancements were inspired by previous work in computer vision where multiscale linear convolutions of a signal/image were modeled via the heat (linear isotropic diffusion) PDE. For multiscale dilations and erosions of 2D signals/images $f(x,y)$ by typical compact flat structuring elements (sets) B at a continuum of scales $s \geq 0$ the generating PDEs have the form [3]

$$\Psi_s = \pm S_B(\Psi_x, \Psi_y) \quad , \quad \Psi(x,y,0) = f(x,y) \tag{1}$$

where we use the notation $\Psi_s = \partial\Psi/\partial s$ and $\Psi_x = \partial\Psi/\partial x$ for the scale and spatial partial derivatives, $\Psi(x,y,s)$ is the dilation or erosion of f by sB, $+/-$ corresponds to dilation/erosion respectively, and $S_B(\cdot)$ is a nonlinear function depending on the shape of B. For example, if B is the unit disk, then $S_B(u,w) = \sqrt{u^2+w^2}$ and hence

$$\Psi_s = \pm\sqrt{(\Psi_x)^2 + (\Psi_y)^2} \tag{2}$$

* This work was supported by the US NSF under Grant MIP-94-21677 and by the JSEP under Contract DAAH-04-96-1-0161.

Even if the initial image/signal f is smooth, at finite scales $s > 0$ the above dilation or erosion evolution may create discontinuities in the derivatives, like 'shocks'. Thus, the dilations $f \oplus sB$ or erosions $f \ominus sB$ are *weak solutions* of (1). Ways to deal with these shocks include replacing standard derivatives with morphological derivatives [3, 8] or replacing the PDEs with differential inclusions [9].

In parallel to the development of the above ideas, there have been some advances in the field of differential geometry for evolving curves or surfaces using level set methods; see [17] for a recent survey. Consider at time $t = 0$ an initial simple, smooth, closed planar curve $C(0)$ which is propagated for $t > 0$ along its normal vector field with speed F. Let this evolving curve $C(t)$ be represented by its position vector $\vec{X}(p,t) = (x(p,t), y(p,t))$ and parameterized by p so that it has its interior on the left in the direction of increasing p. A general propagation law is

$$\frac{\partial \vec{X}(p,t)}{\partial t} = F\vec{N}(p,t) \tag{3}$$

where $\vec{X}(p,0)$ represents $C(0)$, $\vec{N}(p,t)$ is the instantaneous unit outward *normal* vector at points on the evolving curve, and F is the *speed* function which generally depends on local geometrical information such as the *curvature* $K(p,t)$, global image properties, or other factors independent of the curve. If $F = 1$ or $F = -1$, then $C(t)$ is the dilation or erosion of the initial curve $C(0)$ by a disk of radius t.

To overcome the topological problem of splitting and merging and numerical problems with the Lagrangian formulation (3), an Eulerian formulation was proposed in [13] where the original curve $C(0)$ is first embedded in the surface of an arbitrary 2D Lipschitz continuous function $\Phi_0(x,y)$ as its zero level set. (For example, we select $\Phi_0(x,y)$ to be equal to the distance function $\pm d(x,y)$ from the boundary of $C(0)$ where $+$ is for points inside and $-$ is for points outside the curve.) Then, the evolving 2D curve is obtained as the zero level set $C(t) = \{(x,y) : \Phi(x,y,t) = 0\}$ of a 2D function $\Phi(x,y,t)$ that evolves on a (x,y,t) grid according to the PDE

$$\Phi_t = F\sqrt{(\Phi_x)^2 + (\Phi_y)^2} = F||\nabla\Phi|| \quad , \quad \Phi(x,y,0) = \Phi_0(x,y) \tag{4}$$

This function evolution PDE makes all level sets expand at normal speed F. If $F = \pm 1$, it is identical to the flat dilation/erosion PDE (1) by equating scale s with time t.

So far the dilation PDE has played a modeling role, either for morphological scale-space or for curve evolution. However, its practical usefulness is greatly amplified by the existence of a stable and efficient curve evolution algorithm [13] for the numerical implementation of (4). Its main steps (for $F = 1$) are :

- Let $\Phi^n_{i,j}$ be an approximate estimate of $\Phi(i\Delta x, j\Delta y, n\Delta t)$ on a grid.
- $D^+_x = (\Phi^n_{i+1,j} - \Phi^n_{i,j})/\Delta x \ , \ D^-_x = (\Phi^n_{i,j} - \Phi^n_{i-1,j})/\Delta x$
- $D^+_y = (\Phi^n_{i,j+1} - \Phi^n_{i,j})/\Delta y \ , \ D^-_y = (\Phi^n_{i,j} - \Phi^n_{i,j-1})/\Delta y$
- $G^2 = [\min(0, D^-_x)]^2 + [\max(0, D^+_x)]^2 + [\min(0, D^-_y)]^2 + [\max(0, D^+_y)]^2$
- $\Phi^n_{i,j} = \Phi^{n-1}_{i,j} + |G|\Delta t \quad , \quad n = 1, 2, ..., (R/\Delta t)$

where R is the maximum scale (radius) of interest, $\Delta x, \Delta y$ are the spatial grid spacings, and Δt is the time (scale) step. Thus, by choosing fine grids (and possibly higher order terms) an arbitrarily low error (between signal values on the continuous plane and the discrete grid) can be achieved in implementing morphological operations involving disks as structuring elements. This is a significant advantage of the PDE approach and was the main motivation in [16] to propose the implementation of continuous scale morphology using curve evolution.

In this paper, we use two dilation-related PDEs to model the important morphological image analysis tasks of segmentation by watersheds and opening-by-reconstruction and employ curve evolution algorithms for their solutions.

2. Image Segmentation based on the Eikonal PDE

Recently, many tasks for extracting information from visible images have been related to eikonal optics and 2D wave propagation via the **eikonal PDE** [15]

$$||\nabla U(x,y)|| = \eta(x,y) \tag{5}$$

Its solution $U(x,y)$ can provide 3D shape or topographic segmentation of an image $f(x,y)$ by choosing the refractive index field $\eta(x,y)$ to be an appropriate function of the image field $f(x,y)$ [5, 19, 6, 8].

The solution of the eikonal PDE can be viewed as a *weighted distance function* between a point (x,y) and the sources along a path of minimal optical length. The optical length of any path is obtained by integrating the refractive index field $\eta(x,y)$ along this path and is proportional to the time required for light to travel this path. Thus, we can view the solution $U(x,y)$ to the eikonal as a *gray-weighted distance transform (GWDT)* whose values at each pixel give the minimum distance from the light sources weighted by the gray values of the refractive index field.

An efficient morphological approach to *image segmentation* is the *watershed* [11, 20] which transforms an image $f(x,y)$ to the crest lines separating adjacent catchment basins that surround regional minima or other 'marker' sets of feature points. From the work in [14, 10, 12] it has been established that computing the watershed is equivalent to finding a skeleton by influence zones with respect to a weighted distance function that uses the image regional minima as sources and $||\nabla f||$ as the field of indices. (If other markers different than the minima are to be used as sources, then the homotopy of the function must be modified to impose these markers as minima.) Hence, it has been proposed in [10, 12] to use existing efficient watershed algorithms for finding the solution to the eikonal.

We model the problem of watershed segmentation of an image f as an optics problem where solving the eikonal equation

$$||\nabla U(x,y)|| = ||\nabla f(x,y)|| \tag{6}$$

yields as solution a GWDT $U(x,y)$ whose crestlines segment the image $f(x,y)$ into regions whose boundaries are the crestlines of the image function. Of course, robust features have to be extracted first as markers of these regions, and the image function has to be modified (via edge enhancement and homotopy modification) so that its crest lines correspond to the desired region boundaries.

Next we compare three different approaches to segmentation: the digital algorithm for watershed segmentation [20], the eikonal approach solved via a discrete

GWDT based on chamfer metrics [10], and our new approach based on the eikonal PDE and curve evolution.

2.1. Segmentation via GWDT based on Chamfer Distance Transforms

Let $f[x,y]$ be a discrete image to segment and let $\eta[x,y] = ||\nabla f[x,y]||$ represent a refractive index field. A discrete GWDT of $f[x,y]$ (which is an approximation [19] to the solution of the eikonal PDE $||\nabla U|| = \eta$) can be obtained via the following min-plus recursive difference equation

$$\begin{aligned} U_i[x,y] &= \min(U_i[x-1,y] + a\eta[x,y], U_i[x,y-1] + a\eta[x,y], \\ &\quad U_i[x-1,y-1] + b\eta[x,y], U_i[x+1,y-1] + b\eta[x,y], U_{i-1}[x,y]) \end{aligned} \tag{7}$$

where $U_0[x,y]$ is set equal to 0 if $[x,y]$ belongs to a source (marker) or $+\infty$ otherwise. The constants a and b are the local distance steps by which the planar distances are propagated within a 3×3 neighborhood. The above nonlinear recursive difference equation is run over the whole image in forward and backward order, iteratively ($i = 1, 2, 3, ...$) until stability. At convergence, U_∞ is the GWDT of f. The above is a sequential implementation of the GWDT. There are also other faster implementations using queues [19, 10].

The planar propagation of the local distance steps (a, b) starts at the light sources (markers) and moves at speeds proportional to $1/\eta[x,y]$. If f is a binary image, then the propagation speed is uniformly equal to one and the solution of the above difference equation (after convergence) yields the discrete chamfer distance transform of f [2]. To improve the approximation of the eikonal solution via the GWDT, we must find local distance steps (a, b) that make planar distances as close as possible to Euclidean distances. Thus, we can optimize (a, b) by minimizing some norm of the error between the chamfer distances and Euclidean distances. The error can be further minimized if we use larger local neighborhoods (e.g., 5×5) to propagate local distances. Overall, this approach to GWDT (based on chamfer metrics) is fast and easy to implement but it cannot use very large neighborhood masks to achieve very accurate values of the distance transform. Due to the required small neighborhoods, the distance propagation is not isotropic and leads to incorrect segmentation. In our work [4] we have improved the accuracy of this approach by finding optimal local distances for small masks based on a new approach that compares the unit balls of chamfer metrics (chmafer polygons) with the disks of the same size. Using a neighborhood larger than 5×5 can further reduce the approximation error but at the cost of an even slower implementation. However, larger neighborhood masks cannot be used with gray-weighted distance transforms because they give erroneous results since the large masks can bridge over a thin line that separates two segmentation regions.

2.2. Solving the Segmentation PDE via Curve Evolution

In this approach, the gray-weighted distance propagation to approximate the segmentation eikonal PDE is achieved by numerically solving the PDE using the curve evolution algorithm of [13]. The continuous modeling of general weighted distance propagation (or time of arrival) via the eikonal has also been in investigated in [6]. For the problem of watershed segmentation of an image f, we represent the boundary of the sources/markers as a planar curve $C(0)$ from which distance is propagated. Then a family of curves $C(t)$ is obtained by evolving the initial curve $C(0)$ according to the rule (3) by using a gray-weighted speed function $F(x, y) = 1/||\nabla f||$. The value of the resulting GWDT at any pixel of the image is the time it takes for the evolving curve to reach this pixel. To overcome the problem of splitting and merging, the original curve is first embedded as the zero-level set of an initial 2D function $\Phi_0(x, y)$ which is positive in the interior of the curve and negative in its exterior. Then, the evolving curve is obtained as the *zero level set* of a 2D function $\Phi(x, y, t)$ which is evolved, and the evolution time t is assigned as the distance transform value of the point (x, y) as soon as $\Phi(x, y, t)$ becomes positive.

A simple but computationally expensive approach to implement this algorithm would be to evolve the whole surface and then check all the pixels to see whether the surface has become positive at any pixel. Instead, we have used a queue-type data structure to hold the addresses of the pixels which just turned positive. After each evolution step, the neighboring pixels of the pixels on the queue are checked for a sign change. If found, such a pixel is put on the queue. If no neighbor pixel is found, the pixel is taken out of the queue and marked 'processed'. This method reduces the computational complexity of the original algorithm.

Multiple Interfaces and Triple Points: In a simple distance propagation algorithm, the evolving surface divides the domain in two sets: 'processed', where surface is non-negative, and 'unprocessed', where surface is negative. In case of multiple distinct sources with each having different distance propagation rules the domain could be divided into many subsets which interact with each other as the propagating distance wavefronts interact with each other. This problem is more complicated than the problem of simple distance propagation. We have to keep track of multiple wavefronts with multiple queues for tracking the zero-level set. There are special cases that can be solved more quickly than the general cases. One such special case is when evolving distance wavefronts do not penetrate each other.

In the standard watershed algorithm [20, 11], the flooding at each level is achieved by a planar distance propagation that uses the chess-board metric. This kind of distance propagation is highly non-isotropic and could give wrong results, particularly for images with large plateaus, as we found experimentally. Eikonal segmentation using chamfer-based GWDTs improves this situation a little but not entirely. In contrast, segmentation via the eikonal PDE and curve evolution gives results very close to ideal. However, real images may not have many plateaus.

For a better evaluation of the performance of these approaches, a test image has been proposed by Matheron. This test image was used by Meyer [10] to compare different discrete segmentation algorithms. This is a difficult image to segment as it forces the expanding distance wavefronts to meet the watershed lines at a wide range of angles, from perpendicular to almost parallel. Figure 1 compares the results of different segmentation approaches on this test image. The segmentation approach based on continuous (PDE) modeling and curve evolution algorithms outperforms the discrete segmentation results reported in [10] (using either the standard watershed algorithm or chamfer-based GWDTs).

In general, the PDE approach to segmentation is based on a different model than the model used in the standard watershed approach. The continuous model is

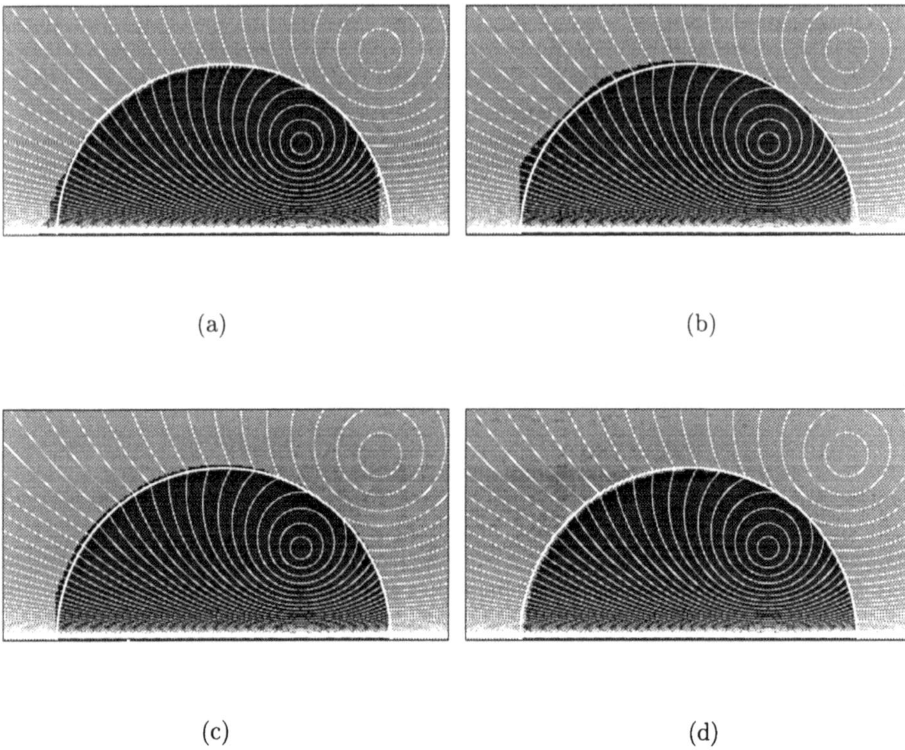

Fig. 1. Performance of different segmentation algorithms on a test image (250×400 pixels). The image is the minimum of two potential functions. The contour plot of the resulting terrain (thin bright curves) is superimposed on all segmentation results. Markers are the two source points of the potential functions. Segmentation results based on: (a) Standard watershed algorithm. (b) (3×3) chamfer-based GWDT. (c) (5×5) chamfer-based GWDT. (d) Eikonal PDE and curve evolution. The thick bright curve shows the correct segmentation.

based on the fact that the gradient magnitude and the GWDT are inverse operators. If we can pre-process the given image in a way such that it represents a gradient magnitude of the terrain we flood using the watershed approach, the more accurate PDE-based distance propagation can be applied to it for a better segmentation of the image.

3. Opening by Reconstruction via Edge-Constrained Dilation Flow

For using curve evolution in shape recovery problems there is often the need to build a mechanism in the flow (4) such that the evolving curve could stop when it reaches the boundaries of desired objects. To model such a constrained propagation of a curve $C(0)$ inside an image $I(x, y)$, an edge-constrained speed F was used in [7] yielding the flow

$$\frac{\partial \Phi}{\partial t} = F_c \left(\frac{M_1 - E(x,y)}{M_1 - M_2} \right) ||\nabla \Phi|| \tag{8}$$

where $\Phi(x, y, t)$ contains the evolving curve $C(t)$ as its zero-level set, $E(x, y)$ is some nonnegative function indicative of the strength of the edges of the original intensity image I, M_1, M_2 are the maximum and minimum value of E, and F_c is some constant speed. Then the speed F assumes values in $[0, F_c]$ and becomes zero at boundary pixels with strong edges where E becomes maximum which stops the flow. One choice for E proposed in [7] is $E = ||\nabla G_\sigma * I||$, where $G_\sigma * I$ is the convolution of I with a Gaussian at scale σ.

Given a digital binary image I consisting of several connected components with some markers in some of them, we modeled the opening by reconstruction of the marked components as an edge-constrained dilation flow of the type in (8) by setting $F_c = 1$, $E = I \oplus B - I$ where B is the 3×3-pixel square, and the initial curve (or curves) $C(0)$ equal to the boundary of the marker(s). Using the dilation residual $I \oplus B - I$ as edges has the advantage in this problem that it builds exact vertical walls which will stop the flow at the exact boundaries of the desired component if there is no noise. An example of implementing such an opening by reconstruction is shown in Fig. 2. In this PDE approach to modeling morphological filtering of the reconstruction type, the final output (i.e., the union of extracted components) is obtained as the steady state of the constrained flow, i.e., at $\partial \Phi / \partial t = 0$.

References

1. L. Alvarez, F. Guichard, P.L. Lions, and J.M. Morel, "Axiomatization et nouveaux operateurs de la morphologie mathematique", *C. R. Acad. Sci. Paris*, pp. 265-268, t.315, Serie I, 1992.
2. G. Borgefors, "Distance Transformations in Digital Images", *Comp. Vision, Graphics, Image Process.*, 34, pp. 344–371, 1986.
3. R. Brockett and P. Maragos, "Evolution Equations for Continuous-Scale Morphological Filtering", *IEEE Trans. Signal Processing*, vol. 42, pp. 3377-3386, Dec. 1994. Also, in *Proc. ICASSP-92*, San Francisco, 1992.
4. M. A. Butt and P. Maragos. "Optimal Design of Chamfer Distance Transforms", *IEEE Trans. Image Processing*, in press.
5. B.K.P. Horn, *Robot Vision*, MIT Press, Cambridge, MA, 1986.
6. R. Kimmel, N. Kiryati, and A. M. Bruckstein, "Sub-Pixel Distance Maps and Weighted Distance Transforms", *J. Math. Imaging and Vision*, 6, pp. 223-233, 1996.
7. R. Malladi, J. A. Sethian, and B. C. Vemuri, "A Fast Level Set Based Alrgorithm for Topology-Independent Shape Modeling", *J. Math. Imaging and Vision*, 6, pp. 269-289, 1996.
8. P. Maragos, "Differential Morphology and Image Processing" *IEEE Trans. Image Processing*, vol. 78, pp. 922–937, June 1996.
9. J. Mattioli, "Differential Relations of Morphological Operators", *Proc. 1st Int'l Workshop on Math. Morphology and its Application to Signal Processing*, Barcelona, Spain, May 1993.
10. F. Meyer, "Topographic Distance and Watershed Lines", *Signal Processing*, vol. 38, pp. 113-125, July 1994.
11. F. Meyer and S. Beucher, "Morphological Segmentation", *J. Visual Commun. Image Representation*, 1(1):21–45, 1990.
12. L. Najman and M. Schmitt, "Watershed of a Continuous Function", *Signal Processing*, vol. 38, pp. 99-112, July 1994.
13. S. Osher and J. Sethian, "Fronts Propagating with Curvature-Dependent Speed: Algorithms Based on Hamilton-Jacobi Formulations", *J. Comput. Physics*, 79, pp. 12–49, 1988.
14. F. Preteux, "On a Distance Function Approach for Gray-Level Mathematical Morphology", in *Math. Morphology in Image Processing*, (E.R. Dougherty, Ed.), Marcel Dekker, NY, 1993.

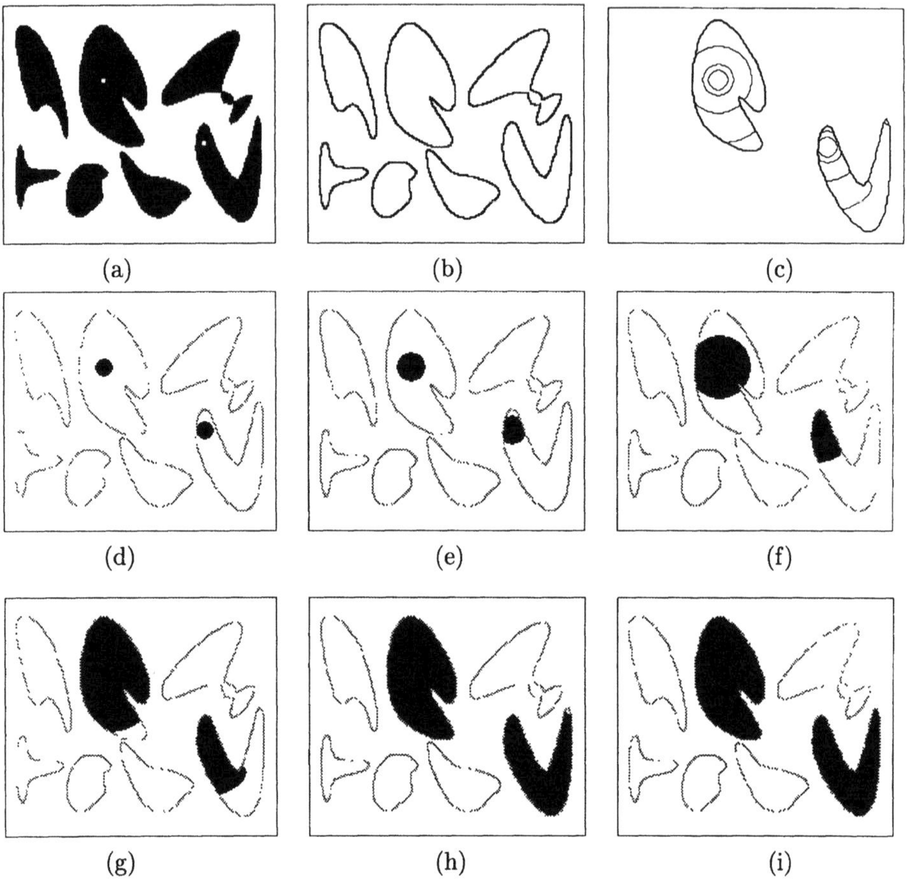

Fig. 2. Binary reconstruction via constrained dilation PDE flow. (a) Original image (170 $\times$ 200 pixels) and two markers. (b) Edge walls. (c) Wavefronts of reconstruction flow at different stages. Reconstructions at: (d) 200 iterations (scale 5 w.r.t. unit disk). (e) 400 iterations. (f) 1000 iterations. (g) 2000 iterations. (h) 4000 iterations. (i) 6000 iterations. ($\Delta t = 0.025$, $\Delta x = 1$.)

15. B.E.A. Saleh and M.C. Teich, *Fundamentals of Photonics*, Wiley, NY, 1991.
16. G. Sapiro, R. Kimmel, D. Shaked, B. Kimia, and A. Bruckstein, "Implementing Continuous-scale Morphology via Curve Evolution", *Pattern Recognition*, 26(9), pp. 1363–1372, 1993.
17. J. A. Sethian, *Level Set Methods*, Cambridge Univ. Press, 1996.
18. R. van der Boomgaard and A. Smeulders, "The Morphological Structure of Images: The Differential Equations of Morphological Scale-Space", *IEEE Trans. Pattern Anal. Mach. Intellig.*, vol. 16, pp.1101-1113, Nov. 1994. Also, Ph.D. Thesis, Univ. of Amsterdam, 1992.
19. P. Verbeek and B. Verwer, "Shading from shape, the eikonal equation solved by grey-weighted distance transform", *Pattern Recogn. Lett.*, 11:618–690, 1990.
20. L. Vincent and P. Soille, "Watershed In Digital Spaces: An Efficient Algorithm Based On Immersion Simulations", *IEEE Trans. PAMI*, vol. 13, pp. 583–598, June 1991.

NUMERICAL SOLUTIONS OF MORPHOLOGICAL PARTIAL DIFFERENTIAL EQUATIONS BASED ON THE MORPHOLOGICAL FACET MODEL

REIN VAN DEN BOOMGAARD
University of Amsterdam
The Netherlands

Abstract. The partial differential equations describing the propagation of (wave) fronts in space are closely connected with the morphological erosion and dilation. Strangely enough this connection has not been explored in the derivation of numerical schemes to solve the differential equations.

In this paper the ***morphological facet model*** is introduced in which an analytical function is locally fitted to the data. This function is then dilated analytically with an infinitesimal small structuring element. These ***sub-pixel dilations*** form the core of the numerical solution schemes presented in this paper.

One of the simpler morphological facet models leads to a numerical scheme that is identical with a well-known classical upwind finite difference scheme. Experiments show that the morphological facet model provides stable numerical solution schemes for these partial differential equations.

Key words: sub-pixel erosions and dilations, PDE, numerical solutions, facet model.

1. Introduction

The partial differential equations describing the propagation of fronts in space are known to be closely connected with the morphological erosion and dilation. These *morphological partial differential equations* (abbreviated as PDE's) known from the work of Alvarez [1], Maragos[5], Matiolli [6] and van den Boomgaard [10], have gained considerable interest in the recent past as canonical descriptions of evolutionary shape deformation (see Osher and Sethian [7], Sapiro [8] and Kimia [4]). Strangely enough this connection has not been explored in the development of numerical schemes to solve these differential equations.

In this paper the *morphological facet model* is introduced, in which a function is locally fitted to the data. This function is then dilated analytically with an infinitesimally small structuring element. These *sub-pixel dilations* form the core of the numerical solution schemes. One of the simpler morphological facet models leads to a numerical scheme that is identical with a well known classical upwind finite difference scheme.

Consider the parameterized shape contour $C(p,t)$ as function of the path parameter p and "time" parameter t. The generic evolution of the shape as a function of time is given by:

$$\frac{\partial C}{\partial t} = \Gamma(\kappa)N$$

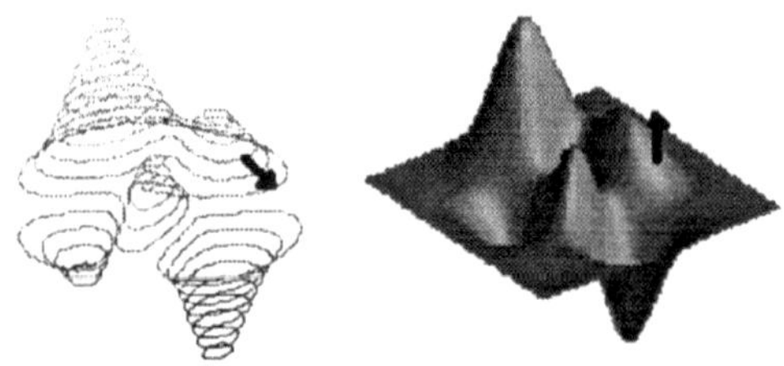

Fig. 1. **Contour versus function evolution.** A function is shown, represented as a collection of levels sets (on the left) and as a surface (on the right). Observe that in the curve evolution description the points are moved in the spatial direction, whereas in the surface (function) description the points are moved in the vertical direction.

where Γ is a function of the curvature κ and N is the inwards pointing normal to the curve. The choice of $\Gamma = -1$ is equivalent with the dilation of the shape with a disk of radius t. A pure local description of the dilation (as in the above PDE) leads to self-intersecting curves. Dorst and van den Boomgaard [2] used this local geometrical description as their definition of the *tangential dilation*. The classical morphological dilation corresponds with the entropy solution of the PDE (i.e. the solution without self-intersections).

A robust way to obtain entropy solutions is to embed the curve as a level set in a function and solve the associated PDE that describes the evolution of the function in time. Let F be a function of space (parameter x) and time (parameter t) and let some level set of F at time $t = 0$ correspond with the original curve C. It can be shown easily that the evolution of the function F, such that the level set behaves in time as if the 'curve PDE' is solved, is given by:

$$\frac{\partial F}{\partial t} = -\Gamma(\kappa)\,\|\nabla F\|\,.$$

Note that the embedding of the curve is chosen in such a way that the shape is characterized with high function values. In that case the gradient vector is indeed the inwards pointing normal. For $\Gamma = -1$ we recognize the PDE that is solved by dilating the initial condition (the function F at time $t = 0$) with a flat disk shaped structuring element of radius t.

When a 2D curve is embedded in a 2D function a remarkable thing happens. The geometry of the curve is not measured in the spatial domain alone, but the smoothness of the embedding function is used to measure the geometry of the curve through the function derivatives. Whereas in the curve representation the points on the curve are moved to a new position in the same (horizontal) plane, in the embedding function the points are moved in vertical direction (i.e. the function value is changed). This is only allowed in case the required smoothness of the embedding function (which is a mathematical construct and not an observation) is guaranteed through out the evolution process.

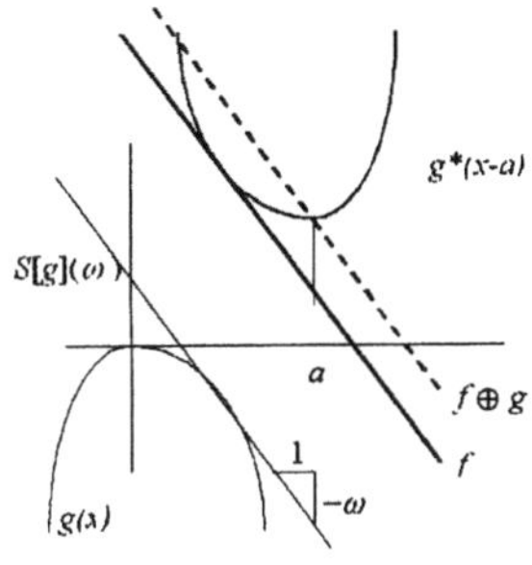

Fig. 2. **Dilating a planar function.** The vertical shift when dilating a planar function is given by the slope transform of the structuring function.

In this contribution we will not look at the numerous applications of these types of morphological PDE's in the computer vision context. Instead we will concentrate on numerical schemes to find solutions. In the mathematical literature, the derivation of robust and stable numerical schemes is complex and relies on the analysis of the conservation law properties of the PDE.

In [10] it was shown that the morphological dilation, out of all possible solutions, selects the unique entropy solution. With this observation in mind, this paper introduces the *morphological facet model* as an elegant method to derive robust and stable numerical schemes to solve these PDE's.

2. Morphological PDE's

In this section a short introduction to the morphological PDE's is given, a more detailed description can be found in [10]. Consider the one-parameter family of images:

$$F(x,t) = (f \oplus g^t)(x)$$

with g a concave function and g^t the umbral scaling of g defined as $g^t(x) = tg(x/t)$. In umbral scaling not only the spatial dimensions are scaled with a factor t but also the grey value dimension is scaled simultaneously. Essentially the graph of the function, interpreted as a geometrical object, is scaled.

In this section it will be shown what happens if we change the scale from t to $t+dt$. Because umbral scaling of any concave function forms a semi-group under dilation (i.e. $g^\lambda \oplus g^\mu = g^{\lambda+\mu}$) we can write $F(x,t+dt) = (f \oplus g^{t+dt})(x) = (F(\cdot,t) \oplus g^{dt})(x)$. As we are interested in the case that dt is infinitesimally small, meaning that g^{dt} becomes sharply pointed and looks like the morphological pulse, we may approximate the function $F(\cdot,t)$ locally around the point x, with its tangent line.

Planar functions are the eigenfunctions of mathematical morphology. Dilating (eroding) a planar function with any structuring function g results in a planar function with the same slope, it is only shifted in space. The vertical shift is equal to the intercept with the function axis (this is illustrated in figure 2) of the tangent plane in the point on the structuring function where the slope equals the slope of the plane to be dilated. This geometrical construction (for all tangent planes to the function g) gives the Legendre transform of the (concave) function g. The generalization of the Legendre transform to arbitrary functions is called the *slope transform*[2].

The dilation of a planar function $e_\omega(x) = \omega \cdot x + c$ is therefore equal to $e_\omega \oplus g = e_\omega + S[g](\omega)$, where $S[g]$ is the slope transform of g. In the case of the tangent plane to the function F in the point x, we obtain $F(x, t+dt) = F(x,t) + S[g^{dt}](\nabla F(x,t))$. Note that $\nabla F(x,t)$ is the 'slope' of the tangent plane. In [2] it is proven that umbral scaling in the spatial domain amounts to grey value scaling (i.e. multiplication with a constant) in the slope domain, therefore $F(x,t+dt) = F(x,t) + dtS[g](\nabla F(x,t))$. Rearranging terms and in the limit $dt \to 0$:

$$\frac{\partial F}{\partial t}(x,t) = S[g](\nabla F(x,t)).$$

This analysis shows that the family of images generated by dilation with the umbral scaling of a concave structuring function is causal in the scale-parameter. I.e. the change in grey value going from scale t to scale $t + dt$ is determined by the (first order) differential structure of the image at scale t.

In summary, the entropy solution of the PDE $\partial F/\partial t = G(\nabla F)$ with initial condition $F(x,0) = f(x)$ is given by the dilation $F(x,t) = (f \oplus g^t)(x)$, such that G is the slope transform of g.

3. Numerical solutions

In this section we look at numerical schemes to solve the PDE

$$\frac{\partial F}{\partial t} = \|\nabla F\| \tag{1}$$

with initial condition $F(x,0) = f(x)$. According to the theory in the previous section, this PDE is solved by the dilation $F(x,t) = (f \oplus \mu_{tB})(x)$, where μ_{tB} is (the logarithm of) the indicator function of the disk with radius t. In this paper, however, only forward Euler differential schemes will be considered to solve the PDE.

Let $F_{i,j,r} = F(i\Delta x, j\Delta y, r\Delta t)$, then the forward Euler finite difference scheme is given by:

$$F_{i,j,r+1} = F_{i,j,r} + \Delta t \sqrt{\left(D^x_{i,j,r}F\right)^2 + \left(D^y_{i,j,r}F\right)^2}$$

where $D^x_{i,j,r}F$ is the finite difference approximation to $F_x(i\Delta x, j\Delta y, r\Delta t)$. The choice of these finite difference operators proves to be the crucial step. Simple central differences like $D^x_{i,j,r}F = \left(F^r_{i+1,j,r} - F^r_{i-1,j,r}\right)/2\Delta x$, do *not* work. Even in case very small time steps Δt are used, stability is not guaranteed.

3.1. Upwind finite difference schemes

Based on the analysis of the PDE (especially the fact that it expresses a conservation law and the fact that we are looking for an entropy solution) several *upwind finite difference schemes* have been presented in the literature. The simplest one is given by Osher and Sethian[7]:

$$F_{i,j,r+1} = F_{i,j,r} + \Delta t\sqrt{\left(D^{-x}_{i,j,r}F\right)^2 + \left(D^{+x}_{i,j,r}F\right)^2 + \left(D^{-y}_{i,j,r}F\right)^2 + \left(D^{+y}_{i,j,r}F\right)^2} \qquad (2)$$

where:

$$D^{-x}_{i,j,r}F = \frac{F_{i-1,j,r} - F_{i,j,r}}{\Delta x} \vee 0, \qquad D^{+x}_{i,j,r}F = \frac{F_{i+1,j,r} - F_{i,j,r}}{\Delta x} \vee 0$$

Equivalent expressions are used for the directed differences in y-direction. A second finite difference scheme to solve the same PDE is due to Rouy and Tourin (as cited in [9]):

$$F_{i,j,r+1} = F_{i,j,r} + \Delta t\sqrt{\left(D^{-x}_{i,j,r}F \vee D^{+x}_{i,j,r}F\right)^2 + \left(D^{-y}_{i,j,r}F \vee D^{+y}_{i,j,r}F\right)^2}. \qquad (3)$$

It is not within the scope of this paper to reproduce the derivation of these upwind difference schemes. Instead in the following section it will be shown that the upwind schemes are in complete accordance with the schemes that implement the sub-pixel morphological operations.

3.2. The Morphological Facet Model

In this section it will be shown that finite difference schemes that are identical to the schemes cited in the previous section, can be easily derived starting from the fact that the PDE is actually solved by a morphological dilation.

In the previous section it was already stated that the operation to derive $F(x, t+\Delta t)$ given $F(x,t)$ is to dilate the function $F(\cdot, t)$ with a disk of radius Δt. In this section we consider the dilation of a function with a disk shaped structuring element of radius $\Delta t < 1$. For these small values of the radius a sampled version of the disk is of no use as only the origin is a grid point. To be able to dilate the sampled function we therefore propose the *morphological facet model*:

1. approximate the discrete data in a small neighborhood with a function, then
2. dilate this function analytically and, finally,
3. sample the dilated function to give the final result.

We use the term morphological facet model because of its resemblance to the facet model which is used to approximate the derivatives of a sampled function (see Haralick[3]).

The function used in the local approximation of the function data has to be chosen in such a way that the desired operation (calculating the derivatives in the classical facet model and dilation in the morphological facet model respectively) can be calculated analytically. Whereas in the classical (linear) facet model the function needs to be differentiable (its sole purpose is to calculate the derivatives), in the morphological facet model a crucial requirement is that the local range of function values is preserved. If this would not be the case then the dilation of the approximated function could result in function values that cannot be "explained" by function values in the sampled data.

The first morphological facet model considered is a degenerated facet model. Instead of fitting a surface to the data points, just the "beams" between the central

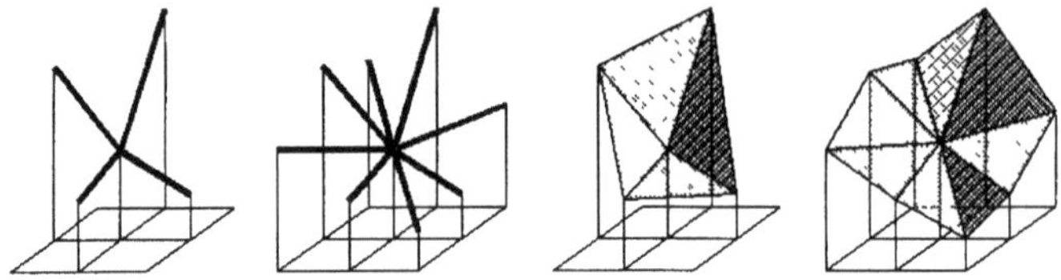

Fig. 3. **Morphological facet models.** a: 4-beam, b: 8-beam, c: 4-plane and d: 8-plane.

data point and the neighboring points are considered. Two beam models are distinguished. The 4-beam model considers only the beams to the 4-connected neighbors. The 8-beam model also considers the beams to the 8 connected neighbors. Both beam models are illustrated in figure 3 a and b. In dilating the facet model, we are only interested in the dilation value in the central pixel. Evidently the final dilation result is the maximum of the dilations of the individual beams. Consider the first beam connecting the central grid point with value $F_{i,j,r}$ with the first neighbor with value $F_{i+1,j,r}$. In case $F_{i+1,j,r} \leq F_{i,j,r}$ the dilation result in the central pixel is $F_{i,j,r}$, i.e. the disk hits the beam in its highest point (the origin). In case the other end of the beam is the highest point (i.e. $F_{i+1,j,r} > F_{i,j,r}$) the dilation value equals $F_{i+1,j,r+dr} = F_{i,j,r} + \Delta t(F_{i+1,j,r} - F_{i,j,r})$. The selection according to the ordering of $F_{i,j,r}$ and $F_{i+1,j,r}$ can be cast elegantly in a maximum operator $F_{i+1,j,r+dr} = F_{i,j,r} + \Delta t\,(0 \vee (F_{i+1,j,r} - F_{i,j,r}))$. An equivalent analysis can be done for all 4-connected neighbors, leading to the following morphological finite dilation scheme:

$$F_{i+1,j,r+dr} = F_{i,j,r} + \Delta t \bigvee_{(k,\ell)\in N_4} (F_{i+k,j+\ell,r} - F_{i,j,r})\,. \tag{4}$$

The beam model is easily extended to take all 8 neighbors into consideration. For the diagonal neighbors a distance correction is needed.

More complex morphological facet models are obtained when interpolating planar surfaces are used as shown in figure 3 c and d. The 4 plane model interpolates the data points with 4 planar function patches, each of them defined in one of the four octants. Dilation of the facet model is then equivalent with the maximum of the 4 dilations of the individual triangular facets. Because in the planar facet model only a small triangular patch of the plane is dilated, we have to make sure that the 'point-of-contact' is indeed within that patch. Let $p^1_{i,j}$ be the planar function in the first quadrant:

$$p^1_{i,j}(x,y) = \begin{cases} (F_{i+1,j,r} - F_{i,j,r})\,x + (F_{i,j+1,r} - F_{i,j,r})\,y & : x \geq 0, y \geq 0, x+y \leq 1 \\ -\infty & : \text{elsewhere} \end{cases}.$$

The dilation result of this patch with a disk of radius t, in the central point is given by:

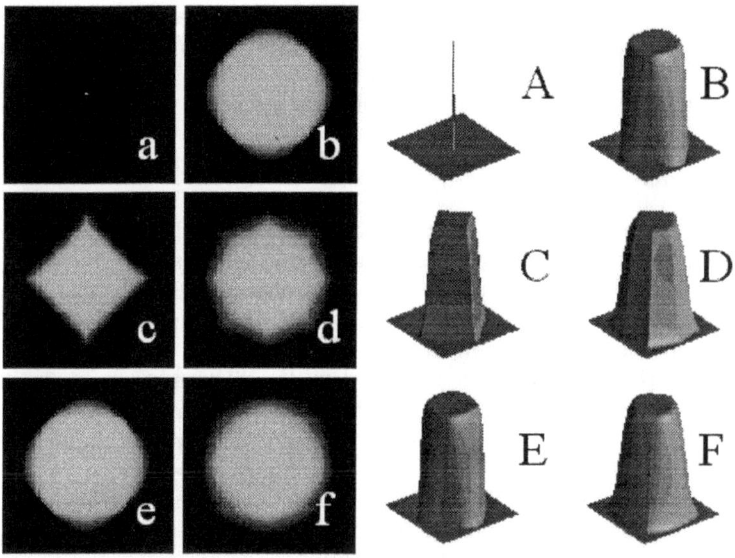

Fig. 4. **Sub-pixel dilation of a pulse.** In a(A) the original 64×64 initial condition of the PDE is shown. The PDE is solved with 5 numerical schemes. In b(B) the R&T scheme is depicted. In c(C) the 4-beam facet model, in d(D) the 8-beam model, in e(E) the 4-plane model and in f(F) the 8-plane model.

$$F_{i,j,r+1} = F_{i,j,r} + \Delta t \sqrt{\left(F_{i+1,j,r} - F_{i,j,r} \bigvee 0\right)^2 + \left(F_{i,j+1,r} - F_{i,j,r} \bigvee 0\right)^2}.$$

The dilation of the entire 4-plane facet model is equal to the maximum of the 4 individual dilations. Careful analysis of the resulting expression reveals that it is equivalent to the Rouy and Tourin scheme (equation 3).

The extension of the 4-plane model to the 8-plane model is straightforward. In this case the 8 planar patches are defined within an octant, making the check to see whether the point-of-contact is within the region of definition more complex.

3.3. Numerical Experiments

The experiment presented in this section is meant to illustrate the concepts developed in previous sections. The morphological numerical schemes are compared with the classical Rouy and Tourin scheme. In figure 4 the experiment is shown in which a pulse (in a 64×64 image) is the initial condition to the PDE given in equation (1). From the figures it is clear that the morphological beam models suffer from severe anisotropy and therefore are of little practical use.

Comparison of the morphological 4-plane and 8-plane models, shows that whereas the 8-plane model is the most isotropic solution, it is so at the cost of being more dissipative (i.e. more 'smoothing' is introduced). An advantage of the 8-plane model is that the scale step Δt can be chosen significantly larger than for the 4-plane model ($\Delta t/\Delta x < \cos\pi/8 \approx 0.92$ compared with $\Delta t/\Delta x < \cos\pi/4 \approx 0.71$ respectively). The 8-plane model therefore can be used with larger time steps, leading to more efficient solutions schemes as fewer iterations are needed.

4. Conclusions

In this paper we have introduced the morphological facet model as an elegant method to implement sub-pixel morphological dilations and erosions and thus to provide a stable numerical scheme to solve the class of morphological PDE's. The morphological numerical scheme based on 4-plane facet model proves to be equivalent with the classical upwind numerical scheme of Rouy and Tourin.

Future research will look at the PDE's where the dilation/erosion is locally controlled by the observed geometry of the shape (i.e. its curvature). The simplest way to use the morphological schemes described in this paper is to use a classical facet model (e.g. bicubic) to observe the local differential geometry (or use Gaussian (fuzzy) derivatives) and calculate the curvature and then to use a morphological facet model to perform the sub-pixel erosion/dilation. A more unified approach is to look for facet models that allow both analytical morphological operations as well as differentiation. It is expected that these facet models will be similar to the ENO schemes known from classical theory.

References

1. L. Alvarez, F. Guichard, P. Lions, and J. Morel. Axioms and fundamental equations of image processing. *Archives for rational mechanics*, 123(3):199–257, 1993.
2. L. Dorst and R. v.d. Boomgaard. The slope transform and morphological signal processing. *Signal Processing*, 38:79–98, 1994.
3. R. Haralick and L. Shapiro. *Computer and robot vision.* Addison and Wesley, 1992.
4. B. Kimia, A. Tannenbaum, and S. Zucker. Shape, shocks and deformations, i. *International journal of computer vision*, 15(3):189–224, 1994.
5. P. Maragos. Max-min difference equations and recursive morphological systems. In J. Serra and P. Salembier, editors, *Mathematical Morphology and its Applications to Signal Processing*, pages 128–133. Kluwer academic publishers, 1993.
6. J. Mattioli. Differential equations of morphological operators. In J. Serra and P. Salembier, editors, *Mathematical Morphology and its Applications to Signal Processing*, pages 162–167, Amsterdam, 1993. Kluwer academic publishers.
7. S. Osher and J. Sethian. Fronts propagating with curvature dependent speed: algorithms based on hamilton-jacobi formulations. *Journal of Computational Physics*, 79:12–49, 1988.
8. G. Sapiro and A. Tannenbaum. Affine invariant scale-space. *International journal of computer vision*, 11:25–44, 1994.
9. J. Sethian. A fast marching level set method for monotonically advancing fronts. *Proc. Nat. Acad. Sci.*, 93(4):1591–1595, 1996.
10. R. v.d. Boomgaard and A. Smeulders. The morphological structure of images, the differential equations of morphological scale-space. *IEEE transactions PAMI*, 16(11):1101–1113, 1994.

AUTO-DUAL CONNECTED OPERATORS BASED ON ITERATIVE MERGING ALGORITHMS

PHILIPPE SALEMBIER, LUIS GARRIDO and DAVID GARCIA
Universitat Politècnica de Catalunya, Barcelona, Spain *
Tel: (343) 401 7404, Fax: (343) 401 6447 Email: philippe@gps.tsc.upc.es

Abstract. This paper proposes a new set of connected operators that are auto-dual. Classical connected operators are analyzed within the framework of merging algorithms. The discussion highlights three basic notions: merging order, merging criterion and region model. As a result, a general merging algorithm is proposed. It can be used to create new connected operators and, in particular, auto-dual operators. Implementation issues are also discussed.

1. Introduction

Connected operators [5, 2, 4] are becoming popular because they simplify images without corrupting the shape information. It has been shown [8] that they interact with the image by merging the zones of the space where the signal is constant (called *flat zones*). As a result, they do not introduce new contours. These filters are extensively used in image analysis and segmentation applications [10, 3, 6, 1].

A large number of segmentation approaches, such as the *Split&Merge*, the *region growing* or the *watershed*, also rely on merging strategies. They have been created in a different context from that of connected operators and for different applications. Connected operators have been designed as a filtering tool whereas the others are devoted to segmentation. However, they all rely on the same fundamental merging process. The objective of this paper is to investigate the basic features of the merging process and to see how the connected operators can benefit from the "segmentation viewpoint". As major output of this study, we will create auto-dual connected operators. Moreover, the operators implementation will be discussed.

The organization of this paper is as follows. Section 2 reviews classical *connected operators* and a general merging strategy is proposed in section 3. An efficient implementation is discussed in section 4. Based on the general merging process, section 5 defines new auto-dual connected operators.

2. Classical connected operators

The most well known connected operators are the *opening by reconstruction* and the *area opening*. They are anti-extensive operators. In practice, this means that they only deal with bright components. Let us describe the filtering process on the Region Adjacency Graph (RAG) of the flat zones partition (see Fig. 1 where the numbers indicate the gray level values of the flat zones).

* This work has been supported by France Telecom under the contract 96ME22.

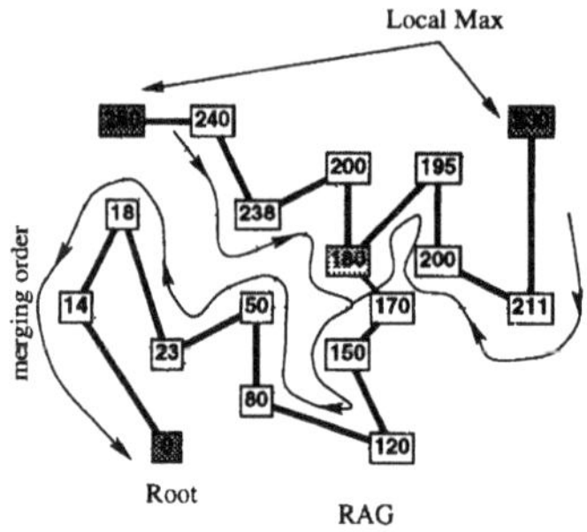

Fig. 1: RAG of flat zones

First, the regional maxima are computed. Then, the operator assesses a given criterion for each connected component of the regional maxima (for an area opening [9], the criterion is the number of pixels). If the criterion value is higher than a given threshold, the connected component is preserved, otherwise it is merged with the flat zone(s) of closest lower gray level value. This procedure is repeated on the new connected component resulting from the merging until the criterion value becomes higher than the threshold.

Note that the merging order is defined by the absolute gray level value of the flat zones. The first pair of flat zones that is studied is the pair involving the highest gray level value. Then, progressively, flat zones of lower gray level values are processed.

Let us now discuss a region growing process. In this case, an initial set of regions is defined. Then, based on a similarity measure such as the mean gray level difference, neighboring pixels are progressively merged to create regions. Note that the merging order is not known at the beginning of the process. Indeed, after each merging, the algorithm has to look for the neighboring pixel of a region that minimizes its distance to the region. However, when one pixel is merged with a region, the distance between this region and the remaining neighboring pixels may change. As a result, the merging order can only be defined step by step during the merging process itself. This merging process is iterated until a termination criterion is reached.

Let us note that the algorithm used for connected operators has strong similarities with a region growing process. It also involves an initialization step defining a set of initial regions (regional maxima) and it also involves merging steps. There are, however, some important differences:

1. The merging order is known a priori in the case of connected operators (gray level value) and not in the case of region growing (on-line distance computation).
2. The second difference is that connected operators make use of a criterion to decide whether a given merging has to be done or not (the size of the connected components in our previous example). In fact, the merging order defines the order in which the pairs of flat zones are analyzed. However, each pair of flat zones can be merged or not. The region growing algorithm decides to merge all pairs of regions that are proposed by the merging order (until a termination criterion is reached). In contrast, the connected operator algorithm does not merge all pairs of flat zones but select some of them. This difference is rather natural, since a filter is a sieving tool, that is, it selects regions or flat zones with a given and specific characteristic.
3. Finally, in the case of classical connected operators, when two flat zones are merged, the lowest gray level value is assigned to resulting connected component. This assignation strategy means that there is an implicit modeling: each flat zone is represented by the lowest gray level value of its pixels in the original image. In the case of a region growing algorithm, the mean is generally used as region model.

3. General merging algorithm

Taking into account the interesting features of region growing and connected operators, let us define a general merging strategy. The proposed algorithm works on the RAG of flat zones which is a set of nodes representing flat zones and a set of links connecting two neighboring nodes. A merging algorithm on this graph is simply a technique that removes some of the links and merges the corresponding nodes. In the sequel, we assume that the merging is done in an iterative way. To completely specify a merging algorithm one has to define three notions:

1. The **merging order**: it defines the order in which the links are studied to know whether or not they should disappear. This order $\mathcal{O}(R_1, R_2)$ is a real value and is a function of each pair of neighboring flat zones R_1, R_2.
2. The **merging criterion**: each time a link is analyzed, the merging criterion decides if the merging has actually to be done or not. It is also a function $\mathcal{C}(R_1, R_2)$ of two neighboring flat zones R_1 and R_2, but it can only take two values ("merge" and "do not merge").
3. The **region model**: when two flat zones are merged, the model defines how to represent the union. Let us denote by $\mathcal{M}(R)$ this model.

In the case of opening by reconstruction or area opening, the merging order and the model are not auto-dual, that is, they do not commute with a modification of the gray level values of the type: $f \rightarrow -f$. In section 5, new auto-dual connected operators will be created by using this algorithm and by selecting an auto-dual merging order and an auto-dual model.

4. Efficient implementation

4.1. Merging algorithm on a RAG structure

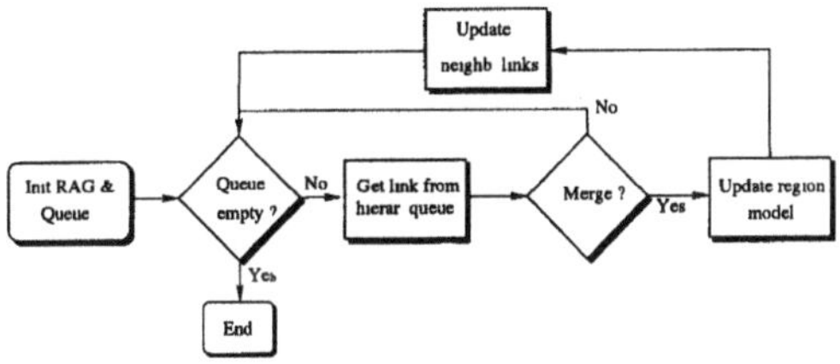

Fig. 2: Merging Algorithm

This section is devoted to the implementation of the merging algorithm described in section 3. The general scheme presented in Fig. 2 can be divided into two stages.

- "Initialization" of the RAG structure and of the hierarchical queue (see section 4.2). The first step consists in calculating the merging order for each pair of neighboring flat zones (called *link* in the following). Then, each link is inserted in the hierarchical queue with a priority defined by the merging order.

- "Merging" process: the algorithm extracts the link of highest priority from the queue. The next step decides whether the pair of flat zones corresponding to this link have to be merged or not. This decision relies on the "merging criterion". If the merging criterion decides not to merge the flat zones, the algorithm returns to the first step of the algorithm[1]. If the merging criterion decides to merge, the next step computes the region model of the union. To avoid redundant computations, a recursive algorithm is needed. "Recursive" here means that the model of the union should be computed from the models of the two initial flat zones. Once the flat zones have been merged, the values of the merging order of the neighboring links are updated. This implies the extraction of the corresponding links from the queue, the computation of the new priority and the insertion of the links in the queue with their new priority. At this point the merged RAG has been computed and updated: the iterative process starts again by checking if the queue is empty.

The key element of the implementation is a queue. In our context, its main features should be: first, fast access, insertion and deletion of an element and, second, no constraints on the dynamic range of the priority (floating point ordering).

4.2. Hierarchical queues and binary trees

Queues and hierarchical queues have been extensively used for fast implementation of morphological operators [10, 4]. For our merging algorithm, the main issue is the merging order that has to be constantly updated. In fact, the hierarchical queue should be updated and re-organized on-line. It can be seen as a dynamic hierarchical queue by contrast to the more classical "static" hierarchical queues. Another drawback of the classical hierarchical queue is that it is generally limited in the range of priority it can deal with and does not allow a floating point priority.

The solution proposed here is based on a binary tree. The basic idea of the hierarchical queues implementation with binary trees is depicted in Fig. 3. Each tree node represents a given priority. Furthermore, the sub-tree hanging on the left (right) side of each node involves nodes of lower (higher) priority. At each node, the links having the same priority are stored in a FIFO structure.

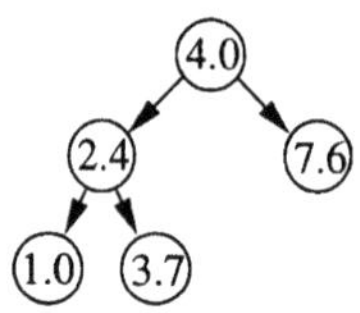

Fig. 3: Binary tree as a queue

To extract the links with highest priority, one begins with the root node and walks down the tree using the right branches until a node with no right branch is found. Search, insertion and deletion of nodes in the queue can be done in $O(log_2 N)$ steps where N is the number of nodes in the tree [11].

One of the drawbacks of using binary trees to implement hierarchical queues is that the tree may degenerate. This problem happens when the merging order is not random. If, for example, the merging order is increasing, the resulting tree "leans" too much to the left and it turns into a simple FIFO queue. In this case, the $O(log_2 N)$ efficiency does not hold anymore. Fortunately, there is a solution to this problem called "balanced trees". Efficient implementations of insertion and deletion preserving the tree balance can be found in [11]. The efficiency of the algorithm turns out to be very high. In practice, the computation time is about the same as the one for an anti-extensive connected operator or a watershed algorithm. For instance, the CPU time is about 1 second for a QCIF image on a 200MHz Pentium.

[1] Note that this decision is final in the sense that the corresponding pair of flat zones will never be merged.

5. Auto-dual connected operators

The objective of this section is to present new connected operators. As discussed in the introduction, one of our motivations is to create auto-dual operators. With the structure defined in sections 3 and 4, it is rather easy to create new connected operators that are auto-dual. To this end, one has to define the merging order $\mathcal{O}(R_1, R_2)$, the merging criterion $\mathcal{C}(R_1, R_2)$, and the region model $\mathcal{M}_R(x)$ in such a way that the resulting operator be auto-dual.

5.1. Region model: $\mathcal{M}_R$

The first choice that has to be made is how each flat zones is going to be modeled. For instance, in the case of an opening (a closing), each flat zone is modeled by the lowest (highest) gray level value of its pixels in the original image. Obviously, this model does not allow the creation of auto-dual operators. Two simple auto-dual models are the mean and the median. In order to allow a fast implementation of the merging process, the model of region $R = R_1 \cup R_2$ should be computed recursively from the models of the two merged regions R_1 and R_2. In the case of the mean model, one has to compute the weighted average of the mean of each flat zone:

$$\mathcal{M}_R = (N_1 \mathcal{M}_{R_1} + N_2 \mathcal{M}_{R_2})/(N_1 + N_2) \tag{1}$$

where N_1 and N_2 denote the number of pixels of each region. In the case of the median, one has simply to select the model of the largest flat zone [2]:

$$\begin{aligned} &\text{if } N_1 < N_2 \Longrightarrow \mathcal{M}_R = \mathcal{M}_{R_2}, \text{ if } N_1 > N_2 \Longrightarrow \mathcal{M}_R = \mathcal{M}_{R_1}, \text{ and} \\ &\text{if } N_1 = N_2 \Longrightarrow \mathcal{M}_R = (\mathcal{M}_{R_1} + \mathcal{M}_{R_2})/2 \end{aligned} \tag{2}$$

The median is known to be a robust estimator and, from our practical experience, it is indeed more robust than the mean. In fact, it rapidly defines areas where the model is stable. These areas can be seen as the core of the final flat zones. Then, small flat zones are progressively merged to the core without modifying the model. This property does not hold in the case of the mean. In the sequel, we assume that the median is used.

5.2. Merging order: $\mathcal{O}(R_1, R_2)$

The merging order defines the notion of objects. It can be seen as a measure of the likelihood that two neighboring flat zones belong to the same object. A natural measure is the squared error between the original image and the model in the region of support defined by $R = R_1 \cup R_2$: $\mathcal{O}(R_1, R_2) = \sum_{x \in R_1 \cup R_2} (f(x) - \mathcal{M}_{R_1 \cup R_2})^2$, if $f(x)$ denotes the gray level value of the original image at position x. This criterion actually allows the extraction of meaningful objects that are homogeneous in gray level, however it does not define precisely the contours. This drawback is related to the size dependence of the criterion. Indeed, small flat zones have a tendency to merge together since the squared error contribution of the union of two small flat zones is small compared to the contribution resulting from the merging with a large

[2] Note that the resulting model is not exactly the median of the original pixels since the median is computed in an iterative way.

flat zone. An alternative solution is to use the difference between the models of the two neighboring flat zones: $(\mathcal{M}_{R_1} - \mathcal{M}_{R_2})^2$ or the Mean Square Error (MSE): $\sum_{x \in R_1 \cup R_2} (f(x) - \mathcal{M}_{R_1 \cup R_2})^2/(N_1 + N_2)$. These criteria are size independent and provide a very good definition of the contours, however, they do not define in a robust way the flat zones themselves. In practice, they produce a few large flat zones surrounded by a very large number of tiny flat zones. To obtain a compromise between the two previous orders, we propose the following merging order:

$$\mathcal{O}(R_1, R_2) = N_1(\mathcal{M}_{R_1} - \mathcal{M}_{R_1 \cup R_2})^2 + N_2(\mathcal{M}_{R_2} - \mathcal{M}_{R_1 \cup R_2})^2 \qquad (3)$$

In the case of the median model and if R_1 is smaller than R_2, the model after merging is $\mathcal{M}_{R_2}$ and the order reduces to $\mathcal{O}(R_1, R_2) = N_1(\mathcal{M}_{R_1} - \mathcal{M}_{R_2})^2$. It is the squared difference between the models weighted by the size of the smallest flat zone.

5.3. Merging criterion: $\mathcal{C}(R_1, R_2)$

The function $\mathcal{O}(R_1, R_2)$ defines the order in which the flat zones have to be processed. Now, the objective of the merging criterion is to select among the set of possible flat zones a few that fulfill a given criterion. In the following, we illustrate two examples: area and PSNR.

Area merging criterion: The objective of this criterion is to remove from the original image all flat zones that are smaller than a given threshold. To this end, the criterion simply states that two flat zones have to be merged if at least one of them is smaller than a given threshold (the size is defined by the number of pixels).

The effect of the operator can be seen in Fig. 4. When the area threshold is set to 10 pixels, the resulting image contains most of the original information except the very small regions. For example, the texture of the fish has been removed. The process is a connected operator and the contour information of the objects that have not been removed is well preserved. Note that the image is made of flat zones of at least 10 pixels. The simplification effect can also be seen in the number of flat zones. When the area threshold is increased to 100 pixels, the same kind of effect can be observed. Note that the simplification effect of this area operator is different from that of an area opening or closing. The operator proposed here not only deals with bright or dark areas but also with transition areas.

Fig. 4: Auto-dual area connected operator

PSNR merging criterion: for simplification or even segmentation, an interesting criterion is the Peak Signal to Noise Ratio (PSNR) between the original image and the modeled one. It is in fact a termination criterion which defines when to stop the merging process. The merging strategy defined in section 3 is particularly useful to compute pyramids of operators. Starting from the flat zone partition, several merging steps are performed. Note that with the implementation of section 4, it is possible to compute the full pyramid with only one run of the merging algorithm. Indeed, one has only to output some of the partitions created at intermediate stages of the merging process. This approach is therefore an efficient way to compute an entire hierarchy of partitions. An example of pyramid is shown in Fig. 5.

PSNR: 30dB (136 fz) PSNR: 28dB (29 fz) PSNR: 26dB (12 fz)

Fig. 5: Pyramid of PSNR connected operators

Let us discuss a potential use of the merging algorithm for *granulometric* analysis of images [7]. The approach consists in using a hierarchical filtering structure (classically, morphological opening) and in measuring an image characteristic (classically the integral of the image) at each filter output. This strategy allows the characterization of what has been removed by each filter. Consider now the PSNR connected operator and assume that PSNR values ranging from 40dB to 25dB are assigned to 16 levels of the hierarchy. Finally, suppose that we measure the number of flat zones of each processed image. The results are shown in Fig. 6. Three typical frames (frames #50, #200 and #270) of the *Foreman* sequence have been processed and the three corresponding curves are shown. These curves define how many flat zones are necessary to achieve a given PSNR. Intuitively, this number is function of the image "complexity": for simple frames with a few objects (homogeneous in gray level), the number of flat zones necessary to achieve a given PSNR is rather low. On the contrary, if the image involves a large number of contrasted objects, a high number of flat zones are necessary to reach the same PSNR. This intuitive discussion is confirmed by Fig. 6: among the three frames, the simplest one is the frame #200. It involves a few objects and the sky covers most of the frame. The corresponding curve is the lowest one. By contrast, the most complex frame is frame #270. It involves a fairly high number of contrasted objects and the corresponding curve is the highest one. Finally, frame #50 is of intermediate complexity and produces a curve in-between the two previous ones. Each curve (or a reduced set of curve points) can be used to characterize the image complexity.

6. Conclusions

This paper has focussed on the class of auto-dual connected operators based on merging techniques. The operator definition relies on three notions: first, the *merging order* that defines the notion of object homogeneity, second, the *merging criterion* that characterizes the set of flat zones we are interested in and, third, the *region model* that define how flat zones are represented.

The efficient implementation of a merging operator relies on a good management of the merging order. At each time instant, one should have, first, a very fast access to the next pair of flat zones to merge, and, second, an easy way to modify the various merging order values as the merging process goes on. The key element for a fast implementation is a dynamic hierarchical queue. A very efficient solution to this queue problem is to use a balanced binary tree.

Using this philosophy and implementation, new auto-dual connected operators based on area and on PSNR have been proposed and illustrated. These operators are particularly useful as simplification, pre-processing and segmentation tools.

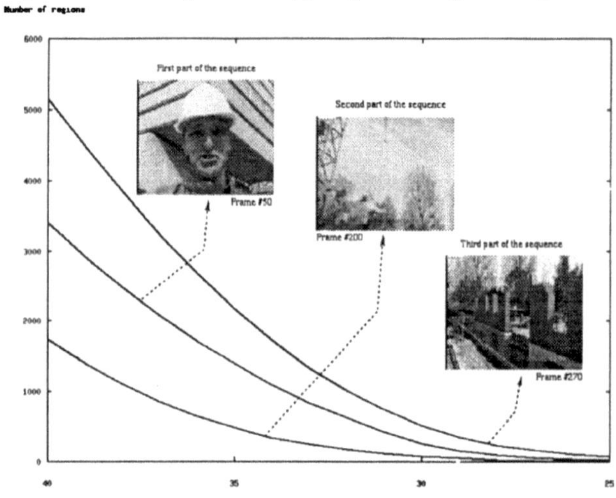

Fig. 6: Granulometric analysis with the PSNR connected operator

References

1. E. Breen and R. Jones. An attribute-based approach to mathematical morphology. In P. Maragos, R.W. Schafer, and M.A. Butt, editors, *International Symposium on Mathematical Morphology*, pages 41–48, Atlanta (GA), USA, May 1996. Kluwer Academic Publishers.
2. J. Crespo, J. Serra, and R.W. Schafer. Theoretical aspects of morphological filters by reconstruction. *Signal Processing*, 47(2):201–225, 1995.
3. B. Marcotegui, J. Crespo, and F. Meyer. Morphological segmentation using texture and coding cost. In I. Pitas, editor, *1995 IEEE workshop on Nonlinear Signal and Image Processing*, pages 246–249, Halkidiki, Greece, June 20-22 1995.
4. P. Salembier, A. Oliveras, and L. Garrido. Anti-extensive connected operators for image and sequence processing. *IEEE Transactions on Image Processing*, April 1998.
5. P. Salembier and J. Serra. Flat zones filtering, connected operators and filters by reconstruction. *IEEE Transactions on Image Processing*, 3(8):1153–1160, August 1995.
6. P. Salembier, L. Torres, F. Meyer, and C. Gu. Region-based video coding using mathematical morphology. *Proceedings of IEEE (Invited paper)*, 83(6):843–857, June 1995.
7. J. Serra. *Image Analysis and Mathematical Morphology*. Academic Press, 1982.
8. J. Serra and P. Salembier. Connected operators and pyramids. In SPIE, editor, *Image Algebra and Mathematical Morphology*, volume 2030, pages 65–76, San Diego (CA), USA, July 1993.
9. L. Vincent. Grayscale area openings and closings, their efficient implementation and applications. In J. Serra and P. Salembier, editors, *First Workshop on Mathematical Morphology and its Applications to Signal Processing*, pages 22–27, Barcelona, Spain, May 1993. UPC.
10. L. Vincent. Morphological gray scale reconstruction in image analysis: Applications and efficients algorithms. *IEEE, Transactions on Image Processing*, 2(2):176–201, April 1993.
11. N. Wirth. *Algorithms & Data Structures*. Prentice-Hall, 1986.

FROM CONNECTED OPERATORS TO LEVELINGS

FERNAND MEYER
Centre de Morphologie Mathématique - Ecole des Mines de Paris
35, rue Saint-Honoré - 77305 Fontainebleau (FRANCE)

Abstract. Applying a connected operator to a grey-tone image may create a regional minimum or maximum. This paper studies two subclasses of connected operators which do not create extrema: the monotone planings and the flattenings. For the monotone planings, each regional minimum of the transformed image contains a regional minimum of the original image. The intersection of these subclasses form the levelings. Flattenings and levelings are basically symmetrical extensions of the openings and closings by reconstruction.

Key words: Connected operators, planings, monotone planings, flattenings levelings, reconstruction openings and closings

1. Introduction

Object oriented compression is based on an extremely important step: the segmentation of an image sequence. There is no preliminary knowledge of the image content available ("generic image sequences" in the MPEG4 requirements) and the segmentation has to be unsupervised. Furthermore, the rate control of the coding scheme imposes to adjust the number of regions of the segmentation to the available bitstream. Hence the classical monolevel morphological segmentation paradigm [5] does not apply. A multiscale segmentation has to take place. In order to facilitate the optimal allocation of coding methods to the various regions of the image, it is best to construct a segmentation pyramid: the lowest level represent the finest mosaic image (with the highest number of regions); from one level to the next, only fusions take place. A good way to obtain such a pyramid is by repeating the watershed construction on the same gradient image, but with a decreasing number of markers: each step will produce a new mosaic image, with as many tiles as there were markers. The problem will be solved, if we know which markers we may skip from one level to the coarser level above. A solution consists in applying to the gradient image a family of reconstruction closings of increasing size: for each new reconstruction closing, some regional minima will be filled and disappear as regional minima. One may then rank all minima of the gradient image by the index of the reconstruction closing for which they vanish for the first time. Using h-closings (the reconstruction closing uses a marker image obtained by adding a constant value to the reference image), we get the ranking by the dynamics introduced by M. Grimaud [1]. Corinne Vachier and Fernand Meyer have extended this concept to other classes of connected filters, namely the areal and volumic closings [11]. Using one class of closings or another yields different pyramids, h-closings stressing the contrast, areal closings the size. If one desires to obtain other types of pyramids, one has to apply a different series of filters to the gradient image. As we have seen, the ranking of the minima in a gradient image works because the reconstruction filters possess an important feature: they never create a regional minimum. To each family of filters having this property, we will be able to associate a segmentation pyramid. Hence the question which is at the origin of the present work: which is the most general family of filters which allow to construct a segmentation pyramid, if one applies the mechanism presented above. Such filters should never create a regional minimum; if this is granted, we will be able to trace the minima all along the filtered versions of the gradient image and label each of them by the index of the filter for which it vanishes. As openings and closings by reconstruction are connected filters (there exists a large literature on connectivity with respect to image processing: for a general analysis of the connectivity: [9], [8] chapter 2 and 7; as binary operators: [6, 2]; for the definition of connected operators for grey-tone images: [10] and for their use in filtering: [7]) one may wonder if the connected operators are not precisely the most general class we are looking for. All connected filters have in common an extremely attractive feature: they preserve the flat zones of the image, but as such they may turn a minimum into a maximum, and hence, we will have to define a subclass of connected operators. This will lead us to the definition of two specialisations of connected operators, first monotone planings, and second flattenings. Combining both specialisation, we will get the levelings, which are by far the most interesting. Levelings have been introduced in [4]; their definition has been enlarged and their algebraic properties extensively studied in [3].

2. General definitions

We will present the notions, in a purely digital framework. For an analysis of the continuous case, see [4, 3]. Let G be the underlying grid, which can be of any type, square or hexagonal in 2 dimensions, cubic, centred cubic or face-centred cubic in 3 dimensions. G is a subset of $Z^n \times Z^n$. We call $N_G(p)$ the set of neighbors of a pixel p, with respect to the grid G. We suppose that $N_G(p)$ is translation invariant and symmetrical versus the point p. In other words:

$$(p, p') \text{ neighbors} \quad \Leftrightarrow \quad (p', p) \text{ neighbors}$$

A path P of cardinal n between two pixels p and q on the grid G is an n-tuple of pixels $(p_1, p_2, ..., p_n)$ such that $p_1 = p$ and $p_n = q$, and for all i, (p_i, p_{i+1}) are neighbors. Based on these paths, we will be able to define connected components: a set X is arcwise connected if any pair of pixels (x, y) of X is linked by a path, entirely included in X.

2.1. Specializing connected operators

Since we are dealing with discrete images defined on a grid, all connectivity relations may be checked just by considering what happens to the function between two adjacent pixels. The connected operators have been introduced by Salembier and Serra in [10].

Definition 1 *A connected operator maps an image f into an image g in such a way that the following relation is verified for all pairs of neighboring pixels :*

$$\forall (p, q) \text{ neighbors: } f_p = f_q \quad \Rightarrow \quad g_p = g_q$$

If a function g and a function f verify this condition for all pairs of pixels, we say that by definition g is a planing of f. The condition in the definition may be formulated in an equivalent manner as:

$$g_p \neq g_q \quad \Rightarrow \quad f_p \neq f_q$$

meaning that if there is a transition between the pixels p and q in the destination image g, a transition also has to occur between the same pixels in the original image f. There is however no coupling of the directions of the transitions: between p and q, the function g may for instance be increasing and f decreasing. As an example, the complementation of a binary image is a connected operator. This shows that a connected operator may turn a regional minimum into a maximum and vice-versa. Of course, complementation is not an increasing operation. One may wonder, if an increasing connected operator would not verify our required property: never to create a regional minimum or maximum. In [4], we have given an example, showing that this is not the case. For this reason we will try to reinforce the condition

$$g_p \neq g_q \quad \Rightarrow \quad f_p \neq f_q$$

We will do it in two independent and complementary ways.

The first specialisation leads to the flattenings.

Definition 2 *An image g is a* flattening *of the image f iff $\forall\,(p,q)$ neighbors:*

$$g_p > g_q \quad \Rightarrow \quad \begin{bmatrix} f_p \geq g_p \text{ and } g_q \geq f_q \\ \text{or} \\ f_q \geq g_p \text{ and } g_q \geq f_p \end{bmatrix}$$

or equivalently as a symmetric expression:

$$g_p \neq g_q \Rightarrow f_p \wedge f_q \leq g_p \wedge g_q < g_p \vee g_q \leq f_p \vee f_q$$

Basically the previous definition means that any transition in the destination image g is bracketed by a larger variation in the source image. Flattenings introduce a coupling between the values of the function f and the function g, whereas simple planings have no such coupling. A different specialisation of planings will ignore the coupling of values but will introduce a coupling of the sense of variation. This leads us to the arcwise connected operators, or monotone planings.

Definition 3 *An image g is a* monotone planing *of the image f iff $\forall\,(p,q)$ neighbors:*

$$g_p > g_q \quad \Rightarrow \quad f_p > f_q$$

The last relation being equivalent to $\forall\,(p,q)$ neighbors: $f_p \leq f_q \quad \Rightarrow \quad g_p \leq g_q$, a property we may describe as "arcwise increasing". Combining the two previous strengthenings, we get the levelings, where we have both a coupling of the values and of the senses of variations.

Definition 4 *An image g is a* leveling *of the image f iff $\forall\,(p,q)$ neighbors:*

$$g_p > g_q \quad \Rightarrow \quad f_p \geq g_p \text{ and } g_q \geq f_q$$

Recalling that the logical meaning of $[A \Rightarrow B]$ is $[notA$ or $B]$ we easily derive a criterion for a function g to be a flattening : $f \wedge \delta\,(f \wedge g) \leq g \leq f \vee \varepsilon\,(f \vee g)$ or a leveling : $f \wedge \delta g \leq g \leq f \vee \varepsilon g$.

2.2. Basic properties

2.2.1. *Binary planings*

In what follows we call binary planing a planing applied to a binary image. We say that a set X is a planing of a set Y iff the indicator function of X is a planing of the indicator function of Y. The following properties are easy to establish. If we restrict ourselves to binary images, we have the following properties:

- every binary planing is a flattening.
- every binary monotone planing is a leveling.

2.2.2. *Examining the thresholds.*

Defining the lower threshold of a function f as: $X^t(f) = \{x \mid f(x) \leq t\}$ we may state the following propositions.

Proposition 1 *A function g is a flattening of a function f if and only if for any threshold value t, $X^t(g)$ is a planing of $X^t(f)$.*

Proposition 2 *A function g is a leveling of a function f if and only if for any threshold value t, $X^t(g)$ is a monotone planing of $X^t(f)$.*

2.3. Preorder and order relations

The notions introduced so far may be considered as binary relations between functions: as an example we say that the leveling relation holds between two functions g and f, if g is a leveling of f. In this section we will consider the planing relation, the monotone planing relation, the flattening relation, and the leveling relation. These relations are reflexive: obviously any function f is leveling, hence a flattening, a monotone planing or a planing of itself. It is easy to check, that they are also transitive.

Proposition 3 *Being reflexive and transitive, all relations under consideration are preorder relations.*

Remark 1 *With the exception of the levelings, they are not anti-symmetric, hence they are not order relations. For the levelings however, we have a stronger property.*

Proposition 4 *The family of levelings, from which we exclude the trivial constant functions, verify the anti-symmetry relation: if f is a non constant function and a leveling of g, and simultaneously g is a leveling of f, then $f = g$.*

3. The monotone planings

The monotone planings are not extremely interesting in practice, as the values of the functions are not coupled, but only their sense of variation. However, they will inherit from this loose form of coupling the property we are looking for: they do not create regional minima or maxima. And as a consequence, levelings, which are far more interesting, being a particular kind of monotone planings will share the same property.

Theorem 5 *A monotone planing does not create regional minima or maxima. In other words, if g is a monotone planing of f, and if g has a regional minimum (resp. maximum) X, then f possesses a regional minimum (resp. maximum) $Z \subset X$.*

Hint of the proof: If X is a regional minimum of g all its neighbors have a higher altitude: if (p, q) are neighbors such that $p \notin X$ and $q \in X$, then $g_p > g_q$. This implies that $f_p > f_q$. It is then easy to show that the lowest pixel for f within X belongs to a regional minimum Z for f included in X.

4. The flattenings

The flattenings on the other hand are not as well behaved as the monotone planings with respect to the regional extrema. They do not however create a regional extremum from nothing, but may turn a regional minimum into a regional maximum:

Proposition 6 *Let g be a flattening of the function f. Then if X is a regional extremum for the function g, there exists a set $Y \subset X$ such that Y is a regional extremum of the function f. It may be a regional maximum or a regional minimum.*

4.1. Extensivity $\{g > f\}$ and anti extensivity $\{g < f\}$ domains of flattenings

If a function g is a flattening or a leveling of a function f, we call the set $\{g > f\}$ its extensivity domain and the set $\{g < f\}$ its anti extensivity domain. Suppose now that two neighboring pixels (p, q) verify $f_p < g_p$ and $f_q < g_q$. For such a pair of pixels the second part of the following implication which defines the flattenings is not verified:

$$g_p > g_q \quad \Rightarrow \quad \begin{bmatrix} f_p \geq g_p \text{ and } g_q \geq f_q \\ \text{or} \\ f_q \geq g_p \text{ and } g_q \geq f_p \end{bmatrix}$$

This means that for such a pair of pixels the hypothesis also is not verified: $g_p \leq g_q$. The pixels p and q playing symmetrical roles we also have $g_q \leq g_p$. This implies that each connected particle on which $f < g$ is necessarily flat. The same holds naturally also for the connected particles where $f > g$. Not only the flattenings does not create new contours, but the zones where they change the values of the function f are necessarily flat.

4.2. Flattenings and openings/closings by reconstruction

The following theorem establishes the link between flattenings, which until now are simply defined by relations between neighboring pixels and two fundamental transformations of mathematical morphology, namely the openings and closings by reconstruction. We recall that a function g is an opening by reconstruction of f iff $g \leq f$ and $g = g \wedge \delta g$; in such a case we write $g = \text{Rec}(g, f)$. By duality, a closing by reconstruction is characterized by $g \geq f$ and $g = g \vee \varepsilon g$; in such a case we write $g = \widetilde{\text{Rec}}(g, f)$.

Theorem 7 *We have an equivalence between the three following properties:*
1) The function g is a flattening of a function f; 2) $f \vee g$ is a closing by reconstruction

of f and $f \wedge g$ is an opening by reconstruction of f; 3) $f \vee g$ and $f \wedge g$ are levelings of f.

This theorem has the following corollaries:

Corollary 8 *Any flattening g of a function f which verifies $g \leq f$ (resp. $g \geq f$) is a leveling of f and an opening (resp. closing) by reconstruction of f*

Corollary 9 *Any opening (resp. closing) by reconstruction g of f is a leveling.*

As leveling, the openings and closings by reconstruction are monotone planings. Hence in each regional minimum of an opening or closing by reconstruction, there is a regional minimum of the original function.

Since the flattenings are closed under supremum and infimum, we may exhibit a rare case where the infimum of two openings still is an opening:

Corollary 10 *Let g and h be two openings by reconstruction of a function f, then $g \vee h$ and $g \wedge h$ also are openings by reconstruction of f.*

Since each flattening g of a function f is a leveling on $\{f \geq g\}$ and on $\{f \leq g\}$, we have to check what happens on the transition between these two domains, in order to find a necessary and sufficient condition for a flattening to be a leveling:

Proposition 11 *Let g be a flattening of a function f. The function g is a leveling of f if and only if the following condition holds:*

$$\forall (p,q) \text{ neighbors: } \begin{matrix} f_p > g_p \\ f_q < g_q \end{matrix} \Rightarrow g_q \leq g_p$$

5. Construction of a few flattenings which are filters

In what follows, a morphological filter is an idempotent and increasing operator ([8], p.115). We also use the activity order relation ([8], p.159) : we say that an operator ψ_1 is more active than an operator ψ_2 and write $\psi_1 \succ \psi_2$ iff $\begin{matrix} f \vee \psi_1 f \geq f \vee \psi_2 f \\ f \wedge \psi_1 f \leq f \wedge \psi_2 f \end{matrix}$

5.1. The most active flattening, given its extensivity and anti-extensivity domain

Let X and Y be two sets such that $X \cap Y = \emptyset$. We associate to any function f a new function $\Phi_Y^X(f)$ defined as follows.We will construct a marker function $\gamma_0 f$ such that by reconstruction we get $\gamma f : \gamma f = \text{Rec}(\gamma_0 f, f)$:

- on $\overline{X}$ we define $\gamma_0 f = f$
- on X we define $\gamma_0 f = -\infty$ (or the minimal value ot the destination lattice)

The reconstruction $\gamma f = \text{Rec}(\gamma_0 f, f)$ verifies:

$$\gamma f \leq f \text{ on } X \quad \text{and} \quad \gamma f = f \text{ on } \overline{X}$$

Let us write $\Phi^X(f)$ for this opening by reconstruction associated to any function f and any set X.

Similarly we construct a marker function $\varphi_0 f$ such that by reconstruction closing, we get $\varphi f : \varphi f = \widetilde{\text{Rec}}(\varphi_0 f, f)$:

- on $\overline{Y}$ we define $\varphi_0 f = f$
- on Y we define $\varphi_0 f = \infty$ (or the maximal value ot the destination lattice)

The reconstruction $\varphi f = \widetilde{\text{Rec}}\,(\varphi_0 f, f)$ verifies:

$$\varphi f \geq f \text{ on } Y \quad \text{and} \quad \varphi f = f \text{ on } \overline{Y}$$

Let us write $\Phi_Y(f)$ for this closing by reconstruction associated to any function f and any set Y.

According to the preceding, the function $\Phi_Y^X(f)$ defined as follows:

$$\Phi_Y^X(f) = \begin{cases} \gamma f & \text{on} \quad X \\ \varphi f & \text{on} \quad Y \\ f & \text{on} \quad \overline{X \cup Y} \end{cases}$$

is a flattening.

Proposition 12 *For a fixed X and Y, the mapping $f \mapsto \Phi_Y^X(f)$ is a flattening and a morphological filter.*

Proposition 13 $\begin{matrix} X_1 \subset X_2 \\ Y_1 \subset Y_2 \end{matrix}$ *implies* $\Phi_{Y_2}^{X_2}(f) \succ \Phi_{Y_1}^{X_1}(f)$.

The following series of pictures illustrates the operator $\Phi_Y^X(f)$. On the left the initial image. In the center, the partition (X, Y), X representing the zone where an opening by reconstruction takes place, and Y the zone where a closing by reconstruction takes place. We first construct in parallel a reconstruction opening γ and a reconstruction closing φ of the initial image f. We call X' the domain $\{\gamma f < f\}$ and Y' the domain $\{\varphi f > f\}$. We obtain the tesselation (X, Y) by applying a morphological segmentation to the initial image : the image to be flooded is the gradient of the original image and the markers are X'/Y' and Y'/X'. On the right, the resulting leveling $\Phi_Y^X(f)$.

5.2. Where the extensivity and anti-extensivity domains are obtained by comparing two functions

Let g be a function. We associate to any function f a new function $\Xi_g(f)$ defined as follows.

We define the anti-extensivity and extensivity sets as:

$$X_f^g = \{x \mid f(x) < g(x)\}$$
$$X_g^f = \{x \mid f(x) > g(x)\}$$

Proposition 14 *For a fixed g, the mapping $f \mapsto \Xi_g(f) = \Phi_{X_g^f}^{X_f^g}(f)$ is a flattening of f and a morphological filter.*

Proposition 15 *For a fixed f, the mapping $g \mapsto \Lambda_f(g) = \Phi_{X_f^g}^{X_g^f}(f)$ is a flattening of f and a morphological filter.*

Remark 2 *For a fixed f, the mapping $g \mapsto \Xi_g(f)$ is not a morphological filter and for a fixed g, the mapping $f \mapsto \Lambda_f(g)$ is not a morphological filter.*

Let f and g be two functions. We associate to (f, g) a new function $\Theta(f, g)$ defined as follows.

$$\Theta(f,g) = \begin{cases} \mathrm{Rec}(f\ ,\ f \wedge g = g) & \text{on} \quad X_g^f \\ \widetilde{\mathrm{Rec}}(f\ ,\ f \vee g = g) & \text{on} \quad X_f^g \\ f & \text{on} \quad \overline{X_g^f \cup X_f^g} \end{cases}$$

Proposition 16 *The mapping $f \mapsto \Theta(f, g)$ is a flattening of f and a morphological filter.*

Proposition 17 *The mapping $g \mapsto \Theta(f, g)$ is a flattening of f and a morphological filter.*

References

1. M. Grimaud. *La Géodésie Numérique en Morphologie Mathématique. Application a la Détection Automatique de Microcalcifications en Mammographie Numérique.* PhD thesis, E.N.S. des Mines de Paris, 1991.
2. H. Heijmans. Connected morphological operators for binary images. Technical Report PNA-R9708, CWI-Amsterdam, 1997.
3. G. Matheron. Les nivellements. Technical report, Centre de Morphologie Mathématique, 1997.
4. F. Meyer. Morphological connected mappings with scale space properties. Technical report, Centre de Morphologie Mathématique, 1997.
5. F. Meyer and S. Beucher. Morphological segmentation. *J. Visual Commun. Image Repres.*, 1(1):21–45, 1990.
6. C. Ronse. Set-theoretical algebraic approaches to connectivity in continuous or digital spaces. *Journal of Mathematical Imaging and Vision*, (to appear), 1998.
7. P. Salembier and J. Serra. Flat zones filtering, connected operators and filters by reconstruction. *IEEE Transactions on Image Processing*, 3(8):1153–1160, August 1995.
8. J. Serra, editor. *Image Analysis and Mathematical Morphology. II: Theoretical Advances.* Academic Press, London, 1988.
9. J. Serra. Connectivity on complete lattices. In M. B. Petros Maragos, Ronald W. Schafer, editor, *Mathematical Morphology and It's Application to Image and Signal Processing*, pages 81–97. Kluwer Academic, 1996.
10. J. Serra and P. Salembier. Connected operators and pyramids. In *Proceedings of SPIE, San Diego*, volume 2030, pages 65–76, 1993.
11. C. Vachier and F. Meyer. Extinction values: A new measurement of persistence. In I. Pitas, editor, *1995 IEEE Workshop on Nonlinera Signal and Image Processing*, pages 254–257, 1995.

THE LEVELINGS

FERNAND MEYER
Centre de Morphologie Mathématique - Ecole des Mines de Paris
35, rue Saint-Honoré - 77305 Fontainebleau (FRANCE)

Abstract. This paper defines and studies the levelings, an interesting subclass of connected operators. The relations with openings and closings by reconstruction are established. Algorithms are derived for constructing them. Extended levelings are introduced through examples.

Key words: Levelings, reconstruction openings and closings, flat zones

1. Introduction

Levelings have been introduced in [2]; Matheron enlarged their definition in the continuous space and studied their algebraic properties in [1]. In [3] we have presented how connected operators may be particularized and we introduced various subclasses, whose definition is recalled in the following table, in which f and g represent functions, and p and q are two neighbouring pixels on a digital grid.

Operator	**Descriptive criterion**
planing	$g_p \neq g_q \Rightarrow f_p \neq f_q$
monotone planing	$g_p > g_q \Rightarrow f_p > f_q$
flattening	$g_p > g_q \Rightarrow [f_p \geq g_p$ and $g_q \geq f_q]$ or $[f_q \geq g_p$ and $g_q \geq f_p]$
leveling	$g_p > g_q \Rightarrow f_p \geq g_p$ and $g_q \geq f_q$

In the present document we keep the same notations as in [3] and we analyse more extensively the family of levelings. After introducing and characterizing lower and upper levelings, we describe algorithms for constructing them. Finally, we show through two examples a way to extend levelings and we will associate to them "pseudo-flat zones", of great interest as segmentation seeds.

2. Lower and upper levelings

2.1. Definitions and characterisation

It will be fruitful to consider the levelings as the intersection of two larger classes: the lower levelings and the upper levelings, defined as follows.

Definition 1 *A function g is a lower-leveling of a function f if and only if for any couple of neighbouring pixels (p,q): $g_p > g_q \Rightarrow g_q \geq f_q$*

Definition 2 *A function g is an upper-leveling of a function f if and only if for any couple of neighbouring pixels (p,q): $g_p > g_q \Rightarrow g_p \leq f_p$*

The name "upper-leveling" comes from the fact that all connected components where $g > f$ are flat: for any couple of neighbouring pixels (p, q): $\left| \begin{matrix} g_q > f_q \\ g_p > f_p \end{matrix} \right| \Rightarrow g_p = g_q$.
Similarly if g is a lower leveling of f, then all connected components where $g < f$ are flat.

Definition 3 *A function g is a leveling of a function f if and only if it is both an upper and a lower leveling of the function f.*

Remark 1 *A function $g \geq f$ is a lower-leveling of the function f. A function $g \leq f$ is an upper-leveling of the function f.*

Recalling that the logical meaning of $[A \Rightarrow B]$ is $[notA \text{ or } B]$ we may interpret $[g_p > g_q \Rightarrow g_q \geq f_q]$ as $[g_p \leq g_q \text{ or } g_q \geq f_q]$ or in a equivalent manner $[g_q \geq f_q \wedge g_p]$. This gives the following criterion (followed by the dual criterion)
Criterion Low1: A function g is a lower-leveling of a function f if and only if for any couple of neighbouring pixels (p, q): $[g_q \geq f_q \wedge g_p]$.
Criterion Up1: A function g is an upper-leveling of a function f if and only if for any couple of neighbouring pixels (p, q): $[g_p \leq f_p \vee g_q]$.
Applying repeatedly the criterion Low1 to all neighbours N_q of the central point q yields: $g_q \geq f_q \wedge \bigvee_{x \in N_q} g_x$ which implies $g_q \geq f_q \wedge \left(g_q \vee \bigvee_{x \in N_q} g_x \right) = f_q \wedge \delta g_q$, where δ represents the elementary morphological dilation with a flat structuring element containing the central point and all its neighbours. Inversely, it is easy to verify that this criterion Low2 implies the criterion Low1. Hence:
Criterion Low2: A function g is a lower-leveling of a function f if and only if : $[g \geq f \wedge \delta g]$
Criterion Up2: A function g is an upper-leveling of a function f if and only if : $[g \leq f \vee \varepsilon g]$, where ε represents the elementary morphological erosion with a flat structuring element containing the central point and all its neighbours
The levelings being both lower and upper levelings are characterized by the two following criteria:
Criterion Lev1: A function g is a leveling of a function f if and only if for any couple of neighbouring pixels (p, q): $[f_q \wedge g_p \leq g_q \leq f_q \vee g_p]$
Criterion Lev2 (Matheron): A function g is a leveling of a function f if and only if: $[f \wedge \delta g \leq g \leq f \vee \varepsilon g]$.

2.2. Openings and closings by reconstruction

We recall that a function g is an opening (resp. closing) by reconstruction of a function f iff $g = f \wedge \delta g$ (resp. $g = f \vee \varepsilon g$). As it verifies the criterion Low2 (resp. Up2), such a function g is then a lower (resp. upper) leveling of f. The reciprocal is also true. Hence:

Proposition 1 *g is an opening (resp. closing) by reconstruction of a function f if and only if g is a lower (resp. upper) leveling of f verifying $g \leq f$ (resp. $g \geq f$).*

Using this characterisation, we may particularize the initial definition of lower levelings in the case where $f \geq g$:

Proposition 2 *g is an opening by reconstruction of a function f if and only if $g \leq f$ and for any couple of neighbouring pixels (p, q): $g_p > g_q \Rightarrow g_q = f_q$.*

Proposition 3 *g is a closing by reconstruction of a function f if and only if $g \geq f$ and for any couple of neighbouring pixels (p, q): $g_p > g_q \Rightarrow g_p = f_p$.*

2.3. Basic properties

Applying the definition of the lower and upper levelings, we obtain:

Lemma 1 *If g and h are two lower (resp. upper) levelings of f, then $g \vee h$ and $g \wedge h$ also are lower (resp. upper) levelings of f.*

Lemma 2 *If g and h are two levelings of f, then $g \vee h$ and $g \wedge h$ also are levelings of f.*

Lemma 3 *The relation "g is a leveling of f" is reflexive and transitive: it is a preorder relation*

Proposition 4 *The family of levelings, from which we exclude the trivial constant functions, verify the anti-symmetry relation: if f is a non constant function and a leveling of g, and simultaneously g is a leveling of f, then $f = g$.*

Remark 2 *If g is a (lower) leveling of f then $g \wedge f$ is an anti-extensive (lower) leveling of f,i.e. an opening by reconstruction. Similarly if g is an (upper) leveling of f then $g \vee f$ is an extensive (upper) leveling of f, i.e. a closing by reconstruction.*

2.4. Algorithms for construction of lower, upper and plain levelings

2.4.1. *Construction of a lower leveling*

Given two functions g and f, we want to transform g into a lower leveling of f. If g is not a lower leveling of f, then the criterion Low1 is false for at least a pair of neighbouring pixels (p, q) : $g_q < f_q \wedge g_p$. Hence the smallest modification of g_q for which the criterion Low1 becomes true is $g'_q = f_q \wedge g_p$. This leads to the algorithm niv^- for producing a lower leveling: for all neighbouring pixels (p, q) for which $g_p > g_q$ and $g_q < f_q$ do $g'_q = f_q \wedge g_p$.

Starting with the criterion Low2 will produce an alternative and equivalent algorithm: niv^δ for producing a lower leveling: for $g < f$ do $g' = f \wedge \delta g$.

We repeat the algorithm until the criterion is satisfied everywhere. The implementation may be made as a parallel or sequential algorithm, it does not matter. For a higher speed a queue implementation can be taken.

2.4.2. *Construction of an upper leveling*

By duality we get two alternative algorithms for upper levelings, equivalent at convergence:

- Algorithm niv^+ for producing an upper-leveling: for all neighbouring pixels (p, q) for which $g_p > g_q$ and $g_p > f_p$ do $g'_p = f_p \vee g_q$.
- Algorithm niv^ε for producing an upper leveling: for $g > f$ do $g' = f \vee \varepsilon g$

We repeat the algorithm until the criterion is satisfied everywhere.

Remark 3 *Each of the algorithms niv^δ and niv^ε may be written in a unified manner (Matheron): $[do\ g' = (f \vee \varepsilon g) \wedge \delta g = (f \wedge \delta g) \vee \varepsilon g]$. Matheron studies the levelings in the continuous space under the following form : g is a leveling of f iff $[do\ g = (f \vee \beta g) \wedge \alpha g = (f \wedge \alpha g) \vee \beta g]$, where α is an arbitrary extensive dilation and β the adjunct anti-extensive erosion [1].*

2.4.3. *Construction of a leveling*

We obtain a complete algorithm by putting both parts together: executing the algorithm niv^+ in the case $g > f$ and the algorithm niv^- in the case $g < f$. We call this parallel execution of niv^+ and niv^- : $niv = (niv^+ \shortparallel niv^-)$.

If g is not modified, while applying this complete algorithm to a couple of functions (f, g), then g is a leveling of f. If on the other hand g is modified, one may repeatedly apply the same algorithm. If the algorithm converges, then at convergence, g clearly will be a leveling of f. Two questions arise then. Will the algorithm converge ? If the algorithm converges, are the various versions of the algorithms equivalent ? In particular, has the scanning order any influence upon the result ?

Convergence of the algorithm In both cases, the algorithm may never converge if some oscillations appear. But this is not possible. Let us examine the simplest version of the algorithm:

- for the pixels q verifying $g_q < f_q$ and for any of their neighbours p for which $g_p > g_q$ do $g'_q = f_q \wedge g_p$. Hence $g_q \leq g'_q \leq f_q$
- for the pixels p verifying $g_p > f_p$ and for any of their neighbours q for which $g_p > g_q$ do $g'_p = f_p \vee g_q$. Hence $g_p \geq g'_p \geq f_p$

This shows that the convergence of the algorithm is pointwise monotonous, in the direction from g_p to f_p. The values taken being discrete, the algorithm must converge, whether it is parallel or sequential and independently of the scanning order.

Importance of the order of operations

Lemma 4 *The operator inequalities $(niv^- \circ niv^+) \leq (niv^+ \shortparallel niv^-) \leq (niv^+ \circ niv^-)$ are always true and become equalities if there exists no couple of neighbouring pixels (p, q) for which $g_p > g_q$ and $f_q > g_q$ and $f_p < g_p$*

It is easy to verify that this condition is fulfilled after one of the following transformations: (niv^+) or (niv^-) or $(niv^+ \shortparallel niv^-)$. Hence we have basically three natural levelings associated to f and g :

- (niv^+) followed by any sequence of (niv^+) and (niv^-) or $(niv^+ \shortparallel niv^-)$ until convergence. This gives the lowest leveling.

- (niv^-) followed by any sequence of (niv^+) and (niv^-) or $\left(niv_f^+ \shortparallel niv_f^-\right)$ until convergence. This gives the highest leveling.
- $(niv^+ \shortparallel niv^-)$ followed by any sequence of (niv^+) and (niv^-) or $(niv^+ \shortparallel niv^-)$ until convergence. This gives an autodual leveling, between the two preceding ones.

3. Extended levelings

3.1. General arcwise connected flat-zones

In this section we want to extend the notion of flat-zones for grey-tone functions. Usually flat-zones are plateaus of uniform altitude:

Definition 4 *Two pixels x, y belong to the same flat-zone of a function f if and only if there exists a n-tuple of pixels $(p_1, p_2, ..., p_n)$ such that $p_1 = x$ and $p_n = y$, and for all i, (p_i, p_{i+1}) are neighbours and have the same grey-tone: $f_{p_i} = f_{p_i+1}$.*

Changing the symmetrical relation $f_{p_i} = f_{p_i+1}$ will modify the flat-zones. As an example, let us consider the following relation: for two neighbouring pixels p and q, $f_p \approx f_q$ is defined by $|f_p - f_q| \leq \lambda$. This relation is symmetrical and defines "flat-zones" with a maximal slope equal to λ.

3.2. A first example of extended levelings

3.2.1. *Lower levelings*

Modifying the order relations appearing in the definition of the lower levelings will produce the definition of extended lower levelings:

Definition 5 *A function g is a lower-leveling of a function f if and only if: for any couple of neighbouring pixels (p, q): $g_p \succ g_q \Rightarrow g_q \sqsupseteq f_q$*

Lower levelings establish a coupling between any internal transition of the function g with a pointwise relation between f and g. To each new definition of $\succ$ and of $\sqsupseteq$ is associated a new leveling. The first example will be treated in the present section 3.2, the second example being treated in section 3.3. In the first example:

- $g_p \succ g_q \Leftrightarrow g_p > g_q + \lambda$ and its negation $g_p \preccurlyeq g_q \Leftrightarrow g_p \leq g_q + \lambda$
- $g_q \sqsupseteq f_q \Leftrightarrow g_p \geq f_q - \mu$ and its negation $g_q \sqsubset f_q \Leftrightarrow g_q < f_q - \mu$

Flat zones: for any couple of neighbouring pixels (p, q): $\begin{vmatrix} g_q \sqsubset f_q \\ g_p \sqsubset f_p \end{vmatrix} \Rightarrow \begin{vmatrix} g_p \preccurlyeq g_q \\ g_q \preccurlyeq g_p \end{vmatrix}$.

We will consider that such pixels, verifying $\begin{vmatrix} g_p \preccurlyeq g_q \\ g_q \preccurlyeq g_p \end{vmatrix} \Leftrightarrow g_q - \lambda \leq g_p \leq g_q + \lambda$, belong to the same"flat-zone" and we write $g_p \approx g_q$. Expressing $[g_p \succ g_q \Rightarrow g_q \sqsupseteq f_q]$ as $[not\,(g_p \succ g_q)$ or $g_q \sqsupseteq f_q]$ yields a criterion for lower levelings:

Criterion Low1: A function g is a lower-leveling of a function f if and only if for any couple of neighbouring pixels (p, q): $g_q \geq (f_q - \mu) \wedge (g_p - \lambda)$

Applying the criterion to all neighbours yields:

Criterion Low2: A function g is a lower-leveling of a function f if and only if

$g \geq (f-\mu) \wedge [g \vee (\delta g - \lambda)]$
Algorithm: on $\{g < f - \mu\}$ do $g = (f-\mu) \wedge [g \vee (\delta g - \lambda)]$

3.2.2. *Upper levelings*
By duality we obtain

Definition 6 *A function g is an upper-leveling of a function f if and only if for any couple of neighbouring pixels (p, q): $g_q \prec g_p \Rightarrow g_p \sqsubseteq f_p$*

Attention, we not necessarily have $g_q \prec g_p \Leftrightarrow g_p \succ g_q$. In our first example however, it is the case:
- $g_q \prec g_p \Leftrightarrow g_q < g_p - \lambda$ and its negation $g_p \nsucceq g_q \Leftrightarrow g_p \geq g_q - \lambda$
- $g_p \sqsubseteq f_p \Leftrightarrow g_p \leq f_p + \mu$ and its negation $g_p \sqsupset f_p \Leftrightarrow g_p > f_p + \mu$

Flat zones: for any couple of neighbouring pixels (p, q): $\begin{vmatrix} g_q \sqsupset f_q \\ g_p \sqsupset f_p \end{vmatrix} \Rightarrow \begin{vmatrix} g_p \nsucceq g_q \\ g_q \nsucceq g_p \end{vmatrix}$.
As previously, we consider that g_p and g_q belong to the same flat zone and write $g_p \backsimeq g_q$. This relation means here that $g_q - \lambda \leq g_p \leq g_q + \lambda$, showing that in our first example $\eqsim$ and $\backsimeq$ have the same meaning. In the second example it will not be the case.
Criterion Up1: A function g is an upper-leveling of a function f if and only if $g_p \leq (f_p + \mu) \vee (g_q + \lambda)$
Applying the criterion to all neighbours yields:
Criterion Up2: A function g is an upper-leveling of a function f if and only if for any couple of neighbouring pixels (p, q): $g \leq (f + \mu) \vee [g \wedge (\varepsilon g + \lambda)]$
Algorithm: on $\{g > f + \mu\}$ do $g = (f + \mu) \vee [g \wedge (\varepsilon g + \lambda)]$

3.2.3. *Levelings*
Following our general definition, a function g is a leveling of a function f if and only if it is both an upper and a lower leveling of the function f.

A criterion for levelings is obtained by regrouping the criteria for lower and upper levelings. Doing in parallel the algorithms for lower and upper levelings will produce a leveling. The flat zones associated to the leveling are based on the relation $g_p \approx g_q \Leftrightarrow [g_p \eqsim g_q \text{ or } g_p \backsimeq g_q]$

In the particular example treated here, we may wonder if the leveling is still a monotone planing: for (p, q) neighbours: $g_p \succ g_q \Rightarrow f_p \succ f_q$? In our case, we know that $g_q < g_p - \lambda \Rightarrow g_p \leq f_p + \mu$ and $f_q \leq g_q + \mu$. It follows $f_q \leq g_q + \mu < g_p + \mu - \lambda \leq f_p + 2\mu - \lambda$. Hence the operator is a monotone planing and as a consequence a planing only if $\mu = 0$. This shows that the extended levelings are not necessarily planings anymore.

3.3. A SECOND EXAMPLE OF EXTENDED LEVELINGS

3.3.1. *Lower levelings*
Adopting a new definition for $\succ$ and $\sqsupseteq$ will lead to a new extended leveling. In what follows γ represents a morphological opening, constructed as a supremum of morphological openings with flat structuring elements:
- $g_p \succ g_q \Leftrightarrow \gamma g_p > g_q$ and its negation $g_p \npreceq g_q \Leftrightarrow \gamma g_p \leq g_q$

– $g_q \sqsupseteq f_q \Leftrightarrow g_p \geq f_q$: it is the ordinary order relation between functions

Flat zones: for any couple of neighbouring pixels (p,q): $\begin{vmatrix} g_q < f_q \\ g_p < f_p \end{vmatrix} \Rightarrow \begin{vmatrix} g_p \not\prec g_q \\ g_q \not\prec g_p \end{vmatrix}$.

"Flat zones" are now defined by $\begin{vmatrix} \gamma g_p \leq g_q \\ \gamma g_q \leq g_p \end{vmatrix} \Leftrightarrow g_p \eqsim g_q$

Criterion Low1: A function g is a lower-leveling of a function f if and only if for any couple of neighbouring pixels (p,q): $g_q \geq f_q \wedge \gamma g_p$

Applying the criterion to all neighbours yields:

Criterion Low2: A function g is a lower-leveling of a function f if and only if $g \geq f \wedge [g \vee \delta\gamma g]$

and the algorithm for constructing such a lower leveling easily follows from these criteria. From criterion Low1, we derive: if $\gamma g_p > g_q$ and $g_q < f_q$ do $g'_q = f_q \wedge \gamma g_p$. Replacing g_q by a higher value g'_q could modify the value of the opening γ at the pixel p. This is not the case, since γ is a flat opening and $\gamma g_p > g_q$. Hence the criterion Low1 becomes true for the function g'.

Algorithm: on $\{g < f\}$ do $g = f \wedge [g \vee \delta\gamma g]$

3.3.2. *Upper levelings*

By duality we define

– $g_q \prec g_p \Leftrightarrow g_p > \varphi g_q$ and its negation $g_q \not\succ g_p \Leftrightarrow g_p \leq \varphi g_q$, where φ represents a morphological closing, constructed as an infimum of morphological closings with flat structuring elements:

– $g_p \sqsubseteq f_p \Leftrightarrow g_p \leq f_p$ it is the ordinary order relation between functions

Flat zones: for any couple of neighbouring pixels (p,q): $\begin{vmatrix} g_q \sqsupset f_q \\ g_p \sqsupset f_p \end{vmatrix} \Rightarrow \begin{vmatrix} g_p \not\succ g_q \\ g_q \not\succ g_p \end{vmatrix}$

"Flat zones" are now defined by $\begin{vmatrix} g_p \leq \varphi g_q \\ g_q \leq \varphi g_p \end{vmatrix} \Leftrightarrow g_p \simeq g_q$

Criterion Up1: A function g is an upper-leveling of a function f if and only if for any couple of neighbouring pixels (p,q): $g_p \leq f_p \vee \varphi g_q$

Applying the criterion to all neighbours yields:

Criterion Up2: A function g is an upper-leveling of a function f if and only if $g \leq f \vee [g \wedge \varepsilon\varphi g]$

Algorithm: on $\{g > f\}$ do $g = f \vee [g \wedge \varepsilon\varphi g]$

3.3.3. *Levelings*

A criterion for levelings is obtained by regrouping the criteria for lower and upper levelings. Doing in parallel the algorithms for lower and upper levelings will produce a leveling.

We will define the flat zones associated to this leveling as the arcwise connected particles based on the symmetrical relation $g_p \approx g_q \Leftrightarrow [g_p \simeq g_q$ or $g_p \eqsim g_q] \Leftrightarrow$ $\begin{vmatrix} g_p \leq \varphi g_q \\ g_q \leq \varphi g_p \end{vmatrix}$ or $\begin{vmatrix} \gamma g_p \leq g_q \\ \gamma g_q \leq g_p \end{vmatrix}$

In the particular example treated here, the obtained leveling is not a monotone planing.

Fig. 1. (a) reference image; (b) marker; (c-f) variouslevelings.

3.4. Conclusion and illustration

The preceding images present various levelings applied to the same reference (a) and marker image (b), obtained by an alternate sequential filter of size 3 based on hexagonal openings and closings. The ordinary leveling (c) is the basic leveling presented in the first part. This leveling reconstructs the most faithfully the initial image, as it creates flat zones which are really flat and which may also be extremely thin. The next leveling (d) corresponds to the first example of extended levelings: it generates "flat zones" with a slope 1. The two last levelings correspond to the last example and are leveling based on $\delta\gamma$ and $\varepsilon\varphi$; in (e) γ and φ are the hexagonal opening and closing; in (f) γ and φ are an opening and a closing with line segments; they also generate "flat zones" which are flat, but without thin details as may be seen within the camera for instance, where many details have been lost. In all cases the contours of the remaining objects are faithfully preserved. The autodual nature of the levelings constitutes one of their major interests. For this reason they are ideal operators to apply to colour images.

References

1. G. Matheron. Les nivellements. Technical report, Centre de Morphologie Mathématique, 1997.
2. F. Meyer. Morphological connected mappings with scale space properties. Technical report, Centre de Morphologie Mathématique, 1997.
3. F. Meyer. From connected operators to levelings. *This volume.*

FACE SEGMENTATION USING CONNECTED OPERATORS *

VERÓNICA VILAPLANA and FERRAN MARQUÉS
Dept. of Signal Theory and Communications
Universitat Politècnica de Catalunya
Campus Nord - Mòdul D5
C/ Gran Capità, 08034 Barcelona, Spain
Tel: (343) 401 64 50, Fax: (343) 401 64 47
E-mail: {veronica, ferran}@gps.tsc.upc.es

Abstract. A connected operator that allows the discrimination of face information is presented. The operator removes from the image the components that belong to a face. The filtering process is implemented by means of a Max-tree. At each node, a criterion based on the likelihood of belonging to a face is computed. The likelihood is estimated relying on the principal components of a set of training images from a face-class. The dual operator is applied to complete the face extraction. This approach opens the door to semantic based connected operators and segmentation techniques.

Key words: connected operators, max-tree, face segmentation, eigenvector.

1. Introduction

Automatic face detection is a very active area of image analysis and several techniques have been proposed to detect and track faces or features in faces [2]. The new functionalities of the future coding standard MPEG4 [6] have even increased the interest in this area. Nevertheless, MPEG4 applications rise a new necessity since the analysis algorithm should, not only detect the position of the face, but really segment it obtaining its actual shape.

One of the basic approaches in the area of face detection is to compute the principal components of a set of training images from a face-class. The most relevant eigenvectors are then used to estimate the probability of an image or a sub-image of being a face [7, 3]. With this technique, the position of a face in an image can be detected. In Figure 1, an example of face detection is presented. The first image shows an original frame whereas the second image illustrates the face detected using the technique proposed in [3]. The third image presents the type of face segmentation that is necessary for MPEG4 applications.

In this paper, we present a new type of connected operators that deal with face information. Connected operators [10] have shown their capability to discriminate very different types of information while correctly preserving the contours in the scene. This is the case of the area [13], dynamic [5], volume [12], moment [1] and motion [9] oriented operators. Here, we extend the face detection algorithm presented in [7, 3] to the case of connected filters. This way, the face is not only detected but correctly segmented. The image is represented by means of a Max-tree [9] and

* This work has been partially supported by the ACTS-057 project (VIDAS) of the European Union and the grant 1997FI-00762 of the Direcció General de Recerca (Generalitat de Catalunya)

Fig. 1. Example of face detection and segmentation

the filtering is directly applied on this structure. The selection criterion used in the filtering process is related to the likelihood of a component of belonging to a face. This is a non increasing criterion and the components to be selected should correspond to the maximum of the likelihood function (analogously, the minimum of the distance function). Therefore, a new specific selection criterion is proposed. The face shape is obtained by combining the result of the two dual connected operators.

This technique is described in the paper as follows: after this introduction, Section 2 presents a new connected operator that is able to discriminate the information related to faces in the scene. The implementation of the operator using the Max-tree representation of the image and the selection criterion that is applied are discussed. Section 3 details the way to apply the previous operator to the problem of face segmentation and, to assess the proposed technique, several results are shown. Finally, Section 4 drives some conclusions.

2. A face-oriented connected operator

Connected operators have been first defined for binary functions and extended to the gray level case using the concepts of *flat zone* and *partition* [11, 10]. A flat zone of a gray level function f is the largest connected component where the function f is constant. All the flat zones that can be defined for a given function f form a partition of the space, the so-called *partition of flat zones.*

Definition [Gray level connected operators]: *An operator Ψ acting on gray level functions is connected if, for any function f, the partition of flat zones of f is finer than the partition of flat zones of $\Psi(f)$.*

Therefore, a connected operator acts by removing connected components while preserving the boundaries of the remaining components. In the sequel, we are going to concentrate in the case of anti-extensive connected operators.

2.1. The use of the Max-tree

The *Max-tree* of an image is a structured representation of the image which is suitable for the implementation of anti-extensive connected operators [9]. A Max-tree contains in each leaf a component being a local maximum of the grey level image. Nodes in the tree are associated to the different grey level values at which, if the image is thresholded, new connected components would appear. Since we are dealing with a tree of maxima, the threshold should start with the highest possible value and vary down to 0.

The filtering step is directly performed in the tree structure. At each node of the tree, a criterion is assessed and, based on the value of this criterion, the node is either preserved or removed (pruning of the tree). The filtered result is obtained by transforming the pruned tree into a gray level image.

2.2. FACE-ORIENTED CRITERION: MEASUREMENT

The criterion used in the filtering process is related to the likelihood of a connected component of belonging to a face. The estimation of this likelihood is based on the distance proposed in [7, 3] for the case of face detection. For each node in the tree, the distance between the connected component (x) and the face class (Ω) is estimated. To be able to detect faces with sizes different from the size of the faces in the data base, this distance should be minimized with respect to a geometrical transformation (w) applied to the connected component.

2.2.1. *Distance definition*

The class membership of the transformed component (x_w) is modeled as a unimodal gaussian density:

$$P(x_w/\Omega) = \frac{\exp[-\frac{1}{2}(x_w - \bar{x})^T \Sigma^{-1}(x_w - \bar{x})]}{(2\pi)^{\frac{N}{2}}|\Sigma|^{\frac{1}{2}}} \tag{1}$$

where the mean ($\bar{x}$) and the covariance matrix (Σ) of the face class are estimated using a training data set. The parameter N is the size of the vectors in the training data set. The Mahalanobis distance is used as a sufficient statistic for characterizing this likelihood

$$d(x_w, \Omega) = (x_w - \bar{x})^T \Sigma^{-1}(x_w - \bar{x}) \tag{2}$$

This distance is defined on the complete vector space $\mathcal{R}^N$. A computationally tractable estimate of this distance, based on the M first principal components of the covariance matrix, is

$$\hat{d}(x_w, \Omega) = \sum_{i=1}^{M} \frac{y_i^2}{\lambda_i} + \frac{1}{\rho}\epsilon^2(x_w, \Omega) \tag{3}$$

where y_i are the projections of x_w over the M principal components and λ_i are the M principal eigenvalues, with $M << N$. Moreover, ρ is the average of the $N-M$ remaining eigenvalues, and the residual reconstruction error is

$$\epsilon^2(x_w, \Omega) = \sum_{i=M+1}^{N} y_i^2 = ||x_w - \bar{x}||^2 - \sum_{i=1}^{M} y_i^2. \tag{4}$$

The estimate of expression (3) divides the complete vector space into a principal M-dimensional subspace and a complementary (N-M)-dimensional subspace. The distance to the principal subspace is exactly computed while the distance to the complementary subspace is approximated.

2.2.2. *Distance computation*

For each node, the distance between its associated connected component and the face class is estimated. For computational reasons, this distance is not minimized with respect to the whole set of possible geometrical transformations. Practically, an extended support is created for each component. This support is defined as the tightest rectangle bounding the original component. In the areas outside the connected component, the data base background is introduced. The set of possible geometrical transformations is reduced to a normalization of the horizontal or the vertical size of this new support with respect to the size of the images in the database.

2.3. Face-oriented criterion: Decision

Relying on the measure that has been computed for every node in the Max-tree, a decision on which components should be removed has to be taken. In previous works on connected operators, the used criteria and resulting operators are increasing [5, 13, 12] or mainly present an increasing behaviour [9]. In the first case, the decision results in a threshold to be applied on the value of the assessed criteria. In the second case, a solution based on the Viterbi algorithm has been proposed [9].

The criterion used in the face-oriented connected operator is based on the estimation of the likelihood of a component of being part of a face. Therefore, the decision should only be a selection of the node leading to the absolute maximum of the likelihood function (analogously, the absolute minimum of the distance function). However, in [3], the authors already reported that the previous distance may lead to false detections in the case of face-like objects. Therefore, the criterion may exhibit local minima. To solve this problem, a decision rule has been adopted that analyzes and compares the criterion values of the various nodes.

2.3.1. *Selection of a minimum*

It has been observed that false minima appear mainly in nodes associated to very large flat components. To solve this problem, the tree is analyzed starting from the leaf nodes down to the root, and consecutive local minima are compared in pairs. This comparison takes into account the maximum value of the intermediate nodes. When analyzing two minima, if there is an intermediate node whose distance is too large compared to the lowest minimum, the first node is selected. Otherwise, the lowest minimum is selected. Formally, let m_i and m_{i+1} be two consecutive minima on the same branch, $\{d_j\}$ the distance values of the intermediate nodes and

$$T = \frac{max_j(d_j) - min(m_i, m_{i+1})}{|m_i - m_{i+1}|}. \tag{5}$$

The function *select*(m_i, m_j) implements the selection process: If $T < Thres$, $min(m_i, m_{i+1})$ is selected; otherwise, m_i. This concept is illustrated in Figure 2 where $d = max_j(d_j)$. The process is iterated using the selected and the next minima.

The idea behind this criterion is that if, in order to go from one connected compo-

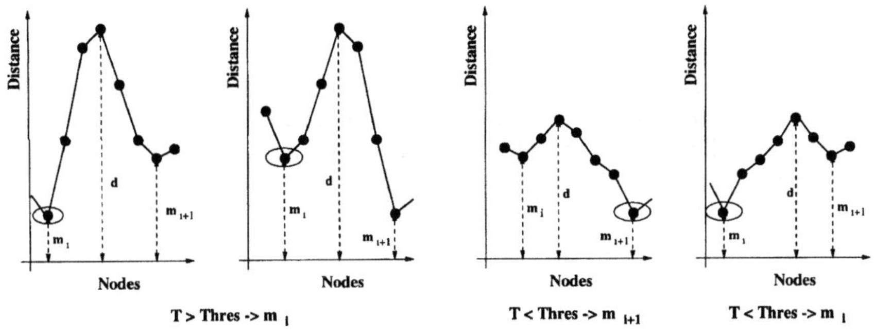

Fig. 2. Selection between two consecutive minima

nent to a larger one both associated to minima, there is an intermediate component whose likelihood of belonging to a face is very low (distance very high), it is very likely that the second minimum leads to a false face detection. Finally, in the intersection of two branches, the rule that is applied is presented in Figure 3. Following this criterion, only one minimum for image is selected and the area associated to this node is the set of bright components belonging to a face (initial estimate of the segmented face).

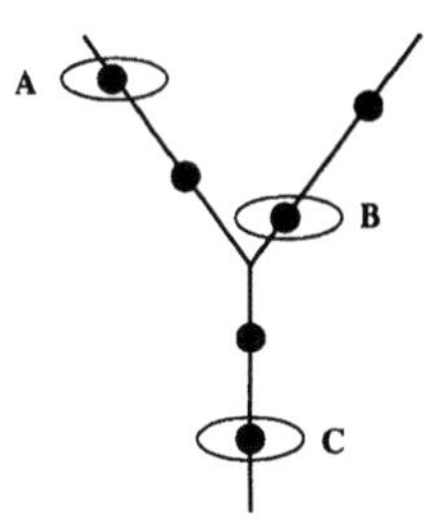

select (A, C)	*select* (B, C)	**Action**
A	B	*select* (A, B)
A	C	A
C	B	B
C	C	C

Fig. 3. Selection of the minimum in the intersection of two branches

2.4. Combining the dual operator

The anti-extensive connected operator is combined with its dual (the extensive one) so that the face dark components are extracted as well. The filtering step is, in this case, driven by the result of the anti-extensive operator. The area covered by the removed component is set to gray level zero before creating the Min-tree. Nevertheless, when computing the criterion on the Min-tree nodes, the information from the original image is used. The fact of zeroing the removed component ensures that the previously removed area corresponds to a leaf node in the Min-tree. The filtering step takes only into account the nodes associated to this leaf node (down to the root of the tree).

An example of the combination of both (anti-extensive and extensive) face-oriented connected operators is presented in Figure 4. The first column shows two original frames of the sequences *Vidas1* and *Vidas2*. The second column contains the result of the anti-extensive operator, where the removed connected component is reconstructed with the mean gray level value of the node it belongs to after filtering. The third column presents the residue of the anti-extensive connected operator where, to better illustrate the result, the original value of the pixels from the removed component are shown. Finally, the last column shows the dark and bright face components that have been extracted by the combination of the dual operators.

Fig. 4. Extraction of the bright and dark face components

3. Face extraction

In the second example of Figure 4, it can be seen that the final result does not contain all the face components. The reason for that is the use of the Max-tree and Min-tree structures for filtering. These filtering approaches allow very efficient implementations but they order in a specific way the set of connected components.

Nevertheless, once the core components have been detected by the connected operators, a refinement step can be applied to completely extract the face information. In the refinement step, the final face information is obtained by merging neighboring regions to the filtered components [4]. Neighboring components are merged relying on the same measure used in the filtering process. A new component is add to the face information if its addition increases the likelihood of belonging to a face.

In Figure 5, the behaviour of the method with four different images (*Vidas1*, *Vidas2*, *Foreman* and *Miss America*) is illustrated. Results have been obtained using the *Olivetti Research Laboratory* data base of face images (400 images with 40 different people and 10 views for person). For computational reasons, the face-oriented operators have not been applied directly on the original images. Images have been first filtered by means of an auto-dual size-oriented connected operator [8]. This allows the face-oriented operators to deal with a reduced set of connected components with a relevant size.

The first column in Figure 5 shows the original images. The second and third columns present the face components extracted by the anti-extensive operator and its combination with the extensive operator, respectively. In the fourth column, the final selected component is shown. In the first example (*Vidas1*), it can be seen how the different steps of the method incorporate new components to the final result. The second example (*Vidas2*) shows a case where the combination of the dual connected operators already achieves the component with maximum likelihood. The third example (*Foreman*) illustrates two different behaviours of the method. First, although the anti-extensive operator has correctly detected a set of components from the face, the dual operator does not incorporate any new component. The reason is that the Min-tree connects the node associated to the bright parts of the face with the darkest area of the building. The addition of this new component decreases the likelihood of belonging to a face and, thus, the decision criterion selects the initial component. Second, although the final step is able to extract the face, the hat is not recognized as being part of the face since the initial data base does not contain any person wearing a hat. Finally, the fourth example (*Miss America*) shows a behaviour similar to the previous one. In this case, the last step is not able to extract any component from the hair due to the similar gray level values of the hair and the background. This leads to connected components covering partially both objects which do not result into a maximum of the likelihood function.

4. Conclusions

We have presented a new connected operator that allows the extraction of the face information present in the scene. The combination of the two dual operators, implemented using the Max-tree and Min-tree respectively, leads to the extraction of the core components of faces. The tree-based implementation of the operators has been chosen in order to speed up the filtering process. To obtain the complete set of components that form the face, a last refinement step is applied.

Although these connected operators have been proposed in the framework of face information extraction, their bases can be applied to other kinds of information. The current work is, therefore, focussed on two aspects. First, to improve the performance of the current approach mainly by improving the used criterion and by looking for recursive techniques for computing it. Second, to extend the technique to other types of semantic information, opening the door to semantic-oriented operators and semantic-based segmentation.

References

1. E. Breen and R. Jones. An attribute-based approach to mathematical morphology. In P. Maragos, R.W. Schafer, and M.A. Butt, editors, *International Symposium on Mathematical Morphology*, pages 41–48, Atlanta (GA), USA, May 1996. Kluwer Academic Publishers.
2. R. Chellappa, C. Wilson, and S. Sirohey. Human and machine recognition of faces: a survey. *Proceedings of the IEEE*, 83(5):705–740, May 1995.
3. F. Davoine, C. Kervran, H. Li, P. Pérez, R. Forchheimer, and C. Labit. On automatic face and facial features detection in video sequences. In *International Workshop on Synthetic-Natural Hybrid coding and three dimensional imaging*, pages 196–199, Rhodes, Greece, 1997.

Original picture Bright components Bright&dark comps. Final segmentation

Fig. 5. Examples of face segmentation

4. L. Garrido, F. Marqués, M. Pardàs, P. Salembier, and V. Vilaplana. A hierarchical technique for image sequence analysis. In *Workshop on Image Analysis for Multimedia Interactive Services*, pages 13–20, Louvain-la-Neuve, Belgium, 1997.
5. M. Grimaud. A new measure of contrast: the dynamics. In SPIE, editor, *Visual Comm. and Image Processing'92*, volume 1769, pages 292–305, San Diego (CA), USA, July 1992.
6. ISO/IEC JTC1/SC29/WG11. MPEG-4 Proposal Package Description (PPD). July 1995.
7. B. Moghaddam and A. Pentland. Probabilistic visual learning for object representation. *IEEE Transactions on Pattern Analysis and Machine Intelligence*, 19(7):696–710, July 1997.
8. P. Salembier, L. Garrido, and D. García. Auto-dual connected operators based on iterative merging algorithms. In *Int. Symposium on Mathematical Morphology*. Submitted, 1998.
9. P. Salembier, A. Oliveras, and L. Garrido. Anti-extensive connected operators for image and sequence processing. *IEEE Transactions on Image Processing*, March 1998.
10. P. Salembier and J. Serra. Flat zones filtering, connected operators and filters by reconstruction. *IEEE Transactions on Image Processing*, 3(8):1153–1159, August 1995.
11. J. Serra and P. Salembier. Connected operators and pyramids. In SPIE, editor, *Image Algebra and Mathematical Morphology*, volume 2030, pages 65–76, San Diego (CA), USA, July 1993.
12. C. Vachier. *Extraction de Caractéristiques, Segmentation d'Image et Morhpologie Mathématique*. PhD thesis, Paris School of Mines, 1995.
13. L. Vincent. Grayscale area openings and closings, their efficient implementation and applications. In J. Serra and P. Salembier, editors, *First Workshop on Mathematical Morphology and its Applications to Signal Processing*, pages 22–27, Barcelona, Spain, May 1993. UPC.

A CONNECTED COMPONENT APPROACH TO THE WATERSHED SEGMENTATION

ANDREAS BIENIEK and ALINA MOGA
Albert-Ludwigs-Universität Freiburg
Lehrstuhl für Mustererkennung und Bildverarbeitung
Prof. Dr.-Ing. H. Burkhardt
Am Flughafen 17, D-79085 Freiburg, Germany
bieniek,moga@informatik.uni-freiburg.de

Abstract. The watershed transformation is a popular image segmentation algorithm for grey scale images. In this paper, we present a definition of the watershed segmentation, based on sets of neighboring pixels. The definition is consistent with the behavior of most implementations of the watershed algorithm which do not construct watersheds lines, namely, to chose one arbitrary label in the case of competing labels. Moreover, different distance metrics to approximate the geodesic distance on plateaus can be incorporated into the definition. The relation to the traditional definition of the watershed segmentation is proven in the paper. The definition leads to a new type of watershed algorithm which is closely related to the connected component algorithm. Our complexity analysis shows that for practical cases, the algorithm has a linear complexity with the number of pixels, independent on the image content and number of grey-levels. The algorithm is simpler with respect to implementation and data structures. Additionally, the memory requirement is small and independent of the number of grey-levels in the input image. Furthermore, our timing results show a significant improvement in the running time, compared against the classical watershed algorithm based on hierarchical queues.

Key words: Watershed Segmentation, Connected Components

1. Introduction

The watershed transformation performs grey scale segmentation. Interpreting an image as a topographical relief, flooding is performed starting from regional minima (connected plateaus of constant altitude surrounded only by pixels of higher grey-levels). Dams (watershed lines) are built when water from different sources meet. The classical algorithm simulates the flooding process using hierarchical queues [4].

The paper is organized as follows. In Section 2, the definition of the watershed segmentation for images without non-minima plateaus is presented. In Section 3, the definition is compared with Meyer's definition of watersheds [4]. Further on, the proposed definition leads in Section 4 to a connected component like watershed algorithm for images without non-minima plateaus. The extension of the definition for images with non-minima plateaus is described in Section 5, whereas the corresponding algorithm and comparing timing results follow in Section 6. The complexity analysis is presented for each step of the algorithm in Section 4 and 6. Finally, conclusions are drawn in Section 7.

2. Segmentation based on local conditions

In this section, we propose a local condition for images without non-minima plateaus, which leads to a watershed segmentation. The reason of considering just such images is that each pixel has a lower neighbor, excepting minima pixels. This property holds also for lower complete images [5, 7], where the flooding order is 1D [7]. The extension to include images with non-minima plateaus is presented in Section 5.

Let $f(p)$ be a function of grey-levels, representing a digital image with the domain $\Omega \subset \mathbb{Z}^2$. Each pixel $p \in \Omega$ has a grey-level $f(p)$ and a set of neighboring pixels $p' \in N(p)$ with a distance function $dist(p, p')$ to each neighbor. In most cases, a 4- or 8-square neighborhood is used with a constant distance of 1 to all neighboring pixels. Before giving our definition of watershed segmentation and catchment basins, some preliminary definitions are introduced:

Definition 1 (Lower slope) *The lower slope of a pixel p is given by the maximal ratio* $\frac{f(p)-f(p')}{dist(p,p')}$ *to all its neighboring pixels of lower gray-level than itself:*

$$LS(p) = \max_{\forall p' \in N(p)} \left(\frac{f(p) - f(p')}{dist(p, p')} \mid f(p') \leq f(p) \right)$$

and is not defined for the case $f(p') > f(p)$, $\forall p' \in N(p)$ *[4].*

We define the notion of watershed segmentation and catchment basin based on the sets of those neighboring pixels that can be part of the path of steepest descent:

Definition 2 (Neighbors on a path of steepest descent) $\forall p \in \Omega$, $NLS(p)$ *is the set of pixels* $p' \in N(p)$ *defined as follows:*

$$NLS(p) = \left\{ p' \,\middle|\, \frac{f(p) - f(p')}{dist(p, p')} = LS(p) \ , \ f(p') < f(p) \right\}.$$

For the case $dist(p, p') = 1$, $\forall p' \in N(p)$, *the set becomes:*

$$NLS(p) = \left\{ p' \,\middle|\, f(p') = \min_{\forall p'' \in N(p)} f(p'') \ , \ f(p') < f(p) \right\}.$$

A similar definition exists also in [4]. Let us note that in an image without non-minima plateaus, $NLS(p) \neq \emptyset \ \forall p \in \Omega$, p not a minimum. In addition, the path of steepest descent from a pixel p down to a regional minimum m_i will pass only pixels of the set $\bigcup_{p'=p}^{m_i} NLS(p')$. In contrast to [4] the following definition of catchment basins and watershed segmentation is here based on the $NLS(p)$ function:

Definition 3 (Watershed segmentation for images without non-minima plateaus) *For any image without non-minima plateaus, a segmentation is called watershed segmentation if every regional minimum* m_i *has an unique label* $L(m_i)$ *and, for every pixel* $p \in \Omega$ *with* $NLS(p) \neq \emptyset$, *the following condition holds:*

$$\exists p' \in NLS(p) \quad \textit{such that} \quad L(p) = L(p').$$

Definition 4 (Catchment Basin) *For the watershed segmentation defined above, a catchment basin* $CBLC(m_i)$ *of the regional minimum* m_i *is the set of pixels with the label* $L(m_i)$*:*

$$CBLC(m_i) = \{p \mid L(p) = L(m_i)\}.$$

$CBLC(p \rightarrow m_i)$ denotes the catchment basin of m_i containing pixel p. The definition of watersheds and catchment basin does not imply uniqueness of the segmentation result; several solutions can be sometimes valid. In the following section the relation between the definition of a catchment basin based on the topographic distance and on the local condition, respectively, is demonstrated.

3. Relation to the traditional definition of the watershed segmentation

In this section, Meyer's formalism [4] is presented and compared with our definition of watershed segmentation. From functions on continuous space, Meyer derived a formal definition of catchment basins for the digital space [4] as follows:

Definition 5 (Cost function based on lower slope) *The cost for walking on the topographical surface from position p_{i-1} to $p_i \in \mathrm{N}(p_{i-1})$ is given by:*

$$\mathrm{cost}(p_{i-1}, p_i) = \begin{cases} \mathrm{LS}(p_{i-1}) \cdot \mathrm{dist}(p_{i-1}, p_i) & : \ \mathrm{f}(p_{i-1}) > \mathrm{f}(p_i) \\ \mathrm{LS}(p_i) \cdot \mathrm{dist}(p_{i-1}, p_i) & : \ \mathrm{f}(p_{i-1}) < \mathrm{f}(p_i) \\ \frac{1}{2}(\mathrm{LS}(p_{i-1}) + \mathrm{LS}(p_i)) \cdot \mathrm{dist}(p_{i-1}, p_i) & : \ \mathrm{f}(p_{i-1}) = \mathrm{f}(p_i). \end{cases}$$

Definition 6 (Topographical distance) *The topographical distance between two pixels p and q of an image is the minimal π-topographical distance among all paths π between p and q inside the image:*

$$\mathrm{TD}_f(p, q) = \inf \ \mathrm{TD}_f^{\pi}(p, q).$$

where $\mathrm{TD}_f^{\pi}(p, q) = \sum_{i=2}^{n} \mathrm{cost}(p_{i-1}, p_i)$ is the π-topographical distance of a path $\pi = (p = p_1, p_2, \ldots, p_n = q)$, such that $\forall i$, $p_i \in \mathrm{N}(p_{i-1})$ and $p_i \in \Omega$.

Definition 7 (Catchment basin based on topographical distance) *A catchment basin $\mathrm{CBTD}(m_i)$ of a regional minimum m_i is the set of pixels $p \in \Omega$ closer to m_i than to any other regional minimum m_j, according to the topographical distance and the grey-levels of the minima:*

$$\mathrm{CBTD}(m_i) = \{p \mid \mathrm{f}(m_i) + \mathrm{TD}_f(p, m_i) < \mathrm{f}(m_j) + \mathrm{TD}_f(p, m_j) \ \forall j \neq i\}.$$

With these definitions Meyer presents the following theorem (Prop. 5 in [4]):

Theorem 1 *The topographical distance between a pixel p and the regional minimum m_i in the depth of its catchment basin is minimal and equal to $\mathrm{f}(p) - \mathrm{f}(m_i)$ and the geodesic line between them is a line of steepest descent.*

The reversal of Theorem 1 states that a path of steepest descent ensures minimal costs. The construction of the catchment basins is reduced to a problem of finding a shortest path between each pixel and a regional minimum. The relation between Definition 4 and Definition 7 is stated in the following theorem:

Theorem 2 *A catchment basin based on the topographical distance, as in Definition 7, is a subset of the catchment basin in Definition 4, based on the local condition given in Definition 3.*

Proof The formal construction of the catchment basin according to Definition 3 can be described as a recursion. The process starts with the set of pixels belonging to the regional minimum m_i. All these pixels are labeled with $L(m_i)$. At each step, unlabeled pixels, whose neighbors of steepest descent are already in the set, are appended to the set. The recursion ends when no more pixels can be incorporated into the set.

$$\begin{aligned} \mathrm{CBLC}^0(m_i) &= m_i \\ \mathrm{CBLC}^{k+1}(m_i) &= \mathrm{CBLC}^k(m_i) \cup \triangle\,\mathrm{CBLC}^k(m_i) \\ \triangle\,\mathrm{CBLC}^k(m_i) &= \left\{p \,\middle|\, \forall j, p \notin \mathrm{CBLC}^k(m_j) \text{ and } \exists p' \in \mathrm{NLS}(p) \text{ with } p' \in \mathrm{CBLC}^k(m_i)\right\} \end{aligned}$$

Each newly inserted pixel p has a neighbor p' being part of the catchment basin $\mathrm{CBLC}^k(m_i)$. Thus, the local condition of Definition 3 is valid for each p. The proof proceeds as follows:

$$p' \in \mathrm{NLS}(p) \Rightarrow \mathrm{LS}(p) \overset{def2}{=} \frac{\mathrm{f}(p) - \mathrm{f}(p')}{\mathrm{dist}(p,p')} \Rightarrow$$

$$\mathrm{LS}(p) \cdot \mathrm{dist}(p,p') = \mathrm{f}(p) - \mathrm{f}(p') \overset{def5}{\Rightarrow} \mathrm{cost}(p,p') = f(p) - f(p').$$

According to Theorem 1, one recursion step adds only those pixels p building paths of steepest descent down to $\mathrm{CBLC}^k(m_i)$ with the minimal cost $\mathrm{f}(p) - \mathrm{f}(p')$, $p' \in \mathrm{CBLC}^k(m_i)$. After the recursion is finished, all paths between pixels of the catchment basin and its minimum are paths of steepest descent. Therefore, it is not possible to construct a steeper path to a different minimum m_j. However, there might exist another steepest path to a different regional minimum m_j. In this case, the pixel is a watershed pixel according to Definition 7. This proves that $\mathrm{CBTD}(m_i)$ is a subset of $\mathrm{CBLC}(m_i)$[1]. ∎

The difference between Definition 7 and 4 is the treatment of the pixels that have steepest paths to more than one minimum. According to Definition 7, these pixels are watershed pixels. Following Definition 4, based on the local condition, such a pixel is assigned to one of the minima, m_i, where it is connected by a steepest path and where the condition $\exists p' \in \mathrm{NLS}(p), L(p') = L(m_i)$ holds. All possible assignments are valid watershed segmentations. A segmentation according to Definition 4 is consistent with most implementations of the watershed algorithm which do not construct watershed lines, including most algorithms destcribed in [4].

4. The algorithm for lower complete images

The idea of the proposed algorithm originates in the connected components problem [2, 3, 6, 9]. Apart from the input image f with n pixels, our design solution makes use of an image l which stores for every pixel its representative, or label. Let us remark that pixel addresses are used for labeling instead of arbitrary integer values [6]. When the algorithm is finished, every pixel in the label image contains the address of the representative for the region it belongs to. The algorithm consists of three steps described below:

1. Raster scan image and for every pixel p do:
   a) Search in $N(p)$ for pixel p' of minimum grey-level and $f(p') < f(p)$
   b) **If** $\exists p'$ then do:
      – Set $l(p) = p'$
      else
      – Set $l(p) = PLATEAU$
2. Raster scan image and for every pixel p do:
   – **If** $l(p) = PLATEAU$ then do:
      a) Set $l(p) = p$
      b) For all pixels $p' \in N(p)$ with $p' < p$ and $f(p) = f(p')$ do:
         i. Find representative $r1$ of p and $r2$ of p' in l
         ii. Set $l(r1) = l(r2) = \min(r1, r2)$
3. Raster scan image and for every pixel p do:

[1] Please note that Meyer's formalism leads to thick watershed lines and areas. Algorithms which avoid thick watershed lines are not covered by the definitions.

- Find representative $r1$ of p in l
- Set $l(p) = r1$

At step 1, the label of each pixel p, which has a lower neighbor, is set to $p', p' \in NLS(p)$. Otherwise, if the pixel has no lower neighbor, it is on a minima plateau and it is labeled $PLATEAU$. The complexity of this step is $O(n)$.

A representative label is computed for every minima plateau at step 2. The connected component operator [2, 3, 6, 9] selects, as a plateau representative, the smallest pixel address in the plateau. Only pixels which have been already accessed in raster scan order ($p' < p$) have to be considered. Path compression [2, 10] is used in each operation which finds the representative r of a pixel p: All labels on the path from p to r are set (shortcut) to r after r has been found. Let $n' \leq n$ be the number of minima plateau pixels. Because we use path compression in combination with naive linking at step 2b, the worst case complexity of this step is $O(n' \log n')$ [10]. The complexity can be reduced to linear for practical problem sizes, if linking by rank or size is used [2, 10]. Nevertheless, for the images we tested, the logarithmic factor could not be observed. Let us remark that performing step 1 and 2 at the same time is also possible, leading to a faster algorithm.

At step 3, all pixel labels are replaced by their representative. In this way, the condition in Definition 3 is true for every pixel, and therefore, the presented segmentation algorithm performs a watershed segmentation. Because of the path compression is performed in each find operation, the complexity of this step is $O(n)$: Each of the n find operations walks along a path of pixels which have not been already accessed before by an other find operation. As soon as it hits a pixel that has been accessed before, the operation finishes. The total number of pixels that have not been accessed before over all find operations is smaller than n.

5. Extension to images with plateaus

Natural images do have non-minima plateaus. Therefore, an extension of Definition 3 and the algorithm is needed. In this section, we will show how to extend the set of lower neighbors NLS on a path of steepest descent to include images with non-minima plateaus.

The basic problem is that the topographical distance in Definition 6 has the same value for any two pixels, on a plateau, which do not have lower neighbors. Therefore, the geodesic distance, or an approximation of it, must be used to ensure that a pixel on a non-minima plateau gets the label from the nearest lower border pixel of the plateau. The geodesic distance between two pixels p and p' on a plateau is equal to the length of the shortest path within the plateau between p and p' [11].

A plateau PL is a connected set of pixels of the same altitude. Let $\partial_{\mathrm{PL}} = \{p' \in \mathrm{PL} \,|\, \mathrm{NLS}(p') \neq \emptyset\}$ denote the set of pixels on the border of the plateau PL which have a lower neighbor; furthermore, let $\mathrm{gdist}_{\mathrm{PL}}(p, p')$ denote the geodesic distance, or an approximation of it, between p and p' within the plateau. The minimal distance between any pixel p on the plateau PL and all border pixels $p' \in \partial_{\mathrm{PL}}$ is $\mathrm{gdist}_{min}(p, \partial_{\mathrm{PL}}) = \min_{\forall p' \in \partial_{\mathrm{PL}}} \mathrm{gdist}_{\mathrm{PL}}(p, p')$. Extending Definition 2, we can further define a watershed segmentation for images with non-minima plateaus:

Definition 8 (Extended neighbors on a path of steepest descent) *The set* $\mathrm{NLS}'(p)$ *contains the pixels of the sets* $\mathrm{NLS}(p')$ *of all border pixels* $p' \in \partial_{\mathrm{PL}}$ *such that:*

$$\mathrm{NLS}'(p) = \left\{ \bigcup_{p' \in \partial_{\mathrm{PL}} \mid \mathrm{gdist}_{\mathrm{PL}}(p,p') = \mathrm{gdist}_{min}(p, \partial_{\mathrm{PL}})} \mathrm{NLS}(p') \right\}.$$

Definition 9 (Watershed segmentation for images with plateaus) *For any image with non-minima plateaus, a segmentation is called watershed segmentation if every regional minimum* m_i *has an unique label* $\mathrm{L}(m_i)$ *and, for every pixel* $p \in \Omega$ *and* $\mathrm{NLS}'(p) \neq \emptyset$, *the following condition holds*[2]*:*

$$\exists p' \in \mathrm{NLS}'(p) \quad \textit{with} \quad \mathrm{L}(p) = \mathrm{L}(p').$$

The case of images without non-minimal plateaus is included, because the equation in Definition 8 possesses the following property: $\mathrm{gdist}_{\mathrm{PL}}(p', p) = 0 \Rightarrow p' = p \Rightarrow \mathrm{NLS}'(p) = \mathrm{NLS}(p)$.

6. The algorithm for images with plateaus

In order to perform a proper watershed segmentation and to fulfill the condition of the extended Definition 9, an additional step is needed for the case of images containing non-minima plateaus. Let us observe that, after step 1 in Section 4, plateaus of minima and non-minima are not distinguishable. For a simple input image illustrated in Fig. 2a), the result of step 1 is shown in Fig. 2c), where the label $PLATEAU$ is -1.

An intermediate step for the treatment of non-minima plateaus, is below described in the frame of the entire general algorithm. This step is similar to the arrowing technique of plateaus described in [4].

1. Do step 1 of the algorithm in Section 4.
2. Raster scan image and for every pixel p do:
   - **If** $L(p) = PLATEAU$ then do:
     a) If $\exists p' \in N(p)$ such that $L(p') \neq PLATEAU$ and $f(p) = f(p')$ then insert p' as a seed into a FiFo queue.
3. Do synchronous wave propagation from the seeds into plateaus (labeled $PLATEAU$); track in l the neighbor where the pixel is reached from.
4. Do step 2 of the algorithm in Section 4.
5. Do step 3 of the algorithm in Section 4.

At step 2, pixels on plateaus which have a lower neighbor are searched and inserted as seeds into a single FiFo queue. Then, a global wave propagation, starting from the seeds in the queue, is performed at step 3. During this process, each seed pixel, accessed in FiFo order, sets its address as a label to all neighboring $PLATEAU$ pixels of the same altitude as itself. The latter become seeds and, at their turn, are introduced into the queue. The result of this step is depicted in Fig. 2d). Let us remark that the condition of Definition 8 is fulfilled for non-minima plateaus, using an approximation of the geodesic distance. The latter is given by the time stamp of the wave propagation process, but it is not actually tracked during the algorithm.

[2] Please note that the definition leaves open the metric used for the geodesic distance. In our and most other implementations, an approximation based on the 4- or 8-square neighborhood is used.

Step 2 and 3 are both linear with the number of plateau pixels n', because each pixel is accessed at most a constant number of times.

After step 3, only minima plateaus are labeled $PLATEAU$ (see Fig. 2d)), because they do not have lower brims. The remaining stages are identical with step 2 and 3 described in Section 4. Thus, pixels on minima plateaus are connected at step 4, using a standard connected component algorithm. The result of this phase is shown in Fig. 2e). Finally, for each pixel, its label is replaced by its representative at step 5. The output image can be observed in Fig. 2f).

In Fig. 1, the presented algorithm is compared with the traditional hierarchical queue algorithm. The time measurements where performed on a Silicon Graphic^{TM} O2 workstation with an R10000 RISC processor. Approach I is the implementation of the algorithm as here presented, while in Approach II, step 2 is performed at the same time with step 1. This saves the overhead of a scan through the image, but many unnecessary seeds are detected, because labels of half of the neighbors are only available at this stage. Therefore, all pixels having a lower neighboring and also neighbors of the same altitude are stored as seeds. The results of both implementations show a significant improvement in running time compared against the classical algorithm. The images and segmentation results are shown in [8].

Running time (sec)	Approach I	Approach II	Hierarchical queues
Cermet (256 × 256)	0.09	0.10	0.15
Lenna (512 × 512)	0.39	0.43	0.77
Peppers (512 × 512)	0.42	0.44	0.74
People (1024 × 1024)	1.91	1.75	3.37

Fig. 1. Timing results.

7. Conclusion

In this paper, we have presented a definition for the watershed segmentation which is consistent with the behavior of most implementations of the watershed algorithm, namely, to chose one arbitrary label in the case of competing labels. Different distance metrics to approximate the geodesic distance on plateaus can be incorporated into the definition.

The definition led to a new type of watershed algorithm which is closely related to the connected component algorithm. We have shown that the algorithm has a linear complexity with the number of pixels, except the connection of minima plateau pixels which introduces, in the worst case, an additional logarithmic factor. For the images we tested, the logarithmic factor could not be observed. The algorithm is simpler with respect to implementation and memory requirement, and independent of the number of grey-levels in the input image. Our timing results also show a significant improvement in the running time, compared against the classical watershed algorithm.

Finally, the connected component like formulation of watersheds exhibits a better parallel potential, allowing the design of efficient and scalable parallel watershed algorithms [1, 7, 8].

References

1. A. Bieniek, H. Burkhardt, H. Marschner, M. Nölle, and G. Schreiber. A parallel watershed

3	5	5	2	8	8	8	11	10	10
5	5	11	11	8	11	11	8	10	10
11	5	11	11	9	9	9	9	8	10
11	11	11	7	7	7	7	9	9	8
11	11	11	11	11	9	7	10	8	10
11	10	11	9	7	7	9	9	10	8
11	10	11	9	11	9	10	10	8	10
11	11	11	8	8	8	8	8	10	10
11	11	11	11	10	10	10	10	10	10
10	10	10	10	10	10	10	10	10	10
11	11	11	11	10	10	10	10	10	10

a)

0	1	2	3	4	5	6	7	8	9
10	11	12	13	14	15	16	17	18	19
20	21	23	23	24	25	26	27	28	29
30	31	32	33	34	35	36	37	38	39
40	41	42	43	44	45	46	47	48	49
50	51	52	53	54	55	56	57	58	59
60	61	62	63	64	65	66	67	68	69
70	71	72	73	74	75	76	77	78	79
80	81	82	83	84	85	86	87	88	89
90	91	92	93	94	95	96	97	98	99
100	101	102	103	104	105	106	107	108	109

b)

-1	0	3	-1	3	-1	-1	6	17	-1
0	0	3	3	3	14	5	-1	17	28
10	-1	21	33	33	34	35	36	-1	28
21	21	21	-1	-1	-1	-1	36	28	-1
51	51	33	33	33	34	-1	46	-1	48
51	-1	53	54	-1	-1	55	46	48	-1
61	-1	73	54	54	54	55	76	-1	68
61	61	73	-1	-1	-1	-1	-1	77	68
90	90	73	73	73	74	75	76	77	-1
-1	-1	-1	-1	-1	-1	-1	-1	-1	-1
90	90	91	92	-1	-1	-1	-1	-1	-1

c)

-1	0	3	-1	3	4	5	6	17	8
0	0	3	3	3	14	5	6	17	28
10	10	21	33	33	34	35	36	17	28
21	21	21	-1	-1	-1	-1	36	28	28
51	51	33	33	33	34	-1	46	39	48
51	-1	53	54	-1	-1	55	46	48	48
61	-1	73	54	54	54	55	76	59	68
61	61	73	74	75	76	77	68	77	68
90	90	73	73	73	74	75	76	77	88
91	92	93	84	84	84	85	88	88	88
90	90	91	92	93	94	97	97	97	98

d)

0	0	3	3	3	4	5	6	17	8
0	0	3	3	3	14	5	6	17	28
10	10	21	33	33	34	35	36	17	28
21	21	21	33	33	33	33	36	28	28
51	51	33	33	33	34	33	46	39	48
51	51	53	54	33	33	55	46	48	48
61	51	73	54	54	54	55	76	59	68
61	61	73	74	75	76	77	68	77	68
90	90	73	73	73	74	75	76	77	88
91	92	93	84	84	84	85	88	88	88
90	90	91	92	93	94	97	97	97	98

e)

0	0	3	3	3	3	3	3	3	3
0	0	3	3	3	3	3	3	3	3
0	0	0	33	33	33	33	33	3	3
0	0	0	33	33	33	33	33	3	3
51	51	33	33	33	33	33	33	3	3
51	51	33	33	33	33	33	33	3	3
51	51	3	33	33	33	33	3	3	3
51	51	3	3	3	3	3	3	3	3
3	3	3	3	3	3	3	3	3	3
3	3	3	3	3	3	3	3	3	3
3	3	3	3	3	3	3	3	3	3

f)

Fig. 2. a)Input Image, b)Pixel addresses, c)After first scan, d)After flooding non-minima plateaus, e)After connecting minima plateaus, and f)After replacing each label with its representative.

algorithm. In *Proc. of the 10th Scandinavian Conf. on Image Analysis (SCIA97)*, Lappeenranta, Finland, June 1997.

2. T. Cormen, C. Leiserson, and R. Rivest. *Introduction to Algorithms*. MIT Press, Cambridge, MA, 1990.
3. R. Lumia, L. Shapiro, and O. Zuniga. A New Connected Components Algorithm for Virtual Memory Computers. *Computer Vision, Graphics, and Image Processing*, 22(2):287–300, 1983.
4. F. Meyer. Topographic distance and watershed lines. *Signal Processing*, 38(1):113–125, July 1994.
5. F. Meyer and S. Beucher. Morphological segmentation. *Journal of Visual Communication and Image Representation*, 1(1):21–46, September 1990.
6. R. Miller and Q. F. Stout. *Parallel Algorithms for Regular Architectures: Meshes and Pyramids*. MIT Press, Cambridge Massachusetts, London England, 1996.
7. A. Moga. *Parallel Watershed Algorithms for Image Segmentation*. PhD thesis, Tampere University of Technology, Tampere, Finland, 1997.
8. A. N. Moga and M. Gabbouj. Parallel image component labeling with watershed transformation. *IEEE Trans. on Pattern Analysis and Machine Intelligence*, 19(5):441–450, May 1997.
9. A. Rosenfeld and J. L. Pfaltz. Sequential operations in digital picture processing. *Journal of the ACM*, 13(4):471–494, October 1966.
10. R. E. Tarjan and J. V. Leeuwen. Worst-case analysis of set union algorithms. *Journal of the Association for Computing Machinery*, 31(2):245–281, April 1984.
11. L. Vincent and P. Soille. Watersheds in Digital Spaces: A Efficient Algorithm Based on Immersion Simulations. *IEEE Trans. on Pattern Analysis and Machine Intelligence*, 13(6):583–598, 1991.

MORPHOLOGICAL SEGMENTATION OF NON-IMAGE DATA

The Closing Function

EDMOND J. BREEN
CSIRO Mathematical and Information Sciences
Locked Bag 17, North Ryde NSW 2113
Australia.

Abstract. In this paper the notion of the closing function and the boundary set for multivariate data analysis is introduced. The concept of centrality and distance is explored in terms of ordering points and a Mathematical Morphological approach to cluster analysis is given.

Key words: Hull peeling, graphs, Voronoi polygons, Delaunay triangulation, cluster analysis, watershed function, centrality, order.

1. Introduction

Mathematical morphology (MM) [11, 17] has provided a collection of tools for working with images as ordered sets under the relation of inclusion. While its theory is highly generalised [16], the majority of the work to date has been in the processing of image data. Although there has been work done on graph-based morphology [8, 20], little attention has been given to the problem of morphological segmentation [13] of non-image or sparse data.

In pattern recognition, segmentation generally applies to the dividing up of a set of points (objects) of some feature space into exclusive subsets according to some partitioning scheme. In MM, image segmentation is carried out using the well-known watershed transform [1, 21]. For sparse data, closings have been used for clustering and extracting marker sets [19, 7], followed by nearest neighbour analysis [14]. Previous approaches for segmenting non-image data has been either via cluster analysis [4], mode detection [19], the use of decision functions [2], statistical and neural networks [15], or more recently, morphological image segmentation of 3D colour histograms [18]. The purpose of this work is to produce a methodology for performing a morphological segmentation of non-image data, and while many of the concepts developed can be applied in any number of dimensions, the attention is restricted to the bivariate case.

2. Definitions

Let X be a set of points $x \in \mathbb{R}^2$, $\#X$ denote the cardinality of X and X^c the complement of X. Let any open disk of radius r and centred at point y be defined as $\overset{\circ}{D}_r(y) = \{ x \mid d(x,y) < r\}$, where $d(x,y)$ is the distance between points x and y. A morphological closing of set X with a disk of radius r will be defined as:

$$\phi_r(X) = \left[\cup_y\{\overset{\circ}{D}_r(y),\ \overset{\circ}{D}_r(y) \cap X = \emptyset\}\right]^c, \tag{1}$$

which is an idempotent, increasing and an extensive transform.

A disc will be defined to be touching point $x \in X$ if $d(x,y) = r$. The radius of the largest open disk touching point x, but not intersecting X will be defined as:

$$r_{max}(x) = \underset{r}{\operatorname{argmax}}\ \overset{\circ}{D}_r(y),\ \forall y \in \mathbb{R}^2,\ d(y,x) = r,\ \overset{\circ}{D}_r(y) \cap X = \emptyset, \tag{2}$$

which is also related to α_{max} defined by Edelsbrunner *et al.* [5], in the field of computational geometry.

The closed Voronoi polygon, V_p, associated with a point $x \in X$ is defined as:

$$V_p(x) = \{ y \mid d(y,x) \leq d(y,x'), x \neq x',\ x,x' \in X\}. \tag{3}$$

From Eq. 3 and as seen from Fig. 1(a), each $r_{max}(x)$ can be equivalently defined as:

$$r_{max}(x) = \max\{ d(x,y) \mid y \in V_p(x)\}. \tag{4}$$

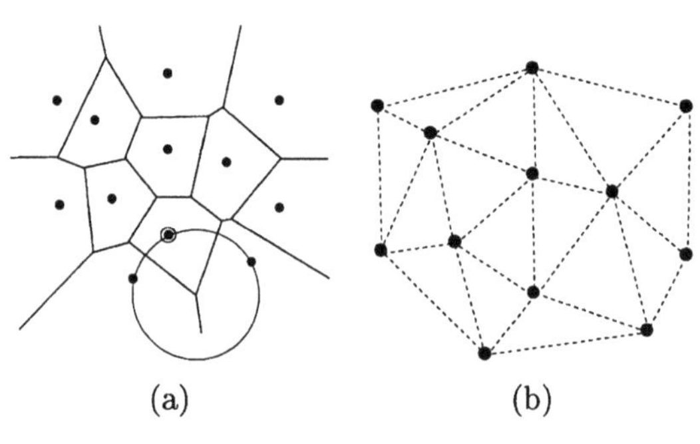

Fig. 1. (a) Voronoi diagram of a set. The large circle overlayed is highlighting the r_{max} of the point encircled with a smaller disc. The Voronoi polygons of X are either unbounded or bounded. $Vp(x)$ is an unbounded polygon if x is a point on the boundary of the convex hull of X. The r_{max} of such points equals infinity. (b) The dual of the Voronoi diagram is the Delaunay triangulation, which is shown in broken lines.

The dual of the Voronoi diagram, Fig. 1(a), is known as the Delaunay triangulation, Fig. 1(b), which forms the structure of a graph, $G = (X, E)$. The set of vertices corresponds to the points in X, and the set of edges E are formed from the arcs linking the Voronoi neighbours of X. Two points $x, x' \in X$ are called Voronoi neighbours if their Voronoi polygons share more than one common point; that is, $\#(V_p(x) \cap V_p(x')) > 1$.

A Delaunay graph is undirected, $(x,x') \in E \Leftrightarrow (x',x) \in E$, and contains no loops, $(x,x) \notin E$. The vertices in the graph can have values associated with them;

the image of values on X is denoted by $f \in fun(X)$. The range of f is $\{0, 1\}$ for a binary graph and $\mathbb{R}$ for a decimal graph.

The neighbourhood of x is then defined through the Delaunay graph of X:

$$N_G(x) = \{ x' \in X \mid (x, x') \in E\} \tag{5}$$

3. Requirements

To apply a morphological segmentation on an arbitrary set of points X, the following are needed: **1**) a segmentation function, which imposes a grey-level topographic surface on the data set – analogous to the distance transform on a binary image. Strictly speaking, a segmentation function should either be a lower or upper complete function [13]; **2**) a neighbourhood set, which defines the neighbours of each point $x \in X$; and **3**) a marker set; that is, a set of inner marker points that roughly identifies each segment or cluster in X. The analogy from image analysis is the set defined by the ultimate erosion [17]. In some instances, an outer marker set may also be required [13].

3.1. Segmentation function

A segmentation function should transform the set X so that interior regions or segments are assigned low values and the regions between and on the outside of segments are assigned higher values. In the binary image case, a segmentation function can be formed from the inverse distance transform, and in the grey-scale case, an edge image is often used.

It is interesting to note that another way of looking at the distance transform, is as an ordering of the points $x \in X$ with respect to their distance to another set. Functions that place an ordering on the points in X may also form segmentation functions.

Sets of points can be ordered with respect to centrality; a measure of how central x is with respect to X. For example, Leavitt introduced a centrality index, β, for each $x \in X$ based on the following ratio (see [10]):

$$\beta_i = \frac{\sum_i \sum_j d(x_i, x_j)}{\sum_j d(x_i, x_j)}, \quad x_i, x_j \in X.$$

Green [6] has suggested using hull peeling techniques for ordering bivariate data. For example, if the boundary points of set X are identified, they can be removed. If this process is repeated, the next set of boundary points will be discovered, and so on, see Fig. 2. However, the major failing of using centrality indices is that they assign high values to points between regions as well as to the interior points. In contrast, a segmentation function will assign low values to interior points and high values to exterior points, which includes those points lying between clusters.

Density functions seem likely candidates to guide the creation of clusters, via a morphological segmentation, Fig. 3. The problem then becomes one of defining the shape and size of the density function. A Parzen window can be used [4, 9], or even simpler, a $n \times n$ summation kernel. If the size of the kernel is too large then it becomes a centrality index, which assigns high values to points lying between

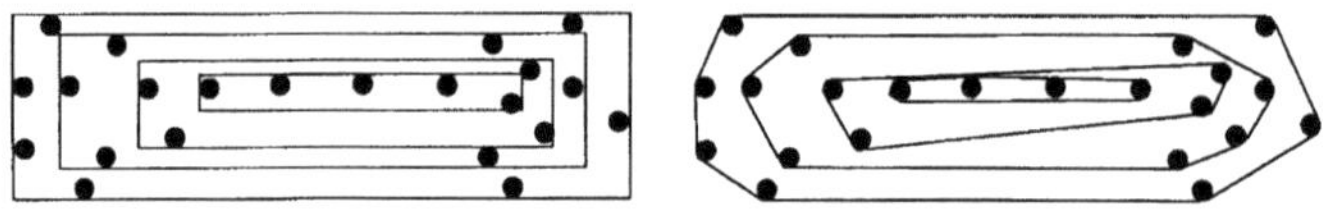

(a) (b)

Fig. 2. Peeling bivariate data. (a) Rectangular peels. (b) Convex-hull peels.

segments and to interior points. However, if there exists an appropriate size, then a reasonable segmentation function can be formed, as shown in Fig. 3(b). This leads to the final segmentation shown in Fig. 3(c).

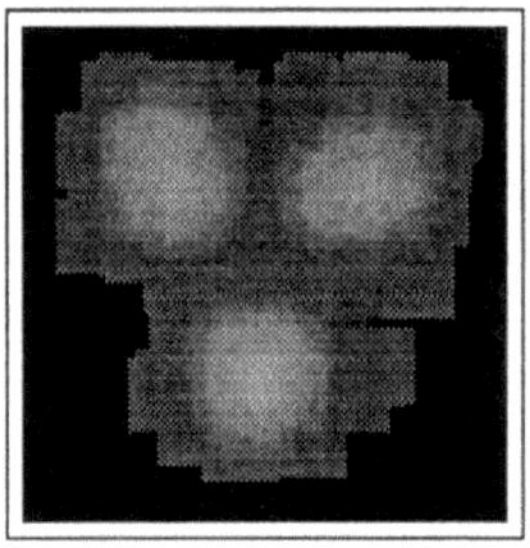

Fig. 3. Density function. (a) Set X. (b) Density distribution of set X formed from a $n \times n$ window. (c) Watershed segmentation. The density function approach essentially transforms the data so that traditional image analysis procedures can be used. Here, the inverse of the density image in (b) was used as the segmentation function, the marker set was formed via thresholding of (b) and the neighbourhood set for each pixel was defined by their 8 connected neighbours.

Closings can be used for clumping, Fig. 4(a), or for forming marker sets. However, care must be taken to detect the creation of new regions not connected to the original set, an example of which is shown in Fig. 4(b), see also [12]. Further, the points in X could be ordered with respect to their distance to $\cup_c \overset{\circ}{D}_r(c)$, given in Eq. 1. The problems with this latter approach are that the appropriate value of r must be defined and that it fails to consider the relations between the points in X.

3.1.1. *The boundary of a sparse set*

If the boundary of X could be defined in some sense, then the points in X could be ordered with respect to this boundary. However, without further information, it is not possible to tell which points of X belong to the boundary and which belong to the interior. Nevertheless, as shown in Fig. 4(a), from a closing of X it is possible to define members of X that intersect with the boundary of the closing hull of X,

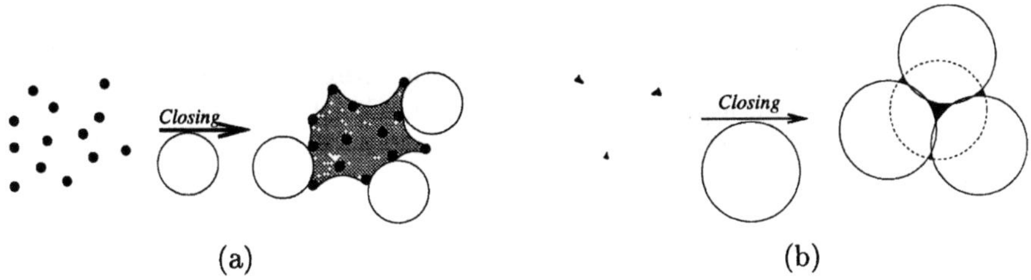

Fig. 4. (a) The closing hull of a set of points, with respect to a disk. The disks overlayed show the relation between a disk of radius, D_r, and $\phi_r(X)$, see Eq. 1. (b) Morphological closings causing the creation of a disconnected set of points (shaded region). The disks overlayed are used to reveal the relationship between the original points and the newly created region. The dotted disk shows that a disk of the specified size can not fit between the points without intersecting them.

$\phi_r(X)$, using the equation:

$$b_r(X) = \{\ x \mid r_{max}(x) \geq r,\ x \in X\}. \tag{6}$$

Here, $b_r(X)$ defines a boundary subset of X with respect to the closing of X using a disk of radius r. The boundary set of X, Eq. 6, has the following interesting properties:

$$\begin{array}{rcll} b_i(b_i(X)) & = & b_i(X) & \text{idempotency,} \\ b_i(X) & \subseteq & X, \quad i > 0 & \text{anti-extensive,} \\ b_i(X) & \subseteq & b_j(X) \ \ i > j & \text{anti-absorption.} \end{array} \tag{7}$$

The anti-absorption property means that $x \in b_i(X)$ iff $i \leq r_{max}(x)$, and that the number of boundary sets that contain x leads to the definition of a closing function:

$$\begin{array}{rcl} C(x) & = & \int_{r=0}^{\infty} I(r \leq r_{max}(x))\, dr,\ x \in X \\ & = & r_{max}(x), \end{array} \tag{8}$$

where I is an indicator function. Thus, the closing function orders the points of X with respect to the radii of the largest open disks that can touch each point without intersecting X, or alternatively expressed, it orders the points in terms of the number of boundary sets that contain each x, see Fig. 5. The closing function can be used to form a segmentation function, as it assigns high values to those points lying between and on the periphery of a clusters, but low values to interior points. One problem however, is that the value of the closing function is infinite for all those points of X that lie on the boundary of the convex hull of X (see Fig. 1(a)). The boundary points of the convex hull are defined by the boundary set, Eq. 6, when $r = \infty$. Therefore, in order to compute the closing function, a further definition is given:

$$C(x) = \begin{cases} r_{max}(x) & \text{if } x \notin b_\infty(X) \\ kR & \text{otherwise} \end{cases} \tag{9}$$

where $R = \max\{r_{max}(x) \mid x \notin b_\infty(X),\ x \in X\}$ and $k \geq 1$.

3.2. Neighbourhoods

In the previous section, the closing function was seen as a candidate segmentation function for MM segmentation of non-image data. Next, we need to be able to define a marker set and a neighbourhood relation. For the purpose of MM-based clustering, the neighbourhood relation between the points of X will be defined by the Delaunay graph $G = (X, E)$, Eq. 5.

3.3. The Marker Set

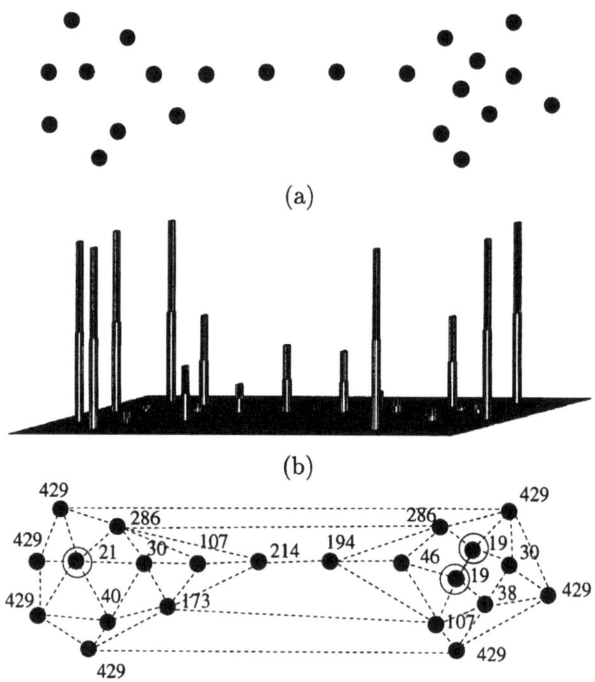

(a)

(b)

(c)

Fig. 5. (a) Set X. (b) The closing function of X. (c) The Delaunay triangulation of X. The value of the closing function at each vertex is given, and the vertices forming regional minima are circled.

Set X in Fig. 5(a), shows (arguably) two identifiable clusters connected by a bridge of points. The closing function of X is given Fig. 5(b). Fig. 5(c) shows the decimal graph, G, formed from $C(X)$. A marker set could be generated for segmentation purposes by thresholding $C(X)$. However, as a more general approach, the use of regional minima will be suggested.

Let $W \subset X$ be a connected set of vertices in G, such that any two vertices of W can be connected by a path of vertices within W. Then W will form a regional minimum if:

$$\forall x \in W,\ C(x) < C(x'),\ \forall x' \in N_G(x)\backslash W,$$

and

$$\forall x \in W,\ C(x) = k,\ \text{where } k \text{ is some constant.}$$

The above defines a set of vertices in G, all of the same value, surrounded by vertices of greater value. In Fig. 5(c), the two regional minima, $\{W_1, W_2\}$, are shown with their vertices circled.

4. Watershed segmentation

Next, the regional minima of G need to be labelled from 1 to M, where M is the number of such minima in G. These labels are then propagated through the topology of G, using flooding: place all unlabelled neighbours of each regional minima onto a stable priority queue [3], and then pop off each vertex, x, in order of increasing $C(x)$; to each vertex, x, popped off the queue, assign the label of the neighbour that pushed them onto the queue; then push the neighbours of x onto the queue that are not already on the queue or labelled. Continue until all vertices have been labelled. The result of this procedure run on G, Fig. 5(c), is given in Fig. 6.

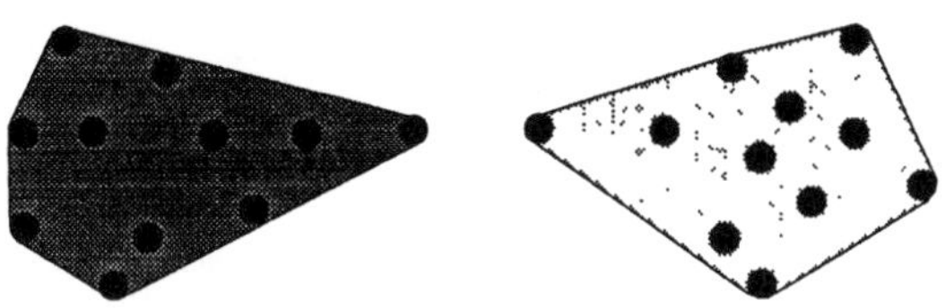

Fig. 6. Watershed segmentation of the data set given in Fig. 5(a). The shaded regions reveals the convex hulls of the two segments automatically detected.

5. Conclusion

In this report, methods for Mathematical Morphological segmentation on non-image data have been presented. Methods for ordering sets of bivariate data were explored and measures of centrality versus distance transforms were presented. In so doing, two new functions were defined: **1**) the closing function Eq. 8 and **2**) the boundary function, Eq. 6. While further treatment of these functions is not possible within the scope of this report, the potential of these transforms is worth exploring in terms of their use for defining shape spectrum information and analysing multivariate data.

6. Acknowledgements

The author would to thank Ronald Jones, Hugues Talbot, Pierre Soille and Michael Buckley for useful comments and suggestions during the course of this work.

References

1. S. Beucher and C. Lantuéjoul. Sur l'utilisation de la ligne de partage des eaux en détection de contours. Technical Report N-598, Ecole des Mines de Paris, Centre de Morphologie Mathématique, May 1979.
2. E. J. Breen. Regression methods for automated colour image classification and thresholding. *J. Microsc.*, 174, 1994.
3. E. J. Breen and D. Monro. An evaluation of priority queues for mathematical morphology. In J. Serra and P. Soille, editors, *Mathematical morphology and its applications to image processing*, pages 249–256. Kluwer Academic Publishers, 1994.
4. R. O. Duda and P. Hart. *Pattern Classification and scene analysis*. John Wiley, New York, NY, USA, 1973.
5. H. Edelsbrunner, D. G. Kirkpatrick, and R. Seidel. On the shape of a set of points in the plane. *IEEE Transactions on Pattern Analysis and Machine Intelligence*, 29(4):551–559, July 1983.
6. P. Green. Peeling bivariate data. In V. Barnett, editor, *Interpreting multivariate data*, chapter 1, pages 3–20. John Wiley, New York, NY, USA, 1981.
7. R. M. Haralick and L. G. Shapiro. *Computer and Robot Vision*, volume 1. Addison-Wesley, Readings Massachusetts, 1992.
8. H. Heijmans, P. Nacken, A. Toet, and L. Vincent. Graph morphology. *Journal of Visual Communication and Image Representation*, 3(1):24–38, March 1992.
9. H. Herbin, N. Bonnet, and P. Vautrot. A clustering method based on the estimation of the probability density function and on the skeleton by influence zones. application to image processing. *Pattern Recognition Letters*, 17:1141–1150, 1996.
10. A. Kaufmann. *Points and Arrows: the theory of graphs*. Transworld student library. Transworld publishers Ltd, 57-59 Uxbridge Rd, London W.5, 1972.
11. G. Matheron. *Random sets and integral geometry*. John Wiley, New York, NY, USA, 1975.
12. G. Matheron and J. Serra. Strong filters and connectivity. In J. Serra, editor, *Image analysis and mathematical morphology: theoretical advances*, volume 2, chapter 10, pages 141–158. Academic Press, New York, NY, USA, 1988.
13. F. Meyer and S. Beucher. Morphological segmentation. *Journal of Visual Communication and Image Representation*, 1(1):21–46, September 1990.
14. J.-G. Postaire, R. Zhang, and C. Lecocq-Botte. Cluster analysis by binary morphology. *IEEE Transactions on Pattern Analysis and Machine Intelligence*, 15(2):170–180, February 1993.
15. B. Ripley. Statistical aspects of neural networks. In O. E. Barndorff-Nielsen, D. R. Cox, J. L. Jensen, and W. S. Kendall, editors, *Chaos and Networks — Statistical and probabilistic aspects*, pages 40–123, London, 1993. Chapman & Hall.
16. C. Ronse and H. Heijmans. The algebraic basis of mathematical morphology. part ii. openings and closings. *Computer Vision, Graphics, and Image Processing: Image Understanding*, 54:74–97, 1991.
17. J. Serra. *Image Analysis and Mathematical Morphology*. Academic Press, New York, NY, USA, 1982.
18. P. Soille. Morphological partitioning of mutlispectral images. *Journal of Electronic Imaging*, 5(3):252–265, 1996.
19. A. Touzani and J.-G. Postaire. Clustering by mode boundary detection. *Pattern Recognition Letters*, 9:1–12, 1989.
20. L. Vincent. Graphs and mathematical morphology. *Signal Processing*, 16:365–388, 1989.
21. L. Vincent and P. Soille. Watersheds in digital spaces: an efficient algorithm based on immersion simulations. *IEEE Transactions on Pattern Analysis and Machine Intelligence*, 13(6):583–598, June 1991.

COLOR SEGMENTATION ALGORITHM USING AN HLS TRANSFORMATION

CLAIRE-HÉLÈNE DEMARTY and SERGE BEUCHER
Centre de Morphologie Mathématique
Ecole des Mines de Paris
35, rue Saint-Honoré 77 Fontainebleau cedex, FRANCE
tel: (19) 01 64 69 47 97, (19) 01 64 69 47 92
fax: (19) 01 64 69 47 07
email: beucher@cmm.ensmp.fr, demarty@cmm.ensmp.fr

Abstract.
A new segmentation procedure for color images is detailed in this article. The segmentation still ends with the watershed function, but the whole series of pre-treatments applied to the color images is of interest. After a HLS transformation of the images, a color reduction is applied, followed by a median filtering. The gradient is then constructed from this resulting image. After this step, two different ways to obtain the boundaries are proposed: either a hierarchical segmentation or a simple watershed line followed by a merging of the bassins according to the color of their minima. All these steps are detailed in the following and the results obtained on test images are discussed.

Key words: Color segmentation, HLS space, Color reduction

1. Introduction

Color images are nowadays materials when dealing with image processing. But, if grey level treatments are numerous and well-known, color image processing is not so developped. In particular, color segmentation is at its beginning. This method presents an algorithm of color segmentation, based on mathematical morphology.

The segmentation produced by this algorithm does not allow to keep the details of the color images. The aim of our algorithm is to obtain the main regions of the image. Such an algorithm could be used in multimedia problems, for example indexation, or, in general, in any application which does not require small details.

As most of morphological segmentation algorithms, it requires a filtering step, before a gradient and a watershed building. But, before any filtering, two pre-treatments are applied to the color images. The first one is based on the color space HLS, and it will be called the HLS transformation, in the following. This pre-treatment is followed by a color reduction, and finally by a median filtering. Except the HLS transformation, which is rather new, the novelty of this algorithm lies in the ordering of these steps. Besides its simplicity, the resulting algorithm presents the advantage of enhancing the main regions of the image.

Each step of the color segmentation is described in the following, and some results are presented on different color images.

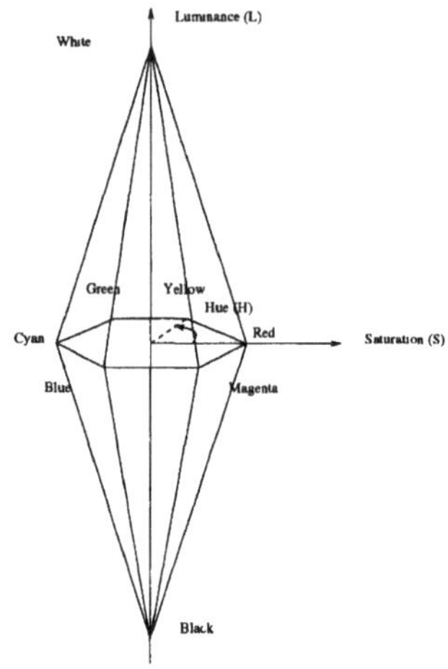

Fig. 1. HLS color space.

2. Segmentation steps

2.1. HLS TRANSFORMATION

Before any description of the HLS transformation, a few reminders of the HLS color space are needed. The three components of the HLS space are respectively the Hue, Luminance and Saturation. The Hue and Saturation components contain the whole color information. Among the Saturation and the Hue components, only the Hue really gives an information of the color domain of a pixel. According to the Hue component, one knows if a pixel is blue, yellow or green ... With a constant Hue, blue for example, different saturation values produce different pixels in the same color domain (with the same hue), from light to dark blue. The HLS space is illustrated in the figure 1. The vertical axis is the Luminance; the horizontal angle represents the Hue and, for a given angle, the radius is the Saturation.

The equations allowing to pass from the RGB color space to the HLS color space will not be detailed in this article. For a description, one should refer to [7].

The HLS transformation is a simple transformation which allows to split the color information from the luminance information. A first transformation was defined as follows:

$$\begin{aligned} \text{HLSTransformation} : \text{HLS} &\longrightarrow \text{HLS} \\ (h, l, s) &\longrightarrow (h', l', s') = (h, 128, 255) \end{aligned} \quad (1)$$

This simple transformation only keeps the Hue information of the images. Because of the HLS space representation, this transformation presents the drawback of producing more or less arbitrary hue values for pixels with small saturation values.

These pixels do not really have any representative hue value, but a grey level. As their saturation value is small, small angle variations lead to critical hue variations. This is the reason why the sky appears green and yellow more than blue, in the HLS transformation (see figure 2, part *b*).

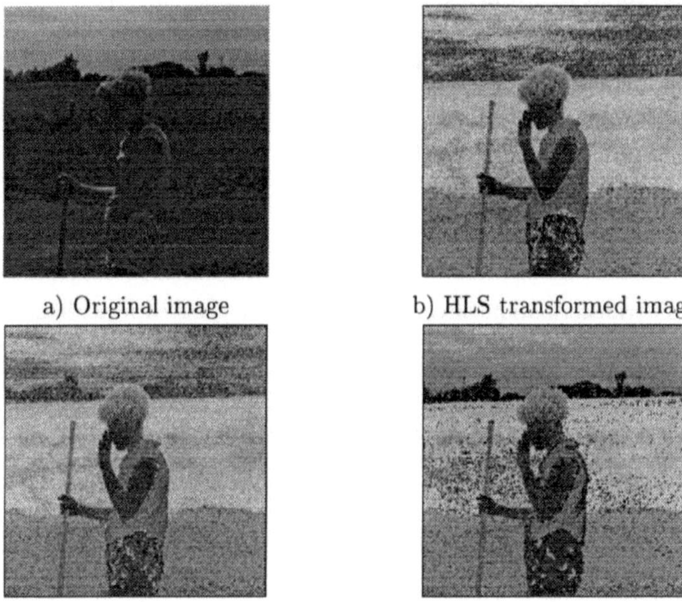

a) Original image b) HLS transformed image

c) HLS transformed image ($s_0 = 10$) d) HLS transformed image ($s_0 = 30$)

Fig. 2. Examples of HLS transformations. The HLS transformation of the original image (a) is successively presented for different thresholds: $s_0 = 0$, $s_0 = 10$ and $s_0 = 30$.

Another HLS transformation is then introduced to take into account those pixels whose color is not significant enough. Two different classes of pixels are considered, according to their saturation value. When the saturation is small enough (less than a threshold s_0), the pixel saturation and hue become zero, whereas the luminance is kept unchanged. For pixels with a high saturation value, the previous HLS transformation is applied.

This new HLS transformation, called ***enhanced HLS transformation***, is then:

$$\text{Enhanced HLS Transformation: HLS} \longrightarrow \text{HLS}$$

$$(h,l,s) \longrightarrow (h',l',s') = \begin{cases} (h,128,255) & \text{if } s \geq s_0 \\ (0,l,0) & \text{if } s < s_0 \end{cases} \quad (2)$$

For different saturation values, the image colors will be transformed either into the nearest grey level, or into the hue only. These changes are illustrated in figure

2, for three different values of s_0: 0, 10 and 30 (the saturation range goes from 0 to 255). One should notice that, when s_0 is increasing, the sky is transformed in grey, and the background trees appear. For $s_0 = 30$, the transformed image contains only the main regions, which are sufficient for the segmentation.

Besides providing initial images for the color segmentation, the enhanced HLS transformation allows to distinguish objects according to their *spectral response* (the hue histogram of the image). For example, the sky and the vegetation have different hues; a simple threshold on the HLS transformation will then separate the sky regions from the vegetation regions. This is also a way to detect people, by analysing the spectral response of the skin. In fact, no matter the original color of the people skin is (white, yellow, black ...), the spectral response will be the same, between red and yellow. This is also illustrated in figure 3.

2.2. Color reduction

After the HLS transformation of the image, another pre-treatment is carried out. The aim of the color segmentation is still to produce the main color regions; with this goal, a drastic color reduction (up to 16) will simplify the image without reducing the information. The reduced HLS image is illustrated in figure 5. From this step, the colors shown on the image have no more significance, they are only labels of the remaining colors. This transformation will obviously not penalize the segmentation.

2.3. Median filtering

Finally, a last filtering (as usual in morphological segmentation) is needed to prepare the image. Three different filters were tested. The first two were morphological filters, an Alternate Sequential Filter and a Filter by Reconstruction. For a definition of these two morphological filters, one should refer to [6, 4]. These two filters were applied on each of the three components of the color image separetely, and the result image was obtained by recombining the three transformed components. This had the major drawback of introducing new colors into the image, which has the inverse effect of the color reduction. Moreover, these new colors do not have any meaning; they do not correspond to any image regions.

Besides, the alternate sequential filter and the filter by reconstruction do not provide a good filtering of the image, as it can be seen in figure 4. The first one modifies the region boundaries, and the second introduces pointillism, which will further produce over-segmentation.

These are the reasons why the filter which is used in this segmentation is a vectorial median filter. In the grey level case, a median filter acts on pixels according to their grey level order. As for color images, no total order can be defined on the colors, which are three-dimensional vectors. The vector median filter replaces then a pixel color by the color, which is at the minimum distance of all its neighbour colors. This is in fact a pseudo-median filter, but it treats the three components together and does not introduce new colors. Furthermore, the result contains no pointillism and the boundaries are still relevant (see figure 5, part *d*). To get more information on such a filter, one can refer to [1].

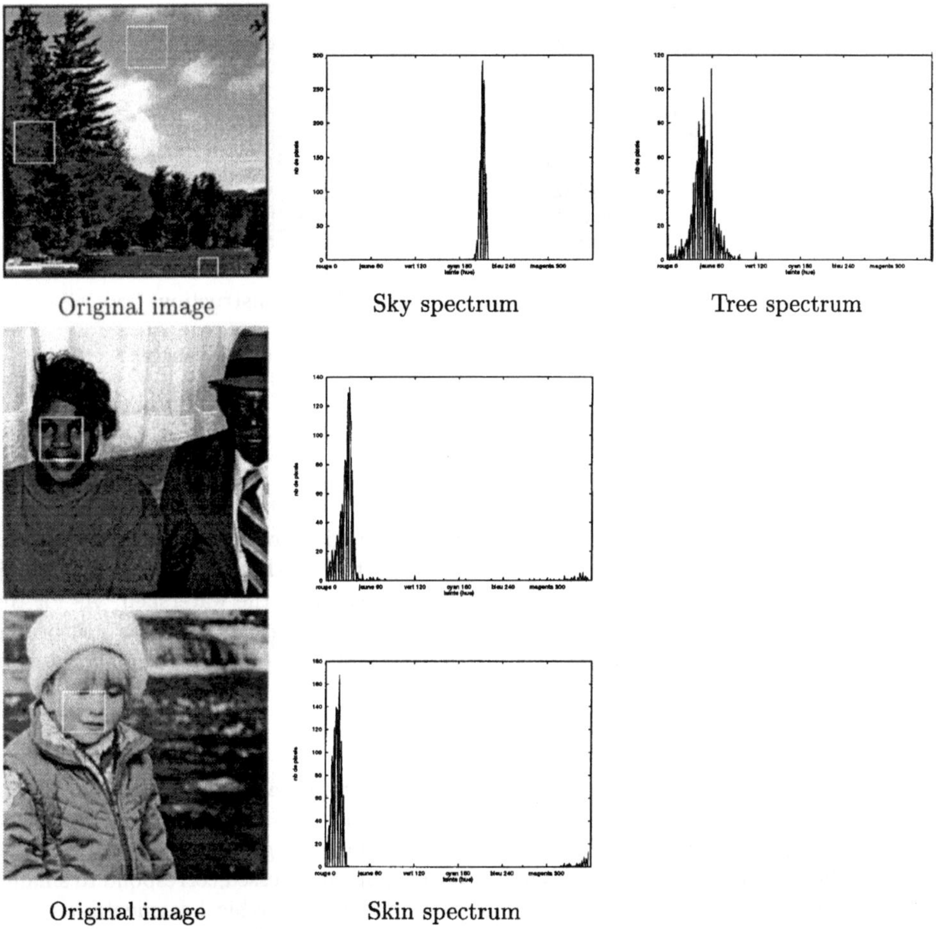

Fig. 3. Study of the spectral response (hue histogram) of different regions.

2.4. Gradient

The gradient is the usual morphological gradient, but as the working image is a color image, one has to build the gradient from the three components again. We chose to construct the gradient separately on each of them, the resulting gradient is then obtained by adding these three gradients. Adding the three gradients allows to enhance the real boundaries which exist in each component, compared to boundaries which are present in one component only, as a suppremum of the three gradients would have led. These last boundaries do not represent real color differences.

The gradient step is shown in figure 5, part *e*.

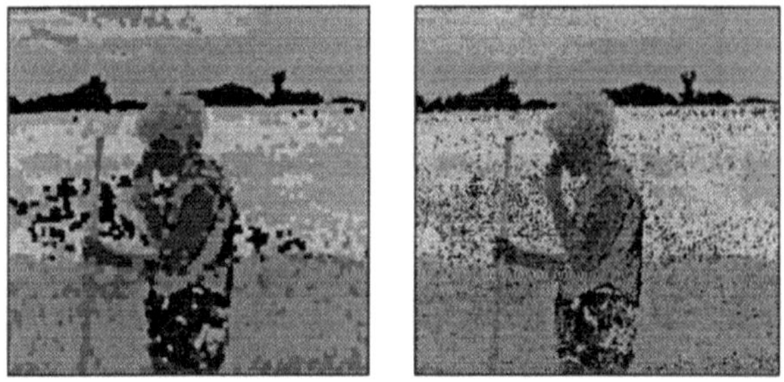

Alternate sequential filter Filter by reconstruction

Fig. 4. Morphological filters applied on the reduced color image.

2.5. SEGMENTATION

In order to obtain the boundaries of the main regions of the image, two different segmentation algorithms were studied. The first algorithm is detailed in [2, 5, 3]; it consists in a hierarchical segmentation, still based on the watershed. As for the second algorithm, a watershed is also build and a merging of the regions, according to the color of their minima, is processed. These two algorihms are successively described in the following.

2.5.1. *Hierarchical segmentation*

The hierarchical segmentation allows to obtain iteratively less detailed segmentations. It starts from the image watershed, which is usually quite over-segmented, and it erases successive boundaries, which are found less significant. At each step, the gradient is modified, and the boundaries which are suppressed correspond to smaller values of the modified gradient. References on the hierarchical segmentation were already given above, but here are a brief description of a hierarchical segmentation algorithm, also called waterfall algorithm.

First the usual gradient of the HLS transformed image is computed. From this gradient, a first watershed is build (see figure 5, part *f*). The gradient modification is then done as follows. One constructs an image, which is at 255 into the bassins, and which has the gradient values on the watershed lines. The modified gradient image is the reconstruction by erosion of the original gradient from this new image. A new watershed is then built from the modified gradient. This operation could be repeated until no more boundaries are left. At each step, the less relevant boundaries, according to the new gradient, are suppressed.

For the kind of color images that were tested, only one step of hierarchical segmentation was necessary. As for the execution time, this algorithm needs then two watershed constructions, and the preparation of the new gradient. This is quite equivalent to a simple watershed segmentation. The result are illustrated in figure 5, part *g*. It should be noticed that all main regions are kept in the final segmentation.

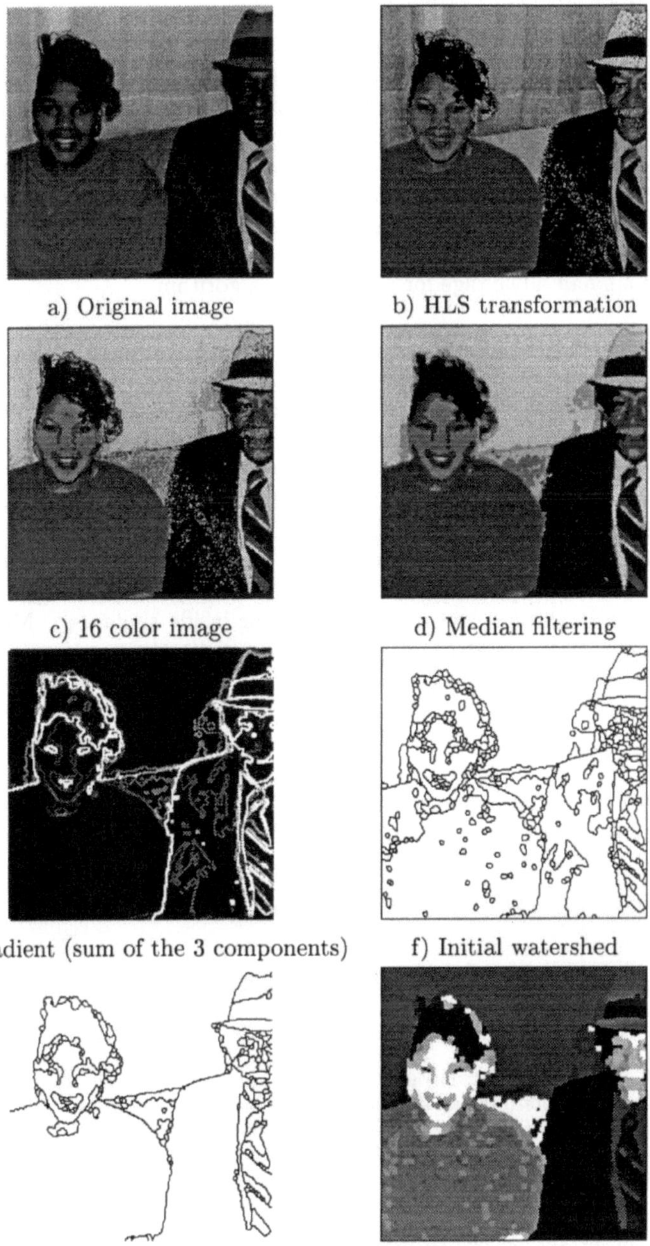

a) Original image
b) HLS transformation
c) 16 color image
d) Median filtering
e) Gradient (sum of the 3 components)
f) Initial watershed
g) Hierarchical segmentation
h) Merging of the bassins

Fig. 5. Different steps of the color segmentation.

2.5.2. *Merging of the bassins*
This algorithm begins with a classical watershed obtained from the gradient, each bassin corresponding to a minimum of the gradient. The resulting image is generally over-segmented. To each minimum of the gradient, one can associate the corresponding color in the reduced image. One merges the bassins whose minima have the same color. The shared boundary is suppressed. This algorithm is rather simple and the result (presented in figure 5, part *h*) is correct: the main regions and their boundaries are well restored. When comparing with the hierarchical segmentation, the results are quite equivalent; as for the execution time, it is also quite the same, with a small advantage for the merging algorithm.

3. Conclusion

In this article, a new color segmentation algorithm was presented. This algorithm contains uses several pre-treatments, in order to simplify the color image, before the segmentation. These pre-treatments reduce dramatically the color information (number of colors, small details, etc), without modifying the important region boundaries. The results on test images prove the validity of the algorithm, for the foreseen applications. These applications can be mainly found in the multimedia domain. One may consider to address such problems as indexing images or sequences, finding people in images, sorting out outside or inside shots, etc. More generally, such a segmentation could be of a great interest for the new image compression standard MPEG7.

Acknowledgements

Parts of this study are supported by the KODAK-PATHE company. Other parts are supported by the CNET-CCETT (France Telecom). The original images are copyrighted by KODAK-PATHE.

References

1. J. Astola, P.Haavisto, and Y. Neuvo. Vector median filters. *Proceedings of the IEEE*, 78(4):678–689, April 1990.
2. S. Beucher. *Segmentation d'images et Morphologie Mathématique*. PhD thesis, Ecole des Mines de Paris, 1990.
3. Serge Beucher. Watershed, hierarchical segmentation and waterfall algorithm. In J. Serra and P. Soille, editors, *Mathematical Morphology and its Applications to Image Processing, Proceedings ISMM'94*, pages 69–76. Kluwer Academic Publishers, 1994.
4. J. Mattioli and M. Schmitt. An efficient algorithm for computing the antigranulometry. *SPIE*, 2030:167–178, july 1993.
5. L. Najman and M. Schmitt. A dynamic hierarchical segmentation algorithm. In *ISMM94*, pages 13–14, september 1994.
6. J. Serra. *Image Analysis and Mathematical Morphology*. Academic Press, London, 1982.
7. T.Y. Shih. The reversibility of six geometric color spaces. *Photogrammetric Engineering & Remote Sensing*, 61(10):1223–1232, October 1995.

COMPUTING THE WATERSHEDS OF LARGE THREE-DIMENSIONAL IMAGES USING LIMITED RANDOM ACCESS MEMORY

COSTAS COTSACES and IOANNIS PITAS
Department of Informatics
University of Thessaloniki
Thessaloniki 540 06 GREECE
Email: pitas@zeus.csd.auth.gr

Abstract. The watershed transform, and in particular its computation by immersion, is probably the most popular morphological segmentation procedure. However, because of its random-access nature, it is inefficient in *large* images, that is those that cannot fit in available random access memory. In this paper, we explore ways to increase its memory efficiency. The algorithm that was developed keeps in memory the propagation fronts and only one greylevel of the image. This algorithm uses much less memory than the one commonly seen in the literature so far. Several techniques have been developed in this work in order to minimize the effect of the set operations required by the algorithm. These include fast search methods, double propagation fronts, directional propagation, and others.

Key words: mathematical morphology,watershed,segmentation,tree-dimensional images ,memory efficiency

1. Introduction

When processing large images, in most cases it is practically impossible to keep the entire image in the computer memory (RAM). Usually, the image is stored in the permanent storage medium and the algorithm reads parts of the image that are needed in the computation. The part of the image that is read each time has to be small enough to fit in available memory. Each point in the image is usually read as many times as the algorithm processes it. However, seek time exceeds read time by orders of magnitude, and therefore it is important to read the data sequentially, when this is possible. Apart from that, naturally it is also important that each point of the image be read as few times as possible. However, in propagation-based methods for the watershed transform there is no predefined order for processing the image, thus making any attempt to read the image in blocks pointless.

So, in order to implement the watershed transform in large images, a way must be found to perform the transformation on a part of the image at a time. The obvious way to achieve this is to take advantage of the fact that the algorithm processes the greylevels of the image sequentially, from the lowest to the highest. After decomposing the image into its constituent greylevels, each one could be processed separately. Thus, only a small part of the image must be kept in memory at a time. The loading of this part can be done in one step, without the need for unnecessary disk seek operations.

In order to store each greylevel in the minimum possible space, it is stored as an array, which is sorted to facilitate fater access. The sorting order of such an array will be called the *basic order*. For reasons apparent later, this order will be the same with the order that images are saved on disk.

There are some other issues in the watershed transform by immersion which require some attention. The first is finding the image minima in each greylevel, from which the propagation fronts emanate. The second is the transition of the propagation fronts from one greylevel to the next, without the storage of unnecessary points. In the following, we shall discuss these issues, as well as the procedure for the decomposition of the image into its constituent greylevels and its recomposition from them. We shall also describe the modifications that need to be made to our algorithm for performing *marker*-based image segmentation.

2. Image decomposition and output image reconstruction

In the decomposition watershed algorithm, the image must be decomposed into separate greylevels. These greylevels need to be stored on disk, each on a separate file so that they can be read independently from each other as the algorithm progresses. At the end of the algorithm, these files will must be merged to give the reconstructed output image.

The decomposition of the image into greylevels poses no particular problems. Since the basic order that is used in the algorithm is the same as the order in which the image is stored on disk, the only thing that needs to be done is to read the input image serially and write each point (in coordinate form) to the file that corresponds to its greyvalue. Points are first written to buffers and when one of them is full its points are written to disk all at the same time. Because this preprocessing happens before the beginning of the actual watershed transform, the buffers can use all available memory.

The reconstruction of the output image is performed in two phases, one during the execution of the algorithm and one after its end. After each greylevel is processed, its points are written (again as coordinates) to intermediate files that correspond to the z-planes of the image (the planes with a constant z coordinate). Because the basic ordering of the image is primarily according to the z coordinate, the disk write head needs to move only as many times as the size of the image along the z-coordinate. When the algorithm finishes, the intermediate files are read again and the points they contain are written at their final position in the output file (which, like the input file, is in bitmap format).

From the above analysis it can be seen that, the algorithm requires reading/writing $3+3+1=7$ bytes per image point, since we need one byte per coordinate. This is not satisfactory and a simple method for the compression of the point coordinates was used in order to reduce the number of bytes that are read/written.

3. Propagation front maintenance

In the classic watershed-by-immersion algorithms, the transition of the propagation fronts from one greylevel to the next does not present any problems. Since all points, irrespective of greyvalue, are directly accessible, it suffices to check whether a point

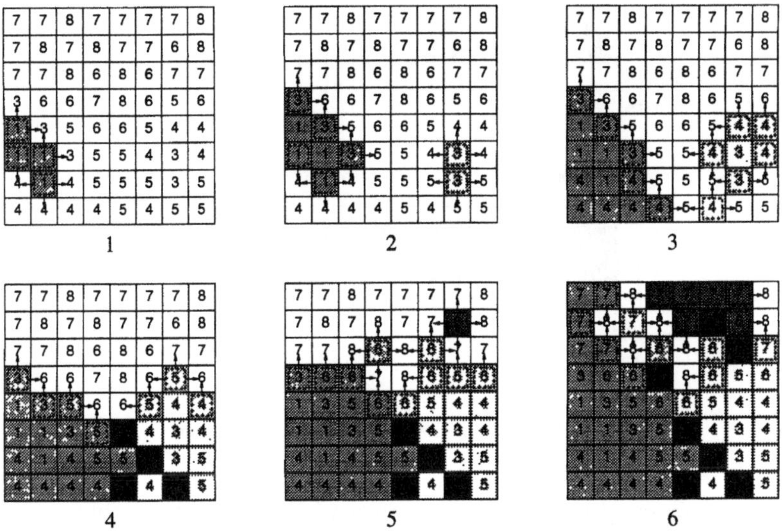

Fig. 1. Example of the operation of the interfront. The various lightly shaded squares are points belonging to catchment basins, the dark squares are watershed points, the dashed borders indicate points belonging to the interfronts, and the arrows show the directions stored in the flags of the labels of the intergrey points.

is adjacent to another of lower greyvalue, in order to determine that the propagation front continues from this point. In the decomposition watershed algorithm we can only access the points of the current greylevel. Thus, in order to perform the transition, we must keep, at the current greylevel and for every propagation front, those points that are adjacent to other points that have a higher greylevel. Because the fronts propagate into all the subsequent greylevels and not only into the next one, such points must be kept in storage until they cease to be adjacent to points of higher greylevels. However, at the same time, the storage of points that are only adjacent to lower greylevels must be avoided for memory minimization. In order to achieve the above objectives, a rather complex data structure is used, consisting of a dynamically managed array of points which also contain information about the position of adjacent points belonging to higher greylevels. For the sake of brevity, this structure will be called an *interfront* (from inter-greylevel propagation front).

The basic functions that must be performed on the interfront are the following:

- Storage of those points of the currently processed greylevel that are adjacent to points belonging to a higher greylevel. These are easily found by checking if the neighboring points of every point that a propagation front goes through, belong to the current greylevel (which is something that the algorithm does anyway). If they do not belong to one, we can conclude that they belong to a higher greylevel and subsequently store the current point in the interfront.
- Reading all the stored points, which will initiate the propagation fronts in the next greylevel. For this initiation, at each new greylevel, every point in the

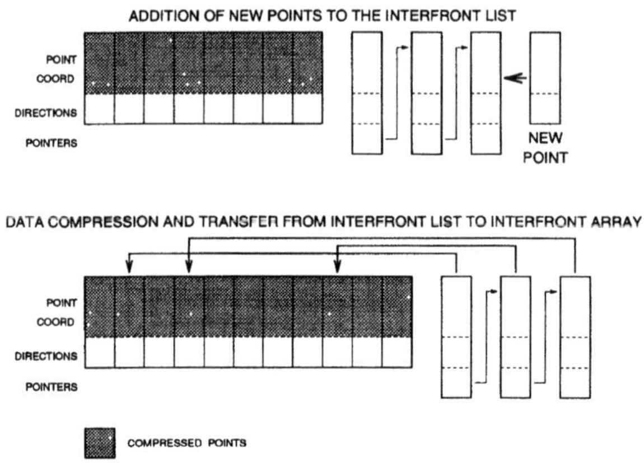

Fig. 2. Overview of the structure of the interfront.

interfront must be checked for adjacency with any points in that greylevel. If such points exist, then they are appended to the propagation front, while the interfront point adjacent to them is appended to the backup level of the propagation front. It must be noted that point reading from the interfront occurs only once per greylevel, that is at the initiation of the propagation.

- Removal of a point from the interfront when it ceases to be adjacent to points of a higher greylevel than that which the algorithm has reached.

In order to avoid keeping more points in the interfront than needed, the position of all adjacent points that the algorithm has not visited are stored for every interfront point (in a fashion similar to [10]). This has two advantages:

- First, that a point can be removed from the interfront when it no longer has such adjacent points.
- Second, at each greylevel we only need to search for points adjacent to an interfront point in the directions we know they exist (which, owing to the size of the interfront, leads to a considerable acceleration of the algorithm).

In order to save memory, the position of these adjacent points is stored as flags (single bits). As mentioned, a point is added to the interfront if some points in its neighborhood are not in the current greylevel. As it is being added, a label is appended to it which contains, in coordinate form, the directions in which no points were found. Thus, when the propagation at the next greylevel starts, the directions where adjacent points may exist are read for each interfront point, and the current greylevel is searched to find these points. If they exist in the current greylevel, it is certain that no other points will exist in this direction, and the corresponding flag in the label byte is cleared. When all flags are cleared, this point is no longer useful and is removed from the interfront.

Because the propagation in each greylevel begins from only a few interfront points, the interfront contains significantly fewer points than the greylevel (considering the fact that the sizes of different greylevels vary significantly). Thus, the greater part of the algorithm's memory requirements is due to the interfront. This means that the reduction of its size is desirable. In order to achieve this, we take advantage of the fact that the addition and readout of points happens separately and consecutively in each greylevel. We also take advantage of the fact that the reading of points takes place in an arbitrary order which can be chosen to be the basic order. The above lead to the compression of the interfront points, which can produce a significant memory gain.

The above compression takes place at the end of the processing of each greylevel, since the number of points added to the interfront during this processing is negligible compared to the total number of interfront points. Because the number of points added to the interfront at each greylevel is not known in advance, appending points to it is implemented using a dynamically managed list. The compression of these new points is performed as this list is merged with the compressed array containing the previous points. All the above are illustrated in Figure 2.

Most of the memory requirements of the algorithm are due to the interfront. This is because it encloses all the area processed by the algorithm, while the current greyscale only adjoins part of it. However, since we took care to compress the interfront, the memory consumed by much smaller than the entire region. If the greyvalue distribution of the image is particularly uneven, the current greylevel can also consume a significant amount of memory. Since the propagation fronts in the current greylevel are much smaller than the greylevel itself, their memory burden is negligible.

4. Marker-based watershed transform

Naturally, the decomposition watershed algorithm does not have access to greylevels other than the one currently being processed. To overcome this limitation the algorithm needs, at each greylevel, to be granted access to all points of previous greylevels that do not belong to catchment basins. This is achieved by scanning each greylevel once after it has been processed, searching for points that have not been labeled as belonging to a catchment basin or to the watersheds. These points are stored in an list, which is subsequently merged with the next greylevel. The merging has a low computational cost because both the current greylevel and the above list are already sorted in the basic order. It should be explained that the list is also on the basic order because it was computed as a part of the previous greylevel using a basic order scan. Because the current list is merged with the current greylevel and from this merged greylevel the next list is derived, each list contains all the non-processed points not only from the current greylevel but also from the previous ones. After the merging, the resulting greylevel is processed as in the regular partial storage algorithm, except of course from finding regional minima and starting new catchment basins.

5. Simulation experiments

In order to verify the algorithm experimentally, we applied it to perform the segmentation of a three-dimensional biomedical image. The volume used was a series of slice CT scans depicted a human head and had a size of $256 \times 256 \times 256$, part of which

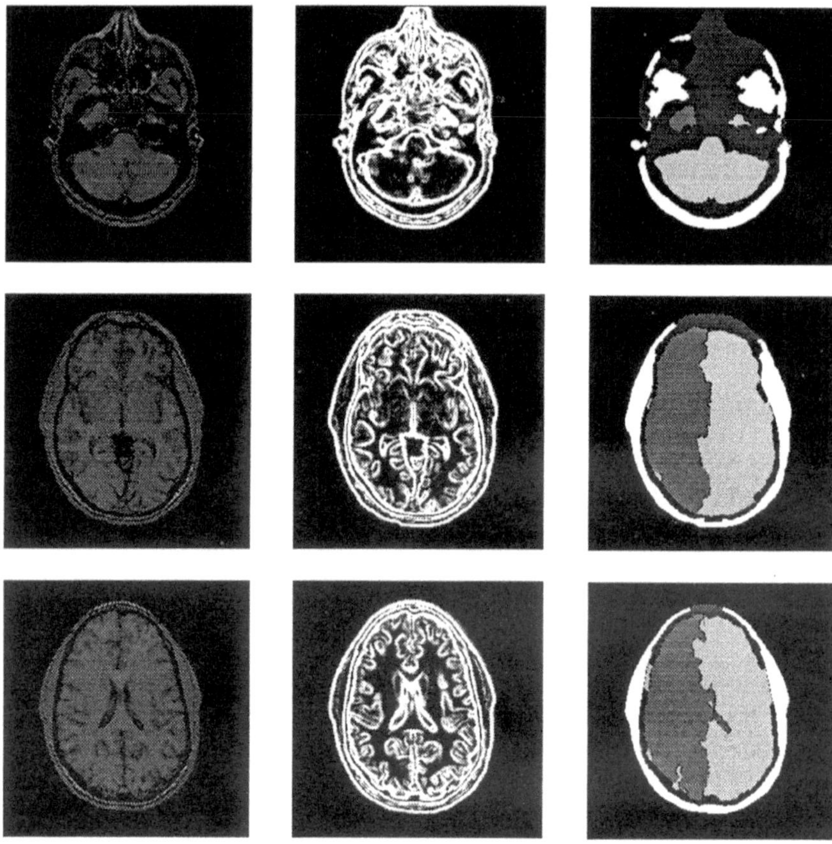

(a) (b) (c)

Fig. 3. Sections of the head volume used for testing the algorithm. On the left is the original image (a), in the center the edge image (b) and on the right the segmented image (c).

is shown in Figure 3a. The purpose of this procedure was to segment as accurately as possible the skull and the two lobes of the brain, as well as the background.

As is common practice when using the watershed transform for image segmentation, an edge detection operation was first applied on the input volume. In this case, Sobel edge detection was used, as shown in Figure 3b. We performed a histogram equalization on the edge image in order to equalize its greylevels. Then, markers were defined interactively for identifying the objects of interest, and the marker-based watershed transform (described in 4) was performed. One marker was defined for each lobe of the brain, two for the skull, three for the region between brain and skull, and one for the background. The machine used to compute the segmentation was a personal computer with an Intel Pentium processor clocked at 133 MHz and

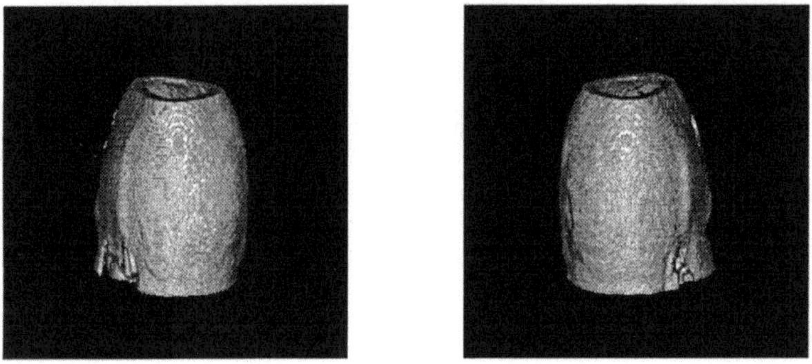

Fig. 4. The skull, as it was segmented by the algorithm.

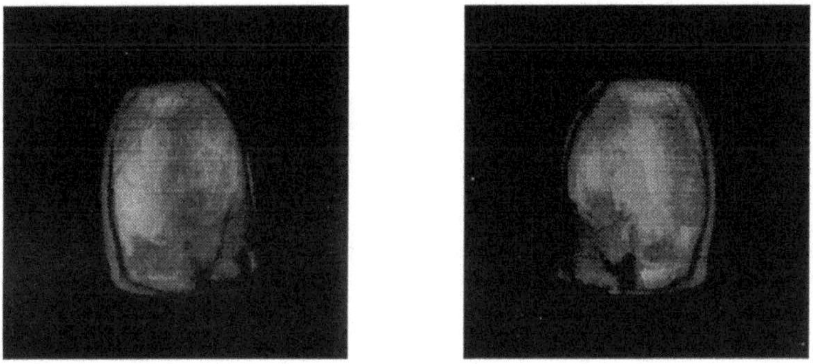

Fig. 5. The segmented brain with the various regions shown in different transparency.

32 Mbytes of random access memory running Microsoft Windows NT Server. Some slices of the segmented volume are given in Figure 3c. The segmented skull is shown in Figure 4 and a rendering of all segmented regions in Figure 5.

The segmentation was correct and its quality satisfactory, since the segmented lobes and skull got the shape they have in reality. There was some distortion in the area of the skull, at the top of the head and at its bottom near the face. This was expected, since the edges in these areas were particularly weak and has nothing to do with the correctness of the algorithm. Naturally, what is of importance is the efficiency of the algorithm in terms of computational cost and memory requirements.

As far as computational cost is concerned, the execution time of the algorithm was on average 2 hours and 25 minutes. If we consider the fact that Vincent's algorithm [1] executes the same transform in approximately 30 minutes (if of course the necessary memory of over 64 Mbytes is available), it can be understood that the execution time is very reasonable. As for the memory requirements, these stem mostly from the interfront since the memory requirements of the propagation fronts are lower by an order of magnitude. The maximum number of points stored in the interfront during the execution of the algorithm never exceeded 12, 000, 00, and thus memory requirements are much lower than in the traditional algorithms.

6. Conclusions

We have presented an algorithm to reduce greatly the memory requirements of the watershed transform, thus making it computable on large images. Memory savings of more than an order of magnitude have been achieved, although with some performance loss. The effectiveness of the algorithm was demonstrated in the task of segmenting a brain MRI image, and produced the expected results.

References

1. L. Vincent and P. Soille. Watershed in Digital Spaces: An Efficient Based on Immersion Simulations. *IEEE Trans. Pattern Anal. Machine Intel.*, vol. 13, no. 6 pp. 583–598, June 1991.
2. P. Soille and M. Ansoult. Automated Basin Delineation from DEMs using Mathematical Morphology. *Signal Processing*, vol. 20, pp. 171–182, 1990.
3. S. Beucher and F. Meyer. The Morphological Approach to Segmentation: The Watershed Transformation. *Math. Morph. in Image Proc.*, E.R. Doherty, Ed., Marcel Dekker, pp. 433–481, 1993.
4. I.E. Pratikakis, H. Sahli and J. Corvelis. Watershed Analysis and Relaxation Labeling. *Math. Morph. and Appl. to Image and Signal Proc.*, P. Maragos *et al.*, Eds. Kluwer, pp. 425–432, 1996.
5. J. Madrid and N. Ezquerra. Automatic 3-Dimensional Segmentation of MR Brain Tissue Using Filters by Reconstruction. *Math. Morph. and Appl. to Image and Signal Proc.*, P. Maragos *et al.*, Eds. Kluwer, pp. 417–424, 1996.
6. I. Pitas. *Digital Image Processing Algorithms.* Prentice Hall, 1993.
7. L. Vincent. Graphs and Mathematical Morphology. *Signal Processing*, vol 16, pp. 365–388, 1989.
8. R. van den Boomgard and A. Smeulders. The Morphological Structure of Images: The Differential Equations of Morphological Scale-Space. *IEEE Pattern Anal. Machine Intel.*, vol. 16, no. 11, pp. 1101–1114, Nov. 1994.
9. P. Salembier and A. Oliveras. Practical Extensions of Connected Operations. *Math. Morph. and Appl. to Image and Signal Proc.*, P. Maragos *et al.*, Eds. Kluwer, pp. 97–118, 1996.
10. L. Vincent. Morphological Transformations of Binary Images with Arbitrary Structuring Elements. *Signal Processing*, vol. 22, pp. 3–23, 1991.

GREY-SCALE IMAGES AND RANDOM SETS

ILYA MOLCHANOV
Department of Statistics, University of Glasgow
Glasgow G12 8QW, Scotland, U.K.

Abstract. This is an outline of several relationships between grey-scale images and random sets. These relationships allow us to define meaningful variants of the distance transform for grey-scale images, new thresholding procedures, and a family of morphological operations with grey-scale images.

Key words: Random set, averaging, grey-scale image, Minkowski addition, sup-measure

1. Introduction

A binary image can be represented as a (deterministic) set in a relevant space W which is typically 2- or 3-dimensional Euclidean space $\mathbf{R}^2$ or $\mathbf{R}^3$ (in the "continuous" setup) or a finite subset of a grid (in the "discrete" setup). If the image is random it is natural to treat it as a random set. For instance, this happens if a sample of images is analysed and treated as a sample of realisations of a random image or the corresponding random set. Such a representation allows us to use tools of random set theory to handle random binary images. Particular examples include filtering [18, 7], averaging of images [2, 20] and models for images [8]. Surveys of relevant statistical methods can be found in [14, 15].

A binary image can be represented as a function f on W with only 2 possible values, so that f is the indicator function of a set $A \subseteq W$:

$$f(x) = \mathbf{1}_A(x) = \begin{cases} 1 \ , \ x \in A \, , \\ 0 \ , \ \text{otherwise} \, . \end{cases}$$

We usually assume the continuous setup and require that the set A is topologically closed (which automatically holds in the discrete case). Note that A is closed if and only if its indicator function is upper semicontinuous.

It was noticed long ago (for example, see [19]) that the *hypograph* (called also subgraph or umbra)

$$\text{hypo} f = \{(x,t) : \ x \in W, \, t \leq f(x)\}$$

is a closed subset of $W \times \mathbf{R}$ if and only if f is an *upper semicontinuous* grey-scale image. Below we always assume that all grey-scale images are upper semicontinuous and the set of grey-levels is $[0, 1]$. This interpretation of grey-scale images as binary sets is very popular and is used to define morphological operations [19, 4, 9] and distances between images [23]. Note that within this framework a *random* image corresponds to a random closed subset of $W \times \mathbf{R}$.

Clearly, the corresponding sets live on the space which is not W but has one additional dimension needed to incorporate the grey values. Morphological operations usually treat this additional dimension in the same manner as operations on the space W. This is not always a desirable property, since this additional dimension has features distinct from those inherent to the position space W. Furthermore, transformations of the range of grey values (anamorphoses or colour-maps) should be treated in a special way, since they change the image is a way, which is different from operations on the position space W. An additional dimension increases computational complexity and makes it more difficult to visualise images.

The current paper aims to outline a new machinery suitable to interpret *deterministic* grey-scale images as *random* subsets of W. Similar methods could be applied within the framework of the theory of *fuzzy* sets, since each grey-scale image (with values in $[0, 1]$) can be interpreted as the membership function of a fuzzy set [24]. Furthermore, a natural generalisation of the presented ideas in the spirit of lattice-based morphology [10] is possible for functions on a given lattice with values in a partially ordered space.

Note that a thresholded set of an image f at a level $t \in [0, 1]$

$$F_t = \{x \in W : f(x) \geq t\} \tag{1}$$

is a closed subset of W. The key idea is that this thresholded set becomes a *random closed set* if the threshold is taken at a random level, so that t becomes a random variable. This allows us to treat deterministic grey-scale images as *random* closed sets in the same carrier space without introducing an additional coordinate. Section 2 elaborates this idea which is utilised in Section 3 to introduce the distance transform for grey-scale images. A thresholding method based on this concept has been discussed in detail in [6], while, in general, a threshold of a grey-scale image is naturally interpreted as an expectation of the corresponding random set. Section 4 discusses morphological operations with grey-scale images based on their representation as random closed sets. Distances between grey-scale images are outlined in Section 5. Finally, Section 6 describes how to use grey-scale images in order to summarise operations on binary images. In particular, it leads to a new filtering procedure based on an "optimal" trade-off between erosion and dilation.

Throughout this paper f, g, h denote upper semicontinuous grey-scale images defined on a compact set $W \subset \mathbf{R}^d$ with grey-scale range $[0, 1]$.

2. Random Thresholds

If f is an upper semicontinuous grey-scale image, then F_t defined by (1) is a non-increasing set-valued function whose values are closed subsets of W. This deterministic function F_t, $0 \leq t \leq 1$, gives rise to a random closed set if t is replaced by a random variable U distributed on $[0, 1]$ (the set of grey values). We call

$$F_U = \{x \in W : f(x) \geq U\}, \tag{2}$$

the *random set model* associated with f. The random variable U describes the type of weighting associated with possible grey values. We usually work with the *uniformly weighted* model where U is uniformly distributed on $[0,1]$. Further useful models appear if U has a distribution related to f. For example, U can be distributed according to the histogram of f (*histogram weighted model*), when

$$\mathbf{P}\{U \geq t\} = \mu(F_t)/\mu(W)$$

with μ being the Lebesgue measure on W. Another option (which yields the so-called *perimeter weighted model*) is to assume that the probability density of U is proportional to the $(d-1)$-dimensional Hausdorff measure of the boundary of F_t (assuming a kind of regularity for f).

The uniformly weighted model yields a simple relationship between the distribution of F_U and the original grey-scale image f, since

$$\mathbf{P}\{x \in F_U\} = \mathbf{P}\{U \leq f(x)\} = f(x)\,, \quad x \in W\,.$$

The histogram weighted model is useful because the distribution of F_U remains unchanged after all increasing continuous transformations of grey values (*anamorphoses*) when $f(x)$ is replaced by $s(f(x))$ with $s : [0,1] \to [0,1]$ being an increasing continuous function. Note also that the histogram weighted model works without any changes for images with arbitrary ranges of grey values. Note as well, that in this model F_U is not empty almost surely, while the uniformly weighted model can lead to random sets with possibly empty values. Clearly, the histogram weighted model corresponds to the uniformly weighted model if f is replaced by its equalised variant. In particular, this means that it is impossible to retrieve the original image from the histogram weighted random set model. It yields only its equalised variant or any other image equivalent up to an anamorphosis.

Furthermore, note that f generates a ***sup-measure***

$$f(K) = \sup_{x \in K} f(x)\,,$$

see [16] for formal discussion of sup-measures. A sup-measure is a particular case of a completely alternating capacity functional [11], so that it defines a distribution of the uniformly weighted random set model F_U. Clearly,

$$\mathbf{P}\{F_U \cap K \neq \emptyset\} = \mathbf{P}\{U \leq f(K)\} = f(K)$$

for all compact sets $K \subset W$. There is the unique correspondence between sup-measures and the introduced uniformly weighted random set models. A *general* capacity functional $\phi : \mathcal{K} \to [0,1]$ defined on the family $\mathcal{K}$ of all compact sets gives rise to the family of sets

$$F_t(\phi) = W \setminus \bigcup_{K \in \mathcal{M},\, \phi(K) < t} K\,, \quad 0 \leq t \leq 1\,, \tag{3}$$

which are called the ***quantiles*** of ϕ for a "suitable" subclass $\mathcal{M} \subset \mathcal{K}$, see [12]. However, the uniqueness is lost and a general capacity cannot be retrieved from its quantiles. Note that if ϕ is a sup-measure and $\mathcal{M}$ is the family of all singletons, then (3) yields $F_t(\phi) = \{x : \phi(\{x\}) \geq t\}$.

As in classical statistics where the expectation is used to summarise a sample, it is possible to "summarise" the random set model F_U (and the corresponding image f) by taking an expectation of this random set. A survey of different expectations of random sets can be found in [14] and [21]. The Vorob'ev expectation is one of the simplest in the current framework, since it is derived from the probabilities of covering a point by F_U. It is shown in [6] that this expectation of the uniformly weighted random set model equals F_t where t is chosen in such a way that $\mu(F_t)$ is the "closest" to the integral of f over W, or, strictly,

$$\mu(F_t) \le \int_W f(x)dx \le \mu(F_q) \quad \text{for all } q > t\,.$$

The Vorob'ev expectation of the histogram weighted random set model is F_t where t is chosen to have $\mu(F_t)$ matching as close as possible $\mu(W)/2$. Thus, the Vorob'ev expectation for these two basic random set models yields naive thresholds. Further thresholds are possible using generalisations of the Vorob'ev expectation outlined in [15].

Below we always work with the uniformly weighted random set model, since other variants are easy to modify to the uniformly weighted case after a suitable transformation of grey values (for example, by means of equalising if the histogram weighted model is chosen).

3. Distance Transform

The distance transform [17] is an established tool in analysis of binary images. If K is a subset of W (or binary image), then the distance transform evaluated at $x \in W$ equals the (Euclidean) distance $d(x, K)$ from x to the nearest point of K. Naturally, for discrete images one replaces the Euclidean distance with a suitable approximation on the grid. We concentrate here only on the Euclidean distance function bearing in mind that it is easy to reformulate the concepts below for other distance functions, discussed in [2], for example, the signed distance function [3].

The distance transform produces a grey-scale image from a binary image. Clearly, the binary image can be retrieved as a thresholded set of its distance function at zero level. If X is a random closed set and $d(x, X)$ has a finite expectation (if it is finite for at least one x, then it is finite for all x), then $\bar{d}(x) = \mathbf{E}d(x, X)$ is naturally called the *expected distance function* of X.

Now let $X = F_U$ be a random set model corresponding to a grey-scale image f. Then

$$\bar{d}(x) = \mathbf{E}d(x, F_U) \tag{4}$$

is called the *distance transform* of f. Therefore, a grey-scale image generates its distance transform being a grey-scale image too. However, the range of grey-values for the distance transform depends on the size of the window and is not necessarily $[0, 1]$ anymore.

Clearly, the result of the distance transform depends on the distribution of U. Note that the distance transform is not defined for empty binary images, so it is essential that F_U in (4) is non-void. This is surely the case for the histogram weighted model, while otherwise, U can be chosen to have a uniform distribution on the effective range of grey values, $[\inf f, \sup f]$. Figure 1 shows the result of the distance transform applied to house image with U uniformly distributed over the effective range of grey values.

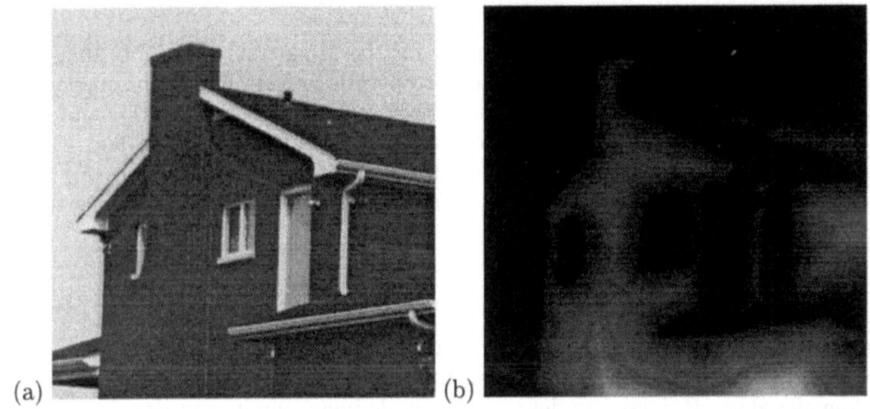

Fig. 1. House image and its grey-scale distance transform

The most interesting question arising in this context is the uniqueness of the introduced grey-scale transform. It is easy to show by induction that $\bar{d}(x)$ uniquely identifies f from the family of step-functions (functions with no more than finite numbers of values). This covers all practical situations appearing in image analysis, where all images are defined on a grid and as such have at most a finite number of possible grey values. However, the relevant mathematical question for the family of all upper semicontinuous functions seems to be open (the conjecture is that it is not unique, so two different images can share the same distance transform). It should be noted that the distance transform is unique if the images are sufficiently "regular", for example, belong to a class of functions $\mathcal{F}_{reg}$ such that $f - g$ does not oscillate too much (has finite number of zeroes) for all $f, g \in \mathcal{F}_{reg}$.

This grey-scale distance transform has been used in [6] to develop a new thresholding procedure for grey-scale images. Then the threshold level t is chosen in such a way that the distance transform of the *binary* thresholded set F_t is the closest in some sense to the grey-scale distance transform $\bar{d}$ of the original image. For example, t can be chosen to minimise the L_2-distance

$$\|\bar{d}(\cdot) - d(\cdot, F_t)\|_{L_2} = \left(\int_W (\bar{d}(x) - d(x, F_t))^2 dx \right)^{1/2}$$

between the functions $\bar{d}(\cdot)$ and $d(\cdot, F_t)$, see [6] for further discussion and examples.

4. Morphological Operations

We have seen that it is easy to retrieve a grey-scale image from the corresponding uniformly weighted random set model. Therefore, by performing binary morphological operations with these random sets, one can interpret their results in terms of grey-scale images.

Let us formulate this idea more formally for the Minkowski sum (note that the same approach naturally works for all other operations from binary morphology). Let f and g be two upper semicontinuous grey-scale images, and let F_U and G_V be the corresponding uniformly weighted random set models, where U and V are two independent random variables uniformly distributed over $[0,1]$. Since F_U and G_V are two random subsets of W (or binary images), it is possible to define their Minkowski sum as

$$H = F_U \oplus G_V = \{x + y : \; x \in F_U, y \in G_V\}.$$

Note that if at least one of the summands is empty, then the sum is empty too. If $W = \mathbf{R}^d$, then $H \subset W$, while in general, the sum may not be a subset of W. The random closed set H is used to define a grey-scale image via its covering function

$$h(x) = \mathbf{P}\{x \in H\}, \quad x \in W,$$

so we write

$$h = f \bar{\oplus} g.$$

By substituting distributions of U and V, we get

$$h(x) = \mathbf{P}\{x \in F_U \oplus G_V\} = \int_0^1 \int_0^1 \mathbf{1}_{x \in F_u \oplus G_v} du dv = \int_0^1 \left[\sup_{y \in x + \check{G}_v} f(y) \right] dv,$$

where $\check{G} = \{-x : \; x \in G\}$. Clearly,

$$\sup_{y \in x + \check{G}_v} f(y) = (f \oplus G_v)(x)$$

appears as the dilation of f by G_v [19, p. 431]. Thus,

$$h(x) = \int_0^1 (f \oplus G_v)(x) dv.$$

Clearly, $f \bar{\oplus} g = g \bar{\oplus} f$. The introduced operation is also associative. Furthermore, a translation of the summands (for example, if f is replaced by $f_1(x) = f(x+u)$ for some u) results in the same translation of the sum. If c is a positive number such that cf has values between 0 and 1, then $(cf) \bar{\oplus} g = c(f \bar{\oplus} g)$.

Let us discuss some particular cases. First, if both f and g are binary images, then F_U and G_V are deterministic sets, so the introduced operation becomes the conventional Minkowski sum of two binary images. Furthermore, if $g = \mathbf{1}_G$ is a binary image (indicator of G), then

$$h(x) = \sup_{y \in x + \check{G}} f(y) = (f \oplus G)(x)$$

is the dilation of f by G.

Example. Let $W = \mathbf{R}^1$ and let

$$f = g = \begin{cases} 0, & x \le 0, \\ x, & 0 < x < 1, \\ 1, & x \ge 1. \end{cases} \tag{5}$$

Clearly, if both f and g are monotone functions on the line, then

$$(f \bar{\oplus} g)(x) = \int_0^1 f(x + g^{-1}(v))dv\,,$$

whence for f from (5) we get

$$h(x) = (f \bar{\oplus} f)(x) = \begin{cases} 0, & x \le 0, \\ x^2, & 0 < x \le 1, \\ x - x^2/4, & 1 < x < 2, \\ 1, & x \ge 2. \end{cases}$$

Let us now find what appears in successive sums of functions are properly normalised. Let W be a compact subset of $\mathbf{R}^d$, and let f be a grey-scale image defined on W. Define

$$h_n(x) = (f \bar{\oplus} f \bar{\oplus} \cdots \bar{\oplus} f)(nx)\,,$$

so that h_n is the sum of n identical summands equal to f with the argument scaled by the factor n. Note that h_n is defined on the set

$$n^{-1}(W \oplus W \oplus \cdots \oplus W)\,,$$

which equals W if W is a convex set. Let $F_{U_1}, F_{U_2}, \ldots$ be independent identically distributed random closed sets obtained as uniformly weighted random set models for the image f and a sequence of independent random variables $U_1, U_2, \ldots$ uniformly distributed on $[0, 1]$. Denote by H_n the uniformly weighted random set model generated by h_n. By definition,

$$H_n = \frac{F_{U_1} \oplus \cdots \oplus F_{U_n}}{n}\,.$$

Note that each random set F_{U_i} is contained in W, so the strong law of large numbers for random closed sets [1] yields

$$H_n \to \mathbf{E}_A F_U \quad \text{almost surely}\,,$$

where the convergence is understood in the Hausdorff metric and $\mathbf{E}_A F_U$ is the *Aumann expectation* (see [1, 22, 14]) of the uniformly weighted random set model F_U associated with f. Remember that $\mathbf{E}_A F_U$ is the (deterministic) convex set K such that

$$h(K,s) = \mathbf{E}h(F_U, s)$$

for all $s \in \mathbf{R}^d$, where $h(K,s) = \sup\{\langle x, s\rangle : x \in K\}$ is the support function of K and $\langle x, s\rangle$ is the scalar product in $\mathbf{R}^d$. Therefore, H_n has an almost sure limit which is a deterministic convex set. This means that the normalised sums h_n converges to a convex binary image as $n \to \infty$. For example, if f is defined by (5), then the (pointwise) limit of h_n is the indicator of $[0.5, 1]$.

It is possible to extend this result for the case when W is unbounded. Then one requires that $\|F_U\| = \sup\{\|x\| : x \in F_U\}$ has a finite expectation.

It is worthwhile to compare this result with the case when grey-scale images are defined through their hypographs (or umbrae). In the latter case the normalised sums of f converge to an image with the hypograph being the convex hull of the hypograph of f.

5. Distances between Images

The elaborated representation of grey-scale images as random closed sets allows us to define distances between grey-scale images by utilising the so-called probability metrics for random closed sets [13]. Probability metrics determine distances between distributions of random sets rather than between their particular realisations.

The full discussion appears in the same volume, see [5].

6. From Binary to Grey-Scale

The conventional Minkowski addition of two sets F and G (representing two binary images) is defined as

$$F \oplus G = \{x : x \oplus \check{G} \cap F \neq \emptyset\}.$$

Based on this definition we can introduce a family of transformations depending on a parameter $p \in [0, 1]$:

$$F \overset{p}{\oplus} G = \{x : \mu((x \oplus \check{G}) \cap F) \geq p\mu(G)\}. \tag{6}$$

If both F and G are regular closed (coincide with the closures of their interiors), then the above defined operation yields standard morphological operations as particular cases

$$\lim_{p \downarrow 0} F \overset{p}{\oplus} G = F \oplus G$$

and

$$F \overset{1}{\oplus} G = \{x : x + \check{G} \subseteq F\} = F \ominus G.$$

The same interpretation holds in the discrete setup, where measures are replaced by numbers of pixels. However, in general case, $F \overset{p}{\oplus} G = F_1 \overset{p}{\oplus} G_1$ if $\mu(F \Delta F_1) = \mu(G \Delta G_1) = 0$ where Δ denotes symmetric difference.

From (6) we can define a grey-scale image h by

$$h(x) = (F\underline{\oplus}G)(x) = \sup\{p \in [0,1] : x \in F \overset{p}{\oplus} G\} = \frac{\mu((x+\check{G}) \cap F)}{\mu(G)}.$$

The maximum value of $h(x)$ is achieved for x from the Minkowski difference $F \ominus G$, while $h(x)$ vanishes for x outside the Minkowski sum $F \oplus G$. Such function can be used to summarise a family of morphological operations applied to F and G. Note that

$$H_t = \{x \in W : h(x) \geq t\} = F \overset{t}{\oplus} G.$$

At the same time, $h(x)$ gives rise to the corresponding random set models $H_U = \{x : h(x) \geq U\}$, which, in its turn, can be used to threshold h. For example, it is possible to use the distance threshold suggested in [6] and outlined in Section 3. The result can be interpreted as a reasonable trade-off between Minkowski sum and difference. Figure 2 shows the stages of this procedure applied to binarised house image. Figure 2(b) shows its noisy variant F obtained by adding a Gaussian white noise with standard deviation 35/255 units to the original grey-scale image on Figure 1(a) and subsequent thresholding at level 0.5. For comparison, (a) shows the original image thresholded at the same level. Figure 2(c) shows the grey-scale image $h = F\underline{\oplus}G$ with the structuring element G being a disk of radius 4 pixels. Finally, (d) displays the thresholded set which appears if the distance thresholding procedure [6] is applied to the grey-scale image h. The "optimal" threshold level is $87/255 = 0.34$, which shows that the "optimal filtering" is achieved by a trade-off between dilation (at level 0) and erosion (at level 1).

Figure 3(a) shows the noisy house image obtained as above in Figure 2(b), but with a stronger noise with the standard deviation of 70 units. The same procedure was applied for G being a disk of radius 8 pixels. Figure 3(b) displays the result of thresholding (at level 126/255) for the corresponding grey-scale image $h = F\underline{\oplus}G$.

Note that similar grey-scale summaries are easy to define for opening and closing applied to binary images. For instance,

$$\sup\{p \in [0,1] : x \in (F \overset{p}{\oplus} G) \overset{1-p}{\ominus} G\}$$

yields closing for $p = 0$ and opening for $p = 1$ (if F and G are regular closed). In the discrete case, the measures must be replaced by the numbers of pixels for each set involved.

7. Conclusions

The described ideas can be generalised for functions with values in a partially ordered space. For example, if $f = (f_1, \ldots, f_m)$ is defined on W and takes values in the unit cube $I = [0,1]^m \subset \mathbf{R}^m$ and $U = (U_1, \ldots, U_m)$ is uniformly distributed over I, then

$$F_U = \{x \in W : f_i \geq U_i, \ i = 1, \ldots, m\}$$

defines the corresponding uniformly weighted random set model. This idea can be applied to handle colour images where different coordinates of f represent three basic colours.

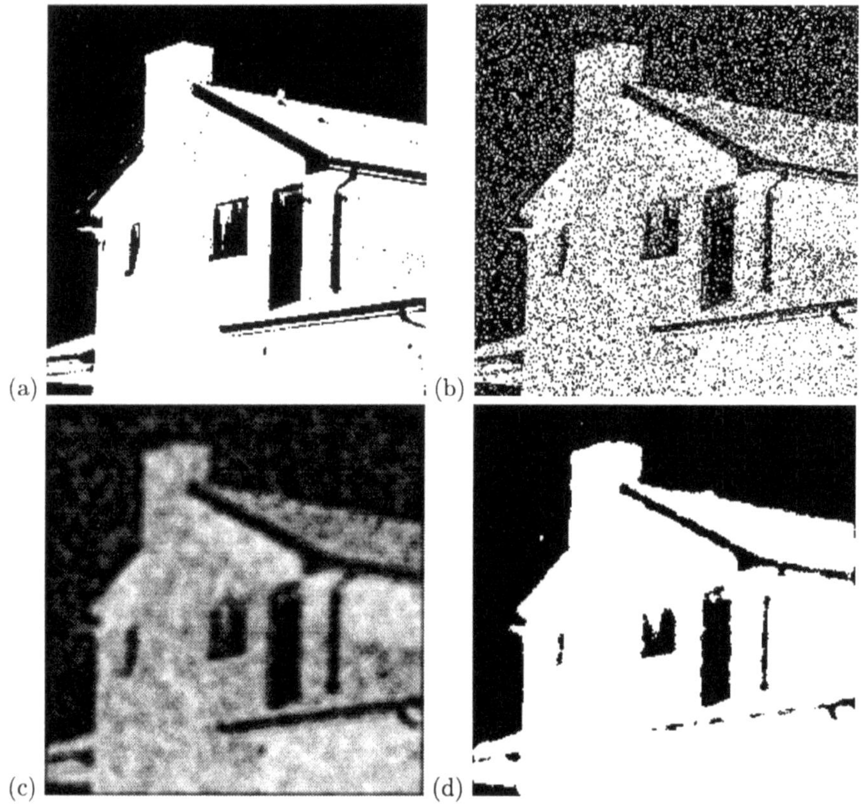

Fig. 2. (a) thresholded house image; (b) its noisy variant; (c) $h = F\underline{\oplus}G$ for G being a disk of radius 4 pixels; (d) the result of thresholding of h at level 87.

Acknowledgements

Sections 2, 3 and 5 survey collaborative research with Nial Friel [6] and [5].

References

1. Z. Artstein and R. Vitale. A strong law of large numbers for random compact sets. *Ann. Probab.*, 3:879–882, 1975.
2. A. Baddeley and I. Molchanov. Averaging of random sets based on their distance functions. *Journal of Mathematical Imaging and Vision*, 8:79–92, 1998.
3. M. Delfour and J.-P. Zolésio. Shape analysis via oriented distance functions. *J. Funct. Anal.*, 123:129–201, 1994.
4. E. Dougherty. Euclidean gray-scale granulometries: representation and umbra inducement. *Journal of Mathematical Imaging and Vision*, 1:7–21, 1992.
5. N. Friel and I. Molchanov. Distances between grey-scale images. In H. Heijmans and J. Roerdink, editors, *Proceedings of the ISSM'98 Workshop*, Dordrecht, 1998. Kluwer.

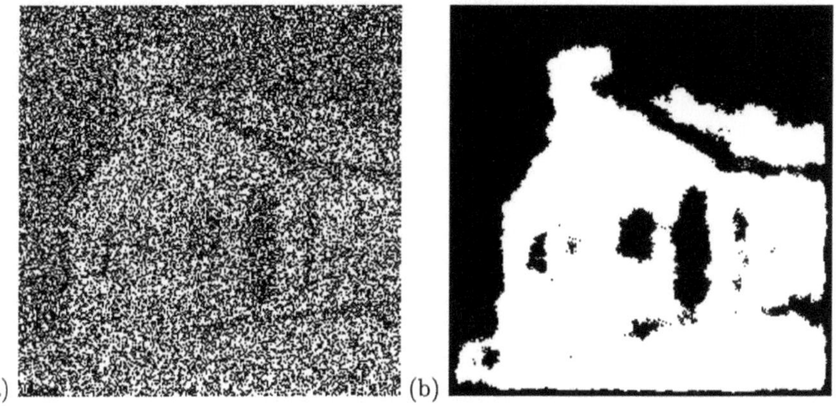

Fig. 3. Noisy variant of **house** image (standard deviation is 70 units) and the result of "filtering".

6. N. Friel and I. Molchanov. A new thresholding technique based on random sets. *Pattern Recognition*, 1998. To appear.
7. J. Goutsias. Morphological analysis of discrete random shapes. *Journal of Mathematical Imaging and Vision*, 2:193–215, 1992.
8. J. Handley and E. Dougherty. Maximum-likelihood estimation for discrete Boolean models using linear samples. *J. Microscopy*, 182:67–78, 1996.
9. H. Heijmans. A note on the umbra transform in grey-scale morphology. *Pattern Recognition Letters*, 14:877–881, 1993.
10. H. Heijmans. *Morphological Image Operators*. Academic Press, Boston, 1994.
11. G. Matheron. *Random Sets and Integral Geometry*. Wiley, New York, 1975.
12. I. Molchanov. Empirical estimation of distribution quantiles of random closed sets. *Theory Probab. Appl.*, 35:594–600, 1990.
13. I. Molchanov. *Limit Theorems for Unions of Random Closed Sets*, volume 1561 of *Lect. Notes Math.* Springer, Berlin, 1993.
14. I. Molchanov. Statistical problems for random sets. In R. M. J. Goutsias and H. Nguyen, editors, *Applications and Theory of Random Sets*, pages 27–45, Berlin, 1997. Springer.
15. I. Molchanov. Random sets in view of image filtering applications. In E. Dougherty and J. Astola, editors, *Nonlinear Filters for Image Processing*, chapter 10. SPIE, 1998.
16. T. Norberg. Random capacities and their distributions. *Probab. Theor. Relat. Fields*, 73:281–297, 1986.
17. A. Rosenfeld and J. Pfalz. Distance functions on digital pictures. *Pattern Recognition*, 1:33–61, 1968.
18. D. Schonfeld. Optimal structuring elements for the morphological pattern restoration of binary images. *IEEE Trans. Pattern Analysis and Machine Intelligence*, 16:589–601, 1994.
19. J. Serra. *Image Analysis and Mathematical Morphology*. Academic Press, London, 1982.
20. D. Stoyan and I. Molchanov. Set-valued means of random particles. *Journal of Mathematical Imaging and Vision*, 7:111–121, 1997.
21. D. Stoyan and H. Stoyan. *Fractals, Random Shapes and Point Fields*. Wiley, Chichester, 1994.
22. R. Vitale. An alternate formulation of mean value for random geometric figures. *J. Microscopy*, 151:197–204, 1988.
23. D. Wilson, A. Baddeley, and R. Owens. A new metric for grey-scale image comparison. *Intern. J. Computer Vision*, 24:5–17, 1997.
24. L. Zadeh. *Fuzzy Sets and Applications: Selected Papers*. Wiley, New York, 1987.

PRIOR INFORMATION IN THE DESIGN OF OPTIMAL BINARY FILTERS

EDWARD R. DOUGHERTY AND JUNIOR BARRERA

Texas Center for Applied Technology.
Department of Electrical Engineering. Texas A & M University

Departamento de Ciencia da Computacao.
Instituto de Matematica e Estatistica. Universidade de Sao Paulo.

Abstract. Except in special circumstances, optimal binary translation-invariant operators are designed statistically from image realizations. A basic question concerns the error increase of a designed filter over that of the optimal filter. The answer depends not only on the random sets themselves, but also on the design methodology. With regard to the latter, if prior information is employed in the design, then the error increase can be substantially less than design without prior information. The present paper provides a coherent examination of prior information in binary filter design by bringing together a number of recently published and yet to be published error measures. It considers precision error, which results from using a finite amount of observations, and constraint error, which results from constraining the class of potential operators to reduce computational complexity and the number of training images. It quantitatively compares to standard-design error the errors occurring when using the following types of prior information: independent constraint, increasing constraint, design from prior filters, and Bayesian design based on prior distributions for the conditional probabilities used in the design process.

Key words: morphological representation, optimization, prior information

1. Introduction

Optimal binary-filtering has evolved in the context of mathematical morphology because optimal filters are naturally represented in terms of hit-or-miss transforms [7] or erosions, if they are increasing[4]. Filter design usually consists of two parts. First, the optimal filter is statistically designed from image realizations and put into the form of a look-up table; second, logic reduction is used to arrive at a minimal morphological representation as a union of hit-miss operators[1] or of erosions [14][10], which also provides hardware architecture for the filter. Use of prior information can have a significant impact on logic minimization; however, this aspect of prior information is not addressed in the present paper, the focus here being on statistical design.

In principle, we prefer using an optimal filter (estimator) to estimate an *ideal* binary image process (random set) from an observed image process; in practice, however, the matter is complicated by the need to observe a sufficient number of process realizations to obtain good estimates. A basic approach to mitigating data requirements is through the use of prior information. The effectiveness of prior information depends on the nature of the information, its relation to the random sets, and the estimation procedure being used; indeed, "bad" prior information can result in poorer estimation.

The effectiveness of prior information can be quantified quite generally, with specifics depending on the particular situation. Consider the following setting: an ideal random set $\mathbf{I}$ is to be estimated from an observed random set $\mathbf{S}$ by a filter from the operator class $\mathcal{F}$; there is an (expected) error measure $Er\left[\mathbf{I}, \Psi\left(\mathbf{S}\right)\right]$ defining the goodness of a filter $\Psi \in \mathcal{F}$; based on realizations of ideal and observed images, a procedure is used to design a filter $\hat{\Psi}_N$, where N is the number of window observations seen during estimation; and the error of estimation is $Er\left[\mathbf{I}, \hat{\Psi}_N\left(\mathbf{S}\right)\right] - Er\left[\mathbf{I}, \Psi_{opt}\left(\mathbf{S}\right)\right]$, where Ψ_{opt} is optimal in $\mathcal{F}$. If a new procedure using prior information is employed, thereby yielding an estimator $\hat{\Psi}_{pri,N}$, then the advantage of using the prior information is

$$\varepsilon_{pri} = Er\left[\mathbf{I}, \hat{\Psi}_N\left(\mathbf{S}\right)\right] - Er\left[\mathbf{I}, \hat{\Psi}_{pri,N}\left(\mathbf{S}\right)\right] \tag{1}$$

This advantage is a function of N and can be negative.

Our concern is with optimal binary W-operators, a W-operator being a translation-invariant operator defined via a window W. Each W-operator Ψ is defined by a Boolean function ψ defined on W: if x_1, x_2, ..., x_m are the observations in W_z, the translation of W to z, then the observations form an m-vector $\mathbf{x}$ and $\Psi\left(S\right)\left(z\right) = \psi\left(\mathbf{x}\right)$. The set of vectors $\mathbf{x}$ for which $\psi\left(\mathbf{x}\right) = 1$ is called the *kernel* of ψ and is denoted by $\mathcal{K}(\psi)$. If Ψ is increasing, then ψ is a positive Boolean function and the set $\mathcal{B}(\psi)$ of the minimal elements in $\mathcal{K}(\psi)$ is called the *basis* of ψ.

Assuming $\mathbf{I}$ and $\mathbf{S}$ are jointly stationary and z is an arbitrary pixel, the *mean-absolute error* of Ψ is the expectation $MAE < \Psi >= E[|\mathbf{I}(z) - \Psi(\mathbf{S})(z)|]$. The optimal MAE filter is the quantized binary conditional expectation,

$$\Psi_{opt}(\mathbf{S})(z) = \begin{cases} 1, \text{ if } P(\mathbf{I}(z) = 1|\mathbf{S} \cap W_z) > 0.5 \\ 0, \text{ if } P(\mathbf{I}(z) = 1|\mathbf{S} \cap W_z) \leq 0.5 \end{cases} \tag{2}$$

Relative to the Boolean function ψ_{opt} for Ψ_{opt}, we define the random variable $Y = \mathbf{I}(z)$ and the random vector $\mathbf{X} = \mathbf{S} \cap W_z$. Realizations y and $\mathbf{x}$ of Y and $\mathbf{X}$, respectively, result from realizations of $\mathbf{I}$ and $\mathbf{S}$. In terms of ψ_{opt} and realizations of $\mathbf{X}$, Eq. 2 takes the form: $\psi_{opt}(\mathbf{x}) = 1$ if and only if $P(Y = 1|\mathbf{x}) > 0.5$. The present paper is confined to MAE (see [3] for design in the context of computational learning).

When using a suboptimal filter Ψ instead of the optimal filter Ψ_{opt}, an error increase can arise in two ways: $\mathbf{x} \in \mathcal{K}(\psi_{opt})$ but $\mathbf{x} \notin \mathcal{K}(\psi)$, or $\mathbf{x} \notin \mathcal{K}(\psi_{opt})$ but $\mathbf{x} \in \mathcal{K}(\psi)$. Either of these result in a MAE increase of

$$|P(Y = 1|\mathbf{x}) - P(Y = 0|\mathbf{x})|P(\mathbf{x}) = |2p_{\mathbf{x}} - 1|P(\mathbf{x}) \tag{3}$$

over the optimal filter, where $p_{\mathbf{x}} = P(Y = 1|\mathbf{x})$ and $P(\mathbf{x}) = \mathbf{P}(\mathbf{X} = \mathbf{x})$. The total error increase from using Ψ instead of Ψ_{opt} is

$$MAE < \Psi > -MAE < \Psi_{opt} >= \sum_{\mathbf{x} \in \mathcal{K}(\psi_{opt}) \Delta \mathcal{K}(\psi)} |2p_{\mathbf{x}} - 1|P(\mathbf{x}) \tag{4}$$

where the last sum is over the symmetric difference between the kernels.

Two kinds of suboptimality interest us here. Suboptimality owing to estimation from realizations results from using a designed filter in place of the filter being estimated. We call this *precision error.* A second form of suboptimality results from constraining the estimation procedure so that the best possible filter that can be designed is not the optimal filter. Hence, the estimation procedure is biased in so far as, even if precision goes to 0 as $N \to \infty$, the limiting filter is suboptimal. We refer to this increase in error as *constraint error.* Whereas precision error is a function of sample size, constraint error is not. Precision error can be so great for the optimal filter that it exceeds constraint error plus precision error for the constrained filter, thereby making constrained design preferable.

We consider several types of prior information and quantify their effects on estimation for binary filters. We take as a basic measure the estimation error resulting from a standard training methodology and consider the effects relative to this method. Focus is on the various kinds of prior information, how they are used, and the quantitative effects. Mathematical derivations are left to full-length references, the purpose here being to provide an overview of approaches and results.

2. Precision for Standard Realization-Based Method

In standard estimation from realizations, filter design involves estimating the probability $p_{\mathbf{x}}$ for each $\mathbf{x}$. The estimate $\hat{\Psi}_{opt}$ of Ψ_{opt} is defined according to Eq. 2 by using an estimate $\hat{p}_{\mathbf{x}}$of $p_{\mathbf{x}}$ in place of $p_{\mathbf{x}}$. Let $u_{\mathbf{x}}$ and $v_{\mathbf{x}}$ denote the number of times, respectively, $y = 1$ and $y = 0$ in $\mathbf{I}$ given $\mathbf{x}$ appears in $\mathbf{S}$ during training, $n_{\mathbf{x}} = u_{\mathbf{x}} + v_{\mathbf{x}}$ be the number of times $\mathbf{x}$ appears, and N be the total number of observations. The usual estimator for $p_{\mathbf{x}}$ is $\hat{p}_{\mathbf{x}} = \frac{u_{\mathbf{x}}}{n_{\mathbf{x}}}$. The Boolean-function estimator of $\psi_{opt}(\mathbf{x})$ is defined by $\hat{\psi}_{opt}(\mathbf{x}) = 1$ if and only if $\hat{p}_{\mathbf{x}} > \frac{1}{2}$. $\hat{p}_{\mathbf{x}}$ is the sample mean of $Y_1|\mathbf{x}$, $Y_2|\mathbf{x}$, ..., $Y_{n_{\mathbf{x}}}|\mathbf{x}$, the observations of Y given $\mathbf{x}$ during sampling. $\hat{p}_{\mathbf{x}}$ has a binomial distribution with parameters $n_{\mathbf{x}}$ and $p_{\mathbf{x}}$ and mean $n_{\mathbf{x}}p_{\mathbf{x}}$.

An estimation error occurs if $\hat{\psi}_{opt}(\mathbf{x}) \neq \psi_{opt}(\mathbf{x})$ and this happens if $\hat{p}_{\mathbf{x}} > 0.5$ and $p_{\mathbf{x}} \leq 0.5$, or if $\hat{p}_{\mathbf{x}} \leq 0.5$ and $p_{\mathbf{x}} > 0.5$. We measure precision error by expected increase in error owing to using the designed filter in lieu of the optimal filter. It is necessary to take into account the possibility that a vector is never observed during training, as well as if it is observed. In our design methodology [3], if $\mathbf{x}$ is not observed during training, then it is placed or not placed into $\mathcal{K}(\hat{\psi}_{opt})$ depending on which choice yields the least logic cost. For the purpose of precision analysis, we assume this placement occurs uniformly randomly. Letting $\delta_{\mathbf{x}} = 2|p_{\mathbf{x}} - 0.5|P(\mathbf{x})$ the precision error using standard design is given by

$$\begin{aligned} \varepsilon_{st} &= E\left[MAE < \hat{\Psi}_{st} > - MAE < \Psi_{opt} >\right] \\ &= \frac{1}{2}\sum_{\mathbf{x}} \delta_{\mathbf{x}} P(n_{\mathbf{x}} = 0) + \sum_{\mathbf{x} \notin \mathcal{K}(\Psi_{opt})} \delta_{\mathbf{x}} P(\hat{p}_{\mathbf{x}} > 0.5) + \sum_{\mathbf{x} \in \mathcal{K}(\Psi_{opt})} \delta_{\mathbf{x}} P(\hat{p}_{\mathbf{x}} \leq 0.5) \quad (5) \end{aligned}$$

where the event $[n_{\mathbf{x}} > 0]$ is part of the events $[\hat{p}_{\mathbf{x}} > 0.5]$ and $[\hat{p}_{\mathbf{x}} \leq 0.5]$ [3] [8]. As $N \to \infty$, $\varepsilon_{st} \to 0$.

3. Constraint

One kind of prior information is constraint of the filter class over which to optimize; rather than optimize over the class $\mathcal{F}$ of all possible W-operators, constrain optimization to some subclass $\mathcal{G}$ of $\mathcal{F}$. While constraint can reduce precision error relative to unconstrained optimization, it introduces a bias (constraint error) unless the optimal filter happens to be in $\mathcal{G}$. If Ψ_{con} is the optimal filter over $\mathcal{G}$, then constraint error is given by Eq. 4 with Ψ_{con} in place of Ψ.

An *independent* constraint is one for which the decision whether or not to place a vector $\mathbf{x}$ in the kernel is constrained by a condition involving only $\mathbf{x}$ itself, and no other vectors. Independent constraints can result from geometric conditions. Letting W be the 3×3 window and reading vectors in the usual raster method, the condition that a pixel be 1-valued if it is interior to a vertical or horizontal line of pixels produces the independent constraints:

$$\begin{aligned} C_1 &: \mathbf{x} \in \mathcal{K}(\psi) \text{ if } \mathbf{x} \geq 010010010 \\ C_2 &: \mathbf{x} \in \mathcal{K}(\psi) \text{ if } \mathbf{x} \geq 000111000 \end{aligned} \tag{6}$$

Independent constraints can arise from algebraic conditions. If we demand that Ψ be anti-extensive, then there exists a single constraint:

$$C_1 : \mathbf{x} \notin \mathcal{K}(\psi) \text{ if } \mathbf{x} \not\geq 000010000 \tag{7}$$

Bounds on the filter produce independent constraints. For instance, $\Psi(S) \supset \Phi(S)$ ($\Psi(S) \subset \Phi(S)$) for all S, meaning $\Psi \geq \Phi$ ($\Psi \leq \Phi$). If Φ is increasing with $\mathcal{B}(\phi) = \{\mathbf{x}_1, \mathbf{x}_2, ..., \mathbf{x}_m\}$, then $\Phi \leq \Psi$ yields m independent constraints:

$$C_i : \mathbf{x} \in \mathcal{K}(\psi) \text{ if } \mathbf{x} \geq \mathbf{x}_i \tag{8}$$

for $i = 1, 2, ..., m$. If Ω is increasing with $\mathcal{B}(\omega) = \{\mathbf{y}_1, \mathbf{y}_2, ..., \mathbf{y}_k\}$ and we desire the bounding constraint $\Phi \leq \Psi \leq \Omega$, then, in addition to the constraints $C_1, C_2, ..., C_m$ we have the independent constraint

$$C_{m+1} : \mathbf{x} \notin \mathcal{K}(\psi) \text{ if there does not exist } \mathbf{y}_j \in \mathcal{K}(\psi) \text{ such that } \mathbf{x} \geq \mathbf{y}_j. \tag{9}$$

For the bounding constraint $\Phi \leq \Psi \leq \Omega$, one can design an unconstrained estimate Ψ' of the optimal filter and then the estimate of the constrained filter is $\Psi = (\Phi \vee \Psi') \wedge \Omega$. This transformation can take place on the truth table defining Ψ', in which case it effects the following changes: if $\psi'(\mathbf{x})$ =0 and $\phi(\mathbf{x})$ =1, then $\psi(\mathbf{x}) = 1$; if $\psi'(\mathbf{x}) = 1$ and $\omega(\mathbf{x}) = 0$, then $\psi(\mathbf{x}) = 0$; if $\psi'(\mathbf{x})$ is undetermined (because $\mathbf{x}$ has not been seen in training), then $\psi(\mathbf{x})$ can be arbitrarily chosen so long as $\phi(\mathbf{x}) \leq \psi(\mathbf{x}) \leq \omega(\mathbf{x})$.

The constraint error resulting from an independent constraint can be computed from Eq. 4 by letting $\mathcal{A}_0^{con}$ and $\mathcal{A}_1^{con}$ be set of vectors deterministically constrained to the kernel complement and kernel, respectively [5]:

$$\varepsilon_{con} = \sum_{\mathbf{x} \in (\mathcal{A}_0^{con} - \mathcal{K}(\psi_{opt})^c) \cup (\mathcal{A}_1^{con} - \mathcal{K}(\psi_{opt}))} |2p_{\mathbf{x}} - 1| P(\mathbf{x}) \tag{10}$$

For the order constraints of Eq. 8, $\mathcal{A}_0^{con} = \emptyset$ and $\mathcal{A}_1^{con}$ consists of all vectors $\mathbf{x}$ such that there exists $\mathbf{x}_i \in \mathcal{B}(\phi)$ such that $\mathbf{x}_i \leq \mathbf{x}$.

Independent constraints can reduce precision error. If prior probabilities for a class C of vectors are negligible, then vectors in C can be arbitrarily independently constrained with $\varepsilon_{con} \leq \beta$, where β is the sum of the prior probabilities in C. In fact, ε_{con} is likely to be much less than β since many vectors may be correctly constrained and the factor $|1-2p_{\mathbf{x}}|$ can be substantially less than 1. Should ε_{con} be small, the designed constrained filter can outperform the designed unconstrained filter because the difference in precision errors between the constrained and unconstrained filters can exceed ε_{con}. In fact, precision error, $E\left[MAE < \hat{\Psi}_{con} > -MAE < \Psi_{con} >\right]$, for the constrained filter takes the same form as precision error for standard, unconstrained design, except that vectors whose filter values are predetermined by constraint do not appear in the sums.

A dependent constraint is one that cannot be applied to each vector independently. There are required relations among vectors that do not reduce to independent constraints. Typically, dependent constraints result from requiring that the filter belong to a given filter class (and it may not be easy to deduce the dependency relations). Moreover, error analysis appears to be complicated.

The most studied dependent constraint is that the filter be increasing[4][11][12]. The relationship among the vectors is given by: if $\mathbf{x} \leq \mathbf{y}$ and $\mathbf{x} \in \mathcal{K}(\psi)$, then $\mathbf{y} \in \mathcal{K}(\psi)$. If $\mathcal{G}$ is the class of all increasing W-operators, then the optimal increasing filter, Ψ_{inc}, is the operator in $\mathcal{G}$ possessing minimal MAE. If Ψ_{opt} is increasing, then $\Psi_{inc} = \Psi_{opt}$; otherwise, the constraint causes an error increase separate from precision error. As originally developed, estimation of the optimal increasing filter is accomplished without estimating the probabilities $p_{\mathbf{x}}$. Since an increasing filter is determined once its basis is determined, we estimate the error resulting from different bases. The basis size is usually restricted and the best basis of size m is estimated by estimating filter MAE for all bases of size m or less, and this is accomplished by applying recursive error representation (the morphological MAE theorem) for increasing filters where recursion is with respect to m. Computational problems remain because the optimal basis of size m need not be a subset of the optimal basis of size $m+1$. Constraint error can be found from Eq. 4.

Precision error for the increasing constraint depends on the manner of estimation. In the simplest scenario, where we take a basis $\mathcal{B}$, consider the filter $\Psi_{\mathcal{B}}$ generated by $\mathcal{B}$, and estimate $MAE < \Psi_{\mathcal{B}} >$ by observing a random sample of M windowed observations and computing the estimator

$$\overline{MAE} < \Psi_{\mathcal{B}} > = \frac{1}{M} \sum_{i=1}^{M} |Y_i - \psi_{\mathcal{B}}(\mathbf{X}_i)| \qquad (11)$$

the variance of the estimator is less than $\frac{1}{4M}$, which is very precise[8]. The problem with this bound is that this estimation method is not used in practice since it requires actual application of the filter on realizations and this is too computationally burdensome. In practice, the morphological MAE theorem is employed and propagation of estimation error through the recursion has not been characterized. The variance bound $\frac{1}{4M}$ for the estimator of Eq. 11 indicates that the increasing constraint should provide better precision than for nonincreasing filters and this is born out in practice when using the MAE theorem. This is a key reason why, even when computationally feasible, derivation of Ψ_{inc} from Ψ_{opt} via switching is not used because estimation of Ψ_{opt} requires so much more data to get a precise estimate.

4. Prior Filters

A different kind of prior information is to begin with a prior filter Ψ_0 and use training images to adjust Ψ_0 to make it a better estimator of the optimal filter[2]. A special case of prior-filter design is the differencing-filter method that has proven useful in digital document processing[9][13]. When using a prior filter, for each vector $\mathbf{x}$, a decision needs to be made based on the sample data: leave $\psi_0(\mathbf{x})$ unchanged or change it. We shall consider only one rule for making this decision.

When using a prior filter, the decision rule $\hat{p}_{\mathbf{x}} = \frac{u_{\mathbf{x}}}{n_{\mathbf{x}}} > \frac{1}{2}$ for $\hat{\psi}_{opt}(\mathbf{x}) = 1$ can be altered according to whether $\psi_0(\mathbf{x}) = 1$ or $\psi_0(\mathbf{x}) = 0$. A non negative *inertial factor* $\alpha_{\mathbf{x}}$ is postulated and the prior-based estimator is defined by

$$\hat{\psi}_{pri}(\mathbf{x}) = 1 \text{ iff } \begin{cases} \hat{p}_{\mathbf{x}} > 0.5 - \frac{\alpha_{\mathbf{x}}}{n_{\mathbf{x}}}, \text{ if } \psi_0(\mathbf{x}) = 1 \\ \hat{p}_{\mathbf{x}} > 0.5 + \frac{\alpha_{\mathbf{x}}}{n_{\mathbf{x}}}, \text{ if } \psi_0(\mathbf{x}) = 0 \end{cases} \tag{12}$$

If $\mathbf{x}$ is never observed, then $\hat{\psi}_{pri}(\mathbf{x}) = \psi_0(\mathbf{x})$. The inertial factor reflects our confidence in the prior filter. The larger $\alpha_{\mathbf{x}}$, the more resistant the prior filter is to being changed. If $\mathbf{x}$ is rarely observed during training, then $\frac{\alpha_{\mathbf{x}}}{n_{\mathbf{x}}}$ resists change, which is consistent with the statistical understanding that $\hat{p}_{\mathbf{x}}$ is a poor estimator for small $n_{\mathbf{x}}$; as $n_{\mathbf{x}} \to \infty$, this weight tends to zero and $\hat{\psi}_{pri}(\mathbf{x}) \to \hat{\psi}_{opt}(\mathbf{x})$.

If $\varepsilon_{pri} = E\left[MAE < \hat{\Psi}_{pri} > -MAE < \Psi_{opt} >\right]$ is the precision error for estimation using a prior filter Ψ_0, then the precision advantage of using Ψ_0 is given by[2]

$$\begin{aligned} \varepsilon_{st} - \varepsilon_{pri} &= \frac{1}{2}\sum_{\mathbf{x}} \delta_{\mathbf{x}} P(n_{\mathbf{x}} = 0) \\ &+ \sum_{\mathbf{x}\in\mathcal{K}(\Psi_{opt})\cup\mathcal{K}(\Psi_0)} \delta_{\mathbf{x}} P(0.5 < \hat{p}_x \leq 0.5 + \frac{\alpha_{\mathbf{x}}}{n_{\mathbf{x}}}) \\ &+ \sum_{\mathbf{x}\in\mathcal{K}(\Psi_{opt})\cap\mathcal{K}(\Psi_0)} \delta_{\mathbf{x}} P(0.5 - \frac{\alpha_{\mathbf{x}}}{n_{\mathbf{x}}} < \hat{p}_x \leq 0.5) \\ &- \sum_{\mathbf{x}\in\mathcal{K}(\Psi_{opt})-\mathcal{K}(\Psi_0)} \delta_{\mathbf{x}} P(0.5 < \hat{p}_{\mathbf{x}} \leq 0.5 + \frac{\alpha_{\mathbf{x}}}{n_{\mathbf{x}}}) \\ &- \sum_{\mathbf{x}\in\mathcal{K}(\Psi_0)-\mathcal{K}(\Psi_{opt})} \delta_{\mathbf{x}} P(0.5 - \frac{\alpha_{\mathbf{x}}}{n_{\mathbf{x}}} < \hat{p}_{\mathbf{x}} \leq 0.5) \end{aligned} \tag{13}$$

The middle two sums apply to observations $\mathbf{x}$ for which $\psi_0(\mathbf{x}) = \psi_{opt}(\mathbf{x})$; the latter two apply when $\psi_0(\mathbf{x}) \neq \psi_{opt}(\mathbf{x})$. Hence, the more ψ_0 agrees with ψ_{opt}, the greater the precision advantage.

5. Prior Conditional Probabilities

In standard design, there exists a probability $p_{\mathbf{x}} = P(Y = 1|\mathbf{x})$. Image pairs are used to form an estimate $\hat{p}_{\mathbf{x}}$ of $p_{\mathbf{x}}$, and $\hat{\psi}_{opt}(\mathbf{x}) = 1$ if and only if $\hat{p}_{\mathbf{x}} > \frac{1}{2}$. A Bayesian approach results from assuming that $R = P(Y = 1|\mathbf{x})$ is a random variable depending on the image-noise model (state of nature) and there is a prior distribution $\pi(r)$ for R. (Note that we have omitted the subscript "x" when writing R to simplify notation, but it needs to be recognized that different vectors can have different prior distributions.) Bayesian estimation yields an estimator $\hat{R}$ of R based on $n_{\mathbf{x}}$ observations of $Y|\mathbf{x}$. The optimal Bayesian estimator is the conditional expectation $\hat{R} = E[R \mid Y_1|\mathbf{x}, Y_2|\mathbf{x}, ..., Y_{n_{\mathbf{x}}}|\mathbf{x}]$. For estimating $\psi_{opt}(\mathbf{x})$, the estimator is

$$\hat{\psi}_{Bayes}(\mathbf{x}) = \begin{cases} 1, \text{ if } \hat{R} \geq \frac{1}{2} \\ 0, \text{ if } \hat{R} < \frac{1}{2} \end{cases} \tag{14}$$

In the Bayesian setting, $\psi_{opt}(\mathbf{x})$ is a random variable (function of R). Hence, filter MAE is also a function of R, as is precision error. Focusing on a fixed vector $\mathbf{x}$, in the nonBayesian case $\mathbf{x}$ is exclusively either an element of or not an element of $\mathcal{K}(\psi_{opt})$; in the Bayesian setting, inclusion in the kernel depends on R. Hence, $\mathbf{x}$ has precision error resulting from being placed correctly in or not in $\mathcal{K}(\hat{\psi}_{opt})$ according to R. For fixed $R = r$, (expected) precision error due to $\mathbf{x}$ is given by

$$\varepsilon_{Bayes,\mathbf{x}}(r) = \begin{cases} |2r-1|P(\mathbf{x})P(\hat{R}|r \leq 0.5), \text{ if } r > 0.5 \\ |2r-1|P(\mathbf{x})P(\hat{R}|r > 0.5), \text{ if } r \leq 0.5 \end{cases} \tag{15}$$

Equation 15 is an expectation relative to the distribution of the conditional random variable $\hat{R}|r$. It must be averaged over the prior distribution $\pi(r)$ to give the Bayesian precision error. Letting $F_{\hat{R}|r}$ denote the probability distribution function of $\hat{R}|r$,

$$\varepsilon_{Bayes,\mathbf{x}} = \left[\int_{0.5}^{1} |2r-1| F_{\hat{R}|r}(0.5)\pi(r)dr + \int_{0}^{0.5} |2r-1|(1 - F_{\hat{R}|r}(0.5))\pi(r)dr \right] P(\mathbf{x}) \tag{16}$$

Total precision error, ε_{Bayes}, is obtained by summing over all $\mathbf{x}$.

The benefit of prior information can be ascertained by comparing the expression of Eq. 16 with a similar expression in which there is no prior information in the Bayesian setting (a uniform prior). Suppose $\pi(r) = 0$ for $r < r_0$, where $r_0 > 0.5$. Then the second integral in Eq.16 vanishes and the first integral is taken over $[r_0, 1]$. But for $r > r_0$, $F_{\hat{R}|r}(0.5)$ is small even for rather small samples ($\hat{R}|r$ is a consistent estimator of r). Hence, the first integral is small. No such simple reduction occurs for a uniform prior ($\pi(r) \equiv 1$).

Many interesting questions arise with Bayesian estimation and many of these have been studied in the context of beta prior distributions, which are excellent for modeling various distributions of R over $[0, 1]$. The questions include: What shaped priors provide the best precision? What is the effect of selecting an inappropriate prior? And, since in the Bayesian setting $\psi_{opt}(\mathbf{x})$ is a random variable depending on R, what is the precision of $\hat{\psi}_{Bayes}(\mathbf{x})$ as an estimator of the random variable

$\psi_{opt}(\mathbf{x})$? Leaving the discussion of these questions to Ref.[6] , we note only that, regarding the last question, this form of precision can be analytically expressed and $E\left[\hat{\psi}_{Bayes}(\mathbf{x}) - \psi_{opt}(\mathbf{x})\right] \to 0$ as $n_{\mathbf{x}} \to \infty$. Although this convergence is very important, the first two questions are crucial to design because they relate to how prior distributions effect estimation error for finite samples.

Acknowledgements

This work has been partially supported by National Science Fundation, grant MIP-9520139 and ProTeM-CC/CNPq through the AnIMoMat project, contract 680067/94-9.

References

1. G. J. F. Banon and J. Barrera. Minimal Representations for Translation-Invariant Set Mappings by Mathematical Morphology. *SIAM J. Appl. Math.*, 51(6):1782–1798, December 1991.
2. J. Barrera, E. R. Dougherty, and N. S. T. Hirata. Automatic design of optimal morphological operators using prior filters. *submitted*, 1998.
3. J. Barrera, E. R. Dougherty, and N. S. Tomita. Automatic programming of binary morphological machines by design of statistically optimal operators in the context of computational learning theory. *Journal of Electronic Imaging*, 6(1):54–67, January 1997.
4. E. R. Dougherty. Optimal mean-square n-observation digital morphological filters - part i:optimal binary filters. *CVGIP: Image Undestanding*, 55(1):36–54, 1992.
5. E. R. Dougherty and J. Barrera. *Non Linear Filters for Image Processing*, chapter Binary Logical Operators. SPIE and IEEE Press, 1998.
6. E. R. Dougherty, J. Barrera, and V. Kamat. Bayesian optimal filter design based on prior distributions for conditional probabilities. *submitted*, 1998.
7. E. R. Dougherty and R. P. Loce. Optimal mean-absolute-error hit-or-miss filter: Morphological representation and estimation of the binary conditional expectation. *Optical Engineering*, 32(4):815–823, 1993.
8. E. R. Dougherty and R. P. Loce. Precision of morphological representation estimators for translation-invariant binary filters: Increasing and nonincreasing. *Signal Processing*, 40(3):129–154, 1994.
9. E. R. Dougherty and R. P. Loce. Optimal binary differencing filters: Design, logic complexity, precision analysis, and applications to digital document processing. *Journal of Eletronic Imaging*, 5(1), 1996.
10. C. R. Giardina and E. R. Dougherty. *Morphological Methods in Image and Signal Processing.* Englewood Cliffs. Prentice-Hall, 1994.
11. R. P. Loce and E. R. Dougherty. Facilitation of optimal binary morphological filter design via structuring-element libraries and observation constraints. *Optical Engineering*, 31(5), 1992.
12. R. P. Loce and E. R. Dougherty. Optimal morphological restoration: The morphological filter mean-absolute-error theorem. *Visual Communication and Vision Representation*, 3(4), 1992.
13. R. P. Loce and E. R. Dougherty. *Enhancement and restoration of digital documents: statistical design of nonlinear algorithms.* SPIE - The International Society for Optical Engineering, 1997.
14. P. A. Maragos and R. Schafer. Morphological filters-part i:their set theoretic analysis and relations to linear shift-invariant filters. *IEEE Trans. on A, S, S. P.*, 35, 1987.

MONTE CARLO SIMULATION AND STATISTICAL INFERENCE OF MORPHOLOGICALLY CONSTRAINED GRFS*

K. SIVAKUMAR
Texas Center for Applied Technology
and Department of Electrical Engineering
214 WERC, Texas A&M University
College Station, TX 77843-3407, U. S. A.

and

J. GOUTSIAS
Department of Electrical and Computer Engineering
Image Analysis and Communications Laboratory
Barton 211, The Johns Hopkins University
Baltimore, MD 21218, U. S. A.

Abstract. This paper is about a new class of random spatial models, known as morphologically constrained Gibbs random fields, that are capable of modeling geometrical constraints in images by means of mathematical morphology. Two issues pertaining these models are considered in this paper: simulation and statistical inference. We first introduce a variant of the Metropolis algorithm, based on a multi-site updating strategy, for simulating morphologically constrained Gibbs random fields, that converges substantially faster than traditional single-site updating algorithms. We then consider the problem of optimally fitting morphologically constrained Gibbs random fields to real data. We show that, under a natural condition, maximum likelihood parameter estimation can be approximately implemented by means of the pattern spectrum.

Key words: Gibbs Random Fields, Metropolis Algorithm, Monte Carlo Simulations, Morphological Constraints, Statistical Inference

1. Introduction

An important subject of statistical image processing and analysis is to design appropriate models for random images. This serves as a tool for incorporating into a problem key knowledge about images in the form of probability distributions. A typical example of such a problem is the specification of a-priori probability distributions in Bayesian image analysis, frequently used for "optimally" extracting images from partially observed and noisy data.

When mathematical morphology is employed for image processing and analysis, it is natural to consider a-priori models capable of modeling knowledge in the form of morphological constraints. For example, in "optimal" morphological filtering problems, it is quite common to assume that the degradation-free image is morphologically smooth, in the sense that is both morphologically open and closed with respect to a given structuring element (e.g., see [1]). If we are interested in

* This work was supported by the Office of Naval Research, Mathematical, Computer, and Information Sciences Division, under ONR Grant N00014-90-1345.
E-mail: `siva@ee.tamu.edu` and `goutsias@mycenae.ece.jhu.edu`

designing a random image model that statistically incorporates such a constraint, we should limit our interest to models for which morphologically smooth images are the most probable realizations.

In this paper, we are considering a new class of random image models, known as morphologically constrained Gibbs random fields, that are capable of modeling a-priori knowledge in the form of morphological constraints. Since these models have been already introduced in [2, 3], we focus our attention here to two important problems pertaining these models, namely simulation and statistical inference. In Section 2, we provide a brief overview of these models and introduce our notation. A Monte Carlo simulation technique is discussed in Section 3. We introduce a variant of the well-known Metropolis algorithm, based on a multi-site updating strategy. The proposed algorithm converges asymptotically to the desirable distribution and is much faster than traditional single-site updating Monte Carlo algorithms. Finally, the problem of statistical inference is discussed in Section 4. It is shown that, under a natural condition, the pattern spectrum is approximately the maximum-likelihood estimator of the morphologically constrained Gibbs random field parameters.

2. Morphologically Constrained GRFs

Recently, a novel statistical model for discrete random images $\mathbf{X}$, defined over a finite rectangular observation window $W = \{w \in \mathbb{Z}^2 \mid w = (m,n), 0 \leq m \leq M-1, 0 \leq n \leq N-1\}$ has been proposed [2–5]. This is used to model images with given size density and is a special case of a *Gibbs random field* (GRF) with (joint) probability mass function $\pi(X)$ of the form:

$$\pi(X) = \frac{1}{Z} \exp\{-\frac{1}{T} V(X)\} , \quad \forall X \in \mathcal{S} , \tag{1}$$

with

$$Z = \sum_{X \in \mathcal{S}} \exp\{-\frac{1}{T} V(X)\} , \tag{2}$$

where

$$V(X) = \sum_{i=0}^{I} a_i \Big| \, ||\gamma_i(X) - \gamma_{i+1}(X)||_W - \nu_i G|W| \, \Big|^p + \sum_{j=1}^{J} a_{-j} \Big| \, ||\phi_j(X) - \phi_{j-1}(X)||_W - \nu_{-j} G|W| \, \Big|^p , \quad \forall X \in \mathcal{S} . \tag{3}$$

In (3), $\{\gamma_s\}_{s=0,1,\ldots}$ is a *granulometry* [6], $\{\phi_s\}_{s=0,1,\ldots}$ is the associated *anti-granulometry*, $I \geq 0$, $J \geq 1$ are two finite integers, $p \geq 1$ is a real-valued parameter, $|W|$ denotes the *cardinality* (or *area*) of W, $||X||_W = \sum_{w \in W} |X(w)|$, $\{a_i, -J \leq i \leq I\}$ are $\{0,1\}$-valued parameters, and $\{\nu_i, -J \leq i \leq I\}$ are real-valued nonnegative parameters such that

$$\sum_{i=-J}^{I} \nu_i \leq 1 .$$

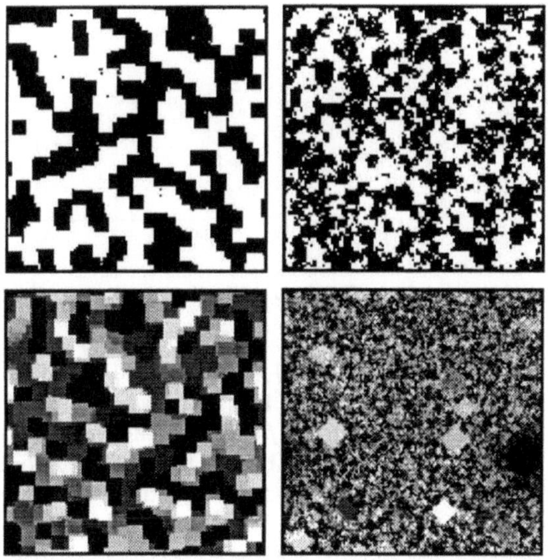

Fig. 1. Binary and grayscale realizations of two morphologically constrained GRFs.

We limit our interest here to discrete-valued random images that take values in the state space $\mathcal{S} = \mathcal{G}^{MN}$ with $\mathcal{G} = \{0, 1, ..., G\}$, $1 \leq G < \infty$. In (1), (2), V is known as the *energy function*, T is a real-valued positive constant known as the *temperature*, and Z is a normalizing constant, known as the *partition function*.

It can be shown that

$$\lim_{T \to 0^+} \pi(X) = \begin{cases} 1/|\mathcal{V}|, & \text{for } X \in \mathcal{V} \\ 0, & \text{otherwise} \end{cases},$$

where

$$\mathcal{V} = \{X^g \in \mathcal{S} \mid V(X^g) \leq V(X), \ \forall X \in \mathcal{S}\}$$

is the collection of global minima of the energy function V (known as the *ground states* of $\mathbf{X}$). Therefore, the probability mass function of a morphologically constrained GRF becomes uniform, over all ground states, in the limit as the temperature decreases to zero. It is the ground states that are most interesting in image modeling problems, since ground states are well structured. It is therefore natural to focus our attention to GRFs at low temperatures.

It can also be shown that

$$\nu_s = \lim_{T \to 0^+} s_X(s), \ \forall s: \ -J \leq s \leq I \text{ such that } a_s = 1, \tag{4}$$

with

$$s_X(s) = \frac{1}{G|W|} \begin{cases} \mathrm{E}[\ \|\gamma_s(\mathbf{X}) - \gamma_{s+1}(\mathbf{X})\|_W\ ], & \text{for } s \geq 0 \\ \mathrm{E}[\ \|\phi_{|s|}(\mathbf{X}) - \phi_{|s|-1}(\mathbf{X})\|_W\ ], & \text{for } s \leq -1 \end{cases}$$

being the *size density* of random image $\mathbf{X}$ [7], where E[ ] denotes expectation. Therefore, and at low enough temperatures, the resulting random image $\mathbf{X}$ may

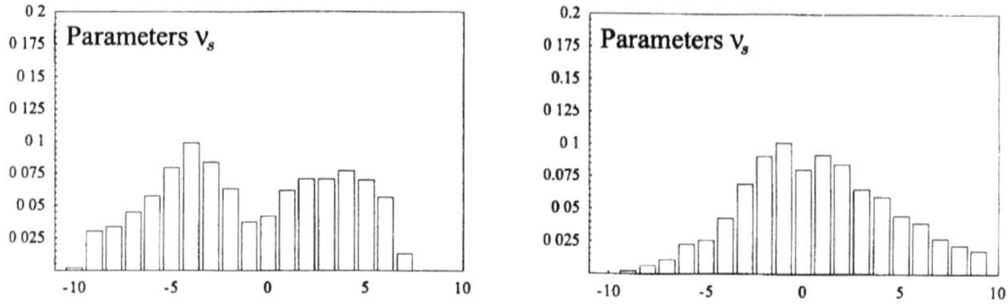

Fig. 2. Morphological size densities of the two realizations in the second column of Fig. 1.

favor realizations whose size density $s_X(s)$ is determined by the particular values of a and ν. A GRF with probability mass function given by (1)–(3) is known as a *morphologically constrained Gibbs random field* (with energy function (3)).

Figure 1 depicts four realizations, two binary (with $G = 1$) and two grayscale (with $G = 15$), of two morphologically constrained GRFs. For the realizations depicted in the first column, we set $a_{-J} = \cdots = a_0 = \cdots = a_I = 1$, $\nu_{-J} = \cdots = \nu_0 = \cdots = \nu_I = 0$, $J = I+1$, $p = 1$, $\gamma_i(X) = X \circ iB$, and $\phi_j(X) = X \bullet j\check{B}$, where $\circ$, $\bullet$ denote opening and closing, respectively, $\check{B} = \{-b \mid b \in B\}$, and $sB = (s-1)B \oplus B$, for $s = 1, 2, ...,$ $0B = \{(0,0)\}$, with $\oplus$ denoting dilation. This leads to an energy function of the form:

$$V(X) = ||X - X \circ (I+1)B||_W \; + \; ||X \bullet (I+1)\check{B} - X||_W \; , \;\; \forall X \in \mathcal{S} \; ,$$

with ground states in $\mathcal{V}_\circ \cap \mathcal{V}_\bullet$, where $\mathcal{V}_\circ = \{X \in \mathcal{S} \mid X = X \circ (I+1)B\}$ is the *invariance domain* of the opening $X \circ (I+1)B$ and $\mathcal{V}_\bullet = \{X \in \mathcal{S} \mid X = X \bullet (I+1)\check{B}\}$ is the invariance domain of the closing $X \bullet (I+1)\check{B}$. At low enough temperatures, the resulting GRF may favor realizations that are approximately both open and closed, with respect to structuring element $(I+1)B$, in the sense that $\pi(X)$ converges (in the limit as $T \to 0^+$) to a uniform distribution over all realizations in $\mathcal{V}_\circ \cap \mathcal{V}_\bullet$. Furthermore, and as a direct consequence of (4), we have that $\lim_{T\to 0^+} s_X(s) = 0$, for $-(I+1) \leq s \leq I$. Such a GRF may serve as an a-priori image model in "optimal" morphological filtering problems, since, in such problems, it is usually assumed that the degradation-free image X is smooth, in the sense that $s_X(s) = 0$, $-(I+1) \leq s \leq I$, for some integer $I \geq 0$ (e.g., see [1]).

For the realizations depicted in the first column of Fig. 1, we set $T = 0.1$, $I = 5$, and $B = B_{sq}$, with B_{sq} being the elementary SQUARE structuring element $\{(0,0),(1,0),(1,1),(0,1)\}$. The realizations depicted in the second column of Fig. 1 are obtained by setting $T = 0.1$, $I = 9$, $a_{-10} = \cdots = a_0 = \cdots = a_9 = 1$, with parameters $\{\nu_s, -J \leq s \leq I\}$ depicted in Fig. 2. For the binary realization, we set $B = B_{sq}$, whereas for the grayscale realization $B = B_{rh}$, with B_{rh} being the elementary RHOMBUS structuring element $\{(0,0),(1,0),(-1,0),(0,1),(0,-1)\}$. These realizations are used in Section 4 to demonstrate the accuracy of the proposed parameter estimation algorithm. The realizations depicted in Fig. 1 have been obtained by means of the Monte Carlo simulation algorithm described next.

3. Monte Carlo Simulation

Although single-site updating *Markov chain Monte Carlo* (MCMC) sampling algorithms are standard tools for simulating GRFs (e.g., see [8]), they tend to converge very slowly when used for simulating morphologically constrained GRFs with large values of I and J at low temperatures. We here propose a variant of the Metropolis algorithm that, at each iteration, allows multi-site updating and converges substantially faster than the single-site updating version. At each iteration, this algorithm chooses a site $w \in W$, uniformly among all sites in W, and a "size" r in $\{0, 1, \ldots, I\}$, randomly with probability

$$g(r) \;=\; \frac{\beta\nu_r + (1-\beta)/(I+1)}{(1-\beta) \,+\, \beta\sum_{r=0}^{I}\nu_r}, \quad \forall\, r = 0, 1, \ldots, I\,, \tag{5}$$

for some $0 \leq \beta \leq 1$. The reason for choosing probability $g(r)$ by means of (5) is the intuitive observation that $g(r)$ should be proportional to the desirable size density, and thus to ν_r. A value of $\beta < 1$ is desirable so as to obtain $g(r) > 0$, even if $\nu_r = 0$.

Given a w, an r, and that $\mathbf{X}_{k-1} = X_i$, we pick-up a realization X_j with probability $q(i,j,w,r)$ (see [4, 5] for a closed-form expression for such a probability) where X_j and X_i are identical at all sites outside of $(rB)_w$, whereas X_j takes, either with probability ϵ, the same value at all sites in $(rB)_w$, this value being chosen at random, uniformly among all values in $\mathcal{G} \setminus \{X_i(w)\}$, or, with probability $1-\epsilon$, value $X_j(v)$, at a site $v \in (rB)_w$, this value being chosen at random, uniformly among all values in $\mathcal{G}$ (except when $v = w$, in which case value $X_j(w)$ is chosen at random, uniformly among all values in $\mathcal{G} \setminus \{X_i(w)\}$). We then set $\mathbf{X}_k = X_j$ with probability $\alpha(i,j,w,r)$, given by

$$\alpha(i,j,w,r) = \min\left\{1, \frac{\pi(X_j)\, q(j,i,w,r)}{\pi(X_i)\, q(i,j,w,r)}\right\}, \quad \text{for } i \neq j\,. \tag{6}$$

It has been shown in [4, 5] that the proposed multi-site updating MCMC sampling algorithm asymptotically converges to the desirable Gibbs distribution π. It has also been noticed that the suggested algorithm converges fast by setting a large value of ϵ. Setting $\epsilon = 1$ should be avoided in general, as explained in [4, 5]. After extensive experimentation, we set $\epsilon = 0.99$.

As we have previously discussed, we are mostly interested in morphologically constrained GRFs at low temperatures. In such a case, a realization X of $\mathbf{X}$ will (approximately) be a ground state, in the sense that $V(X) \approx 0$ (since the energy function V in (3) is non-negative). This observation suggests a simple way of determining the maximum number K of iterations required for the multi-site updating MCMC algorithm (at low enough temperature) to reach equilibrium: take K such that

$$K \;=\; \min_k \left\{ k \mid V(X_k) \leq \delta\, G|W| \right\}, \tag{7}$$

for some small $\delta \geq 0$. Assuming that X is approximately a ground state of $\mathbf{X}$ if $V(X) \leq 1\%$ of $G|W|$, we set $\delta = 0.01$.

The previous discussion leads to the following multi-site updating MCMC sampling algorithm for simulating morphologically constrained GRFs:

Simulation Algorithm: Starting with an i.i.d. image X_0, generate a Markov chain $\{\mathbf{X}_k, k = 0, 1, \ldots, K\}$, where K is given by (7), with $\delta = 0.01$, by means of the following iterative procedure: At each iteration k, choose a site $w \in W$, uniformly among all sites in W, and a "size" r with probability $g(r)$, given by (5); select a realization X_j such that $X_j(v) = X_i(v)$, for every $v \notin (rB)_w$ (where $\mathbf{X}_{k-1} = X_i$), $X_j(w)$ is chosen uniformly among all values in $\mathcal{G} \setminus \{X_i(w)\}$, and, with probability $\epsilon = 0.99$, $X_j(v) = X_j(w)$, for every $v \in (rB)_w$, or, with probability $1 - \epsilon = 0.01$, $X_j(v)$ at a site $v \in (rB)_w \setminus \{w\}$ is chosen uniformly among all values in $\mathcal{G}$; finally, set $\mathbf{X}_k = X_j$ with probability $\alpha(i, j, w, r)$, given by (6).

4. Statistical Inference

A problem of practical interest is to "optimally" fit a morphologically constrained GRF with probability mass function π_ν, indexed by parameters $\nu = [\nu_{-J}, \ldots, \nu_{-1}, \nu_0, \nu_1, \ldots, \nu_I]$, to real data. This can be naturally done by means of a *maximum likelihood* approach. In this section, we limit our interest to a morphologically constrained GRF with energy function V, given by (3), with $p = 2$. We assume that T, I, J, as well as $\{a_i, -J \leq i \leq I\}$ are known a-priori. Given data X in W, the *maximum-likelihood estimate* (MLE) $\hat{\nu}(X)$ of parameters ν is given by $\hat{\nu}(X) = \arg\max_\nu\{\ln \pi_\nu(X)\}$. It can be shown that, under a mild identifiability condition, the maximum-likelihood estimator $\hat{\nu}(\mathbf{X})$ exists, is unique (with probability one), for sufficiently large W, and is always consistent, as $W \to \mathbb{Z}^2$, where $W \to \mathbb{Z}^2$ denotes convergence of the observation window W to $\mathbb{Z}^2$ in the sense of van Hove, provided that the granulometry $\{\gamma_s\}_{s=0,1,\ldots}$ in (3) is translation invariant [4, 5] (see also [9]).

The size density $s_X(s \,|\, \hat{\nu})$ of the morphologically constrained GRF with probability mass function $\pi_{\hat{\nu}}$ satisfies [4, 5]

$$s_X(s \mid \hat{\nu}) = p_X(s)\,, \quad \forall s : -J \leq s \leq I \;\text{ such that }\; a_s = 1\,, \tag{8}$$

where $p_X(s)$ is the *pattern spectrum* of X, given by

$$p_X(s) = \frac{1}{G|W|}\begin{cases} \|\gamma_s(X) - \gamma_{s+1}(X)\|_W, & \text{for } s \geq 0 \\ \|\phi_{|s|}(X) - \phi_{|s|-1}(X)\|_W, & \text{for } s \leq -1 \end{cases}.$$

Therefore, given data X, the size density $s_X(s \mid \hat{\nu})$ of a morphologically constrained GRF with parameter $\hat{\nu}(X)$, obtained by means of maximum likelihood, equals the pattern spectrum $p_X(s)$ of X, for every s such that $a_s = 1$.

The pattern spectrum satisfies some nice statistical properties. For example, it is not difficult to show that the pattern spectrum is a *uniformly minimum variance unbiased* (U.M.V.U.) estimate of the size density $\{s_X(s \mid \nu), -J \leq s \leq I\}$ of a morphologically constrained GRF $\mathbf{X}$ with energy function V, given by (3), with $p = 2$. We here, however, want to discuss the asymptotic behavior of the pattern spectrum $p_X(s)$, in the limit as $W \to \mathbb{Z}^2$. In view of (8), this behavior coincides with the asymptotic behavior of the size density estimate $s_X(s \mid \hat{\nu})$. We consider a sequence $\{W_k = \{(m,n) \in \mathbb{Z}^2 \mid -k \leq m,n \leq k\}, k = 1,2,\ldots\}$ of square observation windows, and assume that $\{\gamma_s\}_{s=0,1,\ldots}$ is a translation invariant granulometry and that parameters ν are independent of k. Let $\mathcal{P}(\nu)$ be the collection of all infinite-volume Gibbs distributions on $\mathcal{G}^{\mathbb{Z}^2}$, associated with parameters ν, and let P_ν be any Gibbs distribution in $\mathcal{P}(\nu)$ (see [9] for more details). For every $k = 1,2,\ldots$, P_ν is a Gibbs distribution on $\mathcal{G}^{\mathbb{Z}^2}$ such that, the conditional probability of $\mathbf{X}_k = \mathbf{X}(W_k) = X_k$, where $\mathbf{X}(W_k)$ denotes the restriction of $\mathbf{X}$ to W_k and $X_k \in \mathcal{G}^{W_k}$, given the value of $\mathbf{X}$ on W_k^c, is given by (1)–(3), with $a_s = 1$, for every $-J \leq s \leq I$, $W = W_k$, $X = X_k$, and $p = 2$. If we assume *identifiability* of the true parameters ν (i.e., if we assume that, for $\nu \neq \nu'$, $\mathcal{P}(\nu) \neq \mathcal{P}(\nu')$) (see also [9]), then we have the following proposition regarding consistency of the maximum-likelihood estimator of the size density.

Proposition 1: *Let the true parameters ν be identifiable, P_ν be any Gibbs distribution in $\mathcal{P}(\nu)$, and for $k = 1,2,\ldots$, let $\hat{\nu}_k = \hat{\nu}(X_k)$ be the MLE of ν, based on a realization X_k of $\mathbf{X}_k$. Suppose that*

$$\lim_{k\to\infty} s_{X_k}(s \mid \nu) = f_s(\nu) \ , \quad \forall s : -J \leq s \leq I \ ,$$

uniformly in a neighborhood $D_\nu = \{u \in \mathbb{R}^{I+J+1} \mid \|u - \nu\|_2 < \delta\}$ of ν, where $\delta > 0$ and $\| \ \|_2$ denotes the Euclidean norm. Then,

$$\lim_{k\to\infty} \mid s_{X_k}(s \mid \hat{\nu}_k) - s_{X_k}(s \mid \nu) \mid \; = \; 0 \ , \quad \forall s : -J \leq s \leq I \ , \quad P_\nu - \textit{almost surely.}$$

As a direct consequence of (8), the previous proposition, and for sufficiently large k, the pattern spectrum $\{p_{X_k}(s), -J \leq s \leq I\}$ is arbitrarily close to the size density $\{s_X(s \mid \nu), -J \leq s \leq I\}$ of a morphologically constrained GRF $\mathbf{X}$, with energy function V, given by (3), with $p = 2$, provided that the assumptions of Proposition 1 are satisfied.

To obtain the MLE of ν, we need to solve a nonlinear system of equations with respect to $\hat{\nu}$. This is clearly a difficult problem whose solution can only be obtained by means of computationally demanding iterative techniques, like the Monte Carlo maximum likelihood estimation technique of Geyer [10]. However, the pattern spectrum can be used to approximate the maximum-likelihood parameters $\hat{\nu}$, provided that data X are at a low enough temperature. Indeed, if we assume that observation X is a sample drawn from $\pi_{\hat{\nu}}$ at a low enough temperature, then (recall (4) and (8))

$$\hat{\nu}_s(X) \cong s_X(s \mid \hat{\nu}) = p_X(s) \ , \quad \forall s : -J \leq s \leq I \ \text{ such that } \ a_s = 1 \ .$$

This leads to an algorithm for approximately fitting morphologically constrained GRFs to real data X, by means of maximum likelihood, without solving a system of non-linear equations.

Estimation Algorithm: Given X, first calculate the pattern spectrum $p_X(s)$, for $-\infty \leq s \leq \infty$. Then, take I and J to be the smallest integers such that $p_X(s) = 0$, for $s > I$ and $s < -J$. Finally, consider a morphologically constrained GRF with energy function (3), where $a_{-J} = \cdots = a_{-1} = a_0 = a_1 = \cdots = a_I = 1$ and $\nu_s = p_X(s)$, for every $s : -J \leq s \leq I$.

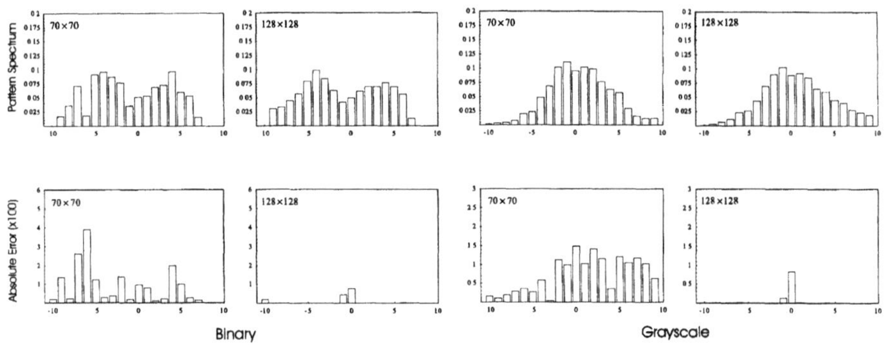

Fig. 3. The pattern spectra (first row) calculated from 70 × 70 and 128 × 128 pixel realizations of the morphologically constrained GRFs depicted in the second column of Fig. 1, and the absolute error (multiplied by 100) in parameter estimation (second row).

Realizations, over a 70×70 and a 128×128 square grid, obtained from the GRFs depicted in the second column of Fig. 1 are used to calculate the pattern spectra depicted in the first row of Fig. 3. The second row depicts the absolute difference (multiplied by 100) between the calculated pattern spectra and the model parameters depicted in Fig. 2. As expected, the pattern spectrum does an increasingly better job in estimating the true model parameters, as the grid size increases. In both cases, data over an 128×128 square grid are sufficient to produce only sparse errors below the 1% mark.

References

1. D. Schonfeld and J. Goutsias, "Optimal morphological pattern restoration from noisy binary images," *IEEE Transactions on Pattern Analysis and Machine Intelligence*, vol. 13, pp. 14–29, 1991.
2. K. Sivakumar and J. Goutsias, "Morphologically constrained discrete random sets," in *Advances in Theory and Applications of Random Sets*, D. Jeulin, Ed., pp. 49–66. World Scientific, Singapore, 1997.
3. K. Sivakumar and J. Goutsias, "Morphologically constrained Gibbs random fields," in *Proceedings of the IEEE International Conference on Image Processing*. Santa Barbara, CA, October 26–29, 1997.
4. K. Sivakumar, *Morphological Analysis of Random Fields: Theory and Applications*, Ph.D. Thesis, Department of Electrical and Computer Engineering, The Johns Hopkins University, Baltimore, MD, 1997.
5. K. Sivakumar and J. Goutsias, *Morphologically constrained Gibbs random fields: Applications to texture synthesis and analysis*, Tech. Rep. JHU/ECE 97-11, Department of Electrical and Computer Engineering, The Johns Hopkins University, Baltimore, MD, 1997.
6. H.J.A.M. Heijmans, *Morphological Image Operators*, Academic Press, Boston, Massachusetts, 1994.
7. K. Sivakumar and J. Goutsias, "Discrete morphological size distributions and densities: Estimation techniques and applications," *Journal of Electronic Imaging*, vol. 6, pp. 31–53, 1997.
8. C.J. Geyer, "Practical Markov chain Monte Carlo (with discussion)," *Statistical Science*, vol. 7, pp. 473–511, 1992.
9. B. Gidas, "Parameter etimation for Gibbs distributions from fully observed data," in *Markov Random Fields: Theory and Applications*, R. Chellappa and A. Jain, Eds., pp. 471–498. Boston, Massachusetts: Academic Press, 1993.
10. C.J. Geyer, "On the convergence of Monte Carlo maximum likelihood calculations," *Journal of the Royal Statistical Society, Series B*, vol. 56, pp. 261–274, 1994.

PARAMETER ANALYSIS OF CONDITIONAL MORPHOLOGICAL OPERATORS FOR IMAGE RESTORATION

ELENA STRINGA AND CARLO S. REGAZZONI
Department of Biophysical and Electronic Engineering (DIBE) University of Genoa, Via all'Opera Pia 11A, I-16145 Genoa, Italy

Abstract. Conditional morphology is a nonlinear method for simultaneous edge detection and image smoothing. It is derived by defining statistical models for the input-output relation as in statistical morphology, with the introduction of a line process taking into account the discontinuities between different regions of the image. It results to be a method presenting the advantages of both morphological (simplicity) and Bayesian (robustness) methods and it allows one to perform image restoration by applying a continuation method. This is performed applying a scheduling of the temperature parameters introduced with the definition of statistical models. In this paper, different kinds of temperature scheduling are analyzed. Experimental results are shown from processing of Synthetic Aperture Radar (SAR) images.

Key words: Mathematical morphology, Bayesian reconstruction, Mean field theory.

1. Introduction

Low level image processing usually involves the problem of coupled edge detection and image restoration. Both edge and restored images are often required by higher level processing modules, for example for pattern recognition tasks. During the last years, two particular methodologies have assumed great interest in the field of image processing: morphological approaches and Bayesian methods.

Mathematical morphology allows one to obtain good results in image restoration taking into account the shape of the objects; this approach is characterized by elementary operators (i.e. erosion and dilation) and their combinations (i.e. opening and closing) [5] which are very fast but not able to remove noise with statistical properties. Bayesian methods allow one to take into account both geometric properties of the object in the scene and noise corrupting the signal, by means of probabilistic models [2]; these operators present good robustness properties but usually involves a very high computational load.

A first attempt to join these approaches to obtain operators presenting the advantages of both methodologies was proposed by Yuille *et al.* with the definition of statistical morphological operators for image restoration [6]; the output of statistical morphology is obtained by applying the Mean Field Theory (MFT), which allows one to define filtering operators presenting robustness and computational simplicity characteristics. They also suggested a solution for edge extraction by appropriately combining the results of basic statistic morphological operators.

Conditional morphology is a new statistical morphological approach proposed in [4]. It was derived by combining statistical morphology [6] and a Markov Random Field (MRF) based method proposed by Geiger and Girosi [1] for coupled edge detection and image restoration. The MRF-based method involves the use of MFT for generating the solving equation. In particular, by using this equation and a gradient descent method, it is possible to iteratively estimate the intensity field and the discontinuity fields of image variables. Moreover, the obtained equation depends on a continuation parameter (equivalent to the reciprocal of temperature in probabilistic models [2] using simulated annealing [3]. Therefore, the fields of variables can be estimated by progressively scheduling the (parameter according to a so called deterministic annealing strategy.

Conditional morphology is an extension of statistical morphology that allows one to solving smoothing and edge detection problems simultaneously [4]; it is derived by using the common theoretical framework of statistical morphology and deterministic annealing given by MFT. The main advantage of conditional morphology with respect to deterministic annealing is that the equation obtained by applying MFT assumption assumes a closed form, i.e. intensity fields solutions for a given model do not depend on variables of the line process. This implies that estimation algorithm must be not recursive, but provides a one-shot estimation method for intensity and discontinuity fields, so being characterized by a grater computational efficiency. Nevertheless, a continuation method has to be applied also for conditional morphology as the obtained equation for the estimation of the intensity field still depends on a (parameter. As in deterministic annealing, the statistical model of conditional morphology involves other parameters that change the performances of the elementary operators if they are chosen with different values.

In this paper an analysis of the results obtained by considering different parameter values is presented. Experimental results are performed by applying the method to synthetic and real Synthetic Aperture Radar (SAR) images that usually are corrupted by speckle noise (noise presenting both statistical and geometric properties).

In the next section, the definition of conditional morphological basic operators (erosion and dilation) is presented. In section III, the behaviour of the conditional morphology is presented by varying the value assumed by the temperature parameter and applying the method to a synthetic SAR image affected by a 2-look simulated speckle noise.

2. Conditional Morphology

Conditional morphology [4] allows one to perform edge detection and image smoothing simultaneously by means of the fusion of statistical morphology [6] and the MRF based method presented in [1].

2.1. Statistical morphology

Statistical morphology [6] is a statistic formulation of mathematical morphology in which morphological operators are redefined under a probabilistic point of view. This methodology involves a temperature parameter T $(1/\beta)$, whose values change the behaviour of the filter: in fact for $T \to \infty$ $(\beta \to 0)$ morphological classic operators can be obtained, while for $T \to 0$ $(\beta \to \infty)$ linear filtering is performed. By considering the input values $I_i, i = 1, \ldots, D_x * D_y$ ($D_x * D_y$ is the image size), the solution at site i is defined by means of a probabilistic model introduced for selecting the winner pixel, which belongs to a neighbourhood set for pixel i (i.e. the structuring element (SE) B_i centred on pixel i). By applying minimum variance estimation to the statistical models proposed in [6], the mean-field equation for statistical dilation is obtained:

$$O_w(I_i) = \sum_{k \in B_i} \frac{e^{\beta I_k}}{\sum_{j \in B_i} e^{\beta I_j}} I_k \tag{1}$$

Eq. (1) shows that the solution for statistical dilation at pixel i depends on the parameter β and the grey-level values of pixels that falling inside B_i. Analogously, Yuille *et al.* [6] derived the solution for statistical erosion: in this case the solution has the form of Eq. (1) in which β assumes negative values.

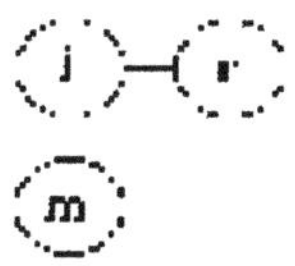

Fig. 1. Neighbourhood set used with the MRF-based method.

2.2. Deterministic annealing solution to MRF models

The method proposed in [1] allows one to perform edge detection and image smoothing simultaneously by means of two coupled MRFs (F, L):
– $F = \{F_i : i = 1, \ldots, D_x * D_y\}$ where F_i represents the image intensity of i-th image pixel (Fig. 2);
– $h = \{h_{i,m} \in [0,1] : i = 1, \ldots, D_x * D_y\}$ where $h_{i,m}$ represents the horizontal line process between i-th and m-th neighbour pixels (Fig. 1);
– $v = \{v_{i,r} \in [0,1] : i = 1, \ldots, D_x * D_y\}$ where $v_{i,r}$ represents the vertical line process between i-th and r-th neighbour pixels (Fig. 1).

The output for (F, L) is obtained by defining probabilistic models for the solution and the noise corrupting the input data; these models depend on a temperature parameter T $(1/\beta)$ which can change the filter behaviour.

By applying minimum variance estimations Geiger & Girosi [1] obtained the following mean-field equations for the line process;

$$h_{i,m} = \frac{1}{1 + \exp\left[\beta(\gamma - \alpha(F_i - F_m)^2)\right]}, \quad v_{i,r} = \frac{1}{1 + \exp\left[\beta(\gamma - \alpha(F_i - F_r)^2)\right]} \tag{2}$$

The mean field equation for the field F depends on noise parameters and the line process values; the solution was obtained by introducing a gradient descent algorithm [1].

2.3. Conditional morphological operators

The basic conditional morphological operation [4], erosion and dilation, are obtained by following the same technique presented in [6]: the possible solution for F_i is selected among the pixels of B_i. This concept is formalized by introducing binary decision unit $V_{i,j}$ as in statistical morphology, and a conditional probability $P(F, h, v/I)$. According to Bayes theorem, we can write:

$$P(F_i, h_{i,m}, v_{i,r}/I) = \sum_{V_i^J} P(F_i, h_{i,m}, v_{i,r}/V_i^J, I] \cdot P[V_i^J/I] \tag{3}$$

where V_i^J represents a possible solution configuration and $P[V_i^J/I]$ has the following expression:

$$P[V_{i,k} = I, V_{i,j} = 0, j \neq k] = \frac{\exp(\beta I_k)}{Z} \tag{4}$$

$\sum_{V_i^J} P(F_i, h_{i,m}, v_{i,r}/V_i^J, I]$ contains the dependence of the solution on discontinuities; in particular for each site $j \in B_i$, we evaluate the interaction among I_j (possible solution) and all possible solutions for F_r and F_m. The F_r and F_m values are chosen in two new neighbourhood sets B_r and B_m presenting the same shape of B_i, but centred on sites r and m respectively.

The output for the image intensity field is obtained by applying the mean field theory:

$$F_i = \sum_{j \in B_i} w_j I_j \tag{5}$$

with

$$w_j = \frac{\sum_{\{p \in B_r, k \in B_m\}} c_{jpk}}{\sum_{j \in B_i} \exp(\beta I_j) \sum_{\{p \in B_r, k \in B_m\}} c_{jpk}} \tag{6}$$

where

$$\begin{aligned} c_{jpk} &= \exp[\beta(I_p - I_k)] \cdot \Big\{ \exp\big[\beta[2\gamma - \alpha(I_j - I_p)^2]\big] + \\ &\quad \exp\big[\beta[\gamma - \alpha(I_j - I_p)^2]\big] \Big\} = \exp\big[\beta[\gamma - \alpha(I_j - I_k)^2]\big] + 1 \end{aligned} \tag{7}$$

with $I_p \in B_r$ and $I_k \in B_m$. By applying the same method we can derive the solutions for the line process:

$$v_{i,r} = \frac{\sum_{j \in B_i} \left(e^{\beta I_j} \sum_{k \in B_m} c_{j,k} \sum_{p \in B_r} e^{\beta I_p} \right)}{\sum_{j \in B_i} e^{\beta I_j} \sum_{\{p \in B_r, k \in B_m\}} c_{jpk}} \qquad h_{i,m} = \frac{\sum_{j \in B_i} \left(e^{\beta I_j} \sum_{p \in B_r} c_{j,p} \sum_{k \in B_m} e^{\beta I_k} \right)}{\sum_{j \in B_i} e^{\beta I_j} \sum_{\{p \in B_r, k \in B_m\}} c_{jpk}} \tag{8}$$

where $c_{j,d} = \exp\Big(\beta I_d \big\{ \exp\big[\beta[\gamma - \alpha(I_j - I_d)^2]\big] + 1 \big\}\Big)$ with $d = \{k \text{ or } p\}$. Eqs. (7), (8) represent the outputs of *conditional dilation* if $\beta > 0$. By considering negative β values, they represent the outputs for *conditional erosion.*

3. Parameter analysis

Conditional morphology allows one to use continuation methods, such as simulated annealing [3], by applying several elementary operation in cascade by using different increasing values of the parameter β. In [4] an analysis of the behaviour of the conditional morphological operators is presented. In this paper their behaviour is analyzed for different values of the β parameter. In particular we analyze the results obtained by applying conditional openings (conditional erosion followed by conditional dilation) in cascade with different kinds of β scheduling.

In order to evaluate the effects off different kinds of β scheduling, statistical morphological operators has been applied to synthetic SAR images (Fig. 2a) corrupted by a simulated 2-look speckle noise (Fig. 2b). Restoration performances are evaluated by means of SNR and the goodness of edge detection is evaluated by counting the percentage of lost and surplus points with respect to the optimal edges; more precisely, to take into account the displacements occurring in edge detection, it is considered that an edge is correctly detected if the discontinuity is present within a cross 3×3 neighbourhood set centred on the optimal edge position point.

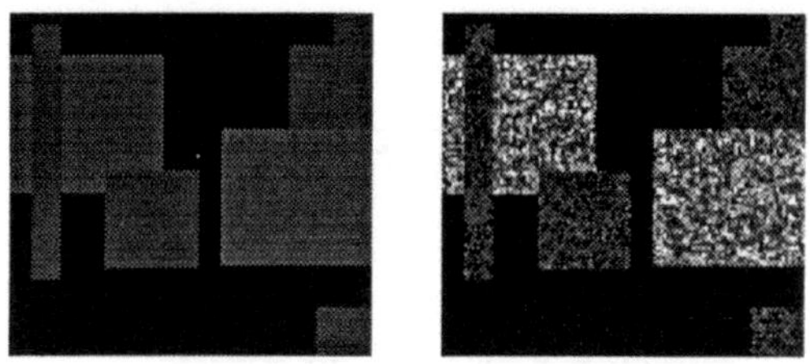

(a) (b)

Fig. 2. Synthetic SAR images used for evaluating the effect of the β parameter.

3.1. Effects of the β values in edge detection from SAR images

Let us consider the scheduling proposed by Zerubia and Chelappa (ZCS) in [7]:

$$\beta_{n+1} = 4 \times \beta_n$$

where n is the considered iteration. By using this method, we have to set the starting β_0 value and the number of the iterations. The starting value has to be considered smaller when the level of noise is higher because conditional morphological operators behaves as linear average operators when $\beta \to 0$ [4]. As we consider a synthetic SAR image corrupted by a 2-look simulated speckle noise, we set $\beta_0 = 0.2$. We have considered six β iterations, corresponding to $\beta_{max} = 204.8$, because in this case the solutions obtained both for the image intensity field and the line process are quite stable or, in other words, increasing the number of the iterations do not introduce a meaningful improvement in both edge and intensity fields. In Fig. 3 the SNR versus the iteration number is shown and it is possible to notice that SNR assumes an asymptotic behaviour.

Starting from this scheduling, several iterations have been added to improve the conditional morphological performances in term of edge detection. The behaviour of the discontinuities has been analyzed to individuate in which range of β values it is convenient to introduce additional iterations. In Fig. 4 a qualitative example of edge detection is presented with different β scheduling. The first raw of pictures represents the image intensity field, the second one represents the line process obtained by using a 3×3 square SE and the third one represents the detected edges (binarized line process). In Fig. 4a the intensity profile of a discontinuity is presented before the application of the conditional morphological operator. By applying conditional morphological operators with ZCS, the discontinuity is modified as in Fig. 4b; it is possible to notice that the discontinuity has been smoothed, in fact when the β value is low a quite linear behaviour is obtained. By introducing two additional iterations with $\beta \in [12.8, 51.2]$ ($\beta = 19.2$ and $\beta = 25.8$, this scheduling will be indicated ZCS1) two extra discontinuities are introduced (Fig. 4c). This situation happened because the conditional morphological operators behave as low pass filters (linear average) for $\beta \to 0$ [4]. This fact is confirmed by results obtained after introducing the two additional iterations in the β range $[51.2, 204.8]$ ($\beta = 82.4$ and $\beta = 104.2$, this scheduling will be indicated ZCS2): Fig. 4d shows that the discontinuity is recovered by using this kind of scheduling.

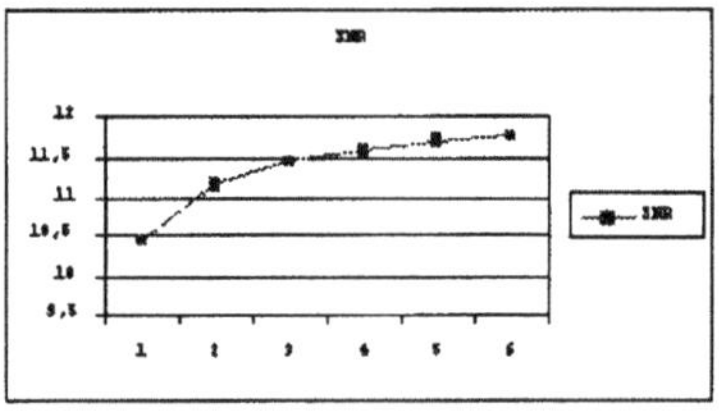

Fig. 3. SNR versus iteration number.

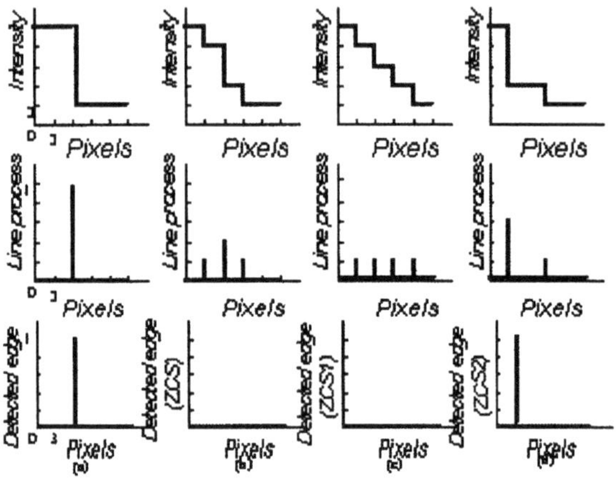

Fig. 4. Smoothing of discontinuity with different β scheduling.

Quantitative results are shown in Table I. It is possible to notice that the introduction of extra iterations with $\beta \in [12.8, 51.2]$ reduces the number of false edge points and increases the number of lost edge points; the introduction of extra iteration with $\beta \in [51.2, 204.8]$ causes the opposite situation. In order to obtain good results in both false edge elimination and lost edge recovering, a modification of the ZCS is proposed by introducing five extra iterations corresponding to the following β values: $19.2, 25.9, 64.8, 82.4, 104.2$ (this scheduling will be named "Conditional morphology scheduling" CMS). The introduction of additional iterations means that the solution is more stable because it is obtained in a slower way. The quantitative results are shown in Table II and they are obtained by using also a cross 5×5 neighbourhood set to take into account that edge point are detected with a little displacement with respect to the original edge position (Fig. 4d).

TABLE I
Quantitative results for edge detection by using different β scheduling

scheduling	Points of lost edges	Points of false edges
ZCS	117 (8.25%)	3173 (4.94%)
ZCS1	295 (20.81%)	1831 (2.85%)
ZCS2	85 (5.99%)	2987 (4.65%)

TABLE II
Quantitative results for edge detection by using different β scheduling and different size of square SE

scheduling	Points of lost edges	Points of false edges
ZCS (Edge points in 3×3 mask)	117 (8.25%)	3173 (4.94%)
ZCS (Edge points in 5×5 mask)	108 (7.62%)	3164 (4.93%)
CMS (Edge points in 3×3 mask)	298 (21.03%)	1942 (3.02%)
CMS (Edge points in 5×5 mask)	82 (5.78%)	1726 (2.69%)

Table II shows that good performances are obtained with the proposed scheduling when the evaluation is made by using a 5×5 mask because edges are detected with a displacement of two pixels with respect to the original position. Similarly, β scheduling for processing images corrupted by 4-look speckle noise is derived and it is given by the following values: 0.8, 3.2, 12.8, 51.2, 104.2, 204.8. In this case, fewer smoothing steps (corresponding to low β values) are used, because of the lower level of noise corrupting the 4-look images with respect to the 2-look images, in this way.

4. Comparisons with statistical morphology and deterministic annealing

Experimental results are obtained on both synthetic and real SAR images. Comparisons are performed with respect to statistical morphology [6] and the deterministic annealing [1]. In the case of statistical morphology, edges are detected by means of a Sobel operator. Conditional morphological operators are used for restoring synthetic 2-look SAR images by using the β scheduling derived in the previous section, $\alpha = 3, \gamma = 305$ and a square 5×5 structuring element. A statistical opening is performed with $\beta = 104.2$ and a square 5×5 structuring element. Deterministic annealing [1] is applied by considering $\gamma = 600, \alpha = 2$, ZCS β scheduling such that $\beta_0 = 0.000125$, 12 βiterations and 5 descent step for each β iteration. Results are shown in Table III.

From experimental results it is possible to notice that conditional morphology presents the advantage of deterministic annealing [1] of being able to perform simultaneously edge detection and image smoothing. The better results are obtained by using deterministic annealing, but conditional morphology presents also the advantage of statistical morphology of using a one-shot expression for computing the output of elementary operations, without using gradient descent methods for obtaining the solutions at each β iteration. Table IV shows the CPU time necessary to perform the operations presented in this section; the algorithm are implemented in a workstation SUN SPARC 2. Results obtained from real SAR images are also satisfactory (Fig. 5). In this case, the best results are obtained by applying the β scheduling derived for cleaning 4-look SAR images.

TABLE III
Quantitative results obtained from a 2-look synthetic image

Evaluated image	Points of lost edges (5×5 mask)	Points of false edges (5×5 mask)	SNR [dB]
Corrupted image (Fig. 2b)	–	–	7.82
Statistical opening (Fig. 5a)	0.49%	33.87%	4.0
Deterministic annealing (Fig. 5b)	2.47	0.76	13.33
Conditional openings (Fig. 8)	3.74%	3.12	11.44

TABLE IV
Computational time

	Statistical morphology (1 iteration)	Conditional morphology (11 iterations)	Deterministic annealing (12 iterations)
CPU time (sec)	7	124	504

(a) (b) (c)

Fig. 5. Results obtained with conditional morphology in a real SAR image.

References

1. D. Geiger and F. Girosi. Parallel and deterministic algorithms for MRFs: surface reconstruction. *IEEE Trans. Patt. Anal. Mach. Intell.*, 13(5):401–412, 1991.
2. S. Geman and D. Geman. Stochastic relaxation, Gibbs distributions, and the Bayesian restoration of images. *IEEE Trans. Patt. Anal. Mach. Intell.*, 6:721–741, 1984.
3. S. Kirkpatrick, C. D. Gelatt Jr., and M. P. Vecchi. Optimization by simulated annealing. *Science*, 220(4598), 13 May 1983.
4. C. S. Regazzoni, E. Stringa, and C. Valpreda. Conditional morphology for image processing. *1997 IEEE Workshop on Nonlinear Signal and Image Processing*, September 7-11, 1997, Mackinac Island, Michigan, USA. In press.
5. J. Serra, editor. *Image Analysis and Mathematical Morphology. II: Theoretical Advances.* Academic Press, London, 1988.
6. A. Yuille, L. Vincent, and D. Geiger. Statistical morphology and Bayesian reconstruction. *J. Math. Imag. Vision*, 1(3):223–238, 1992.
7. J. Zerubia and R. Chelappa. Mean field annealing for edge detection and image restoration. In *Proceedings of EUSIPCO90, V European Signal Processing Conference,* Barcelona, Spain, September 18-21, pages 837–839, 1990.

DISTANCES BETWEEN GREY-SCALE IMAGES

NIAL FRIEL and ILYA MOLCHANOV
Department of Statistics, University of Glasgow,
Glasgow G12 8QW, Scotland, U.K.

Abstract. A grey-scale image gives rise to the corresponding random set model, which appears if the image is thresholded at a random level. This idea allows us to reformulate general definitions of probability metrics for random sets for the special case of random sets generated from images. This yields a number of metrics for grey-scale images, as well as a unified general approach suitable to construct new metrics.

Key words: Error metric, grey-scale image, probability metric, random set, threshold

1. Introduction

It is often required in image processing to find an algorithm which estimates a 'true' image as accurately as possible. This may occur, for example, in image reconstruction, encoding, edge detection etc. In such instances it of interest to formulate a numerical measure, $d(f, g)$, of the distance between two images f and g. In this way we can objectively test the performance of the algorithm.

Many methods used to assess the distance between grey-scale images are based on pixel by pixel differences. An example of such a method is the widely used root mean squared (RMS) distance

$$\|f - g\|_2 = \left(\frac{1}{N} \sum_{x \in W} (f(x) - g(x))^2 \right)^{1/2}$$

where $f(x)$ denotes the grey-level of f at pixel x and W denotes the window of interest with N being the total number of pixels in W. It is widely accepted however that such methods generally don't work well in practice.

Better metrics can be obtained by extending binary error metrics [1] to grey-scale images by applying such metrics to the subgraphs of the corresponding images, see [8].

Vital to this paper is the idea that an image gives rise to a distribution of a random set. Probability metrics are used to numerically assess the distance between random variables, see [6]. Molchanov [5] generalised this theory to the class of closed random sets. We modify this theory to the specific case of random sets generated from images. In this way a measure of error between two images may be formulated via the distance between their corresponding random sets.

The approach we take towards finding a new error metric for grey-scale images is to first find one which performs well when applied to binary images, and then to extend this to grey-scale images. Baddeley's Delta metric [1] for binary images has been shown to work well in practice, while displaying good theoretical properties. In this paper we introduce a class of error metrics and in particular discuss one such metric that is equivalent to Baddeley's Delta metric in the binary case.

2. Preliminaries

A grey-scale image inside a window W can be treated as a function defined on W, where W is usually a subset of a plane or three-dimensional space. In practical imaging W is a collection of pixels within a discretised rectangle, and possible values of f are integers $\{0,\dots,255\}$, so both the domain and the range of f are discrete sets. Below we will work in the continuous setup, where W is a subset of a Euclidean space $\mathbf{R}$ and $f : W \mapsto [0,1]$ is a function with values in $[0,1]$. In fact, most of methods work if W is a general metric space. The continuous setup allows us to refer to analytic tools, while leaving always a possibility to 'discretise' the problem.

Each point $x \in W$ is assigned a grey value $f(x)$ which lies between 0 and 1. We use the notation f to denote both the function and the corresponding grey-scale image. Without further notice we assume that f is upper semi-continuous. This ensures the measurability of f and guarantees that the subgraph (also called hypograph or umbra) of f is a topologically closed set [7, 4].

A grey-scale image can be interpreted as a nested family of sets

$$F_t = \{x \in W : f(x) \geq t\}, \quad 0 \leq t \leq 1 .$$

Observe that F_U becomes a random set, if U is a random variable on $[0,1]$. Throughout we refer to F_U and G_U as the random sets generated by the images f and g respectively.

We could for instance consider the scenario where U is uniformly distributed, we will henceforth call such a random set model 'uniformly weighted'. Another possibility is to adopt the histogram of the image as the distribution. Of course there are many more approaches to take. Other approaches are suitable to emphasise different ranges of grey values.

3. Probability Metrics for Random Sets

Following Rachev [6], a probability metric $d(\xi,\eta)$ between two random variables ξ and η satisfies the following conditions:

1. $d(\xi,\eta)=0$ implies $\mathbf{P}\{\xi=\eta\}=1$.
2. $d(\xi,\eta)=d(\eta,\xi)$.
3. $d(\xi,\eta) \leq d(\xi,\zeta)+d(\zeta,\eta)$ for each random variable ζ.

Probability metrics are used to determine distances between random variables. There are many approaches to generalise probability metrics to the class of random sets, for example, it is possible to generalise some classical probability metrics by replacing distribution functions with capacity functionals of random sets, see [5]. Capacity functionals play the same role in the theory of random sets as probability distribution functions in classical probability theory. The capacity functional of a random set X is defined as

$$T_X(K) = \mathbf{P}\{X \cap K \neq \emptyset\}, K \in \mathcal{K},$$

where $\mathcal{K}$ denotes the space of all compact sets. We see that the capacity functional is the probability that a given compact set K hits the random set X.

The class $\mathcal{K}$ of all compact sets is too large to efficiently define the capacity functional on it. For this reason we need to reduce the class of test sets in order to compute the capacity functional. That is, we need to fix a certain class $\mathcal{M} \subset \mathcal{K}$ on which we define $T_X(K)$.

The *uniform metric* between random sets X and Y is defined as

$$\mathcal{U}(X,Y;\mathcal{M}) = \sup\{|T_X(K) - T_Y(K)| : K \in \mathcal{M}\}.$$

Another example is the *Lévy metric*, which is defined as follows:

$$\mathcal{L}(X,Y;\mathcal{M}) = \inf\{\varepsilon > 0 : T_X(K) \leq T_Y(K^\varepsilon) + \varepsilon, T_Y(K) \leq T_X(K^\varepsilon) + \varepsilon\},$$

where

$$K^\varepsilon = \cup\{B_\varepsilon(x) : x \in K\} = K \oplus B_\varepsilon(0)$$

is the ε-envelope of K with $\oplus$ being the usual Minkowski addition and $B_\varepsilon(0)$ being a ball of radius ε centred at the origin. Both metrics depend on the choice of the class $\mathcal{M}$. This class is often chosen to be the class of all singletons, all balls, or all rectangles (parallelepipeds).

Unfortunately, these two metrics (which are useful in the study of general random sets) are not particularly interesting for random sets obtained as thresholds of grey-scale images. If $X = F_U$ and $Y = G_V$ for two grey-scale images f and g and independent random variables U and V are uniformly distributed over $[0,1]$, then

$$T_{F_U}K) = \sup_{x \in K} f(x), \quad T_{G_V}(K) = \sup_{x \in K} g(x).$$

Assume that $\mathcal{M}$ contains all singletons. Then

$$\mathcal{U}(F_U,G_V;\mathcal{M}) = \sup|f(x) - g(x)|$$

is the uniform distance between f and g. The Lévy metric is reduced to

$$\begin{aligned}\mathcal{L}(F_U,G_V;\mathcal{M}) = \inf\{\varepsilon > 0 :\ & \operatorname{hypo} f \subseteq (\operatorname{hypo} g) \oplus (B_\varepsilon \times [0,\varepsilon]), \\ & \operatorname{hypo} g \subseteq (\operatorname{hypo} f) \oplus (B_\varepsilon \times [0,\varepsilon])\},\end{aligned}$$

where

$$\operatorname{hypo} f = \{(x,t) :\ x \in W,\ t \leq f(x)\}$$

denotes the subgraph (hypograph) of f (respectively of g). Thus, the Lévy metric equals the Hausdorff distance between the subgraphs of the images as subsets of $W \times \mathbf{R}$ if the carrier space is equipped with the metric

$$\rho((x,t),(y,s)) = \max(\|x - y\|, |t - s|), \quad x,y \in W, t,s \in \mathbf{R}.$$

Note however, that both the uniform and Lévy metrics might be more interesting if the random variables U and V have non-uniform distributions, for example, they might be distributed according to the histograms of the images f and g.

4. Integral Metrics

4.1. General Definition

The *integral metric* for random sets X and Y is defined as,

$$I(X,Y) = \sup_{h\in\mathcal{H}} \left| \int h\,dT_X - \int h\,dT_Y \right| \tag{1}$$

where $\mathcal{H}$ is some family of non-negative functions on W. Here the integral with respect $T_X(\cdot)$ is defined as,

$$\int_{\mathbf{R}^d} h\,dT_X = \int_0^\infty T_X(H_t)\,dt, \tag{2}$$

where $H_t = \{x : h(x) \geq t\}$. Note that (2) is called the Choquet integral [2, 3].

Henceforth we concentrate on applying the integral metric to the special case of random sets generated from images. Consider the situation now where the random set $X = F_U$ and $Y = G_V$ are generated from images f and g. Then the capacity functional for a uniformly weighted random set model takes the following form:

$$T_{F_U}(H_t) = \sup_{x\in H_t} f(x). \tag{3}$$

Using (2) and (3) we see that (1) can be reformulated as,

$$I(f,g) = I(F_U,G_U) = \sup_{h\in\mathcal{H}} \left| \int_0^\infty \sup_{x\in H_t} f(x)\,dt - \int_0^\infty \sup_{x\in H_t} g(x)\,dt \right|. \tag{4}$$

4.2. Hausdorff-Type Metrics

Suppose now that the images f and g are binary, where each pixel takes, for example, the values 0 or 1. In this case we define $F = \{x \in W : f(x) \neq 0\}$ and similarly for G, so that the random sets F_U and G_U now become deterministic sets F and G respectively.

It may be seen that,

$$\int_0^\infty \sup_{x\in H_t} f(x)\,dt = \int_0^{\sup\{h(x):\, x\in F\}} dt = \sup_{x\in F} h(x).$$

Therefore,

$$I(f,g) = I(F,G) = \sup_{h\in\mathcal{H}} \left| \sup_{x\in F} h(x) - \sup_{x\in G} h(x) \right|. \tag{5}$$

We now show that by choosing the family of functions $\mathcal{H}$ in a certain manner the integral metric becomes the Hausdorff metric for binary images. Define

$$\mathcal{H} = \{h^a : a \in W\} \tag{6}$$

where

$$h^a(x) = 1 - \|x - a\|/c \tag{7}$$

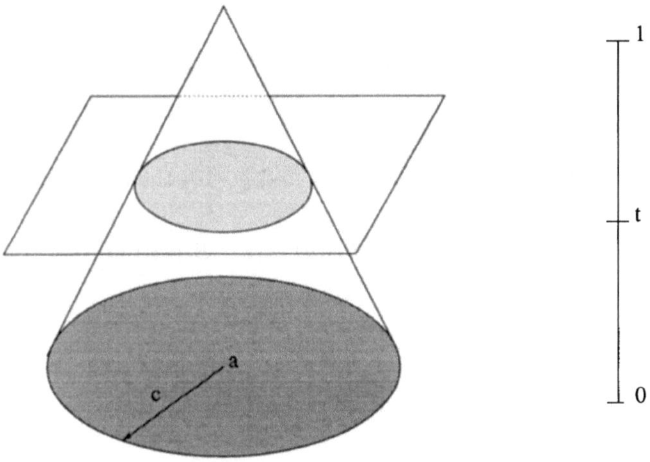

Fig. 1. Diagram showing the function h^a together with the corresponding level set H_t^a.

is the cone of height 1 with the circular base of radius c centred at a, see Figure 1. We set $h^a(x) = 0$ for all points $x \in W$, not contained in the base of h^a. (We allow for the possibility that the base may not be entirely contained in the window.) Observe that,

$$H_t^a = \{x : h^a(x) \geq t\} = B_{(1-t)c}(a)\,.$$

It is well known that the Hausdorff distance between any two nonempty compact sets $F, G \subseteq W$ is

$$H(F,G) = \sup_{x \in W} |d(x,F) - d(x,G)|, \tag{8}$$

where $d(x, \cdot)$ is the Euclidean distance function (the distance from x to the nearest point of the corresponding set).

Theorem 1 *Assume that $\mathcal{H}$ is given by (6) and (7) with c sufficiently large to satisfy*

$$c \geq diam(W) = \sup\{\|x - y\| : x, y \in W\}\,.$$

Then

$$I(F,G) = H(F,G)/c\,.$$

Proof. It may be seen, since c is large enough, that

$$\sup_{x \in F} h^a(x) = \frac{c - d(a,F)}{c}\,,$$

whence

$$|\sup_{x \in F} h^a(x) - \sup_{x \in G} h^a(x)| = c^{-1}|d(a,F) - d(a,G)|\,.$$

Thus,

$$I(F,G) = c^{-1} \sup_{h^a \in \mathcal{H}} |d(a,F) - d(a,G)| = c^{-1} \sup_{a \in W} |d(a,F) - d(a,G)|,$$

as required. □

If we consider general grey-scale images f and g with the above defined family $\mathcal{H}$, then

$$I(f,g) = \sup_{a \in W} \left| \int_0^c \Big[(f \oplus B_t)(a) - (g \oplus B_t)(a) \Big] \, dt \right|,$$

where

$$(f \oplus B_t)(a) = \sup_{x \in B_t(a)} f(x)$$

is the dilation of the image f by the ball of radius t.

4.3. Delta Metric

Let w be a strictly increasing concave function defined on $[0, \infty]$ which satisfies $w(0) = 0$. Baddeley [1] defines the Delta distance between binary images f and g as,

$$\Delta^p(f,g) = \left[\int_W |w(d(x,F)) - w(d(x,G))|^p \, dx \right]^{1/p}. \tag{9}$$

We see that the Delta metric is an L^p average of the differences between the distances, $w(d(\cdot,\cdot))$ from each pixel to the nearest foreground pixel of each image. It is in fact an L^p analogue of the uniform norm on (8).

The Delta metric is implemented using $w(t) = \min\{t, c\}$, so that in (9) pixels greater than a distance c, the so called cut-off value, from the nearest foreground pixel will be given the value c. In this way the effect of background noise is limited. It was shown in [1] that this metric shows desirable theoretical properties. Specifically this metric generates the myopic topology which Serra [7] suggests is the most suitable topology for the medium of binary images.

It is straightforward to extend (5) so that it is equivalent (up to a constant) to the Delta metric. To see this, define $\mathcal{H}$ as before, where the radius c of each cone is now taken to be the fixed cut-off value c, as described above. Then simply replace the outer supremum in (5) with an L^p average over all points $a \in W$. We thus arrive at the following:

$$I^*(F,G) = \left[\int_W |\sup_{x \in F} h^a(x) - \sup_{x \in G} h^a(x)|^p \, da \right]^{1/p}. \tag{10}$$

5. Double integral Metric

Return now to the situation where F_U, G_U are random sets generated from images f, g. Generalising (10) to grey-scale images leads to a new error metric which we call the *double integral metric*, defined as:

$$DI(f,g) = DI(F_U, G_U) \tag{11}$$
$$= \left[\int_W \left| \int_0^1 \sup_{x \in H_t^a} f(x)\,dt - \int_0^1 \sup_{x \in H_t^a} g(x)\,dt \right|^p da \right]^{1/p} .$$

Since

$$\sup_{x \in H_t^a} f(x) = (f \oplus B_{(1-t)c})(a) ,$$

using (12) and changing variables, we get

$$DI(f,g) = \frac{1}{c} \left[\int_W \left| \int_0^c [(f \oplus B_t)(a) - (g \oplus B_t)(a)]\,dt \right|^p da \right]^{1/p} . \tag{12}$$

It is interesting to note that letting $c = 0$ in (12), we find that $DI(f,g) = L^p(f,g)$. That is,

$$DI(f,g) = \left[\int_W |f(x) - g(x)|^p\,dx \right]^{1/p} .$$

It is unclear whether $DI(\cdot,\cdot)$ satisfies the uniqueness axiom of a metric, specifically whether $DI(f,g) = 0$ implies $f = g$. In the discrete case however, this assertion may be seen to be true. First note that (12) can be discretised to get

$$DI(f,g) = \frac{1}{c+1} \left[\sum_{a \in W} \left| \sum_{t=0}^{c} ((f \oplus B_t)(a) - (g \oplus B_t)(a)) \right|^p \right]^{1/p} . \tag{13}$$

Let f achieve its maximum at point $a \in W$ and g at point $b \in W$. Assume that $f(a) \geq g(b)$. Note that $DI(f,g)$ implies in particular that

$$(c+1)^{-1} \sum_{t=0}^{c} (f \oplus B_t)(a) = (c+1)^{-1} \sum_{t=0}^{c} (g \oplus B_t)(a) .$$

If $f(a) > g(a)$, then the left-hand side equals $f(a)$, while the right-hand side is strictly less than $g(b)$. The obtained contradiction shows that $f(a) = g(a) = g(b)$ and so both f and g achieve its maximum at a. By eliminating the point a from W and applying the same argument it is easy to see that both f and g achieve their maximum over $W \setminus \{a\}$ at the same point and their values at this point are identical. Because W is finite (in the discrete setup) we can apply this argument in succession in order to show that f and g coincide at all points inside W.

Note that $DI(f,g) \to |\sup f - \sup g|$ as $c \to \infty$. In general, smaller values of c make the metric more 'local'. Suppose that f and g are both binary chess-board images the first image having alternate pixels black and white and the second having the black and white pixels of the first reversed. Then from (13) $DI(f,g) = n(W)$ if $c = 0$, while $DI(f,g) = \frac{1}{c+1} n(W)$ for each $c \geq 1$, where n(W) equals the number of pixels in the window.

6. Acknowledgements

The first authors research was supported by a TMR Marie Curie research training grant.

References

1. A. Baddeley. Errors in binary images and a L^p version of the Hausdorff metric. *Nieuw Archief voor Wiskunde*, 10:157–183, 1992.
2. G. Choquet. Theory of capacities. *Ann. Inst. Fourier*, 5:131–295, 1953/54.
3. D. Denneberg. *Non-additive measure and integral.* Kluwer, Dordrecht, 1994.
4. J.-B. Hiriart-Urruty and C. Lemaréchal. *Convex Analysis and Minimization Algorithms*, volume 1,2. Springer, Berlin, 1993.
5. I. Molchanov. *Limit Theorems for Unions of Random Closed Sets*, volume 1561 of *Lect. Notes Math.* Springer, Berlin, 1993.
6. S. Rachev. *Probability Metrics and the Stability of Stochastic Models.* Wiley, Chichester, 1991.
7. J. Serra. *Image Analysis and Mathematical Morphology.* Academic Press, London, 1982.
8. D. Wilson, A. Baddeley, and R. Owens. A new metric for grey-scale image comparison. *The International Journal of Computer Vision*, 24(1):5–17, 1997.

SIZE-BIASED RANDOM CLOSED SETS

M. N. M. VAN LIESHOUT
*Centrum voor Wiskunde en Informatica**
POBox 94079, 1090 GB Amsterdam, The Netherlands

Abstract. We indicate how granulometries may be useful in the analysis of random sets. After defining a size distribution function which may be used as a summary statistic in exploratory data analysis, we propose a Hanisch-type estimator and construct new Markov random set models which favour certain sizes above others. The models are illustrated by simulated realisations.

Key words: contact distribution function, empty space function, granulometry, Hanisch-type estimator, Markov random set, size distribution function.

1. Introduction

Stochastic geometry [29] is concerned with the study of random closed sets (rcs). Roughly speaking, an rcs is a mapping X from a probability space into the family of closed subsets of R^d such that $\{X \cap K \neq \emptyset\}$ and $\{X \cap G = \emptyset\}$ are measurable for all compact sets K and all open sets G. The probability distribution of X is completely specified by its *capacity functional*

$$T(K) = P(X \cap K \neq \emptyset) \tag{1}$$

with K ranging over the class of compact sets, which plays a role comparable to that of the distribution function of a real-valued random variable. However, the collection of test sets K is huge, and lower-dimensional summary statistics are called for. Typically these are obtained from the capacity functional (1) by restricting the choice of K. For instance, allowing only singletons results in the coverage probabilities $p(x) = P(x \in X)$, $x \in R^d$.

Below we will assume that the random closed set X is stationary, i.e. its distribution is invariant under translations. In that case, the coverage probabilities $p(x)$ do not depend on the argument x, and $p(x) \equiv p = E|X \cap U|$, the expected volume covered within any set U of unit volume $|U| = 1$. To exclude degenerate cases, it is assumed that $0 < p < 1$.

A summary statistic for assessing the 'size' of pores left open by a stationary rcs is the *empty space function* $F_B(\cdot)$ defined by taking $K = rB$ ($r \geq 0$) in (1), that is

$$F_B(r) = P(0 \in X \oplus r\check{B}). \tag{2}$$

The related *contact distribution function* is defined as

$$H_B(r) = P(0 \in X \oplus r\check{B} \mid 0 \notin X) = \frac{F_B(r) - F_B(0)}{1 - F_B(0)} \tag{3}$$

* This research was partially carried out while the author was at the University of Warwick, Coventry CV4 7AL, UK.

for $r \geq 0$. Both (2) and (3) depend on a structuring element B, usually assumed to be a compact convex set containing an environment of the origin. Typical choices include balls and squares, although non-isotropic structuring elements may be preferred when investigating directional effects. Under these conditions, $H_B(\cdot)$ is a distribution function but this is not the case for general B (see [29]). Note that $1 - H_B(r)$ can be interpreted as the conditional probability that a copy of rB placed at a test point 0 lies entirely in the empty space left by X given that the test point itself does not fall in X. A similar interpretation holds for the empty space function $F_B(r)$.

The particle size operator underlying (2)–(3) is $\psi_r(Y) = (Y^c \oplus r\check{B})^c = Y \ominus r\check{B}$. However, although under our restrictions on B, $\psi_r(\cdot)$ is anti-extensive and increasing, in general the sieving condition

$$\psi_r(\psi_s(Y)) = \psi_s(\psi_r(Y)) = \psi_{\max(r,s)}(Y)$$

for all $r, s \geq 0$ does not hold and hence $(\psi_r)_{r\geq 0}$ is not a granulometry [21]. Thus, in the remainder of this paper, we will consider replacing the dilation in (2)–(3) by a closing.

2. Size distribution functions

In mathematical morphology, a *size distribution law* of a stationary random closed set X may be defined using a granulometry $\psi_r(X) = X \circ rB$ ($r \geq 0$) based on Euclidean openings with a non-empty convex compact structuring element B. Indeed, set [28, p. 335]

$$G_1(r) = 1 - P(x \in X \circ rB \mid x \in X) = 1 - \frac{P(x \in X \circ rB)}{P(x \in X)}, \quad r \geq 0$$

for the size distribution law of the particles, and similarly

$$G_0(r) = P(x \in X \bullet rB \mid x \notin X) = 1 - \frac{1 - P(x \in X \bullet rB)}{1 - P(x \in X)}, \quad r \geq 0$$

for the pores. By the stationarity of X, the definitions do not depend on the choice of $x \in R^d$. Intuitively, $G_1(r)$ will be the conditional probability that a point in X is eliminated by opening with rB ($r > 0$), i.e. the probability that the B-size of X is less than r. Similarly, $G_0(r)$ is the conditional probability that a point from X^c is included by closing with rB, hence that the B-size of the complement is less than or equal to r. By allowing r to vary over $(-\infty, \infty)$, the joint size distribution law

$$G(r) = \begin{cases} 1 - P(x \in X \circ rB) & r \geq 0 \\ 1 - P(x \in X \bullet |r|B) & r < 0 \end{cases} \tag{4}$$

of pores and particles is obtained.

Size distribution laws have been used for a long time in the empirical sciences, and more recently in the analysis of (binary) images. For instance Serra [28] employs size distribution laws for shape and texture analysis, Maragos [20] uses them for multiscale shape representation, [16, 25] apply size distributions to shape filtering and restoration problems, while Sivakumar [27] gives applications in texture classification and morphological filtering. See also [9].

However, from a probabilistic point of view, $G(\cdot)$ (and $G_1(\cdot)$) is not a proper distribution function, as it is semi-continuous from the left rather than from the right. For this reason we prefer the following definition.

Definition 1 *Let X be a stationary random closed set and B a non-empty convex compact structuring element. Define the* size distribution function *of X by*

$$P_B(r) = \begin{cases} P(x \in X \bullet rB) & r \geq 0 \\ P(x \in X \circ |r|B) & r < 0 \end{cases}. \tag{5}$$

The function $P_B(\cdot)$ is called the *granulométrie bidimensionelle* in metallurgi [10]. It is easily verified that $P_B(0) = p$, the coverage fraction of the stationary random closed set X.

The size distribution function $P_B(\cdot)$ is well-defined and does not depend on the choice of $x \in R^d$. In contrast to $G(\cdot)$, it is a proper distribution function. Compared to the empty-space function (2), note that the latter is absolutely continuous except in 0 (see Hansen et al. [15]) but that this property does not generally hold for the size distribution which may have countably many discontinuities.

Explicit expressions for size distributions may be hard to find, being related to covering probabilities [13]. The contact distribution function is available in closed form for Boolean models [29] (but not for most other random set models!) and, if the primary grain is convex, depends only on the moments of a few functionals of the grain. For instance, in R^2 the mean perimeter determines the whole of $H_B(r)$, which, as pointed out in Ripley [23], may result in poor distinguishing power as an exploratory data analysis tool.

3. Estimation and edge effects

In this section we discuss estimating the size distribution function $P_B(r)$ of a stationary rcs X (cf. Definition 1). Since in practice X is only observed within some compact window W of positive volume $|W|$, due to edge effects caused by parts of X outside W, the volume fraction estimator

$$\hat{P}_B(r) = \begin{cases} \frac{|W \cap (X \bullet rB)|}{|W|} & r \geq 0 \\ \frac{|W \cap (X \circ |r|B)|}{|W|} & r < 0 \end{cases} \tag{6}$$

may be biased. To overcome this problem, a minus sampling estimator [23, 29] has been proposed. Briefly, this is just the volume fraction estimator with W replaced by $W \ominus (rB \oplus r\check{B})$. By the local knowledge principle [28], this estimator is pointwise unbiased. However, it is not necessarily monotone in r, nor is all available information used. More refined techniques based on survival analysis ideas have been suggested by Hansen et al. [15] for deriving a Kaplan-Meier type estimator [1] for the empty space function (2). Chiu and Stoyan [8] showed that the ideas underlying the Kaplan-Meier approach are very similar to those involved in the Hanisch estimator [14]. In the remainder of this section, we will derive a Hanisch-type estimator for the size distribution function (Definition 1).

To do so, we need three local size measures: with respect to X, its empty spaces and the boundary. As before, let B be a non-empty convex compact structuring element and X a stationary random closed set observed in a compact observation window W. Set

$$\rho(x,X) = \begin{cases} \sup\{r \geq 0 : \exists h \text{ such that } x \in (rB)_h \subseteq X\} & x \in X \\ 0 & x \notin X \end{cases} \tag{7}$$

$$\eta(x,X) = \begin{cases} \inf\{r \geq 0 : x \in X \bullet rB\} & x \notin X \\ 0 & x \in X \end{cases} \tag{8}$$

$$\zeta(t,W^c) = \begin{cases} \inf\{r \geq 0 : (rB \oplus r\check{B})_t \cap W^c \neq \emptyset\} & t \in W \\ 0 & t \notin W \end{cases} \tag{9}$$

It is easy to see that $X \circ rB = \{x \in X : \rho(x,X) \geq r\}$, $X \bullet rB = \{x \in R^d : \eta(x,X) \leq r\}$ and $W \ominus (rB \oplus r\check{B}) = \{t \in W : \zeta(t,W^c) \geq r\}$.

Discretising over a sampling grid $T = \{t_i\} \subseteq W$, the minus sampling estimator of $P_B(r)$ can be written in terms of ρ, η, ζ as

$$\hat{P}_B(r) = \begin{cases} \frac{\#\{i:\eta(t_i,X) \leq r \leq \zeta(t_i,W^c)\}}{\#\{i:\zeta(t_i,W^c) \geq r\}} & r \geq 0 \\ 1 - \frac{\#\{i:\rho(t_i,X) < |r|; \zeta(t_i,W^c) \geq |r|\}}{\#\{i:\zeta(t_i,W^c) \geq |r|\}} & r < 0 \end{cases}. \tag{10}$$

Note that (10) does not use all information contained in the data. In particular, if $t_i \notin W \ominus (rB \oplus r\check{B})$, but $\eta(t_i,X) \leq \zeta(t_i,W^c)$ the correct void size at t_i is measured. Using this observation, one can define a Hanisch-type estimator for $P_B(r)$ $(r \geq 0)$.

Definition 2 *Let X be a realisation of a stationary random closed set observed in a compact window W. Then for all $r \geq 0$ with $\#\{i : \zeta(t_i, W^c) \geq r\} > 0$, define*

$$\hat{P}_B^H(r) = \sum_{s \leq r} \frac{\#\{i : \eta(t_i,X) = s \leq \zeta(t_i,W^c)\}}{\#\{i : \zeta(t_i,W^c) \geq s\}} \tag{11}$$

and for $r < 0$ with $\#\{i : \zeta(t_i,W^c) \geq |r|\} > 0$, let

$$\hat{P}_B^H(r) = 1 - \sum_{s < |r|} \frac{\#\{i : \rho(t_i,X) = s \leq \zeta(t_i,W^c)\}}{\#\{i : \zeta(t_i,W^c) \geq s\}} \tag{12}$$

The Hanisch-type estimator $\hat{P}_B^H(r)$ is pointwise unbiased for $P_B(r)$. It is increasing and semi-continuous from the right. However, it may be non-negative and exceed 1. If this is undesirable, one can take $R = \sup\{r > 0 : \#\{i : t_i \in W \ominus (rB \oplus r\check{B})\} > 0\}$ and normalise the summands in (11) and (12) by $\hat{P}_B^H(R)$ and $\hat{P}_B^H(-R)$ respectively. The resulting estimator $\tilde{P}_B^H(r)$ is ratio-unbiased. Details can be found in [19], which also provides examples on the use of $\hat{P}_B^H(\cdot)$ in exploratory data analysis.

4. Size-biased Markov random sets

Except from being helpful statistics in exploratory data analysis, size distributions can be used to define new models for random sets observed in a compact window W. Similar ideas for finite random fields have been suggested by Chen and Kelly [6, 7] and Sivakumar and Goutsias [26].

Since the cardinality of the set X is no longer finite, we have to proceed by specifying a density $p(\cdot)$ with respect to some reference process, eg a Boolean model on W. This is a random closed set defined in two steps: first a Poisson process of *germs* is generated (with intensity $\lambda > 0$); then to each of the germs x_i, a random compact *grain* K_i is assigned according to probability distribution $\mu(\cdot)$, independently of other grains. The union $\bigcup_i(x_i \oplus K_i)$ is called a Boolean model. Details on random set densities can be found in [18].

In analogy to [7, 26], set

$$p(X) = \alpha \exp\left[-\int f(s) d\hat{P}_X(s)\right] \tag{13}$$

where $f : R \to R$ is a bounded (measurable) function, $\hat{P}_X(\cdot)$ is an estimator of the size distribution function (5) based on X and α is the normalising constant ensuring that $p(\cdot)$ integrates to 1. It can be shown that for both the naive estimator (6) and the normalised or unnormalised Hanisch-type estimator (Definition 2), (13) is well-defined.

Statistical inference for complex random set models usually relies on iterative procedures making 'local' changes to the random set. First, consider the case were both the germs and the grains of which the random set is composed are fully observable. In that case, $Y = \{(x_i, K_i)\}$ is a germ-grain process and (13) with $X = \cup_i(x_i \oplus K_i)$ is its density with respect to a Poisson marked point process with intensity measure $\lambda\,\mathrm{leb}(\cdot) \times \mu(\cdot)$ (where leb denotes Lebesgue measure, λ is the point intensity and $\mu(\cdot)$ the mark distribution of the grains). Suppose that addition of a grain K at u is considered and the function $f(\cdot)$ is supported on $[-G, G]$. Then the log likelihood ratio depends on X through

$$-\int_{-G}^{G} f(s) d\hat{P}_{X \cup K_u}(s) + \int_{-G}^{G} f(s) d\hat{P}_X(s), \tag{14}$$

writing $K_u = u \oplus K$.

It can be shown that (14) only depends on those $(x_i, K_i) \in Y$ for which

$$(u, K) \sim (x_i, K_i) \Leftrightarrow (K_i)_{x_i} \oplus (GB \oplus G\check{B}) \cap K_u \oplus (GB \oplus G\check{B}) \neq \emptyset.$$

Hence, seen as a grain-marked point process, Y is Markov [24, 3] with respect to the neighbourhood relation $\sim$.

If grains are not individually observable, note that if $X = X_1 \cup \cdots \cup X_k$ is partitioned into its connected components $X_1, \ldots, X_k$, the opening $X \circ B = \cup\{B_h \subseteq X\}$ also partitions, as the convexity of B implies that B_h must fall entirely in one of the X_i. Thus, the naive and Hanisch-type estimators satisfy $\hat{P}_X(s) = \sum_{i=1}^{k} \hat{P}_{X_i}(s)$

for $s < 0$. Similarly, $X \bullet B$ factorises over the connected components of $W \setminus X$ and hence $p(\cdot)$ is of the form

$$p(X) = \prod_{i=1}^{k} \phi(X_i) \prod_{i=1}^{l} \phi(X_i^c).$$

Thus, altering X will only affect the connected components that are modified, a state-dependent Markov property as introduced by Baddeley and Møller [3]. See also [18, 22].

4.1. Morphologically smooth random sets

Let $f(s) = |W| \log \gamma \, \mathbf{1}_{(-1,0]}(s)$. Then, using the volume fraction estimator (6) yields density

$$p(X) = \alpha \gamma^{-|X \setminus (X \circ B)|}, \tag{15}$$

generalising the Chen-Kelly model [7] for binary random fields. Note that for $\gamma > 1$, the most likely realisations X are open with respect to the structuring element B. Thus sets build of approximately convex components are favoured over those with thin or elongated pieces, sharp edges or small isolated clutter.

By duality, taking $f(s) = |W| \log \gamma \, \mathbf{1}_{(0,1]}(s)$ yields

$$p(X) = \alpha \gamma^{-|(X \bullet B) \setminus X|}, \tag{16}$$

favouring for $\gamma > 1$ sets that are approximately closed with respect to B and discouraging small holes or rough edges.

Both models are well-defined for $\gamma \leq 1$ too, for $\gamma < 1$ encouraging morphological roughness.

Note that in (15)–(16), $| \cdot |$ denotes Lebesgue measure restricted to W and hence $p(\cdot)$ is susceptible to edge effects. This can be alleviated by using the Hanisch-type estimator. The resulting model also influences the morphological smoothness of its realisations.

4.2. Morphological area-interaction random sets

Let $f(s) = |W| \log \gamma \, \mathbf{1}\{s \leq -1\}$ and $\hat{P}_X(\cdot)$ given by (6). Then

$$p(X) = \alpha \gamma^{-|X \circ B|},$$

an opening-smoothed version of the area-interaction process in [2, 18]. Similarly, for $f(s) = -\log \gamma \, \mathbf{1}\{s > 1\}$,

$$p(X) = \alpha \gamma^{1 - |X \bullet B| / |W|},$$

a closing-smoothed area-interaction process. Again, Hanisch-type estimators may be employed to better account for edge effects.

Similar ideas may be used if the area measure in the exponent of γ is replaced by the Euler characteristic or other quermass integral [4]. Moreover, since the closing operator removes small holes, the closing-smoothed Euler-interaction process may be integrable when the non-smoothed version is not.

4.3. Size–symmetric random sets

Let f be the indicator function of $(-G, G]$, hence

$$p(X) = \alpha \exp[-\int_{-G}^{G} \gamma d\hat{P}_X^H(s)]. \tag{17}$$

For $\gamma > 0$, particle and pore sizes exceeding G will be favoured, while for $\gamma < 0$ the sizes tend to be smaller than G.

Typical samples from (17) using the Metropolis-Hastings sampler of [11, 12] are displayed in Figure 1. The basic idea is that although the normalising constant α in (17) is not available in closed form, the log likelihood ratios (14) do not depend on the normalisation constant and are 'local'. Hence one can run a Markov chain having equilibrium distribution (17) with transition probabilities based on the likelihood ratio for a sufficiently long time.

For ease of computation, we consider a simple square for our structuring element of three by three pixels. Then $\eta(\cdot, X)$, the size measure of voids, can be computed using the distance transform algorithm [5] for the 'square' metric on R^2 defined by

$$d((p_1, p_2), (q_1, q_2)) = \max\{|p_1 - q_1|, |p_2 - q_2|\}.$$

By duality, $\rho(\cdot, X)$ can be computed by reversing the fore- and background. The reference Boolean model has intensity parameter $\lambda = 1000$ and square primary grains with radius r distributed according to a geometric distribution with parameter $\delta = .2$. Finally, $G = 5$ and $|\gamma| = 2500$ and $\hat{P}_X^H$ is as in Definition 2. It is apparent that the scale in the left image ($\gamma = 2500$) is larger than that in the right image ($\gamma = -2500$).

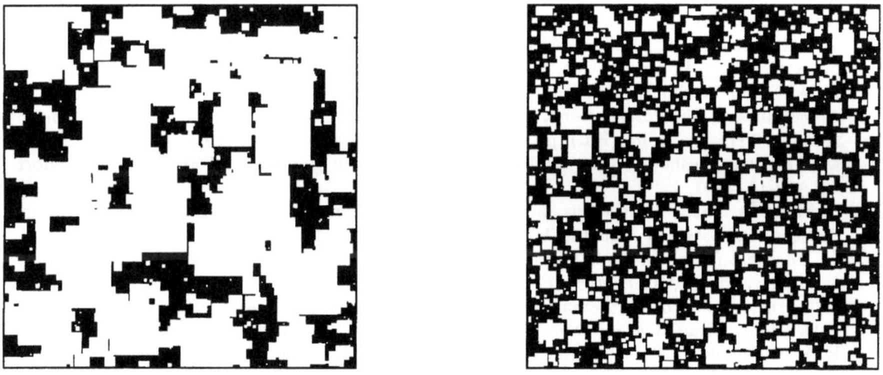

Fig. 1. Samples after 200000 Metropolis-Hastings steps for (17) with $G = 5$ in a 512x512 image for $\gamma = 2500.0$ (left) and $\gamma = -2500.0$ (right).

Exact simulation of size-biased Markov random set models is theoretically possible [17], since its log likelihood ratios are uniformly bounded. However, since (14) is not in general monotone in X (not even for the morphological area-interaction models in Section 4.2), upper and lower bounds based on the current state of the sampler would have to be computed at every iteration. For this reason, we prefer to use the computationally easier Metropolis-Hastings method.

References

1. A.J. Baddeley and R.D Gill. Kaplan-Meier estimators for interpoint distance distributions of spatial point processes. *Annals of Statistics*, 25:263–292, 1997.
2. A.J. Baddeley and M.N.M. van Lieshout. Area-interaction point processes. *Annals of the Institute of Statistical Mathematics*, 47:601–619, 1995.
3. A.J. Baddeley and J. Møller. Nearest-neighbour Markov point processes and random sets. *International Statistical Review*, 57:89–121, 1989.
4. A.J. Baddeley, W.S. Kendall and M.N.M. van Lieshout. Quermass–interaction models. Research report 293, University of Warwick, 1996.
5. G. Borgefors. Distance transformations in digital images. *Computer Vision, Graphics and Image Processing*, 34:344–371, 1986.
6. F. Chen. *Robust and morphologically constrained image segmentation.* PhD thesis, University of Massachusetts at Amherst, 1994.
7. F. Chen and P.A. Kelly. Algorithms for generating and segmenting morphologically smooth binary images. In: *Proceedings of the 26th Conference on Information Sciences*, 1992.
8. S.N. Chiu and D. Stoyan. Estimators of distance distributions for spatial patterns. *Statistica Neerlandica*, 1998. To appear.
9. E.R. Dougherty J.B. Pelz, F. Sand and A. Lent. Morphological image segmentation by local granulometric size distributions. *Journal of Electronic Imaging*, 1:46–60, 1992.
10. A.G. Fabbri. *Image processing of geological data.* Van Nostrand Reinhold, 1984.
11. C.J. Geyer and J. Møller. Simulation procedures and likelihood inference for spatial point processes. *Scandinavian Journal of Statistics*, 21:359–373, 1994.
12. P.J. Green. Reversible jump MCMC computation and Bayesian model determination. *Biometrika*, 82:711-732, 1995.
13. P. Hall. *Introduction to the theory of coverage processes.* John Wiley & Sons, 1988.
14. K.H. Hanisch. Some remarks on estimators of the distribution function of nearest neighbour distance in stationary spatial point processes. *Mathematische Operationsforschung und Statistik, Series Statistik*, 15:409–412, 1984.
15. M.B. Hansen, R.D. Gill and A.J. Baddeley. Kaplan-Meier type estimators for linear contact distributions. *Scandinavian Journal of Statistics*, 23:129–155, 1996.
16. R.M. Haralick, P.J. Katz and E.R. Dougherty. Model based morphology: the opening spectrum. *Graphical Models and Image Processing*, 57:1–12, 1995.
17. W.S. Kendall and J. Møller. Perfect Metropolis-Hastings simulation of locally stable spatial point processes. In preparation.
18. M.N.M. van Lieshout. On likelihoods for Markov random sets and Boolean models. In: *Proceedings of the International Symposium on Advances in Theory and Applications of Random Sets*, D. Jeulin (Ed.), pp. 121–135. World Scientific Publishing, 1997.
19. M.N.M. van Lieshout. Size distributions in stochastic geometry. CWI Research Report, PNA97-15, 1997.
20. P. Maragos. Pattern spectrum and multiscale shape representation. *IEEE Transactions on Pattern Analysis and Machine Intelligence*, 11:701–716, 1989.
21. G. Matheron. *Random sets and integral geometry.* John Wiley and Sons, 1975.
22. J. Møller and R. Waagepetersen. Markov connected component fields. Research Report 96-2009, Department of Mathematics and Computer Science, Aalborg University, 1996.
23. B.D. Ripley. *Statistical inference for spatial processes.* Cambridge University Press, 1988.
24. B.D. Ripley and F.P. Kelly. Markov point processes. *Journal of the London Mathematical Society*, 15:188–192, 1977.
25. D. Schonfeld and J. Goutsias. Optimal morphological pattern restoration from noisy binary images. *IEE Transactions on Pattern Analysis and Machine Intelligence,* 13:14–29, 1991.
26. K. Sivakumar and J. Goutsias. Morphologically constrained discrete random sets. In: *Proceedings of the International Symposium on Advances in Theory and Applications of Random Sets*, D. Jeulin (Ed.), pp. 49–66. World Scientific Publishing, 1997.
27. K. Sivakumar. *Morphological analysis of random fields: theory and applications.* PhD thesis, Johns Hopkins University, 1997.
28. J. Serra. *Image analysis and mathematical morphology.* Academic Press, 1982.
29. D. Stoyan, W.S. Kendall, and J. Mecke. *Stochastic geometry and its applications.* Akademie-Verlag, 1987. Second edition, 1995.

FINDING OPTIMAL SEQUENTIAL DECOMPOSITIONS OF EROSIONS AND DILATIONS

A. JUNIOR BARRERA, B. CARLOS E. FERREIRA and
C. RONALDO F. HASHIMOTO *
Departamento de Ciência da Computação do IME/USP
Rua do Matão, 1010, CEP 05508-900, São Paulo, SP, Brasil.
email: [jb, cef, ronaldo]@ime.usp.br

Abstract. A fundamental problem in the field of automatic design of morphological operators is the transformation of standard morphological representations, that are adequate for operator learning, into other equivalent representations, that permit efficient implementations of the operators learned. In this paper, we study this problem restricted to the families of set translation invariant erosions and dilations. We present a general method for the decomposition of erosions (respec., dilations) by arbitrarily shaped structuring elements into a minimum number of compositions of erosions (respec. dilations) by subsets of the elementary square (i.e.,the 3×3 square centered at the origin). The fundamental idea of the method proposed is the application of Combinatorial Optimization techniques, supported by generic algebraic and geometrical properties of erosions and dilations.

Key words: erosion, dilation, Minkowski addition, decomposition, structuring element.

1. Introduction

The problem of designing automatically morphological operators [1] may be approached by a process composed of two main steps: *i* - learning of the target operator by the estimation of the structuring elements (SE's) that characterize it in a standard representation form; *ii* - transformation of the standard form morphological operator learned into other equivalent representations that permit more efficient implementations.

We find in the literature some works that study the problem of transformation of decomposition structures [3, 2]. However, the general problem of transforming the sup or inf-representation [3] of t.i. set operators into efficient sequential or hybrid (sequential-parallel) representations is extremely hard and practically not studied. In fact, some aspects of this problem have been studied for the families of erosions and dilations.

The speed up achieved by representing erosions and dilations by sequential decompositions, in conventional machines, was quantitatively studied by Maragos [5, p. 77], who showed examples where the complexity of the algorithms that implement erosions and dilations went from quadratic, in the usual representation (i.e., the sup or inf-representation), to linear, in the sequential decomposition.

* The authors are presented in alphabetic order.

In fact, the problem of transforming the usual representation of an erosion or a dilation into a sequential representation is equivalent to find a decomposition for the SE that characterizes it [9, p. 47].

Several researchers [8, 10, 6] have studied the problem of decomposing SE's as sequences of Minkowski additions of smaller subsets and proposed different algorithms to generate them. Xu [10] developed an algorithm for the decomposition of convex SE's in terms of subsets of the *elementary square* (i.e., the 3×3 square centered at the origin). Park and Chin [6] developed an extension of Xu's algorithm for the decomposition of *simply connected* (i.e., an 8-connected set that contains no holes) SE's. Both algorithms are *optimal*, in the sense that they produce decompositions that use the minimum number of subsets.

We should remark that not all SE's have sequential decompositions by Minkowski additions [10]. Furthermore, it is not known an efficient algorithm for determining the existence of such decompositions for an arbitrary SE.

Here, we present a method for the generation of decompositions of arbitrarily shaped SE's as sequences of Minkowski additions of subsets of the elementary square. If there exist such decompositions, the method gives the optimal one, otherwise, it proves that the SE is not decomposable. The formulation adopted is a branch and bound search in a tree that represents the space of all possible sequences of subsets.

Following this introduction, Section 2 presents definitions and properties of Minkowski additions. Section 3 recalls the formulation of Combinatorial Optimization problems by the branch and bound technique. Section 4 joins the information of Sections 2 and 3 for presenting the method proposed. Section 5 presents some experimental results. Finally, Section 6 discusses the main contributions of this paper and some possible next steps of this research.

2. Definitions and Properties

For any $X \subset \mathbb{Z}^2$ and $y \in \mathbb{Z}^2$, X_y denotes the *translation* of X by y, that is, $X_y = \{x \in \mathbb{Z}^2 : x - y \in X\}$.

Let $X, Y \subset \mathbb{Z}^2$. The *Minkowski addition* and *subtraction* of X and Y are the subsets given, respectively, by $X \oplus Y = \cup\{X_y : y \in Y\}$ and $X \ominus Y = \cap\{X_{-y} : y \in Y\}$.

Two properties of Minkowski addition are commutativity (i.e., $X \oplus Y = Y \oplus X$) and translation by unitary sets (i.e., $X \oplus \{y\} = X_y$).

In this paper, we will say that a SE A is *decomposable by Minkowski additions* if there exists a sequence of subsets $B_1, \ldots, B_n$ of the elementary square such that $A = B_1 \oplus B_2 \oplus \cdots \oplus B_n$.

The sequence of subsets that represent a SE can be decomposed in two subsequences: *shape* and *translation*. The shape sequence represents the shape of the SE, while the translation sequence defines its position in the integer plane. The shape sequence $B_1, \ldots, B_k$ and the number k are called, respectively, a *decomposition* of the SE A and the *length* of this decomposition.

Let $A, B \subset \mathbb{Z}^2$. We say that B is an *invariant* of A if and only if (iff) $A = (A \ominus B) \oplus B$. For example, the subset $\begin{bmatrix} 0 & \mathbf{1} & 1 \end{bmatrix}$ is an invariant of the subset $\begin{bmatrix} 1 & \mathbf{1} & 0 \\ 0 & 1 & 1 \end{bmatrix}$, where the 1's represent the points that belong to the subset, the 0's represent points that do not belong to the subset and the bold face character represents the origin (i.e., the point $(0,0)$).

Proposition 1 *Let $A, X, Y \subset \mathbb{Z}^2$. If $A = X \oplus Y$, then X and Y are both invariants of A.*

In the following, we give two immediate corollaries of this proposition.

Corollary 2 *If $A = B_1 \oplus B_2 \oplus \cdots \oplus B_k$, then each B_i is an invariant of A.*

This corollary implies that every subset in a decomposition of a SE is an invariant of it. So, to find a decomposition of a SE, we can restrict our attention only to its invariants.

Corollary 3 *Let $X, A \subset \mathbb{Z}^2$. If X is not an invariant of A, then, for any non empty subset $Y \subset \mathbb{Z}^2$, $Y \oplus X$ is not an invariant of A.*

Let $B \subset \mathbb{Z}^2$ and n be a positive integer. The succession of $n-1$ self Minkowski additions of B is the subset denoted nB, that is, $nB = ((B \oplus B) \oplus \cdots \oplus B)$.

Let $A, B \subset \mathbb{Z}^2$ such that B is an invariant of A. The *multiplicity* of B is the greatest positive integer n such that nB is an invariant of A. For example, the multiplicity of the invariant $B = \begin{bmatrix} 0 & \mathbf{1} & 1 \end{bmatrix}$ of $A = \begin{bmatrix} 1 & \mathbf{1} & 1 \\ 1 & 1 & 1 \end{bmatrix}$ is 2, since $2B = \begin{bmatrix} \mathbf{1} & 1 & 1 \end{bmatrix}$ is an invariant of A and $3B = \begin{bmatrix} \mathbf{1} & 1 & 1 & 1 \end{bmatrix}$ is not. Note that unitary sets have infinity multiplicity.

Proposition 4 *Let $A, B \subset \mathbb{Z}^2$ such that $|B| \geq 2$ and B is an invariant of A. If n is the multiplicity of B, then any decomposition of A contains at most n elements equal to B.*

Let $A \subset \mathbb{Z}^2$ and $\mathcal{B}(A) = \{B \subseteq \{-1, 0, 1\}^2 : B$ is an invariant of A and $|B| \geq 2\}$. Let $B_1, B_2, \cdots, B_k$ be all elements of $\mathcal{B}(A)$ in a fixed order and n_i be the multiplicity of the invariant B_i, $i = 1, \ldots, k$. The *invariant sequence* of A is the sequence $\underbrace{B_1, \cdots, B_1}_{n_1 \text{ times}}, \cdots, \underbrace{B_k, \cdots, B_k}_{n_k \text{ times}}$.

Now, we will study the length of decompositions of an arbitrarily shaped SE from the analysis of measures taken over it.

Let $\vec{u}_1$, $\vec{u}_2$, $\vec{u}_3$ and $\vec{u}_4$ be the Cartesian axis that intersect the origin and have slopes, respectively, 0, 45, 90 and 135 degrees. For a given point $x \in \mathbb{Z}^2$, let $l_1(x)$, $l_2(x)$, $l_3(x)$, $l_4(x)$ be the orthogonal projections of x at the Cartesian axis $\vec{u}_1$, $\vec{u}_2$, $\vec{u}_3$ and $\vec{u}_4$, respectively. Observe that these projections are integer numbers at the directions 0 and 90 degrees and real numbers (proportional to $\frac{\sqrt{2}}{2}$) at the directions 45 and 135 degrees. Let x_1, x_2, x_3 and x_4 denote the *normalized orthogonal projections* of the point x at the Cartesian axis $\vec{u}_1$, $\vec{u}_2$, $\vec{u}_3$ and $\vec{u}_4$, respectively, that is, $x_1 = l_1(x)$, $x_2 = l_2(x) \cdot \sqrt{2}$, $x_3 = l_3(x)$ and $x_4 = l_4(x) \cdot \sqrt{2}$. For example, the normalized projections of the point $x = (-5, 2) \in \mathbb{Z}^2$ are: $x_1 = -5$, $x_2 = -3$, $x_3 = 2$ and $x_4 = 7$.

Let $A \subset \mathbb{Z}^2$. For $i = 1, 2, 3, 4$, let $x_i^{\max}(A)$ and $x_i^{\min}(A)$ be, respectively, the maximum and the minimum normalized orthogonal projection at the Cartesian axis $\vec{u}_i$ of the points in A, that is, $x_i^{\max}(A) = \max\{x_i : x \in A\}$ and $x_i^{\min}(A) = \min\{x_i : x \in A\}$. The *orthogonal projection* of the set A is the vector $\rho(A) \in \mathbb{Z}^4$ such that, for $i = 1, 2, 3, 4$, $\rho_i(A) = x_i^{\max}(A) - x_i^{\min}(A)$. For example, the orthogonal projection of the subset $A = \begin{bmatrix} 1 & \mathbf{1} & 0 \\ 0 & 1 & 1 \end{bmatrix}$ is $\rho(A) = (2, 1, 1, 3)$.

Proposition 5 *If* $A \subset \mathbb{Z}^2$, *then* $\max\{\rho_2(A)/2, \rho_4(A)/2\} \leq \max\{\rho_1(A), \rho_3(A)\}$.

Proof. $\rho_1(A)$ and $\rho_3(A)$ are the length of the sides of the smallest rectangle that contains A. It is easy to see that $\rho_2(A)\frac{\sqrt{2}}{2}$ is less or equal to the length of a diagonal of this rectangle. Thus, $\rho_2(A)^2/2 \leq \rho_1(A)^2 + \rho_3(A)^2 \leq 2(\max\{\rho_1(A), \rho_3(A)\})^2$. Hence, $\rho_2(A)/2 \leq \max\{\rho_1(A), \rho_3(A)\}$. Similarly, $\rho_4(A)/2 \leq \max\{\rho_1(A), \rho_3(A)\}$. Therefore, we have that $\max\{\rho_2(A)/2, \rho_4(A)/2\} \leq \max\{\rho_1(A), \rho_3(A)\}$. ■

The following proposition gives a lower bound for the length of decompositions of a SE by Minkowski additions.

Proposition 6 *If* $A \subset \mathbb{Z}^2$ *is decomposable, then a decomposition of* A *contains at least* $\lceil \max\{\rho_1(A), \rho_3(A)\}/2 \rceil$ *subsets of the elementary square.*

Proof. Let m be the number of subsets of the elementary square of a decomposition of the SE A. Clearly, $A \subseteq mS$, where S is the elementary square, and the orthogonal projection of mS is $\rho(mS) = (2m, 4m, 2m, 4m)$. Hence, since $A \subseteq mS$, we have that $\rho_1(A) \leq 2m$, $\rho_2(A) \leq 4m$, $\rho_3(A) \leq 2m$ and $\rho_4(A) \leq 4m$. Thus, $\max\{\rho_1(A), \rho_2(A)/2, \rho_3(A), \rho_4(A)/2\} \leq 2m$ and, by Proposition 5, we have that $\max\{\rho_1(A), \rho_3(A)\} \leq 2m$. Therefore, $m \geq \lceil \max\{\rho_1(A), \rho_3(A)\}/2 \rceil$. ■

A sequence of subsets of the elementary square $B_1, B_2, \ldots, B_k$ is said an *unfeasible sequence* for the decomposition of a SE A iff the subset $B_1 \oplus B_2 \oplus \cdots \oplus B_k$ is not an invariant of A. The next corollary is an immediate consequence of Proposition 1.

Corollary 7 *If* $B_1, B_2, \ldots, B_k$ *is an unfeasible sequence of a given SE* A, *then* $B_1, B_2, \ldots, B_k$ *is not a subsequence of a decomposition of* A.

In the following, we state some conditions that characterize unfeasible sequences for the decomposition of SE's.

The convex hull $C(A)$ of a set $A \subset \mathbb{Z}^2$ is the intersection of all the half planes that contains A. We will suppose that the subsets are represented in the square grid and consider just the half planes with slopes 0, 45, 90 and 135 degrees to build the convex hull. A set $X \subset \mathbb{Z}^2$ is said *convex* iff $X = C(X)$.

Now, we define the *projection vector* of a SE A as a point in $\mathbb{Z}^8$ such that their coordinates are the lengths of the eight edges of $C(A)$ or, equivalently, the numbers in the chain code of $C(A)$. More formally, for $i = 1, 2, 3, 4$, let $n_i^{\max}(A)$ and $n_i^{\min}(A)$ be the number of points $x \in C(A)$ such that, respectively, $x_i = x_i^{\max}(A)$ and $x_i = x_i^{\min}(A)$, that is, $n_i^{\max}(A) = |\{x \in C(A) : x_i = x_i^{\max}(A)\}|$ and $n_i^{\min}(A) = |\{x \in C(A) : x_i = x_i^{\min}(A)\}|$. The *projection vector* of A is the vector $\nu(A) \in \mathbb{Z}^8$, given by, for $i = 1, 2, 3, 4$, $\nu_i(A) = n_i^{\max}(A) - 1$ and $\nu_{i+4}(A) = n_i^{\min}(A) - 1$. For example, the projection vector for $A = \begin{bmatrix} 1 & 1 & 0 \\ 0 & 1 & 1 \end{bmatrix}$ is $\nu(A) = (0, 1, 1, 0, 0, 1, 1, 0)$.

Note that the projection vector is independent of translation, that is, for any $x \in \mathbb{Z}^2$, $\nu(A) = \nu(A_x)$.

Proposition 8 *Let $A, X, Y \subset \mathbb{Z}^2$. If $A = X \oplus Y$, then $\nu(A) = \nu(X) + \nu(Y)$.*

The ideas for proving this proposition are given in [4].

Corollary 9 *If $A = B_1 \oplus B_2 \oplus \cdots \oplus B_k$, then $\nu(A) = \nu(B_1) + \nu(B_2) + \cdots + \nu(B_k)$.*

Corollary 10 *Let $A \subset \mathbb{Z}^2$ and $B_1, B_2, \cdots, B_k$ be a sequence of subsets of the elementary square, if $\nu(A) = \nu(B_1) + \nu(B_2) + \cdots + \nu(B_k)$ and $A \neq B_1 \oplus B_2 \oplus \cdots \oplus B_k$, then $B_1, B_2, \cdots, B_k$ is an unfeasible sequence for the decomposition of A.*

Given two subsets X and Y, we denote $\nu(X) \leq \nu(Y)$ iff, for any $i \in \{1, \ldots, 8\}$, $\nu_i(X) \leq \nu_i(Y)$.

Corollary 11 *Let $A \subset \mathbb{Z}^2$ and $B_1, B_2, \cdots, B_k$ be a sequence of subsets of the elementary square, if $\nu(A) \not\geq \nu(B_1) + \nu(B_2) + \cdots + \nu(B_k)$, then $B_1, B_2, \cdots, B_k$ is an unfeasible sequence for the decomposition of A.*

Corollary 12 *Let $A \subset \mathbb{Z}^2$ and $B_1, B_2, \cdots, B_m$ be the invariant sequence of A. If $\nu(A) \not\leq \nu(B_1) + \nu(B_2) + \cdots + \nu(B_m)$, then A is not decomposable.*

3. Combinatorial Optimization techniques

The aim of optimization problems is to maximize or minimize a given function over a certain domain. Combinatorial Optimization problems are characterized by the case in which this domain is finite and its elements can be "easily" generated. The difficulty to solve such problems usually stays in the cardinality of the domain. It is not possible to enumerate, in a reasonable time, all possibilities in order to find the best solution.

A strategy to partially enumerate the solutions of a combinatorial optimization problem is known as the branch and bound method [7, p. 40]. The idea is to partition the set of possible solutions into some small sets, originating independent subproblems. When one solves all these subproblems, the best solution found is the optimal solution of the original problem. This partition can be viewed as the construction of a rooted tree, whose nodes correspond to partial solutions, and the root node to the original problem. The search in the tree can be improved if some lower and upper bounds on the value of the optimal solution are known. For example, if it is known a feasible solution whose value is say 25, and we are in a node with lower bound 28, this branch can be pruned, since the best possible solution in this branch cannot be better than a known solution. The art of the strategy is to find good bounds in order to avoid visiting all possibilities.

There are many different ways to visit the nodes of a tree. In the algorithm presented in Section 4, we use the depth first search [7, p. 39], that is described in the following. In the beginning all nodes are marked as "unvisited". In the first iteration, some arbitrary node is selected (if the tree has a root, this is the selected node) and marked as "visited". Its neighbors are marked "reachable", and pushed in a stack. In an arbitrary iteration, if the stack is empty, the search stops. Otherwise, the top element is pop up and marked as "visited". The still "unvisited" neighbors of this node are marked as "reachable", pushed in the stack, and a new iteration begins.

4. Search of Optimal Decompositions

In this section, we describe a combinatorial optimization procedure for the search of decompositions of arbitrarily shaped SE's by subsets of the elementary square. If there exist decompositions for a SE, it gives an optimal sequence of subsets, otherwise, it proves that the SE is not decomposable.

In order to test if a shape sequence represents a given SE, we use a classical matching algorithm. When there is a matching between the SE and the subset generated from the shape sequence, the translation sequence is immediately determined.

The hard part of the problem is the determination of the shape sequence. The formulation adopted is the branch and bound search, described in the last section, in a tree that represents the space of all possible sequences of subsets. In the search we look for feasible sequences, pruning unfeasible ones.

Corollary 2 and Proposition 4 show that the search space for determining decompositions of a SE A can be restricted to the sequence of invariants of A. Thus, given a SE A, we will define a tree that represents the space of all possible subsequences of the invariant sequence of A.

Let $B_1, B_2, \ldots, B_m$ be the invariant sequence of a SE A. The *decomposition tree* of A is a tree such that: *i* - the root is the unitary set that contains the origin; *ii* - a generic node is either the root or a subset $Y = B_{i_1} \oplus B_{i_2} \oplus \cdots \oplus B_{i_j}$, where $B_{i_1}, B_{i_2}, \ldots, B_{i_j}$ is a nomempty subsequence of the invariant sequence of A. *iii* - the direct descendents of a node $Y = B_{i_1} \oplus B_{i_2} \oplus \cdots \oplus B_{i_j}$ are $Y \oplus B_{i_j+1}$, $Y \oplus B_{i_j+2}$, $\ldots$, $Y \oplus B_m$. *iv* - the edge that joins the node $Y = B_{i_1} \oplus B_{i_2} \oplus \cdots \oplus B_{i_j}$ and its direct descendent $Y \oplus B_k$ is labeled B_k $(k = i_j + 1, i_j + 2, \ldots, m)$.

The *level* of a node X in a decomposition tree is the length of the path from the root to X and it is denoted by $level(X)$.

Note that the labels of the edges on the path from the root to a node X forms a subsequence of the invariant sequence and that the subset X is given by Minkowski addition of the elements of this subsequence. So, if the node X is equal to A, then A has a decomposition given by the sequence defined by the path from the root to this node.

Moreover, observe that the nodes of the decomposition tree of A correspond to all possible 2^m subsequences of the invariant sequence of A. Our objective is to find the node A with minimum level. If there is no node A, then the SE is not decomposable.

A crucial task of the procedure is an adequate pruning of the decomposition tree. Corollary 12 gives a necessary condition for the existence of a decomposition for A.

Thus, we should begin the search just if this condition is satisfied. Proposition 6 fixes the lower bound for a decomposition of A, while Corollaries 10 and 11 guaranty important pruning, since they detect unfeasible sequences.

Under this context, after setting the lower bound (with the value given by Proposition 6) and the initial upper bound (with infinite), the search dynamics goes on. In the first iteration, the root is selected to be visited. In an arbitrary iteration, when the node X is being visited: *i* - if X satisfies the constraints of Corollary 11 (i.e., $\nu(A) \not\geq \nu(X)$), or if X is not an invariant of A, or if $level(X)$ is greater or equal to the current upper bound, then this node is pruned; *ii* - if $level(X)$ is greater or equal to the lower bound and the constraint of Corollary 9 (i.e., $\nu(X) = \nu(A)$) is satisfied: *ii.a* - if $X = A$, then the path from the root to X is the present optimal solution and the upper bound is updated to $level(X)$, additionally, if $level(X)$ is equal to the lower bound, then X is the optimal solution; *ii.b* - if $X \neq A$, by Corollary 10, the node X can be pruned; *iii* - if $level(X)$ is less than the lower bound or the constraint of Corollary 9 is not satisfied (i.e., $\nu(X) \neq \nu(A)$), then the node X is an invariant of A, $\nu(X) \leq \nu(A)$ and $level(X)$ is less than the current upper bound, so the next iteration visits one of its unvisited descendents.

Depending on the order chosen to construct the invariant sequence, different heuristic search procedures arise. We have sorted the invariant sequence in decreasing order, according to the sum of the coordinates of the orthogonal projections of each subset.

5. Experimental Results

We have applied our search procedure for the decomposition of digital disks, convex sets, decomposable simply connected sets and decomposable sets that have holes.

The *digital disk* of radius r, centered at the origin, is the subset of $\mathbb{Z}^2$ given by $D(r) = \{(x, y) \in \mathbb{Z}^2 : x^2 + y^2 \leq r^2\}$. For the disks of radius 2 and 4, the procedure spent less than 1 second to find the first decomposition, that were optimal. The disk $D(3)$ and all disks of radius 5 to 16 are not decomposable. The time spent for detecting the non-decomposability of disks $D(3)$, $D(5)$ and $D(6)$ was less than 1 second and for $D(7)$ to $D(16)$ was increasing from 2 seconds to 4 hours.

For convex sets the first solution found was always the optimal one. The time spent for finding the optimal decomposition of subsets of the 7×7 to 61×61 square were increasing from 0.6 to 50 seconds.

For decomposable simply connect sets the time spent, for example, for the first solution of a subset of the 46×46 square was 12 hours. Observing some experiments of this kind, we note that the distance between the optimal and the first solution was at most 1.

For decomposable sets that have holes the time spent for SE's that are subsets of the 9×9, 17×17, 19×19, 27×27, 41×41 and 61×61 square were less than 22 seconds. All the first solutions were optimal.

6. Conclusion

We have presented a combinatorial optimization approach for finding optimal sequential decompositions of erosions and dilations by arbitrarily shaped SE's.The procedure proposed gives an optimal decomposition or proves that the operator considered is not decomposable.

The central idea of the method is walking intelligently on the space of possible decompositions, supported by algebraic and geometrical properties of the SE's.

We have presented a description of the search procedure as well as the Mathematical properties that justify the pruning. As should be expected, the complexity of the search procedure is greatly reduced for convex SE's.

Several examples illustrate the capacity of the approach for finding decompositions of complex SE's in reasonable processing times (compared with the size of the search space). These examples also show that the distance between the optimal and the first solution obtained usually is small. Therefore, in practical applications, it may be a good heuristic to stop when the first solution is obtained.

The model proposed is flexible and open for the addition of other pruning strategies, that should be a consequence of a deeper theoretical understanding of the problem.

The symbiosis between Mathematical Morphology and Combinatorial Optimization seems a powerful approach to solve hard problems as the optimization of the representation of morphological operators in the context of the automatic programming of morphological machines. The problem studied here is just a point in this universe.

References

1. J. Barrera, E. R. Dougherty, and N. S. Tomita. Automatic programming of binary morphological machines by design of statistically optimal operators in the context of computational learning theory. *Journal of Electronic Imaging*, 6(1):54–67, January 1997.
2. J. Barrera and R. F. Hashimoto. Compact representation of w-operators. Submited to International Symposium on Mathematical Morphology' 98.
3. J. Barrera and G. P. Salas. Set Operations on Collections of Closed Intervals and their Applications to the Automatic Programming of Morphological Machines. *Journal of Eletronic Imaging*, 5(3):335–352, July 1996.
4. P. K. Ghosh. Vision, geometry, and minkowski operators. *Contemporary Mathematics*, 119:63–83, 1991.
5. P. A. Maragos. *A Unified Theory of Translation-invariant Systems with Applications to Morphological Analysis and Coding of Images*. PhD thesis, School of Elect. Eng. - Georgia Inst. Tech., 1985.
6. H. Park and R. T. Chin. Decomposition of Arbitrarily Shaped Morphological Structuring Elements. *IEEE Transactions on Pattern Analysis and Machine Intelligence*, 17(1):2–15, January 1995.
7. E. Rich and K. Knight. *Artificial Intelligence*. McGraw-Hill, USA, second edition, 1991.
8. C. H. Richardson and R. Schafer. A lower bound for structuring element decompositions. *IEEE Transactions on Pattern Analysis and Machine Intelligence*, 13(4):365–369, April 1991.
9. J. Serra. *Image Analysis and Mathematical Morphology*. Academic Press, 1982.
10. J. Xu. Decomposition of Convex Polygonal Morphological Structuring Elements into Neighborhood Subsets. *IEEE Transactions on Pattern Analysis and Machine Intelligence*, 13(2):153–162, February 1991.

SETS, PARTITIONS AND FUNCTIONS INTERPOLATIONS

SERGE BEUCHER
Centre de Morphologie Mathématique - Ecole des Mines de Paris
35, rue Saint-Honoré - 77305 Fontainebleau (FRANCE)

Abstract. This paper presents the use of the SKIZ transformation to interpolate sets, partitions and functions. This transformation generates a sequence of intermediary sets which can be used to define a distance between the interpolated sets. Other distances are also introduced. These distances could be useful to generalise the interpolation algorithm. Some examples of use are given.

1. Introduction

Set interpolation and partition interpolation are useful techniques in many domains of image analysis. It is the case for instance in morphing processes, where the design of the deformation grid could be automatised and even replaced by an automatic deformation of homologous regions in the initial and final images. These regions could be built using morphological segmentation tools as watersheds and marker-controlled watersheds. The watershed transform generating a partition of an image (the mosaic image), the problem consists therefore in finding tools to distort the partition corresponding to the initial image and to apply it to the final one. Another efficient use of image and partition interpolation is in image compression. Many morphological compression techniques use mosaic images and interpolation can dramatically reduce the bandwidth when transmitting image sequences. A third domain where interpolation can be applied is for quantifying the distortion or the displacement of objects or regions from a picture to another one. It is the case in particular in stereovision or in motion estimation.

This paper aims at introducing a simple interpolation technique based on SKIZ for sets, partitions and functions. Some extensions will be discussed, in particular the introduction of various distances which can be defined to control these interpolations or to combine displacements and deformations of sets. Some examples of use will also be given.

2. Set Interpolation

All the interpolations described in this paper are based on a sole morphological transform, the skeleton by influence zones or SKIZ [3]. Other approaches using more or less sophisticated geodesic transforms have been studied [4,5]. However, the SKIZ is easy to use, more general (it applies on sets and on functions) and the results it gives are better. We shall not prove in this paper that this transformation induces a geodesic distance and the theoretical aspects will be put aside in favour of a more intuitive and algorithmic approach.

2.1. The basic transformation: the SKIZ

Let, in a space E, X and Y be two sets $\in (E)$ such that $X \cap Y \neq \emptyset$ and $X \cup Y \neq E$. Consider the set W made of the two following components (Fig. 1.):

$$W = \{X \cap Y, (X \cup Y)^c\}$$

The SKIZ of W produces the influence zones (IZ_W) of the two components of W (note that these two components are not necessarily connected). In particular:

$$IZ_W (X \cap Y) = \{x \in E : d(x, X \cap Y) < d(x, (X \cup Y)^c)\}$$

where d is the Euclidean pseudo-distance of a point to a set.

Let us denote $M(X,Y)$ the influence zone of $X \cap Y$. This set is the median set of X and Y. This set is, in the geodesics generated by the SKIZ, at the same distance from X and from Y:

$$M(X,Y) = IZ(X \cap Y)$$

This transformation is auto-dual. We have, in fact:

$$M(X,Y)^c = M(X^c, Y^c)$$

This property provides a good robustness for the set complementation, which is not always fulfilled for any geodesics. It is not true, in particular, for the Haussdorf distance [5].

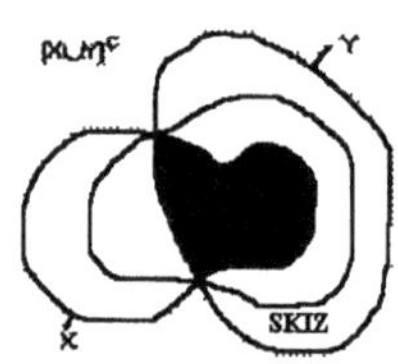

Fig. 1. SKIZ of $X \cap Y$ and $(X \cup Y)^c$

In return, this operator is not associative. If we take three sets X, Y and Z, the median set of these three sets cannot be built by generating $M(X,Y)$ followed by $M[Z, M(X,Y)]$. The result (assuming that it is non simple) will depend on the order of the operations.

The median of n sets X_i, with $\bigcap_i X_i \neq \emptyset$ and $\bigcup_i X_i \neq E$ (all the sets are sharing at least one point and do not cover the entire space) can be written:

$$M(X_i) = IZ_W \left(\bigcap_i X_i\right) \quad \text{with} \quad W = \left\{\bigcap_i X_i, \left[\bigcup_i X_i\right]^c\right\}$$

2.2. Sequences of deformations

By iterating this transformation from the sets X and Y, a sequence of $n+1$ sets representing the n deformation steps from X to Y can be built. We have:

$$K_0 = X; K_n = Y$$

Then: $K_{n/2} = M(K_0, K_n)$; $K_{n/4} = M(K_0, K_{n/2})$; $K_{3n/4} = M(K_{n/2}, K_n)$ and after the $(i+1)$th step:

$$K_{n/2^i} = M(K_0, K_{n/2^{i-1}}); \ldots$$

The number of intermediary sets increasing by 2^{i-1} at each step i, n must be a power of 2 in order that this number corresponds to an integer number of iterations.

We get then a sequence of 2^{i+1} sets, two consecutive ones being at the distance $1/2^i$ if we assume that the distance between the two extreme sets is 1. When n is not a power of 2, intermediary sets can still be built with the same process and be labelled with $n-1$ integer values from 0 to n. However, the distance between two consecutive sets will not be rigorously equal to $1/n$.

The table 1 shows such a deformation sequence. Note that the initial and final sets do not belong to the same class of connectivity.

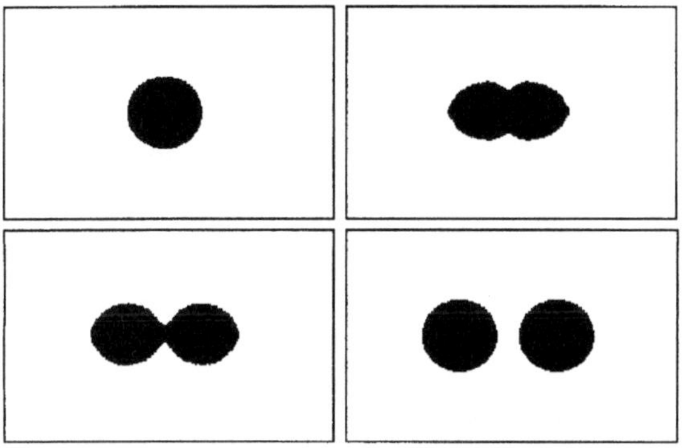

Table 1. Interpolation sequence between sets X (upper left) and Y (lower right).

2.3. The algorithm

Computing $M(X,Y)$ starting from X and Y is simple. Simple dilations are used, instead of thickenings as it is the case in the classical SKIZ. As a matter of fact, the union of $X \cap Y$ and of $(X \cup Y)^c$ can be connected, which prevents the use of thickenings. Moreover, the use of dilations produces a better and unbiased SKIZ because all the connected components grow at the same time. We have:

$$Z_0 = X \cap Y \quad \text{and} \quad W_0 = (X \cup Y)^c$$

We just have to iterate until idempotence the following operations (B is the elementary ball):

$$Z_i = [(Z_{i-1} \oplus B) \,/\, (W_{i-1} \oplus B)] \cup Z_{i-1} \ ; \ W_i = [(W_{i-1} \oplus B) \,/\, (Z_{i-1} \oplus B)] \cup W_{i-1}$$

Then, we have: $M(X,Y) = Z_\infty$

The union at each step with the previous result insures that, despite some parity problems on the digital grid, the operation is extensive. For the same reasons, we generally do not have the equality $Z_\infty = (W_\infty)^c$.

This method does not allow to obtain directly the set at a distance d_1 from X and d_2 from Y with $d_1 + d_2 = 1$. However, this could be possible by modifying the growth speeds (that is the sizes of dilations) of the sets Z_i et W_i. If v_1 is the size of the elementary dilation of Z_i and v_2 the size of the dilation of W_i, and if k is a normation factor, we have:

$$d_1 = k\, v_1 \quad \text{and} \quad d_2 = k\, v_2 \quad \text{with} \quad d_1 + d_2 = 1$$

which gives:

$$d_1 = \tfrac{v_1}{v_1+v_2};\ d_2 = \tfrac{v_2}{v_1+v_2} \quad \text{and} \quad v_1/v_2 = d_1/d_2$$

However, this approach has two drawbacks. Firstly, the induced metrics is different from the one generated by means of the dichotomic technique. Secondly, with dilations of respective sizes v_1 and v_2, the SKIZ thickness may be larger than 1. Therefore, a more sophisticated sequence of dilations must be used to fit with the real ratio v_1/v_2.

3. Interpolation of partitions

3.1. General case

Consider two partitions T and T':

$$T = \{C_i\}, T' = \{C_i'\}$$

We suppose that there exists a one-to-one correspondence between the cells C_i of the first partition and the cells C_i' of the second one. More precisely, we have (Fig. 2):

$$\forall\, C_i \in T,\ \exists\, C_i' \in T'\ :\ C_i \cap C_i' \neq \emptyset$$

Let $W = \{C_i \cap C_i'\}$ be the set made of all the components $C_i \cap C_i'$.

A median partition $M(T, T')$ can then be defined as the influence zones of the sets $(C_i \cap C_i')$:

$$M\,(T, T') = \{IZ_W\,(C_i \cap C_i')\}$$

The definition of a median partition is, to a certain extent, simpler than the definition of the median of two sets.

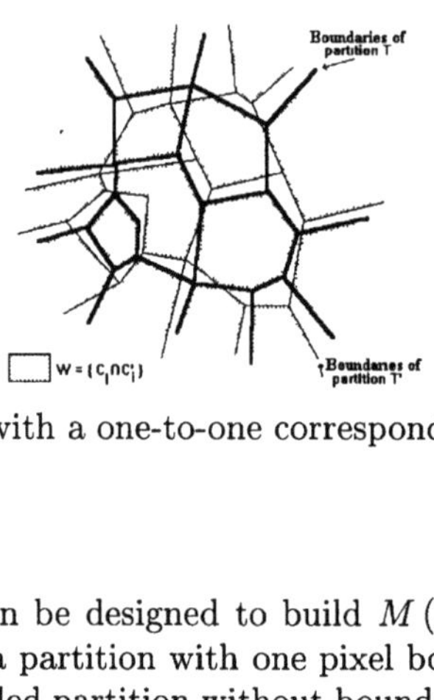

Fig. 2. Partitions with a one-to-one correspondence between cells

3.2. Algorithm

Two kinds of algorithms can be designed to build $M\,(T, T')$. the first one uses a simple SKIZ and produces a partition with one pixel boundaries. The second (and better) one produces a labelled partition without boundaries between the cells. The cell $C_i \in T$ and its corresponding cell $C_i' \in T'$ share the same label. The partitions T and T' can then be considered as multilevel staircase functions f and f':

$$f : E \to N_+\ \ x \to f(x) = i, x \in C_i\ \ \ ;\ f' : E \to N_+\ \ x \to f'(x) = i, x \in C_i'$$

$M(T, T^{'})$ is built by means of simple dilations. A function g can be defined by:

$$g\ : E \to N_+\ \ x \to g(x) = i, x \in IZ_W(C_i \cap C_i')$$

A double labelling of the partitions T and $T^{'}$ is used (they are supposed to contain n cells). The algorithm is as follows:

- g_0 is computed:

$$g_0\,(x) = f\,(x) = f'\,(x) \text{ iff } f\,(x) = f'\,(x) \Longleftrightarrow x \in C_i \cap C_i', \forall i\ ;\ g_0\,(x) = 0 \text{ if not}$$

– then g'_0 is defined:

$$g'_0(x) = n - g_0(x) \quad \text{iff } g'_0(x) \neq 0\ ;\ g'_0(x) = 0 \quad \text{if not}$$

This function is similar to g_0. The only difference is that the labels i of the sets $C_i \cap C'_i$ have been changed to $(n-i)$.

Following this initialisation phase, for each g_i and g'_i, we do:

– calculation of the function h_i:

$$h_i = (g_{ih} \oplus B) + (g'_i \oplus B) \quad (B\text{,elementary ball of } E)$$

– Then, the indicator function h of the points where h_i is less or equal to n is defined:

$$h = 1 \quad \text{iff} \quad h_i \leq n,\ h = 0 \quad \text{if not}$$

– The functions g_{i+1} and g'_{i+1} are calculated:

$$g_{i+1} = \sup\left[(g_i \oplus B) \times h, g_i\right]; g'_{i+1} = \sup\left[(g'_i \oplus B) \times h, g'_i\right]$$

The algorithm is iterated until idempotence. Finally we get $g = g_\infty$.

Table 2. Interpolation of partitions (stereo pair of cleavage facets in steel)

4. Interpolation of functions

4.1. General case

The notion of median set can be generalised to functions by means of the sets over and under the graph of these functions. Let f and g be two functions defined in the space E. Consider first the set under the graph D of $\inf(f,g)$:

$$D\left[\inf(f,g)\right] = D(f) \cap D(g)$$

and then the set over the graph U of $\sup(f,g)$:

$$U\left[\sup(f,g)\right] = \left[D(f) \cup D(g)\right]^c$$

Consider the set W made of the two following components:

$$W = \left\{D\left[\inf(f,g)\right], U\left[\sup(f,g)\right]\right\}$$

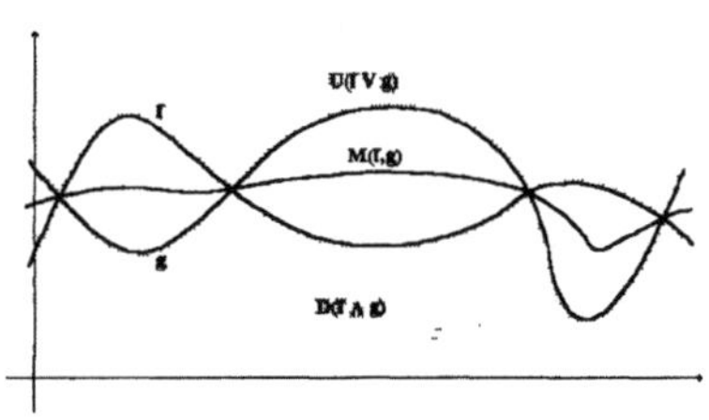

Fig. 3. Median of two functions by SKIZ of $D[Inf(f,g)]$ and $U[\sup(f,g)]$

The median function of f and g is the function $M(f,g)$, its set under the graph being the influence zone of $D[\inf(f,g)]$ in the 3D SKIZ of W (Fig. 3):

$$D\left[M(f,g)\right] = IZ_W\left[D\left[\inf(f,g)\right]\right]$$

4.2. ALGORITHM

As for sets, the algorithm uses 3D dilations of U and D. We have:

$$Z_0 = D\,[\inf(f,g)]\,;\, W_0 = U\,[\sup(f,g)]$$

We can write, with $z_0 = \inf(f,g)$ and $w_0 = \sup(f,g)$:

$$Z_1 = D\,[\sup(\inf(z_0 \oplus B, w_0 \ominus B), z_0)] \;\; ; \;\; W_1^c = D\,[\inf(\sup(w_0 \ominus B, z_0 \oplus B), w_0)]$$

Z_1 is the set under the graph of $z_1 = \sup(\inf(z_0 \oplus B, w_0 \ominus B), z_0)$ and W_1 the set over the graph of $w_1 = \inf(\sup(w_0 \ominus B, z_0 \oplus B), w_0)$.

This is true also at step i:

$$z_i = \sup(\inf(z_{i-1} \oplus B, w_{i-1} \ominus B), z_{i-1})\,; \;\; w_i = \inf(\sup(w_{i-1} \ominus B, z_{i-1} \oplus B), w_{i-1})$$

Finally, we obtain: $\qquad M(f,g) = z_\infty = w_\infty$

The median of two functions is a function.

5. Distance functions

Let us come back to the interpolation of sets. Starting from two sets X and Y with $X \cap Y \neq \emptyset$ and $X \cup Y \neq E$, the intermediary sets $M_\lambda(X,Y)$ depending of the parameter $\lambda(0 \leq \lambda \leq 1)$ can be defined, with $M_0(X,Y) = X$ and $M_1(X,Y) = Y$. This transformation is not monotonous with respect to λ. However $M_\lambda(X,Y)$ can be split into two components, an extensive one and an anti-extensive one with respect to λ. Consider indeed the sets $X \cap Y$ and X^c. The SKIZ of these two disjoint sets can be performed. Let $M^-(X,Y)$ be the influence zone of $X \cap Y$. We see immediately that the obtained SKIZ corresponds to the part of the SKIZ associated to $M(X,Y)$ situated in X/Y. This operation can also be iterated to define a sequence of sets $M_\lambda^-(X,Y)$ with $M_0^-(X,Y) = X$ and $M_1^-(X,Y) = X \cap Y$. The sequence $M_\lambda^-(X,Y)$ is monotonously decreasing with respect to λ. A function d^- defined on X can be associated to this sequence, each M_λ^- corresponding to the threshold at λ of d^-:

$$M_\lambda^- = \{x \in X : d^-(x) \geq \lambda\}$$

In the same way, let us consider now the sets $X \cap Y$ and Y^c. The SKIZ of these two disjoint sets produces the set $M^+(X,Y)$ corresponding to the influence zone of $X \cap Y$. A monotonously increasing sequence of sets $M_\lambda^+(X,Y)$ can then be defined and a function d^+ defined on Y. We have:

$$M_\lambda^+ = \{x \in Y : d^+(x) \leq \lambda\}$$

The sets M_λ^- and M_λ^+ split the deformation of the sets X and Y into an anti-extensive component (an erosion) and an extensive one (a dilation).

The function $d = (d^+ + d^-)/2$ is defined on $X \cup Y$ and is constant on $X \cap Y$ (Fig. 4b).

If the intersection of X and Y is empty, d is not defined. But another distance can be defined on $X \cap Y$ and on the complementary set of their union by means of another intermediary set between X and Y^c. Let us consider two sets X and Y. We can see that, whatever X and Y, the sets $X \cap Y$, X/Y, Y/X and $(X \cup Y)^c$ constitute a partition of the space E. The definition of $M(X,Y)$ is based on the SKIZ of $X \cap Y$ and $(X \cup Y)^c$. But two other sets in the partition could be chosen. Among the possible choices, only the sets X/Y and Y/X are of some interest. Let us define the set $M^*(X,Y)$ as the influence zone of X/Y in the SKIZ of X/Y and Y/X:

$$W = \{X/Y, Y/X\}\;;\; M^*(X,Y) = IZ_W(X/Y)$$

We see immediately that $M^*(X,Y)$ is equal to $M(X,Y^c)$.

The set $M^*(X,Y)$ is therefore the median set of X and Y^c. By iterating the transformation, the sequence M^*_λ of deformations from X to Y^c can be generated. This sequence is not monotonous. However, as for the sequence M_λ, it can be split into two parts, an extensive one and an anti-extensive one. With these two sequences, two new distance functions d^{*-} et d^{*+} can be defined. These two functions are merged to give the function d^* (Fig. 4c):

$$M^{*-}_\lambda = \{x \in X : d^{*-}(x) \geq \lambda\} \; ; \; M^{*+}_\lambda = \{x \in Y^c : d^{*+}(x) \leq \lambda\}$$

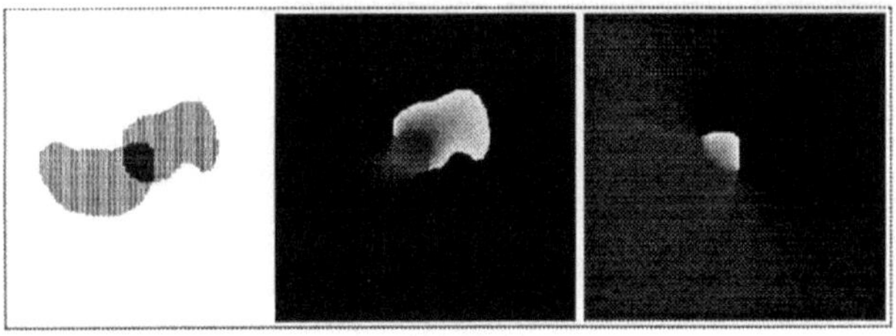

Fig. 4. The sets X and Y, the distances d and d^*

The interest of these functions comes from the fact that they can be used to define deformation fields from a set X to a set Y. The simplest way to achieve this is to derive the functions d and d^*. The gradient vectors represent this deformation field. The use of d and d^* ascertain the existence of at least one such field allowing to go from X to Y or from X to Y^c even when X and Y have no common points. It is also possible to use the field generated by the distance d^* as a displacement field whereas the distance d (when it exists) controls the deformation field.

6. Interpolations in compression of sequences. Use of mosaic images

An interesting application of image interpolation lies in image compression in a moving sequence. Instead of transmitting the entire sequence, one picture over n can be sent, the missing pictures been interpolated. A compression rate of $n : 1$ is then easily obtained. Unfortunately, the interpolation of functions deeply smoothes the intermediary images in the sequence. This is why, instead of using the initial sequence, mosaic images have been built by a watershed transformation [1,2]. This approach reduces the image damaging because, on the one hand, a mosaic image does not present too thin regions and, on the other hand, the calculation of the intermediary mosaic image by means of an interpolation of partition preserves the contrast between the adjacent cells. In the result given in table 3, one picture over 4 is sent, the three others are interpolated. We can see that the general motion in the sequence is well preserved. This amazing result comes mainly from the fact that, as shown before, when two cells in the initial and final mosaic images have some points in common, the interpolation produces a displacement of the cell. So, if there exist a sufficient number of associated cells in the two images, a visual displacement effect will dominate.

Table 3. Sequence interpolations. The 2nd, 3rd and 4th images are interpolated

7. Conclusion

Image interpolation by SKIZ is an efficient and easy to implement tool, in particular when it is associated with segmentation transforms as watersheds and mosaic images. When the direct display of a sequence is the unique goal (videophone, telemonitoring, etc.), we obtain, by these means, good compression rates at low cost.

The use of these interpolations in motion analysis is more complex. The major drawback of the original approach is the necessity for the sets or partitions to be interpolated to share common points (non empty intersection). However, when the motion is important, this correspondence may not exist. But the use of hierarchical segmentations, in the one hand, and the combination, in the other hand, of the distances d and d^* introduced above, which split the displacement and deformation parts of the interpolations, give interesting starting ideas for a better solution of these problems.

8. References

[1] Beucher S., Segmentation d'Images et Morphologie Mathématique, Doctorate Thesis, Paris School of Mines, June 1990.

[2] Beucher S., Meyer F., The Morphological Approach to Segmentation: The Watershed Transformation, in "Mathematical Morphology in Image Processing", E. R. Dougherty Editor, Marcel Dekker, Inc, 1992.

[3] Lantuéjoul C., La squelettisation et son Application aux Mesures Topologiques des Mosaiques Polycristallines, Doctorate Thesis, Paris School of Mines, 1978.

[4] Meyer F. , A morphological interpolation method for mosaic images, in Mathematical Morphology and its application to image and signal processing, Maragos P. et al eds, Kluwer, 1996.

[5] Serra J., Hausdorff distances and interpolations, to be published for ISMM'98.

SEGMENTATION-BASED MOTION COMPENSATION

E. DECENCIÈRE FERRANDIÈRE
Centre de Morphologie Mathématique
École Nationale Supérieure des Mines de Paris
35, rue Saint Honoré
77305 Fontainebleau CEDEX - FRANCE

Abstract.
With the fast growth of the multimedia market, the need for powerful tools for image sequence analysis has become very important. Motion compensation, which increases the correlation between two frames of a sequence, is such a tool. In this article we describe a general backward motion compensation method based on a 2D segmentation of the reference frame. It uses kriging in order to interpolate and filter the motion vector field.

Key words: motion compensation, kriging, vector field interpolation

1. Introduction

With the fast growth of the multimedia market, the need for powerful tools for image sequence analysis has become very important. Motion compensation, which increases the correlation between two frames of a sequence, is such a tool. Image sequence coding and segmentation, as well as object tracking are some examples of applications requiring it.

The most widely used algorithm for backward motion compensation is block-matching. However, it is based on very severe hypotheses, which in practice are often false.

We propose a general backward motion compensation method combining block matching with a 2D segmentation of the reference image. The use of the segmentation allows to treat separately each objet of the scene, and to reduce the problems caused by block matching.

In order to operate the motion compensation we need a displacement vector field. These fields are seldom dense, and often subject to errors. In order to interpolate and filter them we will use kriging, an estimation method used in the field of geostatistics.

Finally we will show its application to motion picture restoration.

2. Backward Motion Compensation

2.1. Definitions

Let I be a grey level image sequence:

$$\{0,\dots,x_{max}\} \times \{0,\dots,y_{max}\} \times \{0,\dots,t_{max}\} \longrightarrow \{0,\dots,255\} \quad (1)$$

$$(x,y,t) \longmapsto I(x,y,t) = I_t(x,y) \quad (2)$$

I_t is the frame at time t of I. We will choose an interpolation method in order to extend the definition set of I. The resulting image sequence is I^*:

$$\mathbb{R} \times \mathbb{R} \times \{0, \ldots, t_{max}\} \longrightarrow \{0, \ldots, 255\} \tag{3}$$
$$(x, y, t) \longmapsto I^*(x, y, t) = I_t^*(x, y) \tag{4}$$

We will only formulate one hypothesis about the chosen interpolation method. We will suppose that it is exact, *i.e.* that it does not modify the original values of the image.

Let V be a displacement vector field:

$$\{0, \ldots, x_{max}\} \times \{0, \ldots, y_{max}\} \longrightarrow \mathbb{R}^2 \tag{5}$$
$$M = (x, y) \longmapsto V(M) = (V_x(M), V_y(M)) \tag{6}$$

If we apply backward motion compensation to a frame I_t of I using the displacement vector field V we obtain the backward motion compensated image I_t^c defined as:

$$\forall M \in \{0, \ldots, x_{max}\} \times \{0, \ldots, y_{max}\} \quad I_t^c(M) = I_t^*(M + V(M)) \tag{7}$$

The compensation is called "backward" because it is done in the opposite direction of the vector field. This has a clear advantage: the construction of the compensated image is very easy.

Let I_{t_1} and I_{t_2} be two frames of the image sequence I. We will say that V is a displacement vector field between frames I_{t_1} and I_{t_2} if each pixel $M = (x, y)$ from I_{t_1} will move to position $M + V(M) = (x + V_x(M), y + V_y(M))$ at time t_2:

$$\forall M \in \{0, \ldots, x_{max}\} \times \{0, \ldots, y_{max}\} \quad I_{t_2}^*(M + V(M)) = I_{t_1}(M) \tag{8}$$

We will call I_{t_1} the **reference image**. As a consequence, if V is a displacement vector field between frames I_{t_1} and I_{t_2}, then:

$$\forall M \in \{0, \ldots, x_{max}\} \times \{0, \ldots, y_{max}\} \quad I_{t_2}^c(M) = I_{t_1}(M) \tag{9}$$

So if we know I_{t_2} and V we can build I_{t_1}. This operation is very useful in image sequence processing and analysis. It allows us to predict one frame given a reference frame and a displacement vector field, or to improve the correlation between frames.

However computing the displacement field is a very difficult problem. Most of the time there is no solution satisfying equation 9, and when there is a solution it is seldom unique. Note also that a displacement field satisfying equation 9 is not necessarily the true displacement field (*i.e.* the projection of the true 3D displacement vector field).

In the next section we will make a short review of the existing displacement estimation methods.

2.2. State of the Art

There are many types of motion estimation methods: differential (optical flow) methods [7], Bayesian methods [9], parametric methods based on a segmentation

[5] and pattern matching methods (among which we can distinguish block-matching methods [8]).

These methods have different fields of application. As far as we know block-matching methods and parametric methods have been the most commonly used for motion compensation.

Parametric methods use a segmentation $\{S_i\}_{i \in I}$ of I_{t_1} in order to estimate the displacement vector field between I_{t_1} and I_{t_2}. For each region S_i of the segmentation a parametric deformation model is estimated (often a linear deformation) [5]. The calculation of a set of parameters for each region can be very long.

The classical block matching motion compensation algorithm, used in video compression standards like MPEG and H-263, divides the reference image into square regular blocks and calculates a translation for each of them. The resulting vector field can model only very simple motions, but can be calculated very fast. New motion compensation methods based on block matching have been proposed, using richer interpolation models [13, 6]. However, these methods, because of the fixed shape of the blocks, have some difficulties with motion frontiers.

We will propose a general structure for a motion compensation algorithm combining the advantages of the previous methods: it uses block-matching for calculating the displacement vectors, and an interpolation constrained by a segmentation of the reference image in order to produce a dense motion field.

3. A General Structure for a Motion Compensation Algorithm

Suppose that we want to operate backward motion compensation on frame I_{t_2} with respect to frame I_{t_1}. The general structure we propose is the following:

1. Segmentation of I_{t_1}. The set of the regions of the segmentation is denoted by $\{S_i\}_{i \in I}$.
2. Estimation through block matching of the displacement vectors between I_{t_1} and I_{t_2} of a set of points $\{P_j\}_{j \in J}$ (**experimental points**). The resulting displacement vector field V_{exp} is called **experimental vector field.**
3. Interpolation constrained by the segmentation. We obtain a dense displacement vector field V^*_{exp}.
4. Backward motion compensation of I_{t_2} using V^*_{exp}.

Note that the classical block matching compensation algorithm can be divided into the same steps: the segmentation is given by regular square blocks; the experimental points are the center points of each block and their displacement vectors are calculated through block matching; finally the interpolation is done inside each block through a simple propagation of the center vector.

This general structure becomes interesting when we accept the following basic hypothesis (which was false in the case of the block matching compensation algorithm):

No region of the segmentation is cut by a motion frontier.

This means that the displacement vector field inside each region of the segmentation is very smooth. Thus the filtering and interpolation steps are made easier.

We need to answer some questions:

- How can we calculate the segmentation?
- How are we going to choose the experimental points and calculate their displacement vectors?
- How are we going to interpolate the displacement vector field?

The following sections answer these questions.

4. Morphological segmentation

Segmentation algorithms based on mathematical morphology have proved to be efficient and robust [15, 12]. Moreover, we expect to have in the near future hardware implementations able to segment an image in real time.

We use a new segmentation algorithm developed by Cichosz and Meyer [1] to calculate the segmentation. It does not matter if the image is slightly over-segmented, as long as our basic hypothesis holds.

We will denote the resulting segmentation by $\{S_i\}_{i \in I}$.

5. Motion estimation

As said before, we use the well known block matching algorithm [8] to estimate the displacement of the experimental points. We only consider integer vector coordinates (no sub-pixel precision).

But how are we going to choose the experimental points?

We want to have as few experimental points as possible inside each region of the segmentation, because calculating displacement vectors is relatively slow. Another reason is that the interpolation method that we are going to use is faster for a small number of experimental points.

On the other hand, block matching is not very robust. It can produce incorrect displacement vectors. The best way to reduce the influence of these vectors is to have a large number of experimental points.

We have chosen a compromise solution using around 10 experimental points inside each region of the segmentation. We will denote $\{P_j\}_{j \in J}$ the set of experimental points.

The resulting displacement vector field is called experimental vector field and denoted V_{exp}.

6. Vector Field Interpolation and Filtering Using Kriging

Kriging is one of the main tools of geostatistics. It is used to interpolate dense fields of data from sparse data. For example, imagine we want to draw the sea bed depth map of a bay. It is practically impossible to obtain all the depth measures that we need, therefore we will perform instead a finite number of measures and, using kriging, interpolate them to the whole bay. The quality of the estimation will depend on the data set and on the hypotheses made about the spatial structure of the sea bed.

We will not describe here the theory of kriging. We refer the reader to [10]. We will just give some results.

Suppose that a function $z : \mathbb{R}^2 \mapsto \mathbb{R}$ is known on a finished set of points $\{M_k\}_{k\in K}$. After making some hypotheses about the structure of z (*i.e.* we associate a covariance model with it), kriging allows to estimate a value of z for any point M. The estimation is noted $z^*(M)$. Its general expression is:

$$z^*(M) = \sum_{k\in K} \lambda_k z(M_k) \tag{10}$$

where the linear coefficients λ_k depend on M and $\{M_k\}_{k\in K}$. In order to calculate them using kriging (and more specifically the theory of Intrinsic Random Functions of Order k [11]) we have to choose a covariance model. In our case, given that we are going to apply this method to displacement vector fields, which are naturally smooth, we will choose the **thin elastic plate model**. The resulting interpolator produces splines.

This interpolation method can be very easily applied to vector values if we consider each vector coordinate separately. The results are very satisfactory [3].

Let us see how we apply this to our case. Suppose that we want to interpolate a vector value $V^*_{exp}(M)$ for a point M of the image. Instead of considering all the experimental vectors, we take only those which belong to the same region S_{i_0} as M: vector value $V^*_{exp}(M)$ is interpolated from the experimental vectors at the points $S_{i_0} \cap \{P_j\}_{j\in J}$.

This interpolation constrained by the segmentation has one important advantage as long as our basic hypothesis holds: the motion frontiers are preserved.

However, some of the experimental vectors, due to the lack of robustness of block matching, can be completely wrong. So, instead of performing directly the interpolation, we will fit an optimal model to the displacement vector field using **inverse kriging** [3].

Suppose that we want to filter the experimental vectors corresponding to region S_{i_0} of the segmentation. Their positions are given by $\{P_j\}_{j\in J_{i_0}} = S_{i_0} \cap \{P_j\}_{j\in J}$ and their values respectively by $\{V_{exp}(P_j)\}_{j\in J_{i_0}}$.

We will place n **information points** $\{Q_l\}_{l=1\ldots n}$ ($n < Card(J_{i_0})$) around the region S_{i_0} and calculate their respective vector values $\{V_{inf}(Q_l)\}_{l=1\ldots n}$ such that when we interpolate vector field V_{inf}, the interpolated values at the experimental points $\{V^*_{inf}(P_j)\}_{j\in J_{i_0}}$ are as close as possible from the experimental values $\{V_{exp}(P_j)\}_{j\in J_{i_0}}$. This is done using Singular Value Decomposition [14]. The complete explanation of the method can be found in [2].

We have chosen $n = 4$ information points, placed on the corners of the image. We call them *information points* because they sum up the information contained in the vector field V_{exp}. The resulting interpolated vector field V^*_{inf} is generally much simpler than V_{exp}. This has two advantages: first this regularization of the field tends to erase false vectors, and second the resulting field is very easy to code (the values $\{V_{inf}(Q_l)\}_{l=1\ldots n}$ are sufficient for that).

7. Results

In figure 1 we show an example of calculation and filtering of a displacement vector field with our method. On the top we have the reference image I_{t_1}, on which we have superposed the experimental vector field. Note that some vectors are erroneous (like the one that we have marked with a white box). On the central image we have the same image, but this time the filtered vector field V_{inf}^{*} is superimposed. It is a dense displacement vector field, but only the values for the experimental points are shown for comparison sake. The false vectors have been smoothed away, but the motion boundaries are left unchanged. Note however that the procedure has not worked well on the right vehicle. There are two reasons for this misbehaviour: first the displacement field was completely false, and second the segmentation did not correspond to the objects of the scene: our basic hypothesis was not true in this case. The bottom image shows the segmentation that was used.

In figure 2 we show the result of the application of our motion compensation algorithm to a restoration problem. In [4] a method for detecting random defects (like dirt and blotches) in old motion pictures is presented. It is based on a connectivity criterion along the time axis. It is implemented using openings and closings by reconstruction. The algorithm was found to be robust and to work well (see [4]). However, small moving objects tend also to get disconnected along the time axis, and therefore, produce false detections. The motion compensation method presented here reduces this problem. In figure 2 true defects are indicated by a circle, while false detections are surrounded by a rectangle; as it can be seen, the application of the motion compensation algorithm has improved the result of the detection.

8. Conclusion

We have presented a general structure for a motion compensation algorithm based on block matching and a segmentation of the reference image.

It was implemented using mathematical morphology for segmentation, kriging for vector field interpolation and inverse kriging for vector field filtering.

The resulting motion compensation algorithm gives good results, as we have shown. Some of the shortcomings of the classical block matching compensation algorithm are avoided.

However, many points can still be optimized, like the placement of the experimental and information points.

Acknowledgements

This work was funded by the European LTR-ESPRIT/NOBLESSE project.

References

1. J. Cichosz and F. Meyer. Morphological multiscale image segmentation. In *Workshop on Image Analysis for Multimedia Interactive Services (WIAMIS'97)*, pages 161–166, Louvain-la-Neuve (Belgium), June 1997.
2. E. Decencière Ferrandière. *Restauration automatique de films anciens.* PhD thesis, Ecole Nationale Supérieure des Mines de Paris, November 1997.
3. E. Decencière Ferrandière, C. de Fouquet, and F. Meyer. Applications of kriging to image sequence coding. *Accepted for publication in Signal Processing: Image Communication*, 1997.

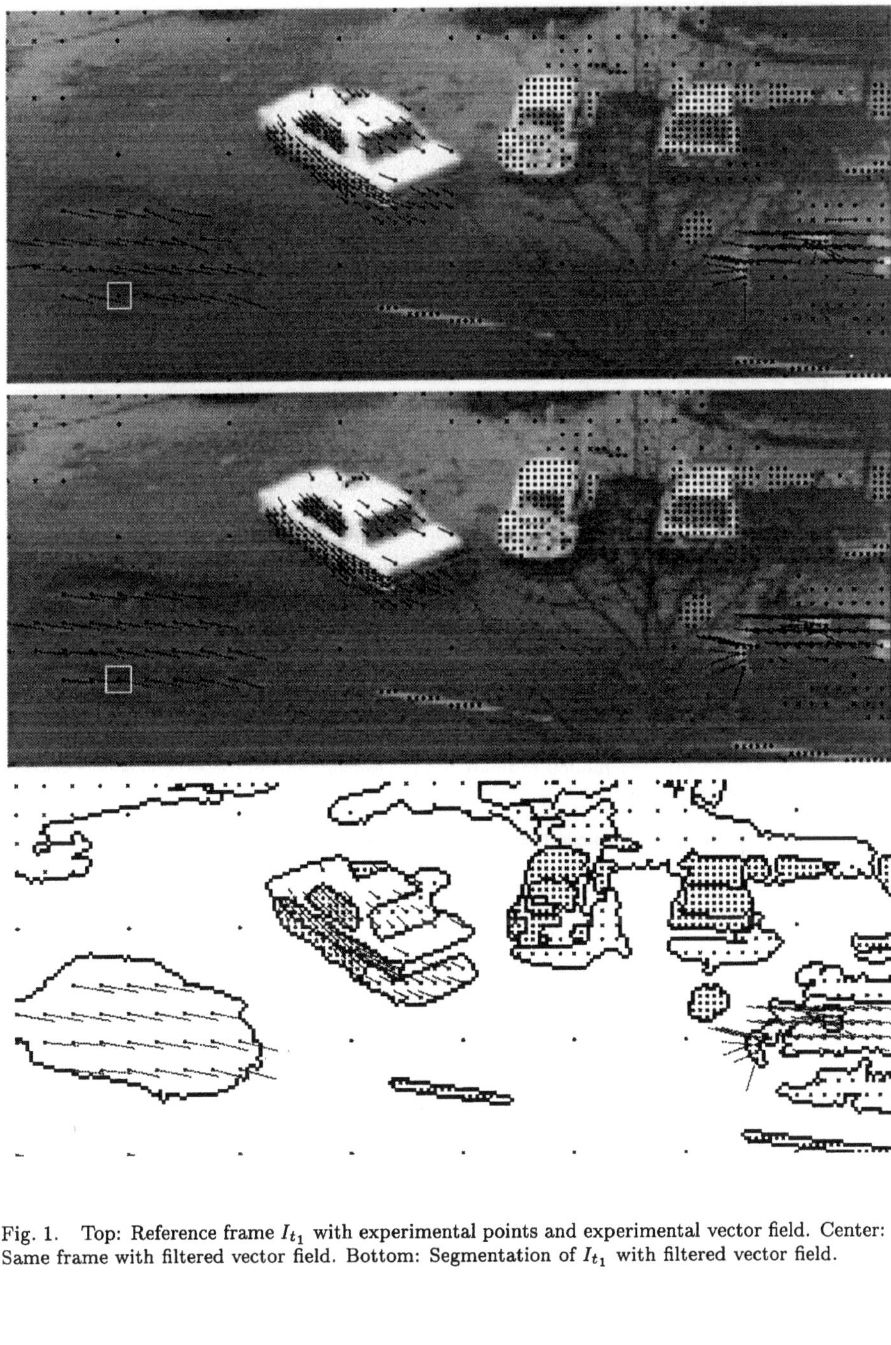

Fig. 1. Top: Reference frame I_{t_1} with experimental points and experimental vector field. Center: Same frame with filtered vector field. Bottom: Segmentation of I_{t_1} with filtered vector field.

Fig. 2. Left: detection of dark defects without motion compensation. Right: detection of dark defects with motion compensation. True defects are surrounded by a circle and false detections by a rectangle.

4. E. Decencière Ferrandière, S. Marshall, and J. Serra. Application of the morphological geodesic reconstruction to image sequence analysis. *Accepted for publication in IEE Proceedings: vision, image and signal processing*, 1997.
5. N. Diehl. Object-oriented motion estimation and segmentation in image sequences. *Signal Processing: Image Communication*, 3:23–56, 1991.
6. M. Ghanbari, S. de Faria, I. Goh, and K. Tan. Motion compensation for very low bit-rate video. *Signal Processing : Image Communication*, 7(4-6):567–580, November 1995.
7. B. Horn and B. Schunck. Determining optical flow. *Artificial Intelligence*, 17:185–203, 1981.
8. J. Jain and J. A.K. Displacement measurement and its application in interframe image coding. *IEEE Transactions on Communications*, 29(12):1799–1808, 1981.
9. J. Konrad and E. Dubois. Bayesian estimation of motion vector fields. *IEEE Transactions on Pattern Analysis and Machine Intelligence*, 14(9):910–927, 1992.
10. G. Matheron. *La théorie des variables régionalisées et ses applications.* Ecole des Mines, Paris, 1970.
11. G. Matheron. The intrinsic random functions, and their applications. *Advances in Applied Probability*, 5:439–468, 1973.
12. F. Meyer and S. Beucher. Morphological segmentation. *Journal of Visual Communication and Image Representation*, 1(1):21–46, September 1990.
13. J. Nieweglowski and P. Haavisto. Temporal image sequence prediction using motion field interpolation. *Signal Processing : Image Communication*, 7(4-6):333–353, November 1995.
14. W. Press, B. Flannery, S. Teukolsky, and W. Vetterling. *Numerical Recipes in C.* Cambridge University Press, 1988.
15. P. Salembier and J. Serra. Multiscale image segmentation. In *Proc. SPIE: Visual communications and image processing*, volume 1818, pages 620–631, Boston, USA, November 1992.

GENERIC ALGORITHMS FOR MORPHOLOGICAL IMAGE OPERATORS

A Case Study Using Watersheds

D'ORNELLAS, M. C.* and VAN DEN BOOMGAARD, R.
Intelligent Sensory Information Systems
University of Amsterdam, Faculty WINS
Kruislaan, 403 - 1098 SJ Amsterdam
E-mail: {ornellas,rein}@wins.uva.nl
FAX: (31)(20) 525-74-90

Abstract.
The aim of the present work is to combine both the advantages of generic programming approach and the wave front propagation interpretation to develop generic algorithms for morphological image operators. The watershed operator is implemented based on this framework and a performance evaluation, using processing time, is provided to compare the generic watershed implementation in contrast to the classical implementation using hierarchical queues.

Key words: generic algorithms, mathematical morphology, priority queues, wave front propagation, and watersheds.

1. Introduction

Mathematical morphology is a powerful and unified approach for geometrical shape analysis and description based on a complete lattice framework. A complete lattice is a partially ordered set $\mathcal{L}$ such that every subset $\mathcal{K}$ has an *infimum*, denoted by $\wedge\mathcal{K}$, and a *supremum*, denoted by $\vee\mathcal{K}$. From a theoretic point of view, a dilation, erosion, adjunction, opening, and closing are the most important algebraic notions in mathematical morphology to characterize operators on lattices.

The present study addresses the generic programming approach to morphological algorithm development. Generic programming reduces the task of creating morphological algorithms available for all combinations of images (i.e., lattices) and data structures. Since pixel values in an image form a partially ordered set and several morphological operators can be most efficiently implemented when this ordering is explicitly used, it seems natural to provide data structures to deal with the ordering. A priority queue structure is used to organize these pixels according to specified priorities while the wave front propagation interpretation is employed to handle the region growing process.

The organization of the rest of this paper is as follows. In section 2, we describe

* Formerly at Universidade Federal de Santa Maria, Departamento de Eletrônica e Computação, Santa Maria-RS, Brazil. Supported by CAPES Foundation under grant BEX 2780/95-0.

the generic programming approach. In section 3 we give a brief introduction to the wave front propagation interpretation of morphological image operators. We also discuss the data structure and the way it can be used to work with every image type. Then in section 4, a watershed segmentation algorithm is implemented based on wave front propagation and following a generic approach. Some code optimizations and enhancements are also provided. The evaluation performance and results of our generic watershed algorithm are shown in section 5. Conclusion and further research are given in section 6.

2. Morphological Algorithm Development

2.1. Generic Programming Approach

The object oriented programming paradigm is so strongly devoted on specifying interfaces for encapsulation that programmers sometimes keep out of sight from data structures and algorithms. Once a class interface has been built, the details of its implementation are said to be of little importance. However, some care must be taken whether or not it has been implemented correctly and efficiently and whether or not the algorithms and data structures have been chosen properly.

Generic programming gives and alternative to this approach. While object oriented programming takes abstract objects and gives them real world representations, generic programming does the same with algorithms. Generic programming also has a number of important advantages:

- it represents efficient algorithms independently of any particular data structure;
- providing an interface to a diverse set of data structures not only gives us the flexibility to choose appropriate ones, but also facilitates the development of more abstract algorithms.

The generic programming paradigm within the C++ context was first proposed by Stepanov and Musser [5] and became popular with the inclusion of their Standard Template Library (STL) into the C++ standard. In C++, the template mechanism is used to build generic algorithms by means of compile-time polymorphism so that the performance overhead is minimal and sometimes, non existent.

Generic algorithms naturally lead to the iterator concepts. Iterators are used in the STL to access data structures, rather than interacting with data structures directly. Moreover, all generic data structures provide the most advanced type of iterator they can implement efficiently. In this way, iterators serve as interfaces to both generic algorithms and generic data structures.

2.2. Working with Generic Images, Algorithms, and Data Structures

Consider software components as a three-dimensional space as shown in figure 1. One dimension represents the image data types (int, float, complex, ...), the second dimension represents the data structures (queue, linked-list, heap, splay, priority queue, ...) and the third dimension represents the morphological algorithms (erosion, dilation, reconstruction, watersheds, ...).

Based on this scenario, $x * y * z$ different versions of code have to be designed - an watershed algorithm for a priority queue of int, a reconstruction algorithm for a priority queue of double and so on. By using template functions that are parametrized

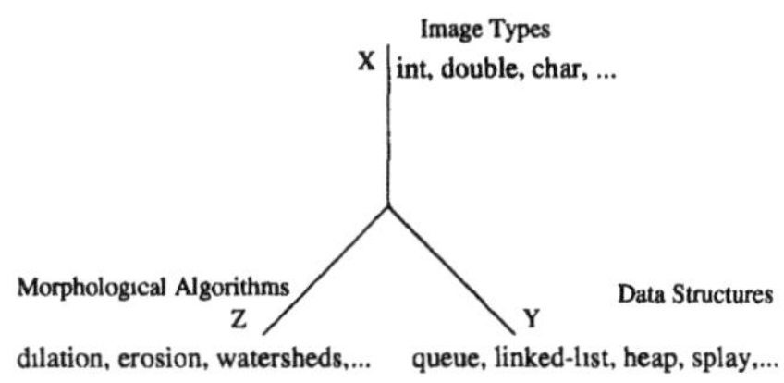

Fig. 1. Software Component Space

by data types, the x-axis can be dropped and only $y * z$ versions of code have to be designed, because there has to be only one priority queue implementation which them can hold objects of any data type. The next step is to make the algorithms work on different data structures - that means that a watershed algorithm should work on hierarchical queues as well as on priority queues, etc. Then, only $y + z$ versions of code have to be created.

With these notions in mind we can define a unique interface between morphological algorithms and the data structures so that the same algorithm implementation can be applied to any number of different image types. Our intention is to assure that one can have algorithms defined as generically as possible without loosing efficiency.

3. The Wave Front Interpretation of Algorithms

3.1. Wave Front Propagation Methodology

Some of the most powerful and often used algorithms in morphological image processing are based on wave front propagation. A wave front propagation interpretation implies that the action of the morphological operator is closely associated with the notion of connectivity. It also implies that the action is based on a kind of growing process, where the information is propagated through the image.

The wave front propagation interpretation of morphological operations was pioneered by Van Vliet and Verwer [8]. In wave front propagation algorithms, the following aspects can be distinguished:

- selection of the initial wave front (i.e., which pixels are selected for processing);
- selection of the points to which a point on the wave front is propagated. (i.e., the elementary wave front);

It should be noted that the main idea of a wave front propagation implies that all points on the wave front act like point sources. The envelope of all infinitesimal wave fronts emanating from all those individual points forms the macroscopic wave front. Also, observe that the morphological selection of minima (or maxima) corresponds with this natural selection of the envelope wave front.

3.2. Wave Front Propagation Algorithms Using Priority Queues

Vincent and Soille [7] were the first to use a priority-like ordered queue in mathematical morphology. Another approach to deal with morphological algorithms using priority queues was introduced by Beucher and Meyer [1] and is known as the hierarchical queue. The inner problem of these two approaches is that they are limited to a finite range of priorities and to discrete values. To overcome this problem a heap structure was used by Vincent [6]. However, a heap is inherently unstable, in the sense that items with the same priority are not handled in FIFO order. Breen and Monro stated that the stability can be achieved at a cost of one integer per item [2].

Recently, Noguet et al. [4] have shown that a priority queue structure can be efficiently used to handle the sorting step if a morphological operator can be treated as a wave front propagation from initial points inside a set of points sharing some features based on the connectivity rules.

A priority queue is an abstract data structure that can deal with an arbitrary number of items with priorities. In our implementation, a priority queue of objects is constructed which every object has its pixel's address and a set of keys $(k1, k2, \ldots, kn)$ that is sorted using the lexicographical ordering. An ordered pair $(k1, k2)$ is lexicographically less than $(k1', k2')$ if either $k1 < k1'$ or $k1 = k1'$ and $k2 \leq k2'$ [9]. Lexicographical ordering can be extended to many keys in a straightforward manner. A priority queue is then specified in table I:

TABLE I
Priority Queue Fundamental Methods

Method	Description
PQ q;	priority queue object declaration.
q.enq(key1, key2, ..., keyn);	put the tuple (key1,key2,...,keyn) on the queue.
q.deq();	get a tuple (key1,key2,...,keyn) from the queue.
q.empty();	checks whether there are items on the queue.

In the next section, we will concentrate in the watershed operator to demonstrate the generic programming approach. However, we should have in mind that the generic algorithm concept is planned to serve as a framework to the development of other morphological operators like reconstruction, skeleton, SKIZ, etc.

4. A Generic Watershed Algorithm Implementation Based on Wave Front Propagation

Watershed analysis has proven to be a powerful tool for many image segmentation problems. In the flooding scheme, water slowly rises within the topographic surface represented by an image, so that all point below water level are immersed. Holes are punched in the regional minima and the topography is flooded from below. As the water rises, more surface minima are pierced, which in turn starts more catchment basins. The catchment basins expand as the water rises and floods more points. When two floods from different catchment basins meet, a dam is built at these points to prevent the catchment basins from merging. After the surface is completely flooded, only the tops of the dams are visible and are treated as dividing lines. These watershed lines separate the surface into catchment basins.

In this paper, we consider images as mappings from $D_I \subset Z^2$ into Z. The grid $G \subset Z^2 \times Z^2$ defines the neighborhood operations (4 or 8 connectivity) and $N_G(p) = \{p' \in Z^2 \,|\, (p, p') \in G\}$ denotes the neighbors of a pixel p, according to a grid G. Also, M is the labeled markers image and f(p) corresponds to the grey-value of pixel p. All the algorithms presented in this paper extend the original image outside the window to cope with borders.

The bare-bones watershed implementation can be composed by two steps: a initialization step, and a data driven propagation step. The pseudocode is given as follows:

```
Initialization Step:
PQ q;
For (every p in domain D) {
      If ( M(p) != 0 and (exists (p') in N_G(p) : M(p') == 0))
          q.enq( p, f(p));
}

Data Driven Propagation Step:
While (!(q.empty())) {
          q.deq();
          For (every (p') in (N_G(p) ∩ D))
                If (M(p') == M(p))
                   q.enq( p', f(p'));
}
```

Fig. 2. Wathershed Algorithm - WS1

The algorithm layout in figure 2 can be found in Noguet et al. [4], Beucher and Meyer [1], and Najman and Schmitt [3]. The data structure used is an ordered queue in the sense that the method *q.deq();* returns the element from the queue that has minimal grey-value. Apart from its well defined structure, there are several drawbacks in this algorithm as stated below:

- the catchment basins are calculated. Although the watersheds themselves are readily calculated from the catchment basins, it is preferred that the watershed algorithm offers at least the possibility to calculate the watershed lines directly;
- in case there are several object tuples on the queue with the same grey-value and associated positions (i.e.,there is a *plateau* in the image), the above algorithm does not specify in what order the pixels on the plateau are processed. It would be preferred in case the pixels on a plateau are processed in accordance to their distance from the regional extremum they belong to, as stated by Beucher and Meyer [1], and Najman and Schmitt [3].

In order to be able to mark the contours of the catchment basins explicitly, the basic algorithm must be extended. This is shown in figure 3:

The extension of this algorithm is potentially rather time consuming because for each pixel p each neighbor p' is checked and also for each p' the neighbors p'' are checked. This increases the number of pixels to be checked from 9 to 81.

It should be noted though that the number of pixels to be checked can be reduced quite significantly. Coming from pixel p that is labeled with $M = M(p)$ we are about

```
Initialization Step:
01: PQ q;
02: For (every p in domain D) {
03:     If ( M(p) != 0 and (exists (p') in N_G(p) : M(p') == 0))
04:         q.enq( p, f(p));
05: }

Data Driven Propagation Step:
06: While (!(q.empty())) {
07:         q.deq();
08:         For (every (p') in (N_G(p) ∩ D))
09:             If (M(p') == 0)
10:               If (exists (p'') in N_G(p') : M(p'') != M(p))
11:                 M(p') = WSHED;
12:               else {
13:                   M(p') = M(p);
14:                   q.enq( p', f(p'));
15:               }
16: }
```

Fig. 3. Wathershed Algorithm - WS2

to label p' with M as well. But we can only do that in case this new pixel p' is **not** connected to a pixel with a label unequal to M. Since p, p' and p'' are connected within a 3×3 neighborhood and because all pixels in the 3×3 neighborhood of p are known to be unlabeled or with label M, we don't have to check all pixels. So, the pixels in $N_G(p)$ do not have to be taken into account when looking for the pixels p'' in the neighborhood of $N_G(p')$. With this optimization, we get a small change reducing the complexity from 81 to at most 45 or 36 on average. This algorithm optimization can be achieved by changing the line 10: in figure 3 by:

```
10: If (exists (p'') in (N_G(p') \ N_G(p)) : M(p'') != M(p))
```

Generic Algorithms can be extended by means of whether or not the wave front propagation is included in the initialization step. By changing the data driven propagation position and some conditional statements one can implement distinct operations like reconstruction, skeleton, and extrema detection easily.

5. Experimental Results

The performance of the watershed algorithms presented in this paper, are compared with the original algorithm by Vincent and Soille [7]. This is illustrated by applying these algorithms to a sample data set in figure 4(a) using a marker image as a reference. Describing the technique used to generate the marker image used would go beyond the scope of this paper.

The result of the first watershed algorithm can be seen in figure 4(c) where just the catchment basins are shown. Figure 4(d) shows both the catchment basins and the watershed lines according to the second implementation. Figure 4(e) shows the results by Vincent and Soille. The original image in figure 4(a) is then superimposed by the watershed lines from figure 4(d), to emphasize the segmentation of the cells in figure 4(f).

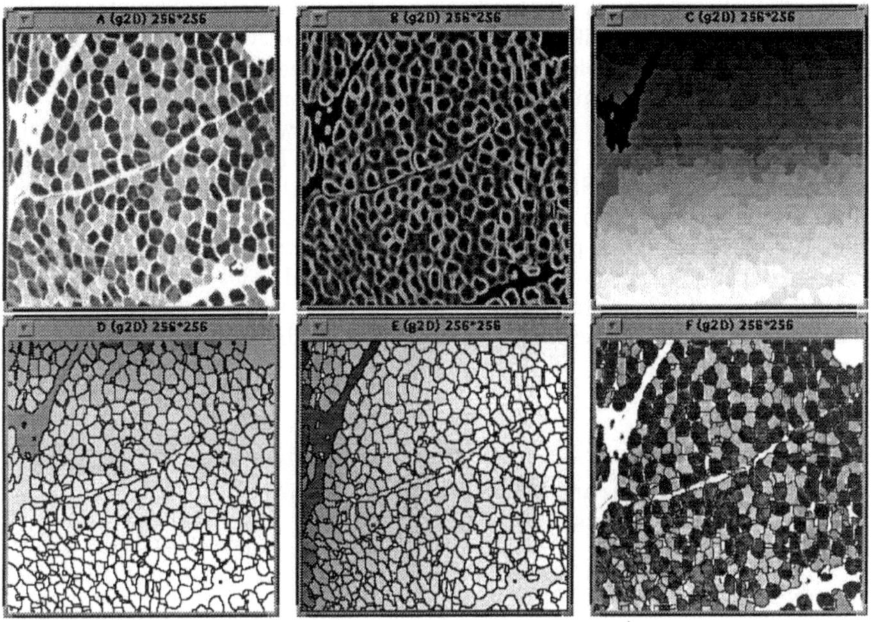

Fig. 4. Input image(a), Morphological gradient(b), Catchment basins(c), Watershed lines(d), Vincent and Soille implementation(e), and Input image with watershed lines(f).

The processing time for the implementations are given in table II for a 256x256 grey-level image based on int type, using 8-connected neighborhood. The watershed implementations are labeled by **WS1,WS2**, and the optimization by **WS3**. The last one ,**WS4**, is the implementation by Vincent and Soille previously cited. All the algorithms were tested on a Sun Sparcstation Ultra-1 running Solaris 2.5 (UNIX), using g++ compiler with optimizations enabled. Note that **WS4** also includes the detection of the regional minima of the image whereas **WS1,WS2** and **WS3** start from this minima.

TABLE II
Performance Evaluation (time in (s))

Algorithms	WS1		WS2		WS3		WS4	
Connectivity	8	4	8	4	8	4	8	4
Regional Minima Detection	0.32	0.32	0.32	0.32	0.32	0.32	—	—
Watershed Algorithm	0.63	0.68	0.53	0.48	0.39	0.32	0.82	0.64
Total Processing time	0.95	1.00	0.85	0.80	0.71	0.64	0.82	0.64

6. Conclusions and Further Research

In this paper we have shown that generic programming approach can be applied to morphological image operators based on wave front propagation like watersheds. Generic programming tools like STL let us apply a more evolutionary and experimental approach to morphological algorithm development. Using the proposed technique, reuse of morphological algorithms should become more easier.

A performance evaluation based on processing time were used to compare our generic watershed algorithms in contrast to the classical implementation. The watershed optimized version **WS3** proved to be faster than the classical one even when the regional minima detection time is considered.

Real images often possess plateaus. If we want thin watershed lines, the unpredictable propagation in the plateaus should be considered. This will be the subject of study for future implementations. One solution is to extend the basic algorithm so that when propagating the labels also the chamfer distance from the regional minimum is propagated.

Acknowledgements

Thanks are due to Niels Nes and Dennis Koelma for making constructive comments on the implementation aspects of our work.

References

1. S. Beucher and F. Meyer. The morphological approach to segmentation: the watershed transformation. In E. R. Dougherty, editor, *Mathematical Morphology in Image Processing*, chapter 12, pages 433–481. Marcel Dekker, New York, 1993.
2. E. J. Breen and D. H. Monro. An evaluation of priority queues for mathematical morphology. In J. Serra and P. Soille, editors, *Mathematical Morphology and Its Applications to Image Processing*, pages 249–256. Kluwer Academic Publishers, The Netherlands, 1994.
3. L. Najman and M. Schmitt. Geodesic saliency of watershed contours and hierarchical segmentation. *IEEE Transactions on Pattern Analysis and Machine Intelligence*, 18(12):1163–1173, 1996.
4. D. Noguet, A. Merle, and D. Lattard. A data dependent architecture based on seeded region growing strategy for advanced morphological operators. In P. Maragos, R. W. Schafer, and M. A. Butt, editors, *Mathematical Morphology and Its Applications to Image and Signal Processing*, pages 235–243, The Netherlands, 1996. Kluwer Academic Publisher.
5. A. Stepanov and D. Musser. Algorithm-oriented generic libraries. *Software - Practice and Experience*, 24(7):623–642, 1994.
6. L. Vincent. Morphological grayscale reconstruction in image analysis: Applications and efficient algorithms. *IEEE Transactions on Image Processing*, 2:176–201, 1993.
7. L. Vincent and P. Soille. Watersheds in digital spaces: An efficient algorithm based on immersion simulations. *IEEE Transactions on Pattern Analysis and Machine Intelligence*, 13(6):583–598, 1991.
8. L. J. V. Vliet and B. J. H. Verwer. A contour processing method for fast binary neighbourhood operations. *Pattern Recognition Letters*, 7:27–36, 1988.
9. D. Wood. *Data Structures, Algorithms, and Performance*. Addison Wesley, New York, 1993.

MINIMAL PATH ALGORITHMS FOR THE ROBUST DETECTION OF LINEAR FEATURES IN GRAY IMAGES

LUC VINCENT
Xerox Scansoft, Inc.
3400 Hillview Avenue, Palo Alto CA 94306, USA

Abstract.
Minimal paths are proposed as a powerful way to extract faint linear structures in noisy gray-level images. A *global* algorithm based on gray-weighted distance transforms is first proposed and shown to sometimes be a compelling alternative to watersheds. A more general *local* minimal path method is also introduced, together with an algorithm for extracting minimal path cost at each pixel location. The proposed algorithms are very efficient, and examples of applications are used to demonstrate their robustness.

Key words: Algorithms, Generalized Distance Transforms, Linear Features, Minimal Paths, Mathematical Morphology, Segmentation, Watersheds.

1. Introduction

A range of image analysis applications are concerned with the extraction of linear features, such as edges, fibers, fault lines, scratches, etc. A common way to approach the problem is to use a morphological *top hat* [7], followed by skeletonization. Traditional edge detection operators can also be used, and in some cases watershed based algorithms provide a powerful alternative [9, 1]. However, in a number of difficult cases, image complexity, faintness of linear structures to be extracted, or high noise level combine to make such approaches fail.

In the present paper, we first examine the difficult problem of extracting correlogram tracks from sonar data [4]. Even watersheds are unable to extract these tracks under low signal-to-noise (SNR) conditions. Analyzing the reasons for this failure leads us to propose a robust "generalized dynamic programming" method based on *minimal paths.* The core of this method is a gray-weighted distance function, a very efficient recursive implementation of which is described in Section 2.

The global nature of this minimal path algorithm is the very reason why it is so powerful, but it is also a limitation: to extract a linear feature L from a grayscale image I, some knowledge about the location of the end-points of L is required, which is not always practical. To address this limitation, a novel *local minimal path* transform is introduced in Section 3, together with an efficient algorithm to compute it. It provides a new and very robust way to detect linear features in images. We demonstrate this by using this method to extract faint and thin glass fibers from noisy grayscale images.

2. Watersheds and Global Minimal Path Algorithms

Let us first examine a difficult image analysis problem: the detection of correlogram tracks in sonar data. Examples of such data are shown in Figs. 1a–b. In each case, the image contains a continuous track, running from the top to the bottom of the image. In image 1a, signal-to-noise ratio (SNR) is relatively high, whereas in image 1b, it is very low and the track is barely visible.

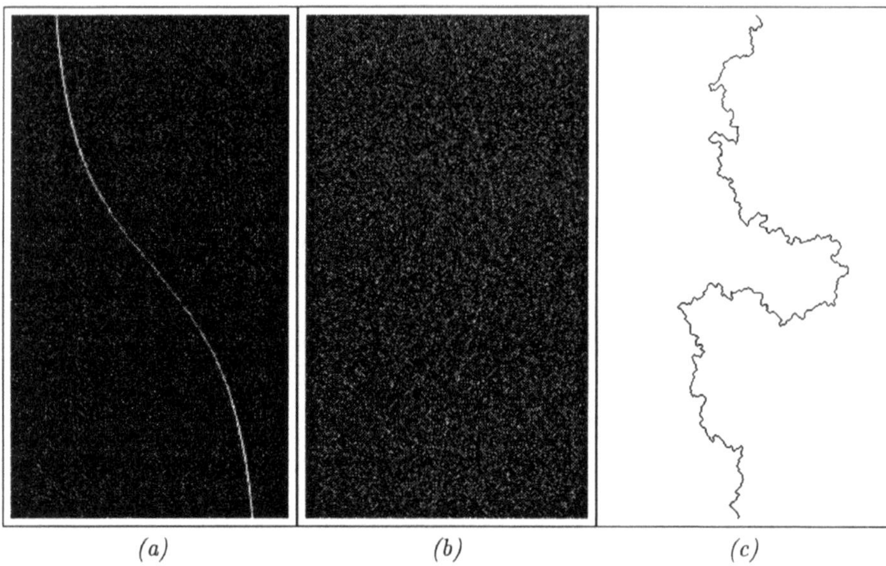

Fig. 1. (a)–(b) Correlogram tracks with high SNR and low SNR respectively. (c) Attempt to use watersheds for extracting track in (b).

In either case, one can think of the track to be extracted as a "crest-line" running from the top of the image to the bottom of the image. The powerful watershed transformation is therefore an obvious candidate for extracting such tracks: following the segmentation methodology exposed in [9, 1], one can constrain the watershed by using as markers the right side and the left side of the image. This is equivalent to saying that we wish to extract a crest-line separating the right side from the left side, that is, running from the top of the image to the bottom of the image.

This approach works extremely well for Fig. 1a, where the track is so well marked that a number of alternative methods would work equally well. However, the watershed approach fails on Fig. 1b: the crest line it extracts (See Fig. 1c) looks fairly "random" and bares little similarity with the desired correlogram track

This apparent contradiction can be explained by the very nature of the watershed algorithm. Examining the algorithm described in [9], or any other derived algorithm, one can get convinced of the following two facts:

1. The watershed line between two markers A and B in an image I highly depends on the position of the ***saddle points*** between these two markers. In other words, considering all the paths that join A to B with the minimal elevation (i.e., gray-

level) gain in I, the highest pixels along those paths are the saddle points, and their location is one of the main factors determining the location of the watershed line between A and B.

2. The criteria used to build the watershed are all solely based on gray level, and length of watershed lines is irrelevant.

In particular, fact #1 above means that a "disconnected" crest-line between A and B will not necessarily be extracted by the watershed operator: it all depends on the gray values of pixels in between the disconnected pieces. In addition, even if the watershed line does end up going over some of these disconnected pieces, there is absolutely no guarantee that it will do it in a way that minimizes its overall length. These two points are illustrated by Fig. 2.

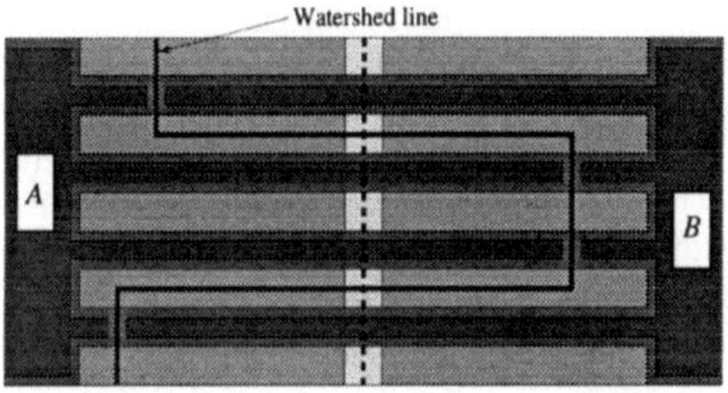

Fig. 2. The watershed line between A and B is very tortuous and does not connect all the pieces of the vertical bright feature in the center of the image. The vertical dashed line represents what one might want to extract as optimal crest line between A and B.

Length constraints can be introduced by using *global minimal paths* algorithms, as already proposed in [4]. Suppose that we wish to *connect* two regions A and B of a grayscale image I using the *shortest path(s) going preferentially over dark pixels.* This can be achieved by considering the neighborhood graph (typically, 4- or 8-connected grid) G of pixels in image I. In this graph, each edge E between two neighboring pixels p and q is assigned value $V_I(E) = V_I(p,q) = I(p) + I(q)$, or any other increasing function of $I(p)$ and $I(q)$, such as $\max(I(p), I(q))$ or $\min(I(p), I(q))$ (See Fig. 3). Each path P in this graph is now associated a *cost* $C_I(P)$, equal to the sum of all its edge values[1]. This provides a new metric in image I, for which the distance d_I between two pixels p and q is given by: $d_I(p,q) = \min\{C_I(P), P \text{ path between } p \text{ and } q\}$.

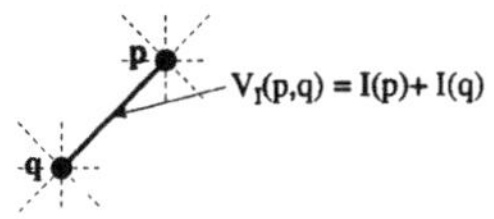

Fig. 3. Valuation of edge E between pixels p and q.

The problem comes down to extracting the path(s) of minimal cost between regions A and B, according to distance d_I. This is almost a classic *dynamic pro-*

[1] $C_I(P)$ should not be mistaken for the *length* $l(P)$ of path P: the latter is equal to the number of edges in P, that is, its number of pixels minus 1.

gramming problem, except that the paths considered here do not necessarily have to be monotonous. One can easily prove that a pixel p belongs to such minimal paths if and only if $d_I(p, A) + d_I(p, B) = d_I(A, B)$. To extract such paths, we can therefore proceed as follows:

- Compute *gray-weighted* distance transform to A, that is, for each pixel p, compute $d_I(p, A)$;
- Compute gray-weighted distance transform to B;
- Sum these two distance functions and threshold the result in order to only keep the pixels p such that $d_I(p, A) + d_I(p, B) = \min_q\{d_I(q, A) + d(q, B)\}$.

Note that the general method of using "forward" and "backward" distance functions is not new, and was already used, e.g., to extract the longest path in a binary set [6], or to detect potential fracture lines in porous media [8].

From an algorithmic point of view, the problem is reduced to computing gray-weighted distance transforms. This can easily be done by modifying the classic two-pass sequential distance function algorithm [5] so that: (1) edge cost is taken into account; (2) raster and anti-raster scans of the image are iterated until stability. More specifically, denoting by $N^+(p)$ (resp., $N^-(p)$) the neighbors of pixel p scanned *before* p (resp. *after* p) in a raster scan of image I, the algorithm proceeds as follows:

Algorithm: Generalized distance function to set A in image I

- `Initialize result image` J: $J(p) = 0$ `if` $p \in A$ `and` $+\infty$ `otherwise;`
- `Iterate until stability:`
  - `Scan image in raster order; For each pixel` p`, do:`
    $J(p) \leftarrow \min\{J(p), \min\{J(q) + V_I(p,q), q \in N^+(p)\}\}$;
  - `Scan image in anti-raster order; For each pixel` p`, do:`
    $J(p) \leftarrow \min\{J(p), \min\{J(q) + V_I(p,q), q \in N^-(p)\}\}$;

This algorithm typically converges in two or three iterations and is therefore extremely efficient. Applying this method to our correlograms, we use as set A the first line of the image, and as set B, the last line. In addition, since the extracted minimal paths are preferentially located on dark pixels, we invert the original image before computing the two generalized distance functions needed. This method, which proved to be very robust, is illustrated in Fig. 4. For more details, see [4].

3. Local Minimal Paths

Unfortunately, in a number of applications, the linear feature extraction problem cannot be formulated as the extraction of a minimal path running between known regions of an image, therefore the global approach described in the previous section cannot be used. A number of authors have proposed techniques for extracting linear structures in difficult images. For example, in [2], directional morphological operations are used to enhance images without destroying linear features and directional dilations are used to combine disconnected lines. In [3], a metric is proposed for line segments and used to provide robust connection criteria.

Such methods still fail to provide satisfactory solutions in some complex situations. In addition, in cases where the problem can indeed be formulated in terms of global minimal path, then the algorithm described in Section 2 proves to be far

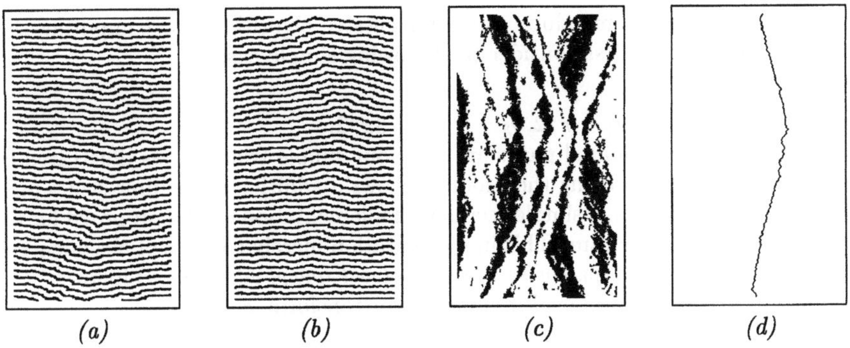

Fig. 4. (a) Gray-weighted distance function to top of image, (b) Gray-weighted distance function to bottom of image, (c) Binarized sum of distance functions, (d) Minimal path extracted between the top and the bottom of the image.

superior. This leads us to propose a *local,* or more exactly, *regional* version of this minimal path algorithm. The principle is as follows:

> *Assign to each pixel p of an image I the minimal cost $C_I(P)$ for all the paths P of a given length l originating in p and whose possible orientation and straighness are within a given range.*

Specifically: (1) the cost $C_I(P)$ is taken to be the sum of the pixel values along path P; (2) let $\theta_1, \theta_2, \ldots, \theta_n$ be a set of possible orientations for the desired paths, and let $\delta\theta$ be the maximal acceptable deviation: the paths P considered must be such that there exists an i such that the dominant orientation of P is within $[\theta_i - \delta\theta, \theta_i + \delta\theta]$.

This orientation constraint is critical: it prevents a candidate minimal path P from looping back on itself or going around pixel p in a circle! It also provides a means to specify the local degree of straightness of desired linear features. Finally, this constraint is key to making the local minimal path algorithm introduced in the present section computationally tractable. In the discrete world, this constraint is implemented as follows: draw a square of size $(2l+1) \times (2l+1)$ pixels centered at pixel p; then draw two discrete lines starting from p, at orientation $\theta_i - \delta\theta$ and $\theta_i + \delta\theta$ respectively. These two lines delineate a section of the square, as shown in Fig. 5. The straightness constraint above is then taken to mean that path P is entirely included in this region, which we refer to as "search cone".

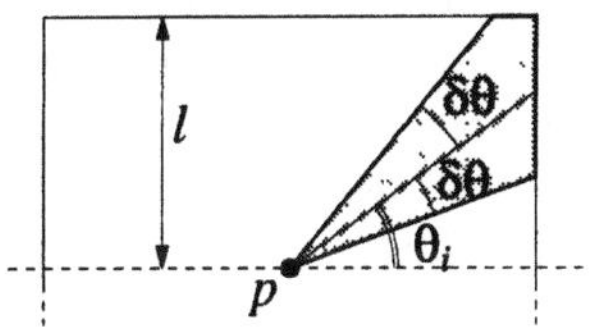

Fig. 5. Search cone: region defining the "orientation tolerance" for paths in direction θ_i.

Implementing this *local minimal path* transformation by considering each pixel p and exhaustively analyzing all the acceptable paths starting at p would be extraordinarily slow. We now introduce efficient recursive algorithms for computing this transformation. Let us first consider the following simple case:

- $\delta\theta = \pi/4$
- $i = 4$, with $\theta_1 = 0, \theta_2 = \pi/2, \theta_3 = \pi, \theta_4 = 3\pi/2$.

The search cones for minimal paths are now four simple triangles, as illustrated by Fig. 6. These structures being the l-fold Minkowski addition of the unit-size 4-pixel triangles with themselves, a simple recursive implementation can be proposed for the local minimal path algorithm. This implementation merely consists of a sequence of *global image shifts, global pixelwise minima, global image additions* as follows:

Algorithm: Local minimal path transform, simple case

- For $\theta = 0$, $\pi/2$, π and $3\pi/2$, do:
  - $I_\theta \leftarrow I$ (image copy); Then repeat l times:
    - For $\alpha = \theta - \pi/4, \theta, \theta + \pi/4$:
      - $J_\alpha \leftarrow I_\theta$ shifted by 1 pixel in direction $-\alpha$;
  - $I_\theta \leftarrow I_\theta + \min_\alpha\{J_\alpha\}$;
- Result: $J \leftarrow \min_\theta\{I_\theta\}$

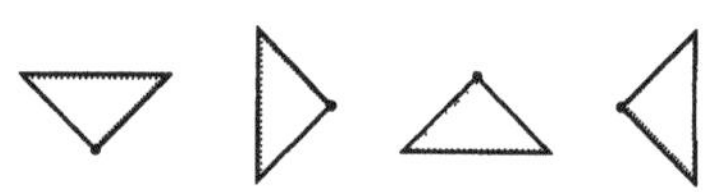

Fig. 6. Triangular "search cones" used when $\delta\theta = \pi/4$, $\theta_1 = 0$, $\theta_2 = \pi/2$, $\theta_2 = \pi$, $\theta_4 = 3\pi/2$.

In this particular case, for a given length l, the algorithm requires $12l$ shift operations, $8l$ pointwise minima operations, and $4l$ image additions. All these steps are very simple and run extremely quickly on any computer. The resulting algorithm is therefore very efficient. It can also be adapted to work with overlapping search cones, or thinner search cones, which often improves the quality of the results.

However, the narrowest search cones that can be used with this method are 45 degree angle triangles. In many cases, this is still not narrow enough. With search cones narrower than 45 degrees, a fully recursive algorithm cannot be used any more. This means that for a search cone of k pixels, the number of image shifts, pixelwise minima, and pixelwise additions needed will be $O(k)$: these simple global image operations being extremely fast, this is still acceptable. However, to minimize the total number of operations needed, we now introduce a *partly recursive* algorithm.

Let us describe this method using the search cone of Fig. 7. In order to extract the minimal path between the pointy side of the cone (left) and the right side, a breadth-first (First-In-First-Out) scan is used. During this scan, each pixel p is assigned: (1) the cost of the minimal path between it and the right side of the triangle; (2) a configuration label config(p). This configuration uniquely characterizes the number of "ancestors" of p in the search cone, as well as the configuration of these ancestors. For example, looking at Fig. 7, all the pixels with configuration "4" have exactly two ancestors in the search cone, each of them with label "2". Each time a new configuration is uncovered, the value of the corresponding pixel is computed from the value of its "ancestors", that is, its (one or two) neigbors that were previously scanned. Since this step is the same for all the pixels in the image, it is also parallelized by means of global image shift, pixelwise minimum, and pixelwise addition operations. However, when the configuration of the current pixel p has been seen previously, it means that the shift, minimum, and addition operations needed to compute the value of this pixel have already been performed! We therefore skip to the next pixel. In the example of Fig. 7, this technique enables us to uncover 10 pixels whose configuration is equivalent to that of a previously scanned pixel. We therefore cut down the total number of global image shifts, minima and additions by about 30% in this particular case.

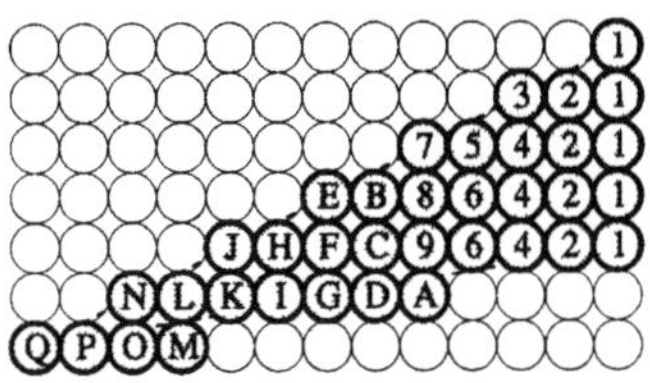

Fig. 7. Successive configurations $1, 2, \ldots, 9, A, B, \ldots, P, Q$ (26 of them) encountered when recursively scanning the pie slice shaped region starting from its right side.

Experimentation was conducted using 512×512 images on a 233 MHz Pentium PC. When using length parameter $l = 15$ pixels, $\delta\theta = 10$ degrees, and $\theta = 0$, 15, 30,..., 345 degrees, several hundred image shifts, pixelwise minima and additions were needed, but the minimal path image was computed in the order of only a minute. This time could still easily be cut by a factor of 5 to 10 by optimizing the speed of the global image operations the algorithm is currently based on.

To illustrate this method, we use the image in Fig. 8a, which represents a thin glass fiber in a rather noisy environment. To extract this fiber, one could carefully design a filter to attenuate background noise, then use a "top hat" followed by skeletonization and pruning... The local minimal path method provides a much more streamlined approach: by applying the algorithm directly on Fig. 8, we combine all the pixels in the fiber and greatly attenuate background noise, as shown in Fig. 8b.

As can be seen on this image, the side effect of this algorithm is that the fiber now appears thicker: we can extract it as one object by simple thresholding, as shown in Fig. 8a, but, extra work is needed to thin it down to unit-pixel thickness. To do this robustly, we compute the ***longest path*** inside the previously extracted object: this can be done using gray-weighted distance transforms geodesically inside the previously extracted object. The excellent result thus obtained is show in Fig. 8d.

4. Conclusion, Future Work

The use of paths of minimal cost for the extraction of linear structures in images was explored. A number of efficient algorithms were proposed to extract such minimal paths as well as to label each pixel p with the cost of the minimal path(s) originating in p. These algorithms are either ***global*** or ***regional*** in nature: to compute the output value of any given pixel, a large neighborhood—sometimes the entire image—is taken into account. As a result, these proposed methods are particularly robust, making it possible to extract even the faintest linear structures.

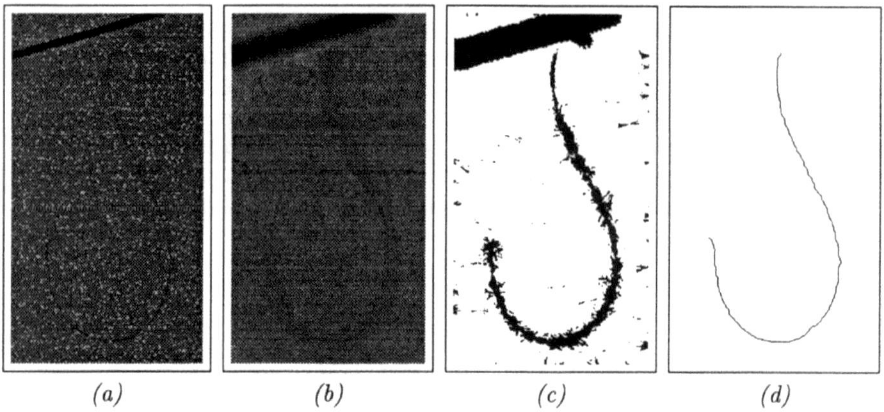

Fig. 8. (a) Original image of thin glass fibers; (b) Result of local minimal path algorithm, using paths of length 15 and overlapping search cones of 15 degrees; (c) Thresholded minimal path image; (d) Extracted fiber.

One of the most novel methods introduced in the paper is the local minimal algorithm described in Section 3. While the proposed technique is a vast improvement over a "brute force" approach, the resulting algorithm remains CPU intensive. Work is currently underway to further improve this algorithm, e.g., through the careful selection of the "search cones" it uses. In addition, relationships between watersheds and minimal paths are being further analyzed. For example, relaxation methods are being explored as a way to constrain minimal path extraction using markers, as is commonly done with watershed segmentation. This work will be reported in subsequent publications.

References

1. S. Beucher and F. Meyer. The morphological approach to segmentation: the watershed transformation. In E. R. Dougherty, editor, *Mathematical Morphology in Image Processing*, pages 433–481. Marcel-Dekker, Sept. 1992.
2. B. Kurdy and D. Jeulin. Directional mathematical morphology operations. In *5th European Congress For Stereology*, pages 473–480, Freiburg im Breisgau FRG, Sept. 1989. Acta Stereologica. Vol. 8/2.
3. P. F. Nacken. A metric for line segments. *IEEE Trans. Pattern Anal. Machine Intell.*, 15:1312–1318, 1993.
4. B. Rosen and L. Vincent. Morphological image processing techniques applied to detection of correlogram tracks. *U.S. Navy Journal of Underwater Acoustics*, 44(2):571–586, Apr. 1994.
5. A. Rosenfeld and J. Pfaltz. Distance functions on digital pictures. *Pattern Recognition*, 1:33–61, 1968.
6. M. Schmitt. *Des Algorithmes Morphologiques à l'Intelligence Artificielle.* PhD thesis, Ecole des Mines, Paris, Feb. 1989.
7. J. Serra. *Image Analysis and Mathematical Morphology.* Academic Press, London, 1982.
8. L. Vincent and D. Jeulin. Minimal paths and crack propagation simulations. In *5th European Congress For Stereology*, pages 487–494, Freiburg im Breisgau FRG, Sept. 1989. Acta Stereologica. Vol. 8/2.
9. L. Vincent and P. Soille. Watersheds in digital spaces: an efficient algorithm based on immersion simulations. *IEEE Trans. Pattern Anal. Machine Intell.*, 13(6):583–598, June 1991.

COMPUTATION OF THE OPENING TRANSFORM

S. SHEYNIN *, A. TUZIKOV and O. KRIVONOS
Institute of Engineering Cybernetics
Academy of Sciences of Republic Belarus
Surganova 6, 220012 Minsk, Belarus
email: {sheynin,tuzikov,oleg}@mpen.bas-net.by

Abstract.
In this report we consider morphological opening transform for binary, grey-level and multi-spectral images. The algorithm introduced by P. Nacken is extended for 3D binary images. It can be also applied for grey-level and multi-spectral images using umbra representation.

Key words: Opening transform, mathematical morphology

1. Introduction

The concept of granulometry was introduced by G. Matheron in 1967 [7] as a new tool for studying size distributions in porous media. The size of porous in such media was characterized using series of morphological openings with structuring elements of increasing size. Granulometry for grey-level images was investigated by Dougherty [3] and used for texture classification [1].

A very much related concept used for shape-size description is pattern spectrum (Maragos, 1989 [6]) which was introduced for both binary and grey-level images. A main problem for implementation of morphological spectrum is the necessity to compute morphological openings for structuring elements of different size and this is a time consuming problem. The problem can be solved efficiently for grey-level images and linear binary structuring elements as it was shown by Vincent [11]. Nacken [8] introduced the morphological opening transform for binary images which gives for every pixel in the images the radius of the maximal ball containing this pixel. He proposed also an algorithm for the computation of the opening transform for the chamfer metric structuring element. It is clear that morphological spectrum can be easily computed from the opening transform. In this note we extend the algorithm proposed by Nacken for the cases of 3D binary images as well as 2D grey-level and multi-spectral images.

Let us denote by $\overline{\mathbb{Z}}$ the set $\mathbb{Z} \cup \{-\infty\}$. A (discrete) grey-level image is a function $f : F \subset \mathbb{Z}^n \to V \subset \overline{\mathbb{Z}}$. Grey-level structuring element g is also a grey-level image $g : G \subset \mathbb{Z}^n \to V \subset \overline{\mathbb{Z}}$ with a small support G. Binary images X and binary structuring elements B are given by subsets of $\mathbb{Z}^n$. The basic mathematical morphology operations [4, 10] dilation $\oplus$ and erosion $\ominus$ are defined as follows

$$X \oplus B = \{x + b \mid x \in X,\ b \in B\},$$

* The authors were supported by the INTAS grant N 96-785

$$X \ominus B = \{y \in \mathbb{Z}^n | \; y + b \in X, \;\text{ for every } b \in B\},$$

$$(f \oplus g)(x) = \max_{z \in G,\; x-z \in F} \{f(x-z) + g(z)\},$$

$$(f \ominus g)(x) = \min_{z \in G,\; x+z \in F} \{f(x+z) - g(z)\}.$$

The opening operation $\circ$ for binary and grey-level images is defined by

$$X \circ B = (X \ominus B) \oplus B, \qquad f \circ g = (f \ominus g) \oplus g.$$

We will further use the following property of the opening operation

$$X \circ B = \bigcup_{x \in X} B_x \subseteq X,$$

where B_x denotes the shift of the set B at the element x.

Given a $k \in \mathbb{N}$ define $B^{\oplus k} = \underbrace{B \oplus \ldots \oplus B}_{k-\text{times}}$. If $k = 0$ then it is supposed that $B^{\oplus 0} = \{0\}$.

2. Binary images

We consider first the problem only for 3D binary images. Then we extend the approach for 2D grey-level and multi-spectral images using a subgraph representation.

Opening transform $O_{X,B}$ of the set $X \subset \mathbb{Z}^3$ by a structuring element $B \subset \mathbb{Z}^3$ is defined by

$$O_{X,B}(x) = \max\{k | \; x \in X \circ B^{\oplus k}\}.$$

It is true that

$$X \circ B^{\oplus k} = \{x \in X | \; O_{X,B}(x) \geq k\}.$$

Let us introduce also the notions of the *normalized pattern spectrum* [6] $p_{X,B}$ and *morphological distribution function* $F_{X,B}$. They are defined as follows for a bounded set X

$$p_{X,B}(k) = \frac{|(X \circ B^{\oplus k}) \backslash (X \circ B^{\oplus (k+1)})|}{|X|},$$

$$F_{X,B}(k) = \frac{|\{x | \; O_{X,B}(x) < k\}|}{|X|}.$$

Here $|X|$ denotes the cardinality of the set X.

The pattern spectrum and morphological distribution function can be easily computed from the opening transform.

Let $X, B \subset \mathbb{Z}^3$ and $0 \in B$. The main idea consists in using a metrical morphology concept. Define a (closed) ball $B(x,r)$ of radius $r \in \mathbb{Z}_+ = \{z \in \mathbb{Z} | \; z \geq 0\}$ centered in a point x by

$$B(x,r) = \{x\} \oplus B^{\oplus r}, \qquad B(x,0) = \{x\}.$$

The function $\rho_{X,B} : \mathbb{Z}^3 \to \mathbb{Z}_+$ defined by

$$\rho_{X,B}(x) = \min\{r \in \mathbb{Z}_+ | \; B(x,r) \not\subset X\}.$$

is called an external *distance transform*.

It is clear that $\rho_{X,B}(x) = 0$ iff $x \notin X$. It is easy to see that the maximal radius of a ball inscribed into X and having the center at $x \in X$ equals $\rho_{X,B}(x) - 1$.

A point $x \in X$ is called a center of a *maximal ball* if there doesn't exist a point $y \in X$ such that $B(x, \rho_{X,B}(x) - 1) \subset B(y, \rho_{X,B}(y) - 1)$.

The idea of the algorithm proposed in [8] and implemented in this paper consists in performing the following steps:

1. Finding distance transform of the set X;
2. Finding a set of points containing maximal ball centers;
3. Opening transform computation using the formula:

$$O_{X,B}(x) = \max\{r|\ x \in B(y,r),\ r = \rho_{X,B}(y) - 1\}.$$

The last formula is obvious since the value of opening transform at the point x equals the radius of the maximal ball containing the point x and inscribed into the set X.

Now we give the detailed description of the algorithm.

The algorithm of finding distance transform for the fixed structuring element is well-known [9] and the description of it does not depend on the image dimension. It performs two scans of the image (in forward and backward directions). Therefore first we choose the forward direction of scanning image voxels (points of the set X). Let us, for example, scan the voxels according to increasing order of the first coordinate then of the second and third one.

For every point $x \in X$ its neighborhood $(\{x\} \oplus B) \cap X$ is partitioned into two parts. The first one consists of points which are scanned before the point x and the second one consists of points which are scanned after x. We will call the points from the first part by preceding neighbors and from the second part by consequent neighbors. The point x itself is not included into the parts.

Algorithm 1 External distance transform of the set X for the fixed structuring element.

1. *Assign* $r_i := \infty$, *if* $x_i \in X$ *and* $r_i := 0$ *otherwise;*
2. *For every point* x_i *under the forward direction scan assign* $r_i := r$ *if* $r = \min\{r_j + 1|\ y_j$*–preceding neighbor of* $x_i\} < r_i$*;*
3. *For every point* x_i *under the backward direction scan assign* $r_i := r$ *if* $r = \min\{r_j + 1|\ y_j$*–consequent neighbor of* $x_i\} < r_i$*.*

The value r_i *is the result of the distance transform at the point* x_i*.*

Let $\rho_{X,B}$ be the computed distance transform of the set X. Find in the set X a subset of points containing maximal ball centers.

Let $\check{B} = \{x|\ -x \in B\}$. If $y \in \{x\} \oplus \check{B}$ and $\rho_{X,B}(y) > \rho_{X,B}(x)$ then it is obvious that $B(x, \rho_{X,B}(x)) \subset B(y, \rho_{X,B}(y))$. Therefore all maximal ball centers are local extrema of the distance transform. Converse in general is not true. All local maxima can be easily found and we will not make additional computations to check which of them are centers of maximal balls.

Algorithm 2 *Maximal ball centers for the fixed structuring element.*

1. *Scan all points $x_i \in X$ in the arbitrary order. For every point x_i check:*
   - *If for some point $y_j \in (\{x_i\} \oplus \check{B}) \cap X$ holds $\rho_X(y_j) > \rho_X(x_i)$ then mark this point as one which is not a maximal ball center.*
2. *Denote the set of all not marked points by M. This set contains all maximal ball centers.*

Let M be the found set of distance transform local maxima. Now it is possible to formulate the whole algorithm of finding the opening transform.

Algorithm 3 *Opening transform for the fixed structuring element.*
1. *Compute distance transform of the set X;*
2. *Find the set M of distance transform local maxima;*
3. *Sort the points from M in the nondecreasing order of distance transform values;*
4. *Assign $O_i := 0$ for all $x_i \in X$;*
5. *For every point $y_j \in M$ (in the nondecreasing order of $\rho_{X,B}(y_j)$) perform:*
   - *For every $x_i \in B(y_j, \rho_{B,X}(y_j) - 1)$ assign $O_i := \rho_X(y_j) - 1$.*

It is difficult to evaluate the exact time complexity of this algorithm. The most time consuming is the step 5. Its time complexity is proportional to the total number of points contained in several maximal balls (each point is countered for every maximal ball which contains this point) and in 3D case is between N^3 and N^6 (N being linear image size). Therefore the exact time complexity depends on the intersection degree of the maximal balls.

The algorithm implemented in this note is valid for binary images of arbitrary dimension.

3. Grey-level images

Now we consider the opening transform for 2D grey-level images.

Given a $k \in \mathbb{N}$ define as before $g^{\oplus k} = \underbrace{g \oplus \ldots \oplus g}_{k-\text{times}}$. If $k = 0$ then suppose that the function $g^{\oplus 0}$ is defined only in a single point 0 and $g^{\oplus 0}(0) = 0$.

The problem addressed here is:

$$\text{find the opening } f \circ g^{\oplus k} \text{ for all } \quad k \in \mathbb{N}.$$

Therefore it is necessary to find the map $k \mapsto f \circ g^{\oplus k}$.

To solve the problem using a finite number of iterations we will suppose that $f, g : \mathbb{Z}^2 \to \mathbb{Z}_+ \cup \{-\infty\}$ are functions with a bounded support. Besides it is supposed that $g(0) \geq 0$ and $G \neq \{0\}$.

Let us formulate now the problem in terms of binary morphology. A grey-level image can be represented as a 3D binary image using its umbra (subgraph). Let $U = \{(x,t)| -\infty < t \leq f(x)\}$, $B = \{(y,\tau)| -\infty \leq \tau \leq g(y)\} \subset \mathbb{Z}^3$, $x, y \in \mathbb{Z}^2$, $t, \tau \in \mathbb{Z}$. Then $U \circ B^{\oplus k}$ – is an umbra representation of $f \circ g^{\oplus k}$. Finding grey-level image from the umbra representation can be obviously done by

$$f(x) = \max\{t \geq 0| \ (x,t) \in U\}.$$

Therefore to compute the opening transform for grey-level images we can first find umbra of the image and structural element and then apply the algorithm described earlier in Section 2 for 3D binary images. It is possible to modify the algorithm in this case considering only bounded from below subgraphs. Another option is to introduce a grey-level opening using binary opening of non-negative subgraphs of the function and the structuring element. It is quite natural to try the same approach for multi-spectral images. However, in this case some additional difficulties arise which are due to the fact that dilation of subgraphs is not a subgraph in general. In fact this situation takes place already for continuous grey-level images and is discussed in [5, Sec.11.6] in the general lattice framework for mathematical morphology. We will treat the multi-spectral case more detailed in the next section.

At the Fig. 1 we show the original grey-level image and the thresholds of the 3D grey-level image (containing the results of the opening transform) at the levels 10,30,50 of the z-coordinate. Here z-coordinate corresponds to the grey-level values in the original image and is used the $3 \times 3 \times 3$ structuring element. The graph of the normalized pattern spectrum is presented at the Fig. 2.

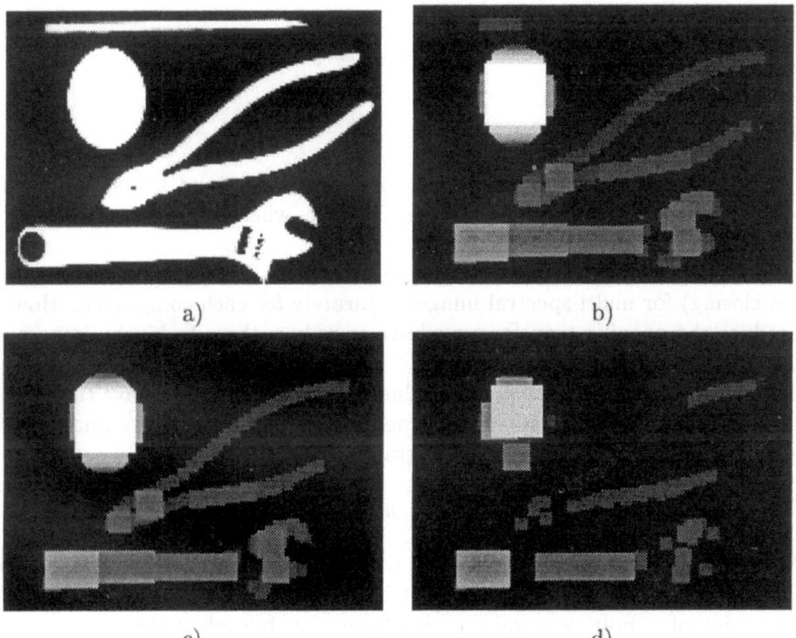

Fig. 1. Original grey-level image a); Thresholds of the 3D grey-level image (containing results of the opening transform) at levels $10, 30, 50$ of the z-coordinate are shown at b),c),d) respectively. Here $3 \times 3 \times 3$ structuring element is used.

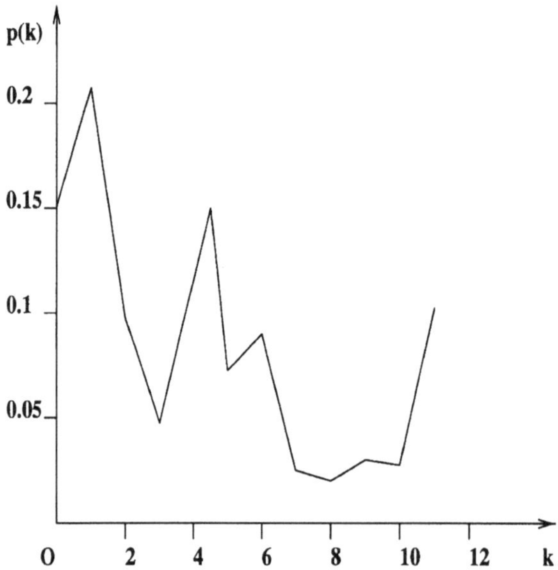

Fig. 2. Pattern spectrum for the grey-level image presented at the Fig. 1 a).

4. Multi-spectral images

By a (discrete) multi-spectral grey-level image we understand a function $\mathbf{f} : F \subset \mathbb{Z}^n \to V \subset \mathbb{Z}^m$. Therefore $\mathbf{f} = (f_1, f_2, \ldots, f_m)$, where each component f_i is a (discrete) grey-level image.

It is possible to introduce basic morphological operations (dilation, erosion, opening and closing) for multi-spectral images separately for each component. However, to introduce the opening transform we have to reduce them to binary images as it was done for ordinary grey-level images.

Further we suppose that all f_i take values from the set $\overline{\mathbb{Z}}$ and have the bounded support. We can now represent a multi-spectral image $\mathbf{f}$ as a binary image $U(\mathbf{f})$ of dimension $m + n$ called further by a subgraph of $\mathbf{f}$:

$$U(\mathbf{f}) = \{(x, u) |\ x \in F,\ u \in \overline{\mathbb{Z}}^m,\ u_i \leq f_i(x)\}. \tag{1}$$

The intersection of a subgraph $U(\mathbf{f})$ with a m-dimensional plane $x = c$ is a parallelepiped (possibly empty) extended by points with a coordinate $-\infty$. The reconstruction of a multi-spectral function from its subgraph is obvious:

$$f_i(x) = \max\{u_i |\ (x, u) \in U(\mathbf{f})\} \text{ for all } i. \tag{2}$$

Let us see how the subgraphs are changed under the morphological operations. Let $\mathbf{f}, \mathbf{g}$ be a multi-spectral image and a structuring element. Their binary representations (subgraphs) are denoted by $X = U(\mathbf{f})$ and $B = U(\mathbf{g})$, respectively. The intersection of the set $X \oplus B$ with a m-dimensional plane $x = c$ is in general the union of parallelepipeds. At the Fig. 3 it is shown the union of parallelograms for the case when $m = 2$. Therefore the set $X \oplus B$ is not a subgraph of a multi-spectral function, and it doesn't define uniquely a multi-spectral function. One of the possibilities to overcome this problem is to find the smallest subgraph U' containing the set $X \oplus B$. Then the multi-spectral function $\mathbf{f}'$ reconstructed from U' by the formula (2) can be called the dilation $\mathbf{f} \oplus \mathbf{g}$. It is not difficult to see that the smallest subgraph U' can be found from all subgraphs $U(f_i \oplus g_i)$ as follows

$$U' = \{(x, u) |\ x \in F \oplus G,\ u_i \leq (f_i \oplus g_i)(x)\}.$$

Here G is the support of $\mathbf{g}$.

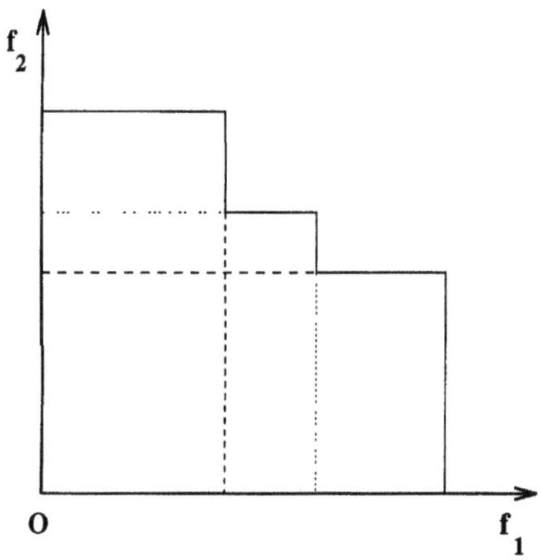

Fig. 3. The intersection of the set $U(\mathbf{f}) \oplus U(\mathbf{g})$ with a plane $x = c$ is represented in general for $m = 2$ as the union of parallelograms and therefore $U(\mathbf{f}) \oplus U(\mathbf{g})$ is not a subgraph of a multi-spectral function.

There are known in the literature other possibilities to introduce morphological operations for multi-spectral images. For example, the case of morphological operation for color images is discussed in [2]. Dilation and erosion operations are introduced there using graphs of functions. The main problem for the case of color images as well as a general case of multi-spectral images is to introduce properly ordering relation on the set of vector values. In [2] the scalar functional is used to order vector values and the maximal value is chosen based on this ordering relation.

For the morphological opening we have even (and what is a bit surprising) a more simple situation, since the set $U(\mathbf{f}) \ominus U(\mathbf{g}) \oplus U(\mathbf{g})$ is a subgraph of a some multi-spectral function. For, we have from the property of the morphological opening

$$\begin{aligned} X \circ B = & \\ & \bigcup_{\xi,\tau} (B_{\xi,\tau} \subset X) = \\ & \bigcup_{\xi,\tau} \Big(\{(y + \xi, v + \tau) |\ v_i \leq g_i(y)\} |\ v_i + \tau_i \leq f_i(y + \xi) \end{aligned}$$

$$\text{for all } y \text{ and } v_i \le g_i(y)\Big) =$$

$$\bigcup_{\xi} \{(y+\xi, v+\tau(\xi)) |\ v_i \le g_i(y)\},$$

where $\tau_i(\xi) = \min_y(f_i(y+\xi) - g_i(y))$. It follows from the last formula that for the computation of the morphological opening for a multi-spectral function by a multi-spectral structuring element it is sufficient to compute it for every grey-level component separately, i.e.,

$$(\mathbf{f} \circ \mathbf{g})(x) = ((f_1 \circ g_1)(x), \cdots (f_m \circ g_m)(x)).$$

Therefore we can define the opening transform for multi-spectral images containing m components of dimension n as a grey-level image of dimension $m+n$ as follows:

$$O_f(x,u) = O_{U(f)}(x,u) = \max\{k|\ (x,u) \in U(f) \circ B_k\}.$$

Here B_k is a subgraph of $g^{\oplus k}$.

Moreover, for multi-spectral images we have the same relation between opening transform and morphological opening as it was for ordinary grey-level images.

$$U(f) \circ B_k = \{(x,t) \in U(f)|\ O_{U(f)}(x,t) \ge k\}.$$

This opening transform can be computed via opening transforms for components of the multi-spectral image:

$$O_{U(f)}(x,u) = \min_i O_{f_i}(x,u_i).$$

The notions of the morphological spectrum and distribution function can be also introduced by analogy with binary images.

References

1. Y. Chen and E. Dougherty. Gray-scale morphological granulometric texture classification. *Optical Engineering*, 33(8):2713–2722, 1994.
2. M.L. Comer and E.D. Delp. Morphological operations for color image processing. Report EDICS: IP 1.7, School of Electrical Engineering, Purdue University (to be published).
3. E. R. Dougherty. Euclidean gray-scale granulometries: representation and umbra inducement. *Journal of Mathematical Imaging and Vision*, 1(1):7–21, 1992.
4. R.M. Haralick, S.R. Sternberg, and X. Zhuang. Image analysis using mathematical morphology. *IEEE Trans. on Pattern Anal. and Machine Intell.*, 9:532–550, 1987.
5. H.J.A.M. Heijmans. *Morphological Image Operators.* Academic Press, Inc, 1994.
6. P. Maragos. Pattern spectrum and multiscale shape representation. *IEEE Trans. on Pattern Anal. and Machine Intell.*, 11(7):701–716, 1989.
7. G. Matheron. *Eléments pour une Théorie des Milieux Poreux.* Masson, Paris, 1967.
8. P. F. M. Nacken. *Image Analysis Methods Based on Hierarchies of Graphs and Multi-Scale Mathematical Morphology.* PhD thesis, University of Amsterdam, The Netherlands, 1994.
9. A. Rosenfeld, J.L. Pfalz. Distance function on digital pictures. *Pattern Recognition*, 1, 33–61, 1968.
10. J. Serra. *Image Analysis and Mathematical Morphology.* Academic Press, London, 1982.
11. L. Vincent. Fast grayscale granulometry algorithms. In: *Mathematical Morphology and Its Applications to Image Processing* (J. Serra and P. Soille, Eds.). Kluwer Academic Publishers, 265–272, 1994.

FAST DILATION AND SKELETONIZATION OF COMPRESSED BINARY PICTURES

V.M. KIIKO and M.I. SCHLESINGER
International Research and Training Centre of Information Technologies and Systems of National Akademy of Sciences of Ukraine
Prospekt Akademika Glushkova, 40,252022,Kiev,Ukraine
E-mail: schles@image.kiev.ua

Abstract. Effective algorithms for dilation (erosion) and skeletonization of binary pictures, represented in compressed form, are described. The proposed line-by-line algorithms are fast and require small memory area, which does not depend on the size of a processed picture. The results of implementation the algorithms for processing of line drawings are presented.

1. Introduction

Let vertices of those polygons, that constitute the contour of a binary picture $v(i,j) \to \{0,1\}$, be called corners. There are eight types (Fig. 1,a) of the corners and the corners of types 1-4 be called convex and of types 5-8 be called concave ones. The set of coordinates of these corners represents one of the lossless compression forms [8,9] of the binary picture ensuring restoring of the initial pixel representation of this picture.

The corner representation (CR) can be effectively used, for example, for translation, scaling or inversion of the binary picture and for adding or subtraction of two binary pictures. For performing the first two operations only coordinates of the corners in the picture must be changed in some way. For inversion of the picture every of four corners in vertices of the rectangle, bounding the picture, must be added to its corner representation in the absence of this corner and removed otherwise.

Let t be a center of some square consisting of 4 pixels (Fig. 1,b), $\rho_1 - \rho_4$ be the values of these pixels and $d_1 = \rho_2 - \rho_3$, $d_2 = \rho_1 - \rho_4$, $dd = (d_1 - d_2)$ be

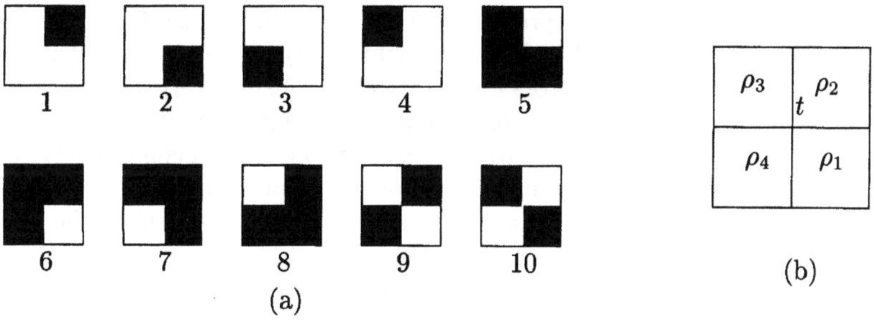

Fig. 1. The types of the corners (1-8) and special places (1-10) in the point t of the binary picture.

correspondingly the first, second and double differences of the pixel's values around this point. Therewith the point t is the corner in the binary picture if $|dd| = 1$ in this point. Some point be called also a special point (SP) in the binary picture, if this point is either the corner or the center of patterns 9-10 in Fig. 1,b and be the SP in the multilevel picture if $|dd| > 0$. A set $\{x, y, dd\}$ of coordinates and double differences of SPs be called a difference representation (DR) of the picture. By sum of two pictures is meant a result of summing of their pixel's values and by sum of two DRs is meant a merging of SPs in these pictures. During this merging instead of some coinciding SPs only one SP with the same coordinates and total double difference must be introduced. It is true also that DR of the sum of two pictures is the sum of DRs of these pictures.

The algorithm for adding of two binary pictures v_1 and v_2 in the corner form consists of the following two parts: 1) summing of DRs of these two pictures and 2) line by line retrieval of the multilevel picture $v = v_1 + v_2$ from its difference representation; binarization of the picture v by the threshold equal to 1 and the corner coding of the result binary picture. In greater detail this algorithm will be considered in Section 2.

An effective algorithm for tracing the contour of the compressed picture and removing the objects with area (number of pixels) less than given threshold is described in [10]. The tracing of the contour is carried out in [10] during one pass scanning of the picture and deletion of the objects is fulfilled by removing their corners from initial data. Considered above examples prove that corner representation can be effectively used in different algorithms but of special interest is the development of the effective algorithms for dilation(erosion) and skeletonization of binary pictures as most time consuming and most often used in the practice.

Dilation of the picture v by mask image x_t with representative point t can be represented as $v_e = \bigcup_{t \subset v_b} x_t$, where v_b is a set of black pixels of the picture v. Usually the number of operations for dilation of the picture depends on the numbers of black pixels in the original and mask images. In Section 2 the effective algorithm for dilation of binary pictures in compressed corner form is described and computational complexity of this algorithm depends on the numbers of the corners in the pictures v and x_t.

Skeletonization [1-7] provides significant geometric features extracted from the binary image data. These features represent the lines that approximate the centers of so called maximal circles or squares, inscribed in the patterns of the picture under processing. Two main approaches are used for skeletonization of the binary pictures: the algorithms [2,5,7] of the first (thinning) approach use different variants of connected deletion operators for iteratively performed shrinking of elongated binary objects to thin line-patterns; on each iteration every boundary(contour) pixel is inspected and those pixels, that are not required to maintain connectivity or ends of lines, are erased. The algorithms [3,4,6] of the second (not iterative) approach define the skeleton points from distance measurements with respect to the opposite boundary points of the regions.

In Section 3 the modified version of the skeletonization algorithm [6] is described. The algorithm implements the second approach and defines the skeleton during one pass sequential scanning of the picture. The computational complexity of the

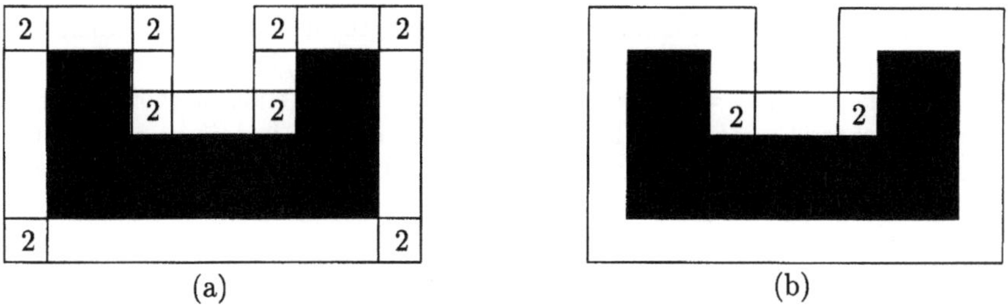

Fig. 2. Two multilevel presentations of the dilated picture.

algorithm depends on the number of the corners in the picture and does not depend, instead of [3], on the size of maximal squares in the picture. Section 4 contains some experimental results and Section 5 some concluding remarks.

2. Description of the dilation algorithm

Let us define corner line as a boundary of two (upper and lower) neighbouring picture's rows that contains two or more corners of this picture. The definition of the line of special places in the picture is similar. The algorithm for dilation of binary pictures consists of three main parts.

The first part of the algorithm consists in transformation of the CR of the picture into its DR, which is carried out during bottom-up and from the left to right scanning of corner lines in the picture. Let us suppose that the most left and bottom point in the picture has coordinates (0,0), and left column and bottom row of the picture are known. Let T_i be a set of points, ordered along some, say, i-th corner line, such that every point is either the corner or the end of a run of pixels (having the same value 0 (white) or 1 (black)) in the lower row of the picture. During processing of this corner line all points $t \subset T_i$ are successively considered and for every of these points the values ρ_1, ρ_2 of to right neighbouring pixels (Fig. 1,b) on the basis of values ρ_3, ρ_4, $|dd|$ and $d_2 = \rho_1 - \rho_4$ are defined: $\rho_1 = \rho_4 + d_2$ and $\rho_2 = 1$, if either a) $|dd| = 1$ and $\rho_1 + \rho_3 + \rho_4$ is the even value or b) $|dd| = 0$ and $\rho_1 + \rho_3 + \rho_4$ is the odd value, and $\rho_2 = 0$ otherwise. The retrieved values $\rho_1 - \rho_4$ are then used for defining parameters of SP (if any) in the current point t and also for run length coding of the upper row of the picture. Before processing of the next $(t+1)$-th point the values ρ_3, ρ_4 are set equal to ρ_1 and ρ_2 correspondingly, and during processing of the next $(i+1)$-th corner line the run length coding data of the upper row are used, being the retrieved data for the lower row during processing of the previous corner line.

The second part of the algorithm consists in retrieval of the DR of some multilevel picture v, which after thresholding by 1 represents the dilated original binary picture. Every connected component of the picture v can be usually represented as a sum of original binary picture's component with a some additional picture. Let us suppose that the mask image represents a black square and its representative point is a center of this square. In this case the additional picture is a strip, which can be represented as a sum of intersected horizontal and vertical rectangles (Fig. 2,a).

Let l_m be the size of mask image, l_c be the length of some line segment of the component's contour and l_r be the length of the rectangle, corresponding to this line segment. Then $l_r = l_c + l_m$, if the corners at the ends of this line segment are convex ones; $l_r = l_c + l_m/2$, if one of these corners is convex and another is concave one, and $l_r = l_c$, if both corners are concave ones.The DR of every of these rectangles consists of four SPs in vertices of this rectangle. In Fig. 2,a the original binary picture and multilevel presentation of expansion of this picture by one pixel are shown. The pixels of the multilevel picture have the value equal to 2=1+1 in intersections of neighbouring rectangles in this figure. For performing the second step of the algorithm it is necessary to define SPs of the additional picture and to merge these SPs with SPs of the initial picture.

Another way for presentation of the additional picture consists in summing pixel's values of every two rectangles, that correspond to the same concave corner, and in joining of all other rectangles. In Fig. 2,b an example of corresponding multilevel representation of dilated picture v is shown. In this figure the pixels in intersected regions of summing rectangles have also the value equal to 2. The DR of the picture v can be received by 1) moving of every SP, that corresponds to convex corner in the original picture, on the distance of mask image's diagonal in certain direction, corresponding to the type of this corner; 2) assigning to every SP: (x, y, dd_0), that corresponds to concave corner of the type t_q, new double difference value $dd_1 = -dd_0$ and introducing of two new SPs: $(x, y+y_1, dd)$ and $(x+x_1, y, dd)$, where a) $dd = -1$, $y_1 = l_m/2$, $x_1 = l_m/2$, if $t_q = 5$; b) $dd = 1$, $y_1 = -l_m/2$, $x_1 = l_m/2$, if $t_q = 6$; c)$dd = -1$, $y_1 = -l_m/2$, $x_1 = -l_m/2$, if $t_q = 7$; and d) $dd = 1$, $y_1 = l_m/2$, $x_1 = -l_m/2$, if $t_q = 8$; 3) ordering of all SPs and summing of the dd values for coinciding SPs. Therewith the SP of the type 9 or 10 (Fig. 1,b) must be considered as a pair of two SPs of the types 1 and 3 or 2 and 4.

Performing of the second part of the dilation algorithm depends in general on the mask image, and for every of these images the algorithm for retrieval of the DR of the multilevel representation of the dilated original image must be defined. For example, if the mask image represents a circle, then concave corners of the original picture will be processed as described above, but processing of every convex corner will result in his replacing on some group of the corners, representing correspondent quadrant of the circle's contour.

The third part of the algorithm is similar to the first one and consists in line by line retrieval of the picture v on the basis of its DR, thresholding of this picture by threshold equal to 1 and corner coding of the result binary picture. This step is carried out during bottom - up and from left to right scanning of the points along the lines of the SPs in the picture, such that every point p is either SP or the end of the run of pixels, having the same value in the lower row of the picture. During processing of every of these points the values ρ_1, ρ_2 of neighbouring pixels (Fig. 2) on the basis of values ρ_3, ρ_4, dd and d_2 are defined: $\rho_1 = \rho_4 + d_2$ and $\rho_2 = \rho_3 + dd + d_2$. The retrieved values $\rho_1 - \rho_4$ are then used for checking if there is the corner in every point p and also for run length coding of the upper row of the picture.

The algorithm of erosion of the binary picture is similar to the considered above algorithm. It can be represented also as dilation of the inversed initial picture, followed by inversion of the dilated picture.

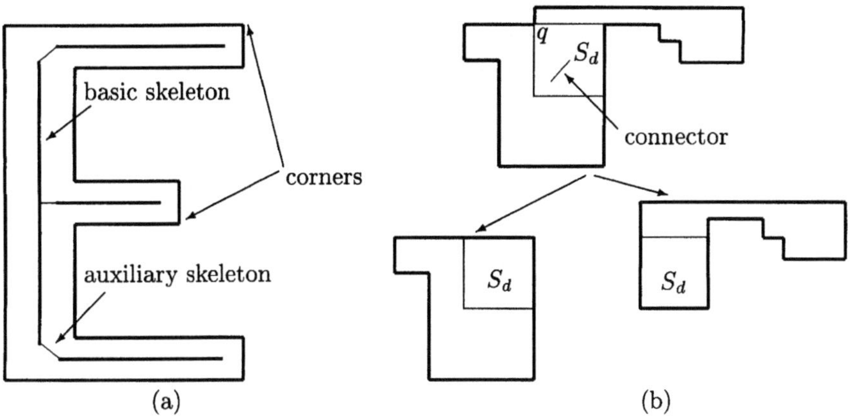

Fig. 3. (a) The basic and auxiliary skeletons of the binary picture. (b) Decomposition of the picture into two subpictures.

3. Description of the skeletonization algorithm

A certain square $S_m \subset R^2$ in the binary picture v is called maximal square [6] if S_m consists of black points and this square can not be increased without violating this feature. A basic skeleton of the picture v is defined as a set of central points of the maximal squares. The basic skeleton represents a set of horizontal and vertical lines, as well as some separate points in the picture and is shown in Fig. 3,a by thick lines. An auxiliary skeleton is a set of additional line segments, ensuring the connectivity of the produced skeleton, and is shown in Fig. 3,a by thin lines. Corner's coordinates represent the compressed form of the picture and also important data for skeleton detection, because to every such corner corresponds (in appropriate diagonal direction) the end of skeleton line segment.

A certain square S_d is called dissective square (Fig. 3,b) in the picture if (a) the vertice of the square S_d coincides with the concave corner q , (b) the diagonal of this square coincides with the bisectrix of the corner q and (c) the square S_d consists of black points and can not be increased without violating this feature.

Let us suppose that intersection $C_v \cap S_d$ belongs to the same connected component of the contour C_v of the picture v. It means, that there is no in the picture v the white component or pattern (the connected set of white pixels), such that its contour contains the concave corner q. In this case splitting the picture v into two connected pictures v_1 and v_2 (Fig. 3,b) is called decomposition if (a) $v_1 \cap v_2 = S_d$, (b) $v_1 \cup v_2 = v$ and (c) $v_1 \neq S_d$, $v_2 \neq S_d$. Connector is the line segment (Fig. 3,b) linking centers of those two maximal squares S_1 and S_2 such that $S_d \subset S_1 \subset v_1$ and $S_d \subset S_2 \subset v_2$. The following two statements are valid: (a) the basic skeleton of the picture v is the union of the basic skeletons of the pictures v_1, v_2; and (b) the auxiliary skeleton of the picture v is the union of auxiliary skeletons of the pictures v_1, v_2 and a third set containing only the connector. That is why skeletonization of the picture v can be reduced to decomposition the picture v into two pictures v_1 and v_2, followed by skeletonization of these two pictures.

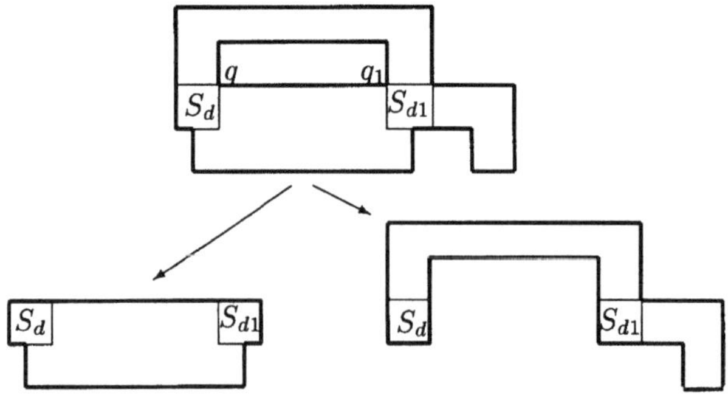

Fig. 4. Decomposition of the picture in concave corners q and q_1.

Otherwise if the picture v contains the white component with the concave corner $q(x,y)$ of the type 5, but below this corner there is no another corner of such type, the picture v has the concave corner $q_1(x_1,y)$ of the type 8, neighbouring to q, and decomposition of the picture v into two pictures can be carried out after defining dissective squares in the corners q and q_1 (Fig. 4).

Let us define also a notion of a residual picture. Let L be a horizontal line with ordinate y dividing the picture v into two parts. The operation of decomposition of lower part $v_b(y)$ of the picture is called permissible, if ordinate of upper side of dissective square, corresponding to this operation, is less than y. After carrying-out all possible decompositions of the picture $v_b(y)$ some set of rectangles below line L and also some set (it may be empty) of the parts of the picture touching to this line will be received. The union of these parts represents the residual picture $v_r(y)$ of the picture $v_b(y)$.

Some picture that does not contain the corners of types 5 and 8 be called a simple picture. At first let us consider the algorithm of skeletonization of the simple picture with the contour which does not intersect vertical sides of dissective squares corresponding to concave corners of types 6-7 in this picture [Fig. 5,a]. If the simple picture does not contain concave corners, then it is rectangle and its skeletonization is obvious. Otherwise the skeleton of the simple picture can be detected in the following manner. The concave corner q with the least coordinate x and the dissective square in this corner are detected. Then decomposition of the picture into two parts is carried out: one of these parts is the rectangle and another is the simple picture, which does not contain the concave corner q. These operations are repeated to complete removal of concave corners in this picture. The skeletonization algorithm of other simple pictures [Fig. 5,b] only somewhat differs from considered above. The computational complexity of skeletonization of any simple picture linearly depends on the number of corners in these pictures.

The algorithm for skeletonization of any (not only simple) picture processes the picture during its bottom-up scanning, so that the boundary of the scanned part successively occupies positions coinciding with corner lines in this picture. In doing so the algorithm processes some, say, i-th corner line with ordinate y_i using not

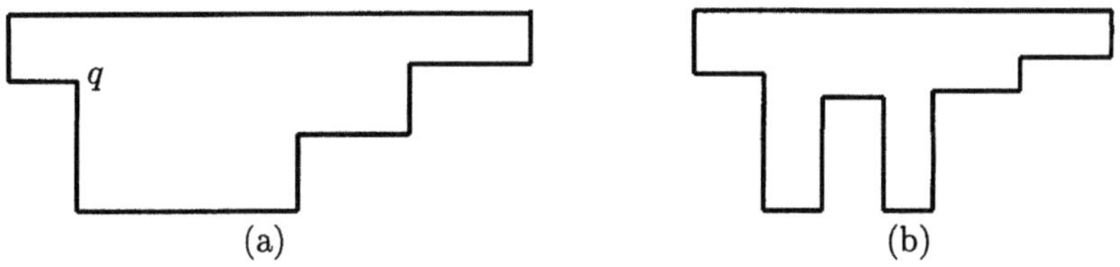

Fig. 5. Simple pictures.

all picture $v_b(y_i)$ below of y_i , but only the residual part $v_r(y_i)$ of this picture. The residual picture $v_r(y_i)$ consists of one or some connected components. During processing each of these components $v_{rc}(y_i)$ it can be defined whether this component has continuation above i-th corner line or no. In the second case the picture $v_{rc}(yi)$ is the simple one and its decomposition must be carried out as considered above.

Otherwise the decomposition of the component $v_{rc}(y_i)$, correspondent either to the corner q of the type 5 or 8 with the least absciss and ordinate y_i or to two such neighbouring corners, is carried out. If there is no such corner in the picture $v_{rc}(y_i)$ this picture is joined to the residual picture $v_r(y_{i+1})$. Otherwise, two following pictures as a result of decomposition $v_{rc}(y_i)$ in this corner are produced. One of these pictures does not contain corners of types 5 and 8 and hence it is the simple picture, which must be skeletonized as described above. Another picture does not contain the corner q and it must be processed in the same way as the initial picture $v_{rc}(y_i)$: carrying out the decomposition in the next corner with the least absciss and so on.

4. Performance

The dilation(erosion) and skeletonization algorithms have been implemented in the software system for automatic input, processing and semantic interpretation of line drawings and maps. Fig. 6 shows a 1248x1700 original image on the left, its dilation (isotropic expansion) by two pixels on the middle and its skeletonization on the right. Size of the original picture is 270x170 mm. After scanning this picture with the resolution 200 dpi the volume of noncompressed TIFF file is 570 kbyte and the volume of corner representation of the picture is 97.1 kbyte. Dilation and skeletonization of the picture take 0.6 and 1.5 sec. on IBM PC Pentium 120 The volume of output (postprocessed) skeleton is 31.2 kbyte. The developed software processes the picture part by part and that is why it can be used for processing of large-size line drawings.

5. Concluding remarks

Fast dilation and skeletonization algorithms have been developed and presented in this paper. The proposed dilation algorithm is fast because its computational complexity depends on the number of corners in the picture and does not depend on the number of the pixels in this picture. This algorithm processes the picture line by line and requires small memory area for its performance.

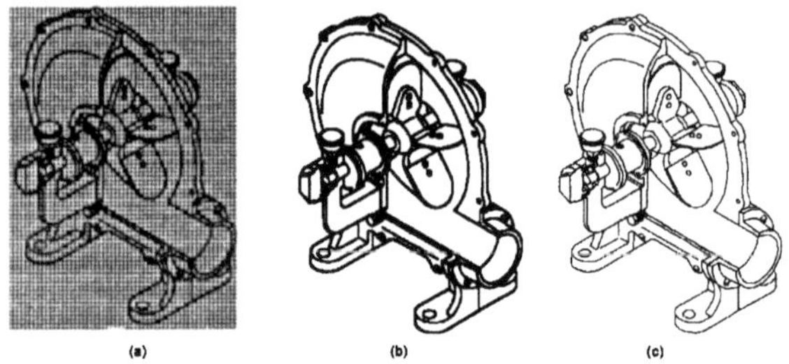

Fig. 6. Results of expansion by two pixels (b) and skeletonization (c) of the original picture (a).

Comparing developed skeletonization algorithm to algorithms [2-5] we find the following.

1. The proposed algorithm is essentially faster, taking in account, that different computers were used, in that it is a) width independent and b) it does not require each pixel of the pattern to be tested except only some points (corners) of its contour.
2. It is structure preserving and is oriented toward a structure-descriptive skeleton representation, because the set of detected maximal squares represents thin lines instead of isolated pixels or points.
3. It requires small memory area, which does not depend on the size of input data.

References

1. H.Blum, "A transformation for extracting new descriptors of shape,"in Models for the Perception of Speech and Visual Form, edited by W.Whaten-Dumn (MIT Press, Cambridge, 1967), pp. 362-380.
2. P.S.P. Wang and Y.Y. Zhang, "A fast and flexible thinning algorithm," IEEE Trans. Comput. 38(5), 742- 745 (1989).
3. T.Wakayama,"A core-line tracing algorithm based on maximal square moving,"IEEE Trans. Pattern Anal. Mach. Intell. PAMI-4(1), 68-74 (1982).
4. C. Arcelli and G.S. Di Baja, "A width-independent fast thinning algorithm," IEEE Trans. Pattern Anal. Mach. Intell. PAMI-7(4), 463-474 (1985).
5. S.V. Ablameyko, B.S. Beregov and A.N.Kryuchkov, "Computer-aided cartographical system for map digitizing," Proc. 2nd ICDAR, Tsukuba, Japan, pp. 115-118, 1993.
6. V.M.Kiyko and M.I.Schlesinger, "Width-independent fast skeletonization algorithm for binary pictures,"Intern. Journal of Imaging Systems and Technology, vol. 3, 222-226(1991).
7. L.Lam, S-W. Lee, C.Y. Suen, "Thinning methodologies - A comprehensive survey", IEEE Trans. Pattern Analysis and Machine Intelligence, vol. 14, No. 9, Sept. 1992, pp. 869-885.
8. M.I.Schlesinger. Matematiceskie sredstva obrabotki izobrazenij, in Russian. (Mathematic tools for image processing). Naukova Dumka, Kiev, Ukraine, 1989.
9. V.Hlavac, J.Fojtik," Adaptive non-linear predictor for lossless image compression. Computer analysis of images and patterns," 7th Int. Conference, CAIP'97, 279-288, 1997.
10. V.V.Matsello, "The algorithm for noise removing in line drawings," in Russian. Avtomatica, No. 2, pp. 86-89, 1986.

THE CREATION OF COMPUTER GRAPHICS WITH MATHEMATICAL MORPHOLOGY

J.S. PLOEM
Leiden University, The Netherlands

Abstract. This paper desrcibes the use of transformations from mathematical morphology to generate computer graphics. Some black and white reproductions representing small parts of large colour graphics are included for illustration.

1. Introduction

Medical image analysis has been one of the main research topics in the Laboratory for Cytochemistry and Cytometry (Leiden University) during the last 20 years. In 1972, we started with a PDP-12 (DIGITAL) computer having only 8000! words of memory (still using paper tape input), in combination with an automated microscope for image analysis with the purpose to detect cancer cells in cervical smears. The first programs we used, were written for conventional computers by the group of Wied [13] in Chicago. In 1974, information reached us about the mathematical morphology (MM) methods developed by Matheron and Serra at the "Centre de Morphologie Mathmatique" (Fontainebleau, France). A prototype machine, the "Texture analyser", which is a special purpose computer optimised for Hit or Miss transformations, openings, closings and Boolean methods, was then developed by Serra and collaborators [4, 9] to efficiently execute the MM methods. A commercial version ("TAS") of the Fontainebleau "Texture Analyser" was subsequentely manufactured by Leica (Wetzlar, Germany); this instrument became available in the Leiden Cytometry laboratory in 1975. Having no mathematical programming experience, we were surprised to be able to compose a reasonably effective procedure for cancer cell detection in a few weeks using this machine. By comparing the cytopathologist's visual and subjective evaluation of cancer cell images with the results obtained with the "TAS", it became clear that image simplification and transformation obtained with MM was a powerful tool for the detection of cancer cells. Also, it became evident that machine operators using MM can in fact compete with, or at least add, significant data to classical human visual interpretations. Subsequently Meyer [7] wrote a much more effective cancer cell detection program for cervical cytology.

VISUAL OBSERVATIONS COMBINED WITH THE RESULTS OF IMAGE TRANSFORMATIONS

Ever since I followed, during my last grades of high school, courses in painting at the college of Art in Maastricht (The Netherlands), I maintained an interest in painting. As I had been impressed with the charm of some transformed cell images obtained with mathematical morphology, I could not resist to include skeleton-like lines in many of my aquarelles and acrylic paintings. In 1992 at the official occasion of ending my professorship at the University of Leiden, I could show an exposition of aquarelles and acrylic paintings presenting classical visual impressions combined with skeleton-like white lines at the gallery of Pulchri Studio in The Hague (The Netherlands). I was now convinced that observations by an artist could potentially be combined with MM image transformations. Then, following my leave of the Leiden Cytometry laboratory in 1992, I could obtain a commercial version of an extensive, versatile and user-friendly mathematical morphology program ("QWin") from Leica Imaging Systems Ltd. in Cambridge (England), running on a regular PC under Windows. Also, QWin provides an attractive macro programming facility that makes it possible to assemble a large variety of image transformations in one program.

After transferring photographs to the computer memory, I could make MM transformations from for example, flowers, human figures, landscapes, buildings, etc., which resulted in interesting computer graphics. The next step was to combine some computer graphics - made with conventional programs as used by many designers (e.g. Corel Draw and Adobe Photoshop) - with the graphics generated by MM transformations. In short, this combines the structures selected by the two eyes of an observer with the result of the analysis by a third "mathematical" eye, into one computer graphic.

2. Computer graphics

Fig. 1. Black and white photograph of a bass scanned into computer memory.

No attempt will be made here to give a comprehensive review of computer art or computer graphics. However, a short description of 5 examples of computer programs, used for making computer graphics, could illustrate aspects of different experimental approaches.

(1) A first type of image software allows to use the computer as a direct technical tool for drawing or painting. There is no data input into the computer by scanned photographs or drawings. Using for example a mouse, the artist-operator makes a drawing not unlike an artist in front of his easel. He has a direct control and responsibility for any computer graphics that he may produce. There is maximum interaction of the operator with the computer. Randomisation is mostly not a dominant factor in such activities.

(2) A next example of software produces the famous Mandelbrot "Fractals" [2]. There is again no image data input beforehand into the computer. Complex fractal graphics can be derived from relatively simple arithmetics, but require millions of iterative operations to produce an almost infinite variety of images. Sometimes pictures are obtained that resemble natural objects like flowers, trees, landscapes etc.. Such results have been described as "fractal art". The pro-

grams use an inherent strong randomisation. The user is only responsible for the choice of the fractal generating software (large program libraries are now available). However, once the program is launched, it operates generally on its own without human interaction.

Fig. 2. Skeletonization of the bass depicted in Fig. 1.

(3) An important and early contribution to image making by computers has been made by the English artist Cohen [3]. He created the computer program "AARON", which after its start was continuously improved and "educated" by Cohen. In 1973 "AARON" made already some interesting sketches. It used no data input in the form of photographs. By the late 1984's, "AARON" had been "educated" to know about the basic visual attributes of people, plants, and rocks and knew how to place them relative to each other. By 1992 Cohen enabled the program to select and apply colours itself. There's no way of predicting what "AARON" will draw or which colours it will choose and its creator never interferes with what the program chooses to do in individual paintings. "AARON" 's work sells well at auctions and has been exhibited in major galleries in London, New York and elsewhere. But is this genuine computer creativity? Cohen is cautious: ' "Creative" is a word I try hard never to use' [11].

(4) Scha [8] in The Netherlands, has developed an innovative program named "Artificial". It uses data input into a computer such as photographs or drawings. From this image data base, the program first randomly selects the pictures with which it will work. Then it begins to cut up and move around the image components according to certain rules designed by Scha, but not always in the same manner, because the choice of such rules is also controlled by the process of random selection. This leads to a fully automated generation of new and often interesting pictures on the screen. Scha describes his program as an investigation into the process of image making without predetermined goals, and as a principle, he does not require aesthetic qualities of the output.

(5) A fifth mode of image software focuses on morphological structures in a scene that are not, or not easily noticed by a human observer. Data input into the computer is by by scanned photographs, black and white drawings or coloured sketches made by the operator-artist. Powerful and versatile image transformations as available through an extensive library of MM operators contribute to creative image making [10]. To direct the transforming operators at fine or coarse detail, they must be chosen by the artist-operator in relation to the size of the structures that are of special interest in a scene. By using a great variety of operators, the random process in image making is enhanced. When a great variety and number of transformed images is generated, the operator-artist has, however, to spend a large

amount of his energy to select only a few interesting images. More details of this approach to the creation of computer graphics using MM are described in the sections about mathematical morphology (below) and the section on MM programming.

Fig. 3. The same music instrument as in figures 1 and 2. With logical operations thin skeletons have been subtracted from thick skeletons in a random way, resulting in combinations of channel-like "skeletons" of multiple sizes. This picture is a black and white copy (2-bit grey depth) of a part of the painting (8-bit colour depth) computer print "Hommage á Fernand Meyer".

3. Mathematical morphology

The purpose of this paper is to show that mathematical morphology [10] cannot only be used for image analysis in science and technology, but also for image making and under certain circumstances for creating computer graphics.

Image scaling

At which level of detail the picture, stored in computer memory, should be analysed using MM? An artist almost intuitively chooses the level of detail in a scene he wants to depict before he paints on canvas or paper. If the size of interesting structures in the scene is wrongly chosen he ends up with irrelevant small details or he will ignore important structures that would have been characteristic for a scene. The artist using a computer for image making with mathematical morphology has to perform a similar task.

Fig. 4. Portrait of a woman scanned into computer memory.

Using the finest details of a high resolution photograph of a scene will not necessarily lead to a satisfactory result. In analogy to the human visual cognition process, where the detection of "zero-crossings" [6] helps to segment a scene, morphological gradients using erosions and dilations can be used for edge detection in images stored in computer memory. To avoid excessive image details they should often be thresholded and smoothed prior to the application of the gradient. The problem of having to determine the right size of the structures to be transformed in a scene directly at the beginning of a computer session, can be avoided by writing a larger comprehensive program that is designed to analyse automatically, sequentially and at multiple image scaling levels with a large variety of transforming operators. This increases the desired randomisation of the procedure. Only at the end of the computer session, the artist-operator can then from the many images that were generated, select the few images that may be interesting for further processing.

Image transformation

It seems not unreasonable to describe the mental process of analysing a scene - which an artist wishes to bring unto paper or canvas - as the process of mentally carrying out multiple image transformations. "To perceive an image is to transform it" (Serra [10]). Artists have an intuitive idea about the spatial structure of e.g. a landscape. But "we see only what we want to look at" (Serra [10]). Simplifying images of complicated scenes to be used for image making only makes sense when the artist is capable of choosing structuring elements for transforming structures of the right size in the image. The selectivity achieved by the human mind is characteristic for a certain painter. For great painters this is part of their genius. It is difficult or impossible to reproduce this process in the computer analysis of a scene. But one can try to analyse a few components of this process. By using various structuring elements in a computer program for MM, an artist-operator can interact with certain pictorial elements, transforming them to a sort of "caricatures" which may be more expressive than the initial morphological elements. Serra (1982) used the word "caricatures" in the context of image analysis, but this term is also appropriate for the results of image making. The creation of caricature-like elements have played a role in the work of many painters (e.g. Picasso, Klee, Appel and many others). Some details of image transformations used for creating my computer graphics are described in the section on MM programming.

4. Randomisation in computer graphics

Fig. 5. Same picture as in Fig. 4 after numerous random transformations. This is a black and white copy (2-bit grey depth) of a part of one of the (8-bit colour depth) pictures selected from about 25 that were generated with MM.

Several artists have deliberately and consciously employed randomising procedures like tossing coins in the creation of art [3]. It is told that van Gogh pulled coloured threads from his pocket, in order to help to choose and vary the coloured pigments to be used in a particular part of his painting. There is no question that, in the paintings of Pollock (e.g. made by dripping paint on a canvas) and many of the other abstract expressionists, the element of random procedures has played a large and deliberate part [1]. For the human artist, randomisation, although in principle unlimited, is in practice constrained. Especially because of the considerable physical effort it takes, and the time required for working out the numerous possible variations (e.g. random colour variations). One function of randomness in image making is, according to Cohen [3], the generation of a variation of images by the program without requiring the artist to constantly "invent" such images. Also Scha [8] stated that computers are particularly effective for randomisation. The theory of mathematical morphology allows that a very extensive and varied set of transformations can be constructed by combining only a small number of basic operators. Considering the speed of modern computers a problem soon arises. What to do with the large number of new images?

Randomisation with subsequent visual selection

If a computer has carried out a random process producing numerous results on the screen, the artist-operator has to decide what to do with such images. Scha [8] decided for principal reasons to perform no subsequent image selection. Van Weelden [12] supports this idea. However, in mathematical morphology - using an already highly interactive programming environment - it seems to me a defendable philosophy to continue this interactivity, by letting the artist-operator finally select the most interesting images. Consequently the artist has then to spend a considerable amount of energy on the intensive mental process of selecting the few images that should in his opinion be saved. Instead of image production, image selection becomes his main artistic effort. Through his decisions, he can use his artistic intuition to the full extent and thus become responsible for his choice.

Fig. 6. Graphic result (2-bit grey depth) of a "QWin" binary transform labelling procedure of a picture of a woman sitting at a table. The distance function in this procedure has been varied by randomly choosing different binary mask planes containing images of various channel-like "skeletons" of multiple sizes.

Since new image combinations can always be created from the various transformed images, he can use his creative talents by active use of the technique of computer operated image cloning. He can in that case use some interesting small parts of transformed images as building blocks for new images.

In summary, the artist uses the computer as a tool to simulate the visual cognitive process followed by image transformation and finally for intensive human interaction in creative image making. Possibly an artist can also use the final image obtained with the computer only as an inspirational input or as design for a manual painting on paper or canvas.

Fig. 7. Mathematical transformations applied to a photograph of the village Sauto, in the French Pyrenees. This is a black and white (2-bit grey depth) reproduction of part of a (8-bit colour depth) painting (computer print).

5. Future specialisation of artists

Artists often fear that the use of computer programs for art making will curtail human intuition and creativity and therefore art itself. This should be seen as an opinion that should truly concern us. It would indeed be undesirable if human creativity would be obstructed by the unwanted limitations of rigid or uncreative computer programming. Mathematical morphology programming is, however, used in so many fields where only human visual analysis dominated, that it appears to have the transformational creativity and interactive accessibility to support artistic creativity. The remarkable transparency of MM operators makes it likely that non-professional programmers can effectively use them with a sufficient predictability of the graphic results. Due to the explosive use of computer graphics on the INTERNET, young artists may already have or will acquire in the near future the abilities to handle computer programs for graphic work.

Van Maarseveen [5] - who won the prize of "CS-essay-contest" by writing a review on a recent opinion poll held in the Netherlands about the state of modern art - predicts a future for computer graphics. He concludes that a certain loss of expertise in the field of art has occurred, that the role of artists with a "nice talent" will diminish, and that in the future well trained artists-experts will make significant contributions to creative art. The artist-specialist is, according to him, capable of producing more interesting or innovative creations. In his opinion, computer art and more specifically computer graphics will be an example of a reinstated combination of specialistic skills and creativity of an artist. Weber [11] - in a review he wrote on computer creativity in 1996 - believes that the increasing use of computers as a medium for artistic expression will also produce computer artists, contributing ideas as well as executing them. He considers computer art as a very dynamic and fertile area in the art world and thinks that a revolution is taking place with shaking up a lot of conservative values. Encouraging is certainly that established galleries such as the Museum of Modern Art, The Guggenheim and the Whitney in New York have taken to show computer art alongside traditional oil paintings.

6. Material

The core of the system is a Compaq PC with a Pentium 233 Mhz processor, 64 MB RAM memory, hard disks providing a total of 8 GB of disk space, a Matrox Millenium Graphic (8 MB) card, supporting an 1600×1200 pixel 24 bit colour image on a Mitshubishi Diamond Pro 21 inch TX screen.The used image editing programs are: Adobe Photoshop and Corel Draw. Mathematical Morphology programming consists of : Leica QWin (Leica Imaging Systems, Cambridge, England), running under Win 95. Digital data input into the computer is through a Nikkon LS-1000 35 mm film scanner. Photographs are brought into the computer as 1575×1050 pixel full colour files. QWin transformations are carried out on a 1600×1082 pixel image with storage in 96 binary planes and 7 colour planes. Graphic files are combined into a 2814×1959 pixel full colour format for printing (after unsharp mask sharpening) on a Tally 7070, A2 format ink-jet printer.

7. Mathematical morphology (MM) programming

Only a few examples of the main MM operations that have been used in image making will be described here. Pre-editing of the image is carried out by drawing a few white lines around certain structures in the original image to provide relatively isolated zones in the picture before transformation is performed. Morphological gradients - formed by various combinations of structuring elements and a number of erosions and/or dilations - provide edge detection facilities for the imported photographic images. Various transforms on grey images for smoothing (obtained by a series of openings and closings) provide images with more or less detail. This is of importance for the selection of rightly sized structures in the image for operators to work on.

Infrequently "QWin" watershed functions have been used to produce a grey (colour) watershed image, as well as a binary image depicting the watershed divide lines. Tophat filters are used with images having large variations in the background.

For the creation of computer graphics, the Hit-or-Miss transforms we most frequently have used are skeletonizations. An advantage is that they preserve the relation of the features to each other; exhaustive skeletonization provides very characteristic images resembling sometimes examples of early prehistoric art painted on rocks, or drawings from very young children. Depending on the picture, the skeletonization may be carried out by zone of influence (SKIZ). Now and then pruning of the skeletons is employed. Skeletonizations are also carried out with only a limited number of cycles providing e.g. networks of skeletons with differently sized free spaces. Skeletons are then dilated 2, 3, 5 and 11 times and copied into 6 different colours. Subsequently thin skeletons are subtracted with logical operations from thick skeletons in a random way, resulting in combinations of channel-like "skeletons" of multiple sizes and colours. This great variety of structures is routinely stored on disk in about 80 binary images.

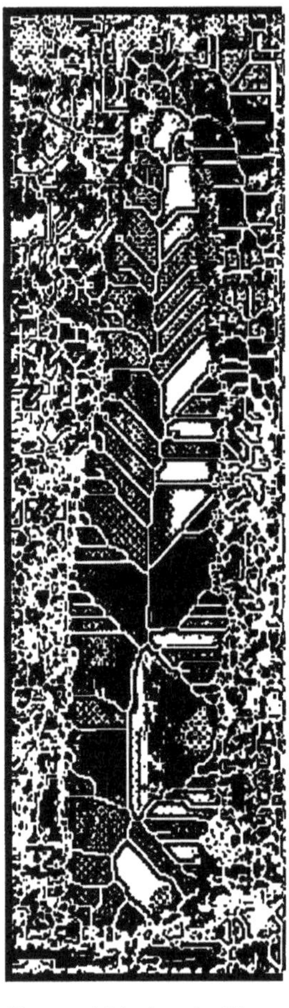

Fig. 8. A black and white reproduction (2-bit grey depth) of a cypress tree as part of a colour painting (computer print) of the abbey Sant Antimo (Italy). The mathematical morphology program has recognized the elongated shape of a cypress and generated the skeleton lines that well depict this type of tree.

A further randomisation is obtained by a binary identify (find) holes operation. This is followed by a binary transform labelling procedure where binary images are converted into grey (colour) images. In this labelling procedure each feature is labelled with different grey (colour) levels . In the "QWin" program for labelling, a distance function creates a grey (colour) image from the binary input image by setting the grey level of each point to a value that represents its distance from the nearest edge. This distance function can be varied by randomly choosing different binary mask planes (from the numerous binary images with different sizes and colours of the channel-like subtracted "skeletons" obtained before, and stored on disk). Fortunately the rather harsh colours that were originally allocated by the "QWin" colour labelling process described above, are then replaced by random colour families related to the colour of one of the randomly chosen 80 binary images. A screen capture procedure has to be used to store these coloured graphics in computer memory, since only 7 colour images (and 96 binary images) of 1600x1062 pixels can be stored by the QWin program configuration that was chosen for our PC. The coloured graphics thus obtained show an impressive variation, but still display a significant relation with the original photograph (this was our purpose). Finally the randomisation of colours can still be significantly increased by treating the images with the powerful random (chaos) colour generation method described below.

Fig. 9. Black and white reproduction (2-bit grey depth) of a picture, obtained after random transforms of an abstract drawing that was composed beforehand with an image editing program.

8. Randomised (and chaotic) colour generation

Simultaneously opened windows with colour images can be activated or non-activated. If a windows screen display of only 256 colours is chosen with the (Matrox Millenium) display card, the active and non-active window may show different types of colour palette support. The system's colour palette for 256 colours is often not sufficient to support the images in both types of windows with an optimal number of colours. Some image editing programs permit to choose one (active) image to be displayed with an optimal colour palette, by having it preferentially supported by the system (256) colour palette. The second, (non active) image can then only use the remaining colour choices from the system colour palette. It shows a more limited colour palette with sometimes an artistic character, especially if in the image editing program is chosen for a dithered colour display. In my system I have miniaturised 40 pictures - obtained from my earlier aquarelles - to about 50 × 50 pixels and placed them in a row on the low side of my screen. The large 1575 × 1050 pixel window of a (target) image - of which the colour has to be randomised - is placed full size on the screen above this row of minute images . By now clicking with the mouse on one of the 40 miniaturised colour images in the row, this small (50 × 50 pixel) window becomes the active window, and the large (1575 × 1050) window of the target image on the screen becomes the non-active window. This image shows a more limited, "impressionistic" palctte with "pointillistic" (dithered) colour spots. If this image is attractive, it can be saved to memory or disk by a screen-grabbing procedure. A click on the next miniaturised picture in the row, results in another "impressionistic" colour palette of the same large target image. In general, there is a fixed relationship between the colour palette of the activated miniaturised picture and the target image. Surprisingly however, sometimes a completely new colour palette is generated in the large target image. From these observations I had to conclude that the windows display algorithm or/and the display card are not always stable, giving sometimes rise to a "chaos-like" colour display process. In fact I could generate more than 80 "impressionistic" colour variations in the one large (target) image, with only 40 different miniaturised pictures. From this, often overwhelming multitude of images, a few have then to be selected by the artist-operator to be saved. In a potential cloning session, interesting parts of several of such images can then be combined into one computer graphic.

9. Results shown in graphics

Computer graphics generated with mathematical morphology methods can only be shown optimally as larger sizes (A2, A1 or A0) colour computer prints. As an illustration to this paper, I include however, some black and white reproductions representing small parts of such large colour graphics. They are chosen only to demonstrate some of the MM transformations described in this paper.

References

1. H. H. Arnason. *History of Modern Art (3rd ed.)*. Harr N. Abrams, New York, 1986.
2. J. Briggs. *Fractals*. Thames and Hudson, 1992.
3. H. Cohen. What is an image. In *I. J. C. A. I., Proceedings, Tokyo*, 1979.
4. J. C. Klein and J. Serra. The texture analyser. *J. of Microscopy*, 95(2):349–356, 1973.
5. H. Maarseveen. NRC Handelsblad (July 11, 1997).
6. D. Marr. *Vision*. W. H. Freeman and Company, New York, 1982.
7. F. Meyer. *Cytologie Quantitative et Morphologie Mathématique*. Thèse Docteur Ingénieur, Ecoles des Mines, Paris, 4 Mai 1997.
8. R. Scha. Computer/Art/Photography. *Perspectief*, 37:4–10, 1989.
9. J. Serra. Patents of the texture analyser. First application in France, no. 1-449-052 (July 2, 1965).
10. J. Serra. *Image Analysis and Mathematical Morphology*. Academic Press, 1982.
11. J. Weber. Computer Creativity. But is it art? *PC PRO*, pages 177–181, June 1996.
12. D. v. Weelden. Remko Scha, An Ideal, a Name, an Explorer. *Perspectief*, 47/48:42–51, June 1994.
13. G. L. Wied, G. F. Bahr, and P. H. Bartels. Automated analysis of cell images by TICAS. In G. L. Wied and G. F. Bahr, editors, *Automated Cell Identification and Cell Sorting*, pages 195–360. Academic Press, New York, 1970.

MORPHOLOGICAL SEGMENTATION APPLIED TO CHARACTER EXTRACTION FROM COLOR COVER IMAGES

LIXU GU and TOYOHISA KANEKO
Toyohashi University of Technology, 1-1, Hibarigaoka, Tempaku-cho, Toyohashi 441, Japan. E-mail: ko(or kaneko) @mmip.tutics.tut.ac.jp

and

NAOKI TANAKA
Kobe University of Mercantile Marine, 1-1, 5-Chome, Fukae-minami-machi, Higashinada-ku, KOBE 658, Japan. E-mail: tanaka@athena.ti.kshosen.ac.jp

Abstract. Mathematical morphology provides us the theory and tools to capture geodesic information. Hence structure segmentation and shape representation have been popularly operated by morphological approaches. A primary morphology-based approach — top-hats transformation (TT) provides an excellent tool for extracting bright or dark objects from uneven background. But to many complicated segmentation problems, the TT alone cannot provide satisfactory solutions. When a series of structure elements with various sizes have been employed in our study, the gray-scale particle segmentation for some complex cases is realized successfully in an iterative manner based on the TT. This paper describes a new morphological segmentation algorithm which was derived from the TT idea. In addition, a complete system for extraction of characters from color cover images using our algorithm is presented. 1,210 characters out of 1,326 characters included in six source images are correctly extracted, indicating a 91% success rate.

Key words: Mathematical morphology, Top-hats, Particle segmentation, Reconstruction

1. Introduction

Image segmentation via the methods of mathematical morphology may be divided into two essential categories: texture segmentation and particle segmentation[1]. Texture segmentation partitions an image into several regions with particular textural features. Applications of this approach include segmentation of document images; medical image processing and aerial photograph analysis.

Particle segmentation is concerned with images of particles (or objects). When the textural information in an image is either not or cannot be used simply as a discriminating factor, extraction of particles in the image would be more reasonable. A typical approach called top-hats transformation (TT), which was originally proposed by Meyer [4], provides an excellent tool for extracting bright or dark objects from gray-scale images. Furthermore investigation of image skeleton and shape representation [5]–[6] provides us many good ideas in morphological iterative operations. Recently considerable research efforts have been addressed to the watershed analysis [7].

This paper presents a new approach for particle segmentation based on the TT idea and morphological iterative operation. We apply this approach to deal with a character extraction problem for color cover images.

Even if a cover image contains many complicated information, the information carried by characters is considered to be the most essential. The importance of this field has attracted a number of researchers and there have been several studies in the character extraction[2]–[3], which are proposed to be optimal solutions for extraction characters, but are limited to simpler background or other restrictive conditions.

Characters in cover images may be characterized by the words, "they float on the background or sink under it with a flat surface". This characteristic underlines the basic idea of our new method. The process includes two distinct stages:

a) Primary processing (particle segmentation)
b) Extraction processing
   - feature emphasis
   - character extraction
   - noise reduction

In the first stage, a new particle segmentation filter based on the TT is implemented. This filter decomposes a gray-scale input image into a set of sub-images according to the size of characters. In the second stage, we first employ a new morphological filter to emphasize characters' features in the sub-images and remove most noises out of them, and then, the characters are extracted directly from the gray-scale sub-images by a histogram method. Lastly, a morphological image cleaning algorithm based on morphological conditional dilation is introduced to make the extracted character region distinct from noises. The resulting sub-images are composed together to generate the final result in binary.

The organization of this paper is as follows: in section 2, after terminologies and notations are defined, a brief overview of typical morphological operations and definition of structure elements are given. Section 3 describes a new algorithm based on the TT. The details of our character extraction implementation for color cover images are presented in section 4. Experimental results are discussed in section 5, and the concluding remarks in section 6.

2. Conception and notation

2.1. Morphological operation

Mathematical morphology is a set theoretical methodology for image analysis[8]. Images are considered as sets of points where set operations can be performed. By convention, we let $X - B, X \oplus B, X \ominus B, X \circ B$ and $X \bullet B$ denote the set difference, dilation, erosion, opening and closing in binary operation, when $X \oplus_g B, X \ominus_g B, X \circ_g B$, and $X \bullet_g B$ denote the dilation, erosion, opening, closing in gray-scale respectively, where X and B denote sets in Z^2.

Definition 1: The recursive dilation is defined as

$$X \overset{i}{\oplus} B = \begin{cases} \{0\} & if\ i = 0 \\ ((\ldots(X \overbrace{\oplus B) \oplus B) \cdots \oplus B}^{i}) & if\ i = 1, 2, 3, \cdots \end{cases} \tag{1}$$

Definition 2: The conditional dilation is defined as

$$(X \overset{i}{\oplus} B)_c = (X \overset{i}{\oplus} B) \bigcap C, \tag{2}$$

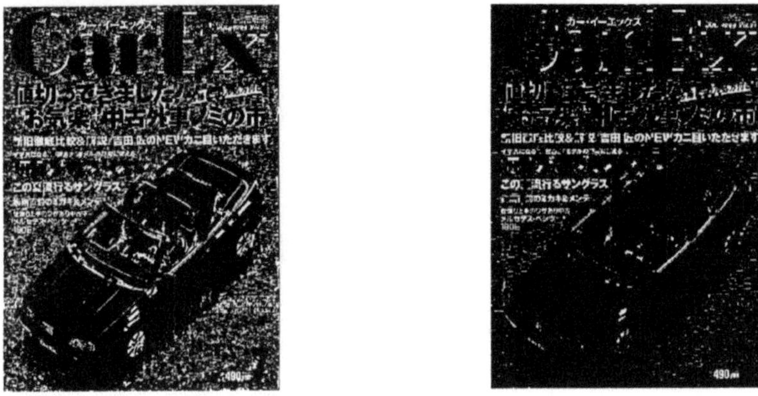

(a) Result of top-hat (b) Result of presented algorithm

Fig. 1. The segmentation result of top-hat (a) contrasts with our presented algorithm (b). They are resulted from the source image of Fig.2(a) which is closed by the same structure element $r_4 B_{disk}$ and thresholded by the same gray level B=31.

where C denotes a mask and the index i is the smallest one satisfying:

$$(X \overset{i}{\oplus} B)_c = (X \overset{i-1}{\oplus} B)_c. \tag{3}$$

Definition 3: The top-hats transformation is denoted by $T_i^{(i)}$ and defined as

$$T_i = X_0 - X_0 \circ_g r_i B \tag{4}$$

$$T^i = X_0 \bullet_g r_i B - X_0, \tag{5}$$

where $r_i B = B \oplus B \oplus \cdots \oplus B$ (i times).

2.2. Structure elements

Two standard shaped structure elements DISK and LINE are selected in our method. A DISK structure element is employed to detect objects by scale when a LINE structure element is utilized to detect edges in different directions. They are described in the following:

- A DISK structure element with its origin at the center and radius i is denoted by $r_i B_{disk}$.
- A LINE structure element with sizes of $4i \times i$ (length $\times$ width) and a α angle direction is denoted by $L^{\alpha}_{4i \times i}$.

3. Segmentation algorithm

3.1. Implementation of segmentation algorithm

For some complicated images, especially those in which the target objects are combined in a uneven background, it's difficult to segment interested particles satisfactorily. Clues for detecting features were discovered when we concentrated on the

(a) CarEx (b) Gems

Fig. 2. Examples of source images

TT with different sizes of disk structure elements. The difference between $T_i^{(}i)$ and $T_{i-1}^{(i-1)}$ includes our interested objects, and that image can be easily thresholded to make features stand out.

$$
\begin{aligned}
X_i &= |T_i^{(i)} - T_{i-1}^{(i-1)}|_B - X'_{i-1}, \;\; X'_i = \bigcup_{1 \le j \le i} X_j, \;\; X'_1 = \emptyset \\
T_i &= X_0 - X_0 \circ_g r_i B; \;\; T^i = X_0 \bullet_g r_i B - X_0,
\end{aligned}
\tag{6}
$$

where X_0, $|D|_B$, and X'_i denote original image, threshold operation by the gray level B and segmented images which hold different sizes of objects respectively. The differences of the neighbor TT results up to i are united together in X'_i with a certain size of features.

An example of the performances of this segmentation algorithm contrasted with the TT is shown in Fig.1.

3.2. Computational complexity

The computational cost of the segmentation algorithm is measured by counting the required number of set unions, set translations and set differences. We assumes implementation on special purpose image processing hardware, where the morphological operation can be implemented with a series of shift and arithmetical operations, and where an entire image can be processed in parallel.

We assume that the dilation and erosion of a set by a structure element B requires $2(Card(B) - 1)$ operations, where Card(.) denotes the cardinality of a set, and B is a 3×3 structure element. When we consider dilation of a set X by B as $X \oplus B = \cup_{b \in B}(X + b)$, the dilation requires $(Card(B) - 1)$ set translations and $(Card(B) - 1)$ set unions. This conclusion can be found in erosion operation.

The opening and closing operation in (6) can be shifted to another statement implemented only by dilation and erosion for speed up purpose:

$$\begin{aligned} X_0 \circ r_i B &= (X_0 \ominus r_i B) \oplus r_i B \\ &= ((\ldots(X_0 \overbrace{\ominus B \ominus B \cdots \ominus B)}^{i} \overbrace{\oplus B) \oplus B) \cdots \oplus B)}^{i} \\ &= (X_0 \overset{i}{\ominus} B) \overset{i}{\oplus} B \end{aligned} \tag{7}$$

Therefore, the cost of X_i is

$$O_{X_i} = O_{T_i} + O_{T_{i-1}} + 3 = 4S(Card(B) - 1)(2i - 1) + 5, \tag{8}$$

where S denotes the area of X_0.

As a result, the cost of the whole algorithm is

$$O_{X_i'} = \sum_{i=1}^{N} 4S(Card(B) - 1))2i - 1) + 5(N - 1) \tag{9}$$

The larger the original image is, the heavier is the computational load. Because the maximal shape of the 3 × 3 structure element B is a square, the worst case of our algorithm is:

$$O_{X_i'} \leq 32SN^2 + (32S + 5)N - (32S + 5) \tag{10}$$

In this case, the order of the segmentation algorithm is $O(N^2)$.

4. Extraction of characters from color cover image

The characters in cover images are intricate with a complicated background showing various sizes, shapes and directions. This situation makes the extraction a difficult task. In this section, a complete system to extract characters from color cover images using the presented segmentation algorithm is described.

4.1. Segmentation processing

The original color image is transformed to RGB images and the three gray-scale images are implemented independently. The source image is difficult to deal with in a general view. Thus we decompose it into simpler ones in this processing stage.

$$\begin{aligned} X_{ki} &= |T_{ki}^{(ki)} - T_{K(i-1)}^{k(i-1)}|_B - X'_{k(i-1)}, \quad X'_{ki} = \bigcup_{1 \leq j \leq i} X_{kj}, \quad X'_{k1} = \emptyset \\ T_{ki} &= X_{k0} - X_{k0} \circ_g r_i B_{disk}, \quad T^{ki} = X_{k0} \bullet_g r_i B_{disk} - X_{k0} \\ X_0 &\Rightarrow X_{c0} = \{X_{R0}, X_{G0}, X_{B0}\}; \quad c = \{R, G, B\}, \quad k \in c, \end{aligned} \tag{11}$$

where X_0, X_{k0} denote source color image and RGB gray-scale images.

The procedure of this processing stage is started with $r_1 B_{disk}$ and a series of sub-images X'_{ki} are generated in a recursive manner. The WTT T_{ki} extracts bright areas from a dark background when the BTT T^{ki} does the opposite operation. For

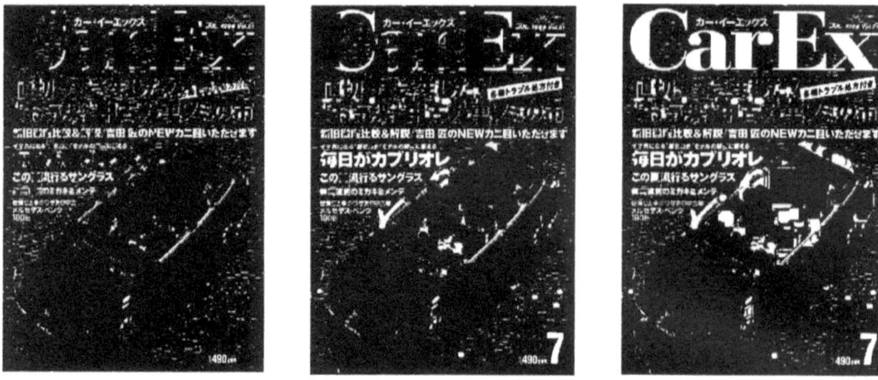

(a) $i = 4$ (b) $i = 10$ (c) $i = 30$

Fig. 3. Examples of segmented sub-images

256 gray level source images, the threshold operation is implemented by a value of $B = 31$ which was determined experimentally. The processing is stopped when $T_{kj}^{(kj)} = T_{k(j-1)}^{(k(j-1))}$ holds. The result of segmentation of "CarEx" (Fig.2(a)) is shown in Fig.3.

4.2. Extraction processing

The decomposed sub-images are processed by a morphological filter to emphasize the character region and suppress the small islands of noises.

$$E_{ki} = \begin{cases} ((X'_{ki} \circ r_{i-1}B_{disk}) \oplus r_{i+1}B_{disk}) \circ r_{2i}B_{disk} & (i \leq 10) \\ (X'_{ki} - X'_{k10}) & (i > 10) \end{cases} \tag{12}$$

Since the character regions are the main component in E_{ki}, they hold the peak values in the histogram. The peak values which are bigger than the average of all peak values are searched and $E_{ki} \times X_{m0}$ (see (13)) is thresholded by the selected peak values to extract characters standing out as:

$$H_{ki} = \bigcap_{m \in c \cap k^c} |E_{ki} \times X_{m0}|_B, \tag{13}$$

where, " $\times$ " denotes an arithmetic multiplication between two same size images for transforming binary image to gray-scale. The noise region attaching with the character candidate region may not hold the same gray level in all the three R,G, and B images. This characteristic is used to distinguish the characters from noises.

4.3. Refinement processing

The extracted characters in H_{ki} are broken and there remain some noises. A morphological filter derived from conditional dilation is implemented to refine the character candidate regions to get more satisfaction. Where the gray level B is the same with

(a) CarEx

(b) Gems

Fig. 4. Examples of result images

the extraction processing stage.

$$\begin{aligned} R_{ki0} &= ((H_{ki} \circ L^{0}_{i\times 4i}) \cap (H_{ki} \circ L^{\pi/2}_{i\times 4i})) \cup ((H_{ki} \circ L^{\pi/4}_{i\times 4i}) \cap (H_{ki} \circ L^{3\pi/4}_{i\times 4i})) \\ R_{kin} &= (R_{ki(n-1)} \oplus r_5 B_{disk}) \cap |X_{k0}|_B \qquad (14) \\ & if \;\; R_{kin} = R_{ki(n-1)} \; then \; stop \end{aligned}$$

Finally, subimages R_{kin} are united to obtain the resulting image X_r, which is denoted by

$$X_r = \bigcup_{\substack{1\le i\le j \\ k\in\{R,G,B\}}} R_{kin}. \qquad (15)$$

5. Experimental result

Color cover images of various magazines are used in our experiment to demonstrate the performance of our system. They are scanned in 100dpi with 256 gray levels. A SUN SPARC STATION 20 is used. Two examples out of six experimental images are shown in Fig.2.

The result for the two examples shown in Fig.4 appears encouraging. In the cover image "CarEx" (Fig.4(a)), 180 characters extracted from total 230 characters, showing a 78% extraction rate(ERA), when extracted characters(ECH) of "Gems" (Fig.4(b)) are 48 from total 52 characters with a 92% ER. The same problem which exists in the experiment is that the very small characters(VSC) can't be extracted so far. If the source images were scanned in a high resolution, a 99% ERA could be obtained.

It's not easy to estimate the extraction error precisely. In this paper, we introduce an "error extracted region"(ERE) to evaluate the performance of our approach in a opposite way. The "error extraction rate"(EER) shows the same mean in percent. The detail can be found in Table 1.

TABLE I
The results of characters extraction

source	sort	totals	VSC	ECH	ERA	ERE	EER
Carex	cover	230	50	180	78%	5	2%
Gems	cover	52	12	48	92%	2	4%
Magagin	cover	303	31	274	90%	12	4%
Jump	cover	267	12	257	96%	9	3%
Spirits	cover	243	10	235	97%	8	3%
Comic	cover	231	17	216	94%	4	2%

6. Summary

In this paper, a new approach for particle segmentation using mathematical morphology has been presented. We described a new idea based on the TT and recursive morphological operations, analyzed its computation complexity, and then described a complete system to extract characters from cover images. This new method can deal with more difficult segmentation problems than the other known algorithms, especially with the case where objects exist in a complicated background with irregular sizes and directions.

The proposed approach to detect characters in cover image was tested to be efficient in our experiment. We intend to proceed with this study for construction of a character recognition system on cover image using mathematical morphology.

References

1. L.Vincent and E.R.Dougherty, "Morphological segmentation for textures and particles", Digital Image Processing Methods, pp. 43-102, Marcel Dekker, 1994.
2. S.Liang and M.Ahmadi, "A Morphological Approach to Text String Extraction from Regular Periodic Overlapping Text/Background Images", Computer Vision, Graphics, Image Processing, VoL.56, No.5, Sep, pp.402-413, 1994.
3. S.Shuji, M.Michihiko, and I.Katsuo, "A method of extraction of character pattern from a color image based on the uniformity of the character color of a string", IEICE Technical Report, PRU94-29, pp.17-24, 1994.
4. F.Meyer,"Contrast Feature Extraction, "Quantitative Analysis of Micro-structures in Material Sciences, Biology and Medicine, J.-L. Chermant, ed., Special issue of Practical Metallography, Riederer Verlag, Stuttgart, Germany, 1978.
5. I.Pitas and A.N.Venetsanpoulos, "Mathematical shape decomposition", IEEE Trans. on Pattern Anal. Machine intell., Vol. 12, No.1, pp.38-45, Jan. 1990.
6. Hochong Park and Roland T.Chin, "Decomposition of arbitrarily shaped morphological structure elements", IEEE Trans. on Pattern Anal. Machine Intell., Vol.17, No.1, pp.2-15, Jan. 1995.
7. L.Vincent and P.Soille, "Watersheds in Digital Spaces:An Efficient Algorithm Based on Immersion Simulations", IEEE Trans. on Pattern Anal. Machine Intell., Vol.13, No.6, pp.583-598, June 1991.
8. R.M.Haralick, S.R.Sternberg, and Zhang, "Image analysis using mathematical morphology", IEEE Trans. on Pattern Anal. Machine Intell., Vol.9, No.4, pp.532-550, July 1987.

THE POWER OF MORPHOLOGICAL FILTERS ALONE AND WHEN COMBINED WITH LINEAR FILTERING

M.H. SEDAAGHI and Q. H. WU
Department of Electrical Engineering and Electronics
The University of Liverpool, Liverpool, L69 3GJ, U.K.

Abstract. This paper presents a robust algorithm for low-pass and band-pass filtering of a periodic 1-D (one-dimensional) and 2-D (two-dimensional) signal buried in noise using a combination of morphological and linear filters. The proposed method also illustrates how efficiently morphological operators filter aperiodic 1-D and 2-D signals compared with linear methods. Designing a suitable structuring element based on the characteristics of the input signal is also investigated.

Key words: Morphological filtering; Joint morphology and convolution; Structuring element design.

1. Introduction

Mathematical morphology ([4, 8]) has become an important formalism in image processing and computer vision as a geometrical approach [10] and is considered as a powerful tool for geometrical shape analysis [1, 3]. The original theoretical work of Matheron and Serra as a quantitative description of shape and size was initially used for binary images, but now it can be applied to grey-scale images and, generally speaking, for multi-scale images [2], and can be realized by special hardware [5, 7], and there are a lot of fast algorithms for implementing morphological filters as in [6, 11]. Serra ([9]) says: "Mathematical morphology is not only a theory but also a powerful image analysis technique. It is mathematical in the sense that the analysis is based on the set theory, topology, lattice algebra, random functions, etc. and it is called 'Morphology' since it aims at analysing the shape and form of the objects". Mathematical morphology and convolution methods have advantages and disadvantages by themselves. One of the disadvantages of linear filtering is that it blurs the image where morphological filters (MF) provide image enhancement without blurring the features (see Figures 1 and 2). Moreover they only use addition/subtraction accompanied with comparison where linear methods use multiplication and addition. On the other hand there is no strong analytical methods for MF as for linear methods due to their nonlinear operations, and the recovery of the original image is impossible after morphological operations (idempotency, i.e. $\psi(\psi(f)) = f$) where linear methods are reversible using both of time and frequency domain for analysis and design. The proposed method joins morphology with convolution to solve the deficits of each method when used individually, and obtains a robust filter using the advantages of both of them.

2. Morphological operators

This section introduces the definition of the operators used in MF. More details can be found in literature as in [1], etc. Let f and b denote respectively a 1-D grey-scale (GS) signal and a 1-D grey-scale structuring element (GSE) of length L, D_f and D_b denote their domains, and $f(\alpha) \mid \alpha \in D_f$. Erosion, dilation, opening, closing of f by b denoted respectively by $f \ominus b$, $f \oplus b$, $f \circ b$ and $f \bullet b$ are defined as follows.

$$(f \ominus b)(k) = \min_v \{f(k+v) - b(v)\}, \tag{1}$$

$$(f \oplus b)(k) = \max_v \{f(k-v) + b(v)\}, \tag{2}$$

$$(f \circ b)(k) = ((f \ominus b) \oplus b)(k) = \max_u \{\min_v \{f(k-u+v) + b(u) - b(v)\}\}, \tag{3}$$

$$(f \bullet b)(k) = ((f \oplus b) \ominus b)(k) = \min_u \{\max_v \{f(k+u-v) - b(u) + b(v)\}\}. \tag{4}$$

where $u, v \in D_b$. Open-closing is simply an opening followed by closing. Similar definition exists for close-opening.

3. Our approach towards joining morphology with convolution

Let $in(n)$ be the input signal defined as

$$in(n) = A_1 \cos(2\pi \frac{f_1}{f_s} n) + A_2 \cos(2\pi \frac{f_2}{f_s} n) + e(n), \tag{5}$$

where f_s is the sampling frequency, $0 < f_1 < f_2 < \frac{f_s}{2}$, and $e(n)$ is white noise. We want to filter the signal with the frequencies f_1 and f_2 separately from the noisy signal (see Figure 3). The result of the low-pass and band-pass linear filtering (see Figures 3-b,i), despite clearing unwanted frequencies (Figures 3-f,l for the normalised spectra) satisfactorily, is accompanied by a delay at the start of the signal which is not desirable. This is one of the shortcomings of the linear methods occured on most of the well-known linear filters (as an example Butterworth low-pass and band-pass filters of the order 5 have been applied for illustration). On the other hand we have experienced that the average of open-closing and close-opening is a desirable morphological operator for filtering. The next steps have been taken as the proposed algorithm:

1. Choose the structuring element SE_1 having the shape of half of a sinus. The length of SE_1 equals to $\frac{f_s/2}{f_2}$ with its peak at its origin. Therefore it will remove the noise which has smaller size than SE_1. The following operation removes noise leaving a residue signal containing the frequencies f_1 and f_2.

$$residue(n) = [((in \circ SE_1) \bullet SE_1)(n) + ((in \bullet SE_1) \circ SE_1)(n)]/2. \tag{6}$$

2. The structuring element size should be increased from $\frac{f_s/2}{f_2}$ to a larger size of length $SE_2 = \frac{f_s/2}{f_1}$ (with sinusoidal shape having its peak at its origin) to extract the low frequency (f_1). As the size of the structuring element is larger than $\frac{f_s/2}{f_2}$, the following equation will remove f_2 from the residue of the Equation 6 preserving f_1:

$$lowpass(n) = [((residue \circ SE_2) \bullet SE_2)(n) + ((residue \bullet SE_2) \circ SE_2)(n)]/2. \quad (7)$$

The result is shown in Figure 3-c with the normalised spectrum shown in Figure 3-g.

3. The following operation will extract f_2 as a band-pass filtering (see Figures 3-j,m for time and frequency domain respectively).

$$bandpass(n) = residue(n) - lowpass(n). \quad (8)$$

Figures 3-c,j show that morphological low-pass and band-pass filters cause no time delay and the original shapes have been preserved. However the normalised spectra shown in Figures 3-g,m for morphological low-pass and band-pass filters respectively are worse than the corresponding spectra of the linear low-pass and band-pass filters shown in Figures 3-f,l.

4. To take the advantage of linear methods besides MF we apply the convolution

$$out(n) = morph(n) * h(n), \quad (9)$$

where $*$ denotes the convolution in time domain, $morph(n)$ is $lowpass(n)$ or $bandpass(n)$ defined in Eqs. 7, 8, $out(n)$ is the desired output, and $h(n)$ is a transfer function that will be defined below. We develop the above idea in frequency domain as:

$$OUT(f) = MORPH(f) \,.\, H(f) \quad (10)$$

where $MORPH(f)$ is the frequency response of the morphological filters (LOWPASS(f) or BANDPASS(f)), $OUT(f)$ is the frequency response of the desired output and $H(f)$ is a transfer function defined as $H(f) = \exp(-\alpha(f - f_0))$ where $f_0 = \pm f_1$ for low-pass, and $f_0 = \pm f_2$ for band-pass filtering, and α is a constant that determines the decay ratio of the frequencies away from the central frequency f_0. Figures 3-d,k show the desired low-pass and band-pass outputs and Figures 3-h,n show their normalised spectra respectively. $Out(n)$ conveys both the advantages of morphological and linear filtering as mentioned above. Figure 4 shows the above results for a periodic 2-D signal.

For aperiodic signals or the signals containing a lot of main harmonics in frequency domain (except noise), we can trust the morphological operators. Figures 1 and 2 show how reliably MF dominate over linear filtering for medical images and a aperiodic square signal respectively. The average of open-closing and close-opening has been used as the morphological operator. The structuring element used is flat and its size is as mentioned under the figures. The linear filter applied for Figure 1 is: $y(n) = .85 * y(n-1) + .05 * x(n) + .05 * x(n-1) + .05 * x(n-2)$. The kernel applied for each pixel of the linear filtering in Figure 2 is the summation of the eight neighboured pixels and the central pixel divided by 9.

4. Structuring element design

We have empirically shown that the following rules exist for the design of the structuring element (SE).

1. For periodic signals (with a limited spectrum), the size of SE should approximately equal the number of samples building the corresponding frequency in time domain. For example if the sampling frequency $f_s = 128$, and $f_1 = 5Hz$, the length of SE should equal to 11 ($\frac{f_s/2}{f_1} = 12.8 \rightarrow 11$ to have a symmetric SE with the origin at its centre, the size of SE should be odd). After getting the size of SE, its shape will be constructed similar to the shape of the desired signal (i.e. sinusoidal in the above example).
2. For aperiodic signals we need to find the approximate length of the desired shapes to be extracted and design a similar shape and length for SE. However if it is difficult to decide, the first and easiest assumption is to use a flat SE of the appropriate size starting from short sizes and leading to longer sizes and to apply a criterion to determine the best result.

5. Conclusion

We have shown the efficiency of the MFs when combined with linear methods for periodic signals in 1-D and 2-D that have a clear spectrum of the desired components to be extracted. For other cases when requiring non-linear filtering, the MFs seem to be completely acceptable.

References

1. R. M. Haralick, S. R. Sternberg, and X. Zhuang. Image analysis using mathematical morphology. *IEEE Trans. Pattern Anal., Machine Intell.*, PAMI-9(4):532–550, July 1987.
2. H. Heijmans. *Morphological Image Operators.* Academic Press, Boston, 1994.
3. P. Maragos and R. W. Schafer. Morphological systems for multidimensional signal processing. *Proceedings of the IEEE*, 78(4):690–710, Apr. 1990.
4. G. Matheron. *Random sets and integrated geometry.* Wiley, New York, 1975.
5. A. M. S. J. Ko and K. H. Lee. Fast recursive algorithms for morphological operators based on the basis matrix representation. *IEEE Trans. Image Proc.*, 5(6):1073–1077, June 1996.
6. M. H. Sedaaghi. Direct implementation of open-closing in morphological filtering. *Electronics Letters*, 33(3):198–199, Jan. 1997.
7. M. H. Sedaaghi and Q. H. Wu. Real-time implementation of grey-scale morphological operators. *Electronics Letters*, 33(21):1761–1763, Oct. 1997.
8. J. Serra, editor. *Image analysis and mathematical morphology.* Academic Press, New York/London, 1982.
9. J. Serra and P. Soille. *Mathematical morphology and its applications to image processing.* Kluwer Academic Publishers, The Netherlands, 1994.
10. R. van den Boomgaard. *Mathematical Morphology: Extensions Towards Computer Vision.* PhD thesis, University of Amsterdam, 1992.
11. D. Wang and D. C. He. A fast implementation of 1-D grayscale morphological filters. *IEEE Trans. Circuits and Systems*, 41(9):634–636, Sept. 1994.

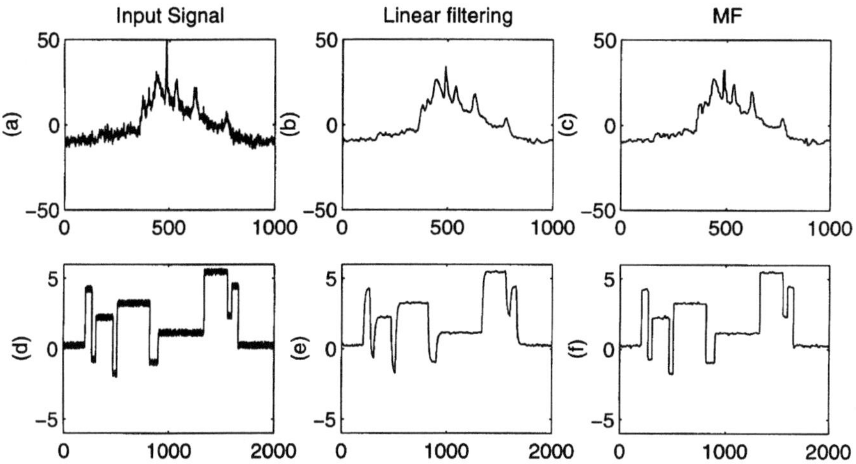

Fig. 1. The effect of MF on aperiodic signals: (a): Input NMR signal buried in noise; (d): Input aperiodic square signal; (b,e): linear filtering; (c,f): MF with a flat SE of size 7.

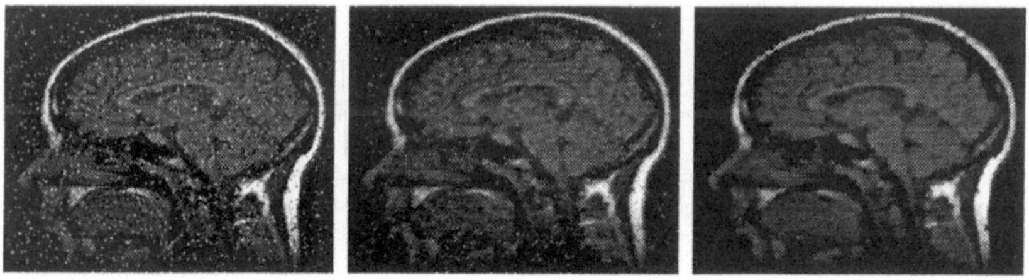

(a) Input (b) Linear (c) MF

Fig. 2. The effect of MF on MRI image: (a): Input image buried in noise; (b): linear filtering; (c): MF with a flat SE of size 3.

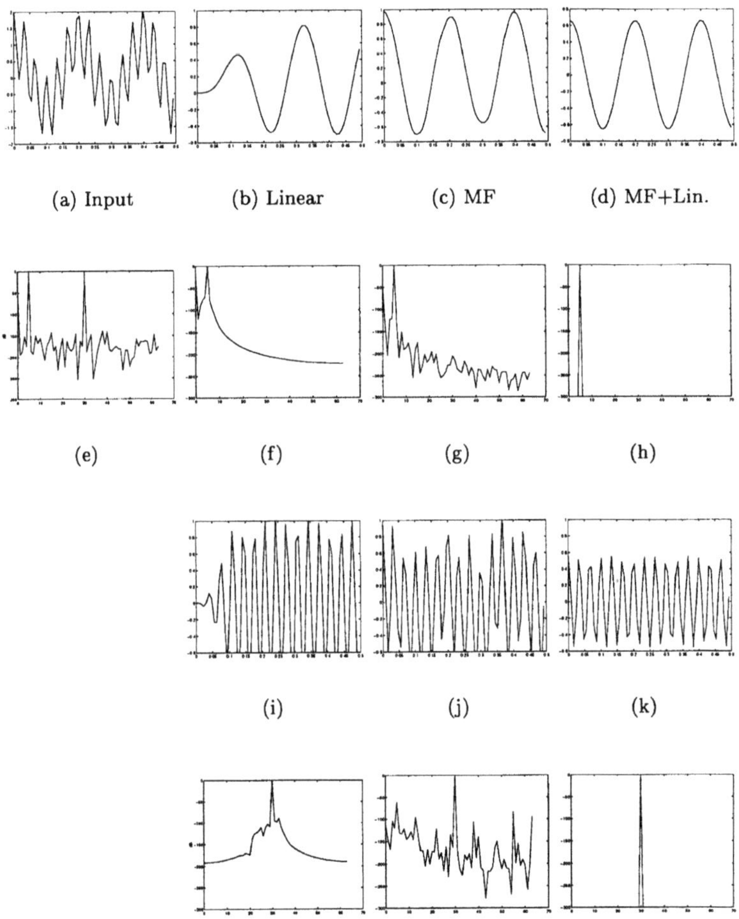

(a) Input (b) Linear (c) MF (d) MF+Lin.

(e) (f) (g) (h)

(i) (j) (k)

(l) (m) (n)

Fig. 3. Frequency analysis of a periodic signal: (a): Input signal; (b,c,d): low-pass filtering with linear, MFs, and combination of them respectively; (e,f,g,h): normalised spectra of a,b,c,d respectively; (i,j,k): band-pass filtering with linear, MFs, and combination of them respectively; (l,m,n): normalised spectra of i,j,k respectively.

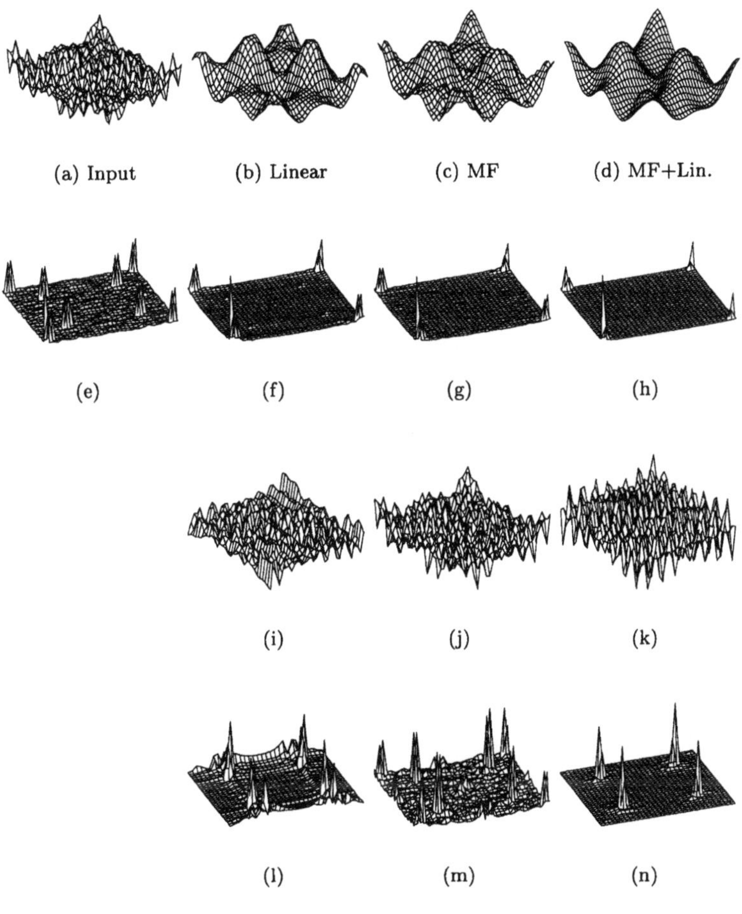

Fig. 4. Frequency analysis of a periodic 2-D signal: The definitions of a-n similar to of the Figure 3.

MORPHOLOGICAL UNWRAPPING OF INTERFEROMETRIC PHASE-MAPS

PIERRE SOILLE
Ecole des Mines d'Alès & EERIE
Laboratoire de Génie Informatique et d'Ingénierie de la Production
Parc scientifique G. Besse, F-30000 Nîmes Cedex, France
Email: Pierre.Soille@eerie.fr

Abstract.
Phase-sampling interferometry outputs a phase-map wrapped into a $[-\pi, \pi]$ interval. The unwrapping of these phase-maps is essential in automatic quantitative quality control and visual inspection applications. The algorithm we propose in this paper, first extracts a region of interest by applying morphological filters to the input interference images. High and low fringe boundaries are then obtained using morphological gradients and watersheds. Finally, the image is unwrapped by a fast propagation algorithm using queue data structures.

Key words: interferometry, quality control, fringe analysis, unwrapping, segmentation.

1. Introduction

Applications of phase-measurement interferometry can be divided into three major types of measurement [5]: surface shape, surface roughness and metrology. Common techniques for the calculation of the phase associated with a fringe pattern imply the shifting of a reference phase through known increments (phase-sampling or quasi-heterodyne method) or by addition of a substantial tilt to the wave front that produces carrier fringes and then by a further Fourier transformation of the fringe pattern (Fourier transform method). In all these cases the phase is calculated by means of an inverse trigonometric function. This function return only principal values, i.e., values in the $[-\pi, \pi]$ interval, generating a discontinuous phase-map wrapped into a $[-\pi, \pi]$ interval [13]. For example, the phase-shifting interference techniques require 4 interferograms corresponding to the reference phases $\phi_r = 0°$ (f_1), $90°$ (f_2), $180°$ (f_3) and $270°$ (f_4). The wrapped phase s' is then given by[1]

$$s' = \arctan \frac{f_2 - f_4}{f_3 - f_1},$$

and the unwrapped phase s by

$$s = s' + n\,2\pi, \tag{1}$$

[1] Further details about the generation of wrapped phase-maps using interferometry can be found in [12] and [7]. The use of optical measurement techniques such as interferometry for industrial applications is detailed in [4].

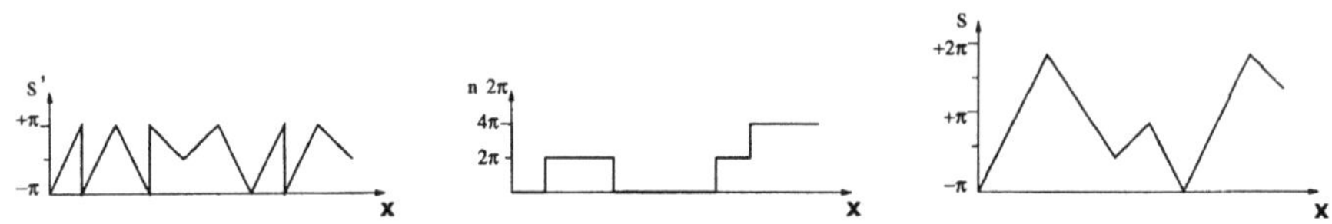

(a) Wrapped phase s' of a 1-D signal.

(b) Bias signal corresponding to *(a)*.

(c) Unwrapped phase of *(a)*.

(d) Wrapped phase s'

(e) Bias signal corresponding to *(d)*.

(c) Unwrapped phase of *(d)*.

Fig. 1. Unwrapping of 1-D and 2-D wrapped phase-maps by adding a bias signal to the original signal (see Eq. 1).

where $n\,2\pi$ is the 'bias' signal which must be added to the wrapped phase for its unwrapping. This principle is illustrated in Fig. 1 on a 1-D wrapped signal as well as a 2-D wrapped image.

The aim of an unwrapping algorithm consists therefore in finding the correct bias signal. For ideal fringe patterns it is enough to apply the following basic algorithm to the first image line:

1. Forward scan of the pixels p of the line, starting with the second pixel
2. if $[s'(p) - s'(p-1)] \bmod 2\pi < \pi$
3. $s'(p) \leftarrow s'(p-1) + (\text{int}\{[s'(p) - s'(p-1)]/2\pi\} + 1)2\pi$
4. else if $[s'(p) - s'(p-1)] \bmod 2\pi > \pi$
5. $s'(p) \leftarrow s'(p-1) + (\text{int}\{[s'(p) - s'(p-1)]/2\pi\} - 1)2\pi$

where $\text{int}(x)$ returns the integer part of the number x. This algorithm is then applied to the first image column and finally to all unprocessed image lines. However, actual fringe patterns are are usually affected by several error sources which make this algorithm unsuitable to most practical applications. Error sources that frequently arise are:

- Electronic and speckle noise.
- Objects of arbitrary shape not covering the whole image plane (holes, corners, etc.).
- Points with low modulation. These points are due to areas of low visibility or areas outside the phase range of the acquisition system.
- Aliasing due to inappropriate sampling frequency.
- Object discontinuities such as adjacent regions having very different depth values.

Consequently, the unwrapping of real wrapped phase-maps is not a trivial problem. In this paper, we propose a robust algorithm based on mathematical morphology [16]. It handles the two first error sources listed above and, in some situations, the three last.

The paper is organized as follows. A brief overview of current algorithms for unwrapping wrapped phase-maps is proposed in Sec. 2. The proposed algorithm is detailed in Sec. 3. Before concluding, results on various images are presented in Sec. 4.

2. Current algorithms

A review of phase unwrapping techniques in fringe analysis can be found in [14, 7, 8]. Most available algorithms belong to one of the following categories:

- Search for local inconsistencies. For example, [6] describe a cellular automaton which looks for inconsistencies in a closed loop around each block of 4 pixels in the phase-map. The sum of the differences of the wrapped phases along the loop is computed. This sum modulo 2π should be equal to 0 except for inconsistent blocks where it equals ± 1. This technique detects only local logical inconsistencies and cannot detect invalid data points.
- Cut methods. The basis of this technique is to place cuts, on a scan line by scan line basis, between points of phase discontinuity in order to minimize the length of the propagated discontinuity.
- Search for regions without any inconsistency. This technique attempts to segment the fringe field into areas of inconsistency and then to relate these areas to one another.
- Minimum spanning tree methods. In this latter approach, the wrapped phase-map is first divided into tiles which are considered as the nodes of a tree. The procedure starts by unwrapping the phase within the tiles. An edge is defined for each pair of adjacent tiles. The weights of the edges is a measure of the validity of phase unwrapping across the boundaries of the tiles. The corresponding minimum spanning tree produces the paths that should be followed for unwrapping the whole set of tiles.

3. Proposed algorithm

We propose a new algorithm based on a morphological processing of the wrapped phase-map. This algorithm can be divided into three main steps:

1. Masking of the regions not carrying relevant phase information.
2. Detection of the internal and external fringe boundaries.
3. Fast unwrapping using an integration procedure based on queue data structures.

3.1. Masking

A mask of the regions carrying relevant phase information is obtained by processing the input interference images. A morphological closing with a square whose size is greater than the gap separating successive interference fringes allows us to fill the gaps between the regions where interference occurs. The arithmetic difference between the closing and the original image outputs an image of these gaps. This operation is called a top-hat by closing or black top-hat [9, 10]. An image of the interference regions themselves is determined by computing the difference between the interference image and its opening by a square larger than the interference regions (i.e., a top-hat by opening or white top-hat). The sum of the white and black top-hats outputs an image of the regions carrying relevant information. In fact, the sum

(a) Interference image (b) Self-dual top-hat of *(a)* (c) Threshold of *(b)*

Fig. 2. Detection of the region of interest or ***masking*** of an interference image using a sef-dual top-hat using a square slightly larger than the dark and bright strips as structuring element.

of the white and black top-hats of an image comes down to the arithmetic difference between its closing ϕ and its opening γ. Due to its self-duality property, we call this operator ***self-dual top-hat*** and denote it by ϱ:

$$\varrho = \phi - \gamma.$$

A threshold T of the self-dual top-hat for values greater than t outputs a mask of the regions carrying relevant phase information:

$$\text{Mask}(f_i) = T_{[t,t_{\max}]}[\varrho(f_i)].$$

Figure 2 illustrates the mask detection starting form on an input interference image. In practice, this masking operation is applied in parallel to all interference images and the union of the resulting masks is used as the actual mask:

$$\text{Mask} = \bigcup_i \text{Mask}(f_i).$$

If necessary, the union can be replaced by an intersection and the resulting mask can be further processed by an opening or even an erosion. These operations increase the noise immunity by masking additional discontinuity sources. In any case, the mask is used as region of interest when detecting the fringe boundaries and unwrapping the phase-map.

3.2. Fringe boundaries

Our fast unwrapping procedure detailed in Sec. 3.3 requires the prior detection of the fringe boundaries. In addition, when going from one fringe to another, we need to know whether we go up or down. This information will be used to subtract or add a multiple of 2π when calculating the bias image.

The high and low boundaries of each fringe are obtained from the internal and external gradients[2] of the wrapped phase-map (see Fig. 3). A mask of the internal and external gradients is obtained by thresholding the gradient images. However,

[2] The internal gradient is also called gradient by erosion and the external gradient, gradient by dilation [1, 15]. They are both referred to as half-gradients because their sum equals the morphological gradient.

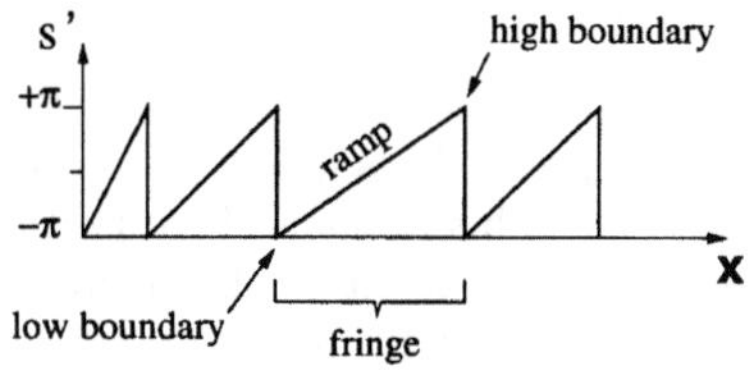

Fig. 3. One-dimensional wrapped phase-map: a fringe with its high and low boundaries plus the corresponding ramp region.

due to the error sources detailed in the introduction, it may be that the detected fringe boundaries are not connected (this is a classical problem of edge detection techniques based on gradient thresholding). A better solution is to threshold the wrapped phase-map for values for the range $[-\pi, 0]$ and use the resulting image as a marker for the watershed segmentation [2, 11, 19, 22, 3] of the internal gradient image. Denoting by ρ^- the internal gradient, WS the watershed transformation and $R^\star$ the morphological reconstruction by erosion [1, 21] used for minima imposition, we obtain the following equation:

$$\text{High-Boundaries}(s') = WS[R^\star_{\rho^-(s') \wedge \text{Marker}}(\text{Marker})],$$

where Marker, is the marker image obtained by thresholding the wrapped phase-map s' for values greater than 0 and setting the resulting object pixels to $t_{\max}$:

$$\text{Marker}(x) = \begin{cases} 0, & \text{if } s'(x) \leq 0, \\ t_{\max}, & \text{otherwise.} \end{cases}$$

When markers of two adjacent fringes are connected, it is necessary to further process the marker image so as to disconnect all connected fringes. A union of openings with long line segments in all possible discrete directions is best suited to this task because fringes are usually elongated and thin regions. This procedure can be fasten using the recursive procedure for line erosions and dilations at arbitrary angles [18].

The low boundaries of the wrapped fringes are following the high boundaries along their side which has very low wrapped-phase values. A mask of the low boundaries is therefore obtained by first calculating the external boundaries of high boundaries. The mean grey scale value along each connected component of the resulting external boundaries is used as a decision criterion to select those edges that correspond to low boundaries (they have a very low mean grey scale value). Multiple points of the network of watershed lines are removed beforehand so as to avoid connections between low and high sides of the original high boundaries.

An image is then created with low boundary pixels at value `low`, high ones at value `high`. Remaining pixels of the region of interest are set to `ramp` and pixels falling outside the mask of the regions carrying phase information are set to `bckg` value.

3.3. Fast unwrapping

Our method for unwrapping is based on the calculation of the bias image, the unwrapped phase-map being defined as the sum of the wrapped phase-map and the bias image (see Eq. 1).

The bias image is calculated from the boundary image generated at the end of the previous section. The algorithm proceeds as follows. The boundary image is scanned until a `ramp` pixel is encountered. This point is then used as the reference pixel for the processing of the connected component of the region of interest it belongs to. The bias of the reference pixel is set to 0. A queue data structure is then used to fill the connected region of `ramp` pixels it belongs to, while setting the bias image to zero within this region. When boundary pixels are encountered, they are inserted into a queue. In fact, there are two queues: one for low and one for high boundary pixels. When no more pixels are queued, one of the queues holding boundary pixels is considered. Assume that it concerns low boundaries. Pixels from this queue are extracted from the queue and set to the current bias value. However, the bias of the neighbour pixels of the queue which are high boundaries is set to the current bias value minus 1. The same bias value is given to all pixels belonging to the `ramp` region adjacent to these high boundaries. The same procedure is applied to the queue holding high boundary pixels, but the current bias values is incremented rather than decremented by 1. The procedure terminates when the connected component of the region of interest marked by the reference pixel is fully processed. The boundary image is then further scanned until all connected components of the mask of the regions carrying phase information are processed. The resulting image is multiplied by 2π so as to generate the bias image (see Figs. 1b and e). It is then added to the input wrapped phase-map.

4. Results

Table I shows five wrapped phase-maps that have been unwrapped using our algorithm. Except for large discontinuities and aliasing, all sources of errors can be found in these images.

Processing time for a 670×560 image is about 30 seconds on a 75 MHz computer. Four images of the same size as the input image are required to run the algorithm.

When large discontinuities (or inappropriate sampling rates) are present, the algorithms fails to unwrap the phase-maps. Examples are provided in Fig. 4.

5. Concluding remarks and perspectives

The proposed algorithm is fast and reliable for a wide range of wrapped phase-maps. It deals with source errors such as speckle and electronic noise, objects that partly cover the image area, small- to medium-sized discontinuities, and, to a certain extent, points with low modulation. Aliasing and large scale discontinuities are not supported by the algorithm.

Current research concentrates on the automatic detection of large discontinuities. We also found that the automatic extraction of elevation contour lines on digitized topographic maps and the interpolation of digital elevations models from these contour lines [17] is closely linked to the unwrapping problem. The automatic extraction of growth rings [20] is another application related to the automatic fringe detection problem. First results show that these research topics will benefit each other. The

TABLE I
Examples of wrapped phase-maps using the proposed morphological unwrapping algorithm.

	Wrapped phase-map	Unwrapped phase-map
Surface with small- to medium-discontinuities		
Arbitrary shaped, high precision, mechanical tool	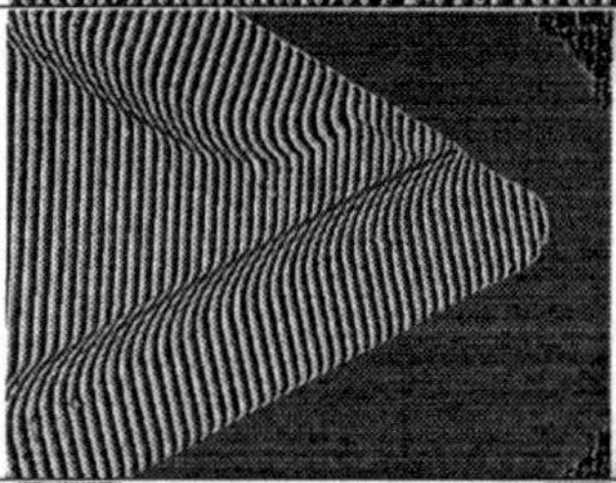	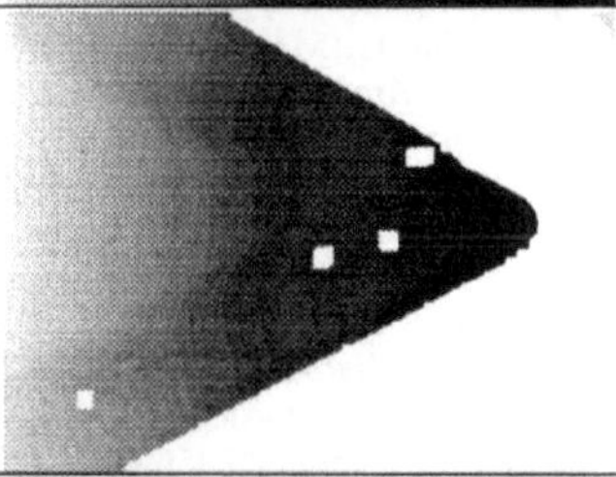
Tool with 2 pixel-wide fringes		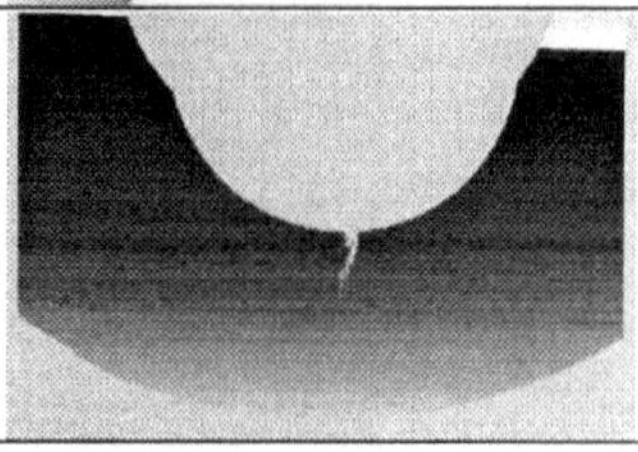

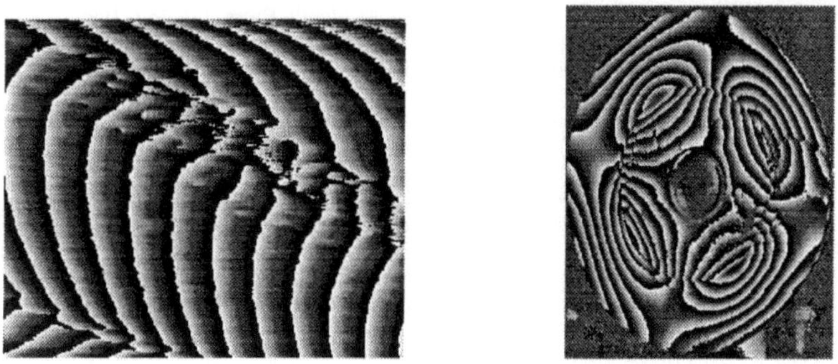

Fig. 4. Two wrapped phase-maps showing large discontinuities that cannot be resolved by the proposed algorithm.

design of a unifying algorithm is also currently investigated.

References

1. S. Beucher. *Segmentation d'images et morphologie mathématique.* PhD thesis, Ecole des Mines de Paris, June 1990.
2. S. Beucher and C. Lantuéjoul. Use of watersheds in contour detection. In *International Workshop on Image Processing*, Rennes, September 1979. CCETT/IRISA.
3. S. Beucher and F. Meyer. The morphological approach to segmentation: the watershed transformation. In E. Dougherty, editor, *Mathematical morphology in image processing*, chapter 12, pages 433–481. Marcel Dekker, 1993.
4. B. Breuckmann. *Bildverarbeitung und optische Meßtechnik in der industiellen Praxis.* Franzis Verlag GmbH, München, 1993.
5. K. Creath. Phase-measurement interferometry techniques. In E. Wolf, editor, *Progress in Optics*, volume 26, pages 349–393. Elsevier Science Publishers, 1988.
6. D. Ghiglia, G. Mastin, and L. Romero. Cellular-automata method for phase unwrapping. *J. Opt. Soc. Am.*, 4:267–279, 1987.
7. P. Hariharan and D. Malacara, editors. *Interference, interferometry, and interferometric metrology*, volume MS 110 of *SPIE Milestones Series*, 1995.
8. P. Judge and P. Bryanston-Cross. A review of phase unwrapping techniques in fringe analysis. *Optics and Lasers in Engineering*, 21(4):199–239, 1994.
9. F. Meyer. Iterative image transformations for an automatic screening of cervical smears. *Journal of Histochemistry and Cytochemistry*, 27:128–135, 1979.
10. F. Meyer. Automatic screening of cytological specimens. *Computer Vision, Graphics, and Image Processing*, 35:356–369, 1986.
11. F. Meyer and S. Beucher. Morphological segmentation. *Journal of Visual Communication and Image Representation*, 1(1):21–46, September 1990.
12. W. Osten. *Digitale Verarbeitung und Auswertung von Interferenzbildern.* Akademie Verlag, Berlin, 1991.
13. J. Quiroga and E. Bernabeu. Phase-unwrapping algorithm for noisy phase-map processing. *Applied Optics*, 33(29):6275–6731, October 1994.
14. G. Reid. Automatic fringe pattern analysis: a review. *Optics and lasers in engineering*, 7:37–68, 1986.
15. J.-F. Rivest, P. Soille, and S. Beucher. Morphological gradients. In T. Szoplik, editor, *Selected Papers on Morphological Image Processing: Principles and Optoelectronic Implementations*, volume MS 127 of *SPIE Milestones Series*, pages 295–305. Society for Photo-Optical Instrumentation Engineers, 1996. [Reprinted from J. Electron. Imaging, (3)2, 137–146, 1993].
16. J. Serra. *Image analysis and mathematical morphology.* Academic Press, London, 1982.
17. P. Soille. Spatial distributions from contour lines: an efficient methodology based on distance transformations. *Journal of Visual Communication and Image Representation*, 2(2):138–150, June 1991.
18. P. Soille, E. Breen, and R. Jones. Recursive implementation of erosions and dilations along discrete lines at arbitrary angles. *IEEE Transactions on Pattern Analysis and Machine Intelligence*, 18(5):562–567, May 1996.
19. P. Soille and L. Vincent. Determining watersheds in digital pictures via flooding simulations. In M. Kunt, editor, *Visual Communications and Image Processing'90*, volume SPIE-1360, pages 240–250, Lausanne, 1990.
20. S. Som, J. Davis, P. Wells, and I. Svalbe. Morphology methods for processing tomographic images of wood. In K. Fung and A. Ginige, editors, *Proc. DICTA'93: Digital Image Computing: Techniques and Applications*, pages 564–571, 1993.
21. L. Vincent. Morphological grayscale reconstruction in image analysis: applications and efficient algorithms. *IEEE Transactions on Image Processing*, 2(2):176–201, April 1993.
22. L. Vincent and P. Soille. Watersheds in digital spaces: an efficient algorithm based on immersion simulations. *IEEE Transactions on Pattern Analysis and Machine Intelligence*, 13(6):583–598, June 1991.

SHOALING WAVES AND WATER DEPTH DETERMINATION USING MATHEMATICAL MORPHOLOGY

SUZANNE M. LEA
Department of Mathematical Sciences
University of North Carolina at Greensboro
Greensboro, NC 27402

and

MATTHEW LYBANON
Remote Sensing Applications Branch
Naval Research Laboratory
Stennis Space Center, MS 39529

Abstract. There is substantial interest in rapid depth determination near coastal areas using remotely sensed images. We have investigated a method which uses measured wavelengths (crest separation) and wave speeds (crest motion from a time sequence of images) to find depths. Data are generated by a simulator program. Mathematical morphology accurately locates crests. The data quality is uneven across the time stack; the variation provides a method of bounding reasonable values for the bottom slope from above and below. For good data in this time stack, predicted depths agree with actual depths to within 0.13 m on the average; the largest difference is -0.52 m, with actual depth shallower than predicted for this time slice

1. Introduction

There is substantial interest in rapid depth determination near coastal areas using remotely sensed images. The problem is of particular interest for coastal areas where the depth can change rapidly because of dredging or sedimentation. Among the methods proposed to find depths is direct measurement of wavelengths ($\lambda = 2\pi/k$, where k is wave number) and wave speeds ($c = \omega/k$, where $\omega = 2\pi f$ and f is frequency), followed by solving the dispersion relation,

$$\omega^2/g = k\, tanh(k\, h) \tag{1}$$

for the depths h. (g is the acceleration of gravity.) An initial approach to the problem, which increases the processing speed dramatically through the use of mathematical morphology, has been reported (Lea *et al.*, 1996). The aim is to produce an operational Navy semi-automated system to process time sequences of images of wave crests. Presently such processing is done manually.

Wavelengths may be estimated by measuring the distance between wave crests in a single image; wavelengths decrease as the waves approach the shore. Estimating wave speeds requires a time sequence of images; the distance a particular crest travels between images can be measured to determine the speed. Clearly, the images must be close enough together in time that motion of a single crest can be unambiguously determined, and far apart enough in time that the wave crests move by visible amounts. For most ocean coasts, a time separation of 1 to 5 s is adequate.

Fig. 1. An example image from the time sequence used. The image is 512 x 512 pixels at 3.048 m/pixel. The contrast has been enhanced.

2. Data

The lack of good images with known depths and the presence of noise in real images cause problems when trying to evaluate methods of analysis. Consequently, we developed a wave simulator which can generate a time sequence of images of wave crests (sea surface elevation). The simulator uses the linearized wave equations for an inviscid irrotational liquid (Weiss, 1997; Lea *et al.*, 1998) and a standard deep-water wave model (Hasselmann *et al.*, 1976; Donelan *et al.*, 1985; Banner, 1991; Wyatt *et al.*, 1997) to produce wave components of different frequencies traveling in different directions. Each component is allowed to approach the shore over a specified piecewise linear bottom of decreasing depth. Each component is also followed forward in time. Superposition of components at a single time yields an image of wave crests at the time; the collection of images forms a time sequence.

The simulator has been extensively tested against model predictions and actual wave images or data (Weiss, 1997; Lea *et al.*, 1998). It provides results in good agreement with the data for wave heights and profiles outside the breaking region and for bottom slopes between 0.005 and 0.2. The advantage of simulator image sequences is that the bottom is precisely known and there is no noise. Consequently, the accuracy of processing methods can be determined.

The data (Fig. 1) produced by the simulator for this study are a time sequence

Fig. 2. The image of Fig. 1 after the initial processing stages, immediately before its incorporation into the time stack image. The lines are wave crest positions.

of images of wave crests moving directly toward a shoreline slightly above the top of the image. The images are 512 x 512 pixels, at a scale of 3.048 m/pixel. There are 11 images in the time sequence, separated by 1 second. The bottom depth used has a constant slope of 0.0065 perpendicular to the shoreline, and a depth of 0.1 m at the top of the image.

Use of simulated data allows evaluation of the accuracy of depth results in the absence of noise. Tests with actual data are planned to determine the effects of noise.

3. Processing

Accurately locating the wave crests in each image of the sequence is essential to successful depth determination. The waves consist of a mixture of frequencies and include directional spreading (lower amplitude components traveling at an angle to the shore). However, refraction causes waves traveling over a sea bottom with a gradient perpendicular to the shoreline to exhibit wave crests parallel to the shoreline. Consequently, the wave crests are linear features in an image (Fig. 1), and mathematical morphology provides a reliable method of locating the crests.

The wave crests are closed using a 3 $\times$ 3 structuring element. Horizontal por-

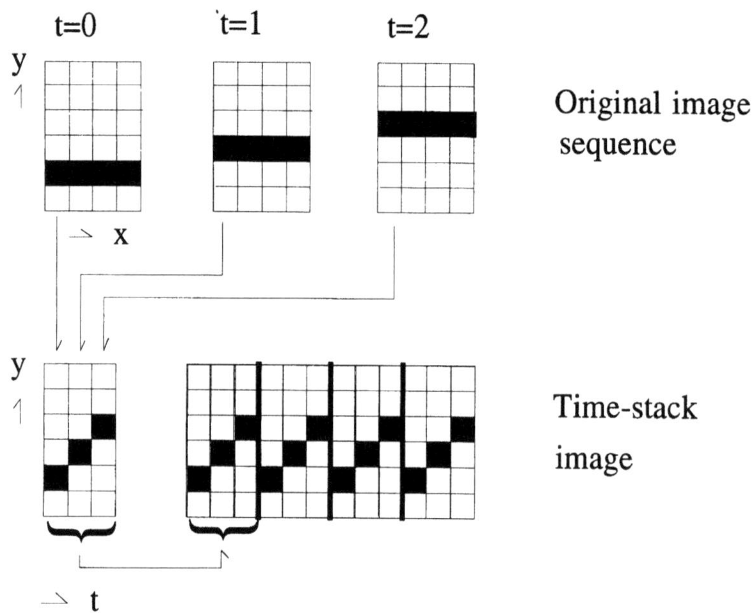

Fig. 3. Construction of a time stack image from a time sequence of images. The dark horizontal line in the sequence represents a wave crest moving upward. Corresponding columns from each image are made adjacent to form a time slice, and time slices are made adjacent to form a time stack image. Vertical motion of wave crests between images appears as diagonal lines in the time slices.

tions of the crests (parallel to the shore) are extracted by four successive horizontal erosions followed by six successive horizontal dilations using a 1×3 structuring element. Successive erosions remove short horizontal portions belonging to crests at an angle to the shore. The horizontal dilations restore the surviving horizontal crests and extend them slightly to fill data gaps. The resulting horizontal crest portions are then closed once more and thinned, resulting in a binary image. The result of processing Fig. 1 is shown in Fig. 2. The minimum depth at which crests can reliably be detected using this method is 0.7 m, well shoreward of the breaking region for real waves.

The time sequence of images is then converted into a time stack image by placing corresponding columns from each image adjacent to one another, forming a time slice, and placing the slices adjacent to each other to form the time stack image (Fig. 3). Since the images are binary images, each row of a time slice in the time stack image is encoded as a single (short) integer to decrease the memory requirements and the time required to read or write the image.

Motion of wave crests in the original sequence appears as diagonal lines in the time slices of the time stack image. Since the waves are moving up toward the shoreline (above the top of the image in Fig. 1), the diagonal lines will have positive slopes. Since waves slow down as they approach the shore, the slope of the diagonal lines will decrease at image points closer to the shore. The slope of a diagonal line provides a value for the speed of a wave crest, whereas the spacing between lines provides a set of values for the wavelengths.

Because the time slices are binary images, the full power of directional mathematical morphology (Kurdy and Jeulin, 1989; Jeulin and Kurdy, 1991; Agam *et al.*, 1996; Soille *et al.*, 1996) is not required or the simulated data (although it may be necessary for actual data). In essence, our method involves identifying and aligning the pixels in each diagonal line. We assume the slice is Y × T and that pixels are on or off. To accomplish the identification and alignment, the following method is used:

1. For each column, extract the y (vertical) coordinate values for the pixels which are on, and save them in the corresponding column of an M × T array. M is the maximum over all columns of the number of coordinate values extracted; the number of values extracted in each column may be different, because of partial or missing lines or noise. The result of this step is an array where the y coordinate values for each pixel that was on in column t are listed in consecutive positions of column t of the M × T aray, for $0 \leq t < T$.
2. Move each column down so that the bottom-most value in each column is aligned in the bottom row of the M × T array.
3. For each column t except the last and for each y-value y_i in column t, find the y-value y_k in the next column t+1 which is closest in value to and greater than y_i. (The *greater than* criterion arises because our lines have positive slopes.) Place y_k in the same row as y_i, moving the values in column t +1 above y_k as a group with y_k. This procedure may result in removing values which are not part of a diagonal line that starts in the leftmost column, or in introducing holes where values are missing. Missing values are replaced by the value in the same row from column t, slightly diminished.
4. At the conclusion of this process, the points in each diagonal line lie in a row across the M × T array. Lines where too much interpolation has occurred are eliminated. (These lines are evident because the maximum difference between the y coordinate values of two adjacent pixels constituting the line is much larger than the average difference.) The slopes are then found by taking the difference between the endpoint coordinates and dividing by the time interval between the first and last image in the sequence. (Clearly, the points in a row could also be fit to a straight line.) Each speed value is assigned to the average y value of its diagonal line.

Finding the wavelengths is straightforward: the distances between consecutive y values in a column are determined. Wavelengths are assigned to the y coordinate midway between the crests. The speeds and wavelengths are then cleaned of outliers and fit to y in the form

$$c = (A_c\, y \,+\, B_c\, y^2)^{1/2}, \qquad \lambda = (A_\lambda\, y \,+\, B_\lambda\, y^2)^{1/2}, \tag{2}$$

which is the third order prediction of the dependence on y based on a power series expansion of the tanh (kh) function in the dispersion relation (Eq. 1), assuming

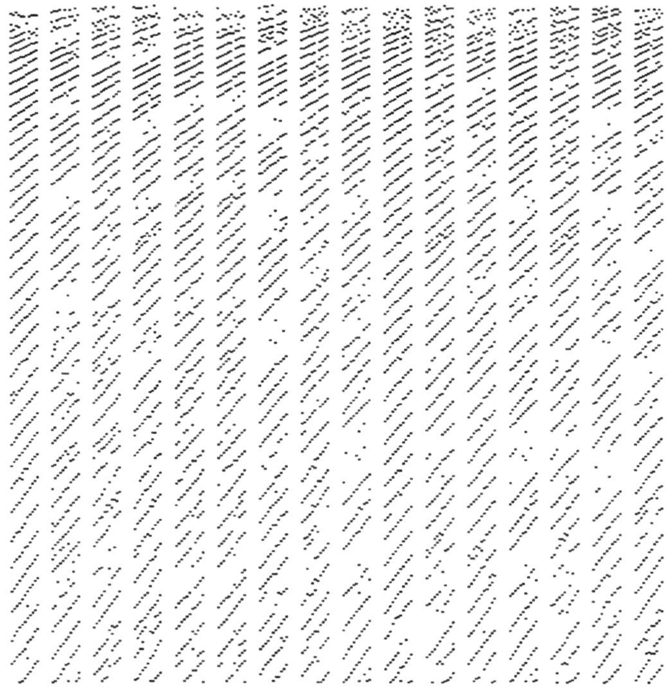

Fig. 4. Time slices from the sequence of images used. The slices have been separated by blank areas for ease of viewing. From left to right, the slices generated depths with a maximum difference from the actual depth (in m) of -0.01, 0.10, 0.21, 0.31, -0.39, -0.51, 0.40, -0.50, 0.60, -0.72, 0.79, 0.87, 0.97, 1.17, -0.53, and 1.26. Positive numbers represent predicted depths shallower than actual depths; negative numbers represent predicted depths deeper than actual depths. The first 6 slices are accepted as good data using the criterion discussed in the text.The next 8 slices are rejected. The next-to-rightmost slice shows data which cause B_c to be positive, and the rightmost slice shows data which cause B_λ to be positive. The slices in this image are not in the same order as in the time stack image.

depth is a linear function of y. The resulting fit values for c and λ can be used to find the frequency of the waves: $f = \langle c / \lambda \rangle$, where $\langle\ \rangle$ denotes an average value. A separate frequency is calculated for each time slice.

The fit values for c, f, and λ are then used in the dispersion relation to derive values of the depth h. The depth can be expressed as a function of any two of the three quantities. Since we do not initially know which pair is the most accurate, we calculate depth using all three possible pairs, resulting in three sets of depth values. These sets may not agree well with one another. Each set is fit to y (distance from the shore), using a linear least squares fit, which yields a slope s_1 and an intercept m for each set. Since the depths should be zero at the shore, a second slope for each set is calculated as $s_2 = \langle h/y \rangle$. The slope s_2 forces an intercept of zero.

The values of m and information from the fits of c and λ to y can be used to bound the values of s_2 which yield depth values close to the actual depths. Figs. 1 and 2 show that, looking horizontally at the crests in an image, there is an alternation between good data and poor data. The good data gradually get less good until they reach the worst slice, and then the trend reverses. The data in the worst slices are so poor that the fits to c and λ may not have the negative B value (Eq. 2) expected from the expansion of the dispersion relation. Fig. 4 shows two time slices with positive B values (furthest right).

Data with positive B_c values have very high c values. These correspond in the depth fit to negative intercepts and high slopes. Consequently, we select the slices with $B_c > 0$ (5 in this time stack image) and collect from these slices the s_2 values which have $m < 0$. The minimum s_2 value from the collection, s_{ub}, is an upper bound for s_2 : $s_2 \geq s_{ub}$ produces depths far from the actual depths, and hence values of $s_2 \geq s_{ub}$ in the slices must be caused by poor data.

Likewise, data with positive B_λ have very large wavelengths. These correspond in the depth fit to positive intercepts and small slopes. We select the slices with $B_\lambda > 0$ and collect from these slices the s_2 values which have $0 < m < 0.01$. The maximum s_2 value from the collection, s_{lb}, is a lower bound for s_2: $s_2 \leq s_{lb}$ produces depths far from the actual depths, and values of $s_2 \leq s_{lb}$ in the slices must be caused by poor data. The restriction $m < 0.01$ selects slices where the slope of the depth fit is in error, rather than its intercept.

For this time stack, $s_{lb} = 0.00631$ and $s_{ub} = 0.00684$. Slices with all s_2 values below s_{lb} or above s_{ub} are considered to have bad data; the further from the bound s_2 is, the larger the error in the predicted depths. Furthermore, the average $s_{av} = (s_{lb} + s_{ub}) / 2$ represents an expected value for the bottom slope. For this time stack, $s_{av} = 0.00657$, and the actual (constant) bottom slope is 0.00650. The predicted depth fit with s_2 closest to s_{av} is selected from the three depth fits as the best depth prediction. Fig. 4 shows 6 slices with good data and 8 slices with poor data, as well as 2 slices with $B > 0$.

4. Results

To evaluate the accuracy of the method, tests were made on 3 time stack images. The first used wave components having only one frequency and all moving directly toward the shore. Predicted depths agreed with actual depths to within 0.08 m. The second time stack used wave components having only one frequency, but allowed waves of smaller amplitude traveling at angles to the shore (directional spreading), in addition to the main component traveling directly toward the shore. Predicted depths for this time stack agreed with actual depths to within 0.08 m on the average, with the largest difference being 0.14 m.

The third case used wave components with frequencies from 0.02 Hz to 0.5 Hz and included directional spreading. (Fig. 1 shows an image from this case.) 339 slices (66%) predict depths which agree with actual depths to within 0.13 m on the average, with the largest difference being -0.52 m. The remaining 171 slices consist of 61 slices with $B > 0$ and 110 rejected because $s_2 \leq s_{lb}$ or $s_2 \geq s_{ub}$. Fig. 4 shows that the diagonal lines in rejected slices are of poorer quality than those in accepted slices.

The method finds depths that are on the average quite accurate, and identifies areas of poor data reliably. Since each slice is analyzed independently of the others, variations in depth parallel to the shore can be detected. We plan to investigate non-constant bottom slopes using simulated data. We also plan to use actual data to investigate whether the method of bounding the bottom slopes can be extended to non-simulated images and to determine depth accuracy in the presence of noise.

Acknowledgements

This work was partially supported by an ASEE-Navy Summer Faculty Fellowship to SML and was sponsored by the Space and Naval Warfare Systems Command, Dr. S. Payne, Program Manager.

References

Agam, G., Luo, H., and Dinstein, I.: 1996, Morphological approach for dashed lines estimation, in *Graphics Recognition - Methods and Applications*, ed. R. Kasturi and K. Tombre, Lecture Notes in Computer Science Vol. 1072, Springer-Verlag.

Banner, M. L.: 1991, On the directional behavior of the equilibrium wave number spectrum: implications for the equilibrium frequency spectrum, in *Directional Ocean Wave Spectra*, ed. R.C. Beal, Johns Hopkins University Press, Baltimore, pp. 39-45.

Donelan, M. A., Hamilton, J., and Hui, W. H.: 1985, Directional spectra of wind-generated waves, *Philos. Trans. R. Soc. London*, Vol. A 315, 509-562.

Hasselmann, K., Ross, D. B., Mueller, P., and Sell, W.: 1976, A parametric wave prediction model, *J. Phys. Oceanogr.* 200-228, 1976.

Jeulin, D. and Kurdy, M. B.: 1991, Directional mathematical morphology for oriented image reconstruction and restoration, *Acta Stereol.* 11 (Suppl. 1), 545-550.

Kurdy, M. B. and Jeulin, D.: 1989, Directional mathematical morphology operators, *Acta Stereol.* 8 (2), 473-480.

Lea, S. M., Lybanon, M., and Peckinpaugh, S. H.: 1996, Water depth determination using mathematical morphology, in *Mathematical Morphology and Its Applications to Image and Signal Processing*, ed. P. Maragos, R. W. Schafer, and M. A. Butt, Kluwer Academic Publishers, Boston, pp. 451-458.

Lea, S. M., Lybanon, M., and Smith, P. M.: 1998, Shoaling waves simulator, to apear, *Proceedings of 1998 Symposium on Applications of Computing.*

Soille, P., Breen, E. J., and Jones, R.: 1996, Recursive implementation of erosions and dilations along discrete lines at arbitrary angles, *IEEE Trans. on Patt. Anal. and Machine Intell.* 18 (5), 562-567.

Weiss, J. M.: 1997, Three-dimensional linear solution for wave propagation with sloping bottom, *IEEE J. Oceanic Engineering* 22(2), 203-210.

Wyatt, L. R., Ledgard, L. J., and Anderson, C. W.: 1997, Maximum-likelihood estimation of the directional distribution of 0.53-Hz ocean waves, *J. Atmospheric and Oceanic Technology* 14 (3, part 2), 591-603.

AN APPLICATION OF MATHEMATICAL MORPHOLOGY TO ROAD NETWORK EXTRACTION ON SAR IMAGES

JOCELYN CHANUSSOT AND PATRICK LAMBERT
LAMII/CESALP - BP 806 - 74016 Annecy Cedex - France
(CNRS - G1047 - ISIS - Information Signal ImageS)
tel : (33) 450 666 074 - FAX : (33) 450 666 020
email : Jocelyn.Chanussot@univ-savoie.fr

Abstract. In this paper, we present an application of morphological operators to the automatic extraction of the road network in SAR images. The proposed method is simple and fast, un-supervised and almost non parametric, and leads to results comparable to those found in the literature. Basically, roads are considered as linear, thin objects, that are darker than their surroundings. As a consequence, the key-point of the processing is a combination of directional closings using flat linear structuring elements (oriented segments). The whole processing chain is meant to deal with the noisy (speckled) structure of SAR images.

Key words: road detection, synthetic aperture radar, directional morphology, remote sensing.

1. Introduction

Since the first Earth-viewing spaceborne synthetic aperture radar (SAR) was put into orbit in 1978, numerous radar sensors have been launched [15]. The images provided by these satellites are now very useful in many remote sensing applications. This paper focuses on the problem of automatic line detection, with application to road network extraction.

Most of the existing methods are organized in two separate steps (Figure 1). The first one consists in the local detection of linear features in the image, whereas the second one, a post-processing step, tries to fill in the gaps existing between the detected segments in order to fully reconstruct the road network (grouping, or reconnection step). The method presented in this paper intends to achieve the first step of the processing.

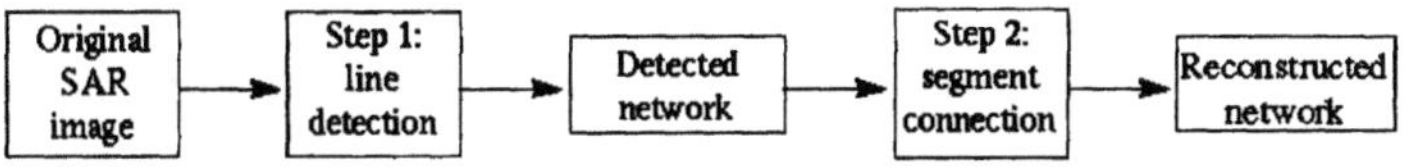

Fig. 1. Typical processing synopsis.

A lot of methods have already been proposed to solve this detection problem (step 1), in particular in the visible domain, for instance on SPOT [9] or aerial [2] images. But these methods are seldom very suitable to SAR images since these images are usually highly corrupted by a multiplicative noise, called the speckle. The use of the Hough transform for lineaments detection [14] and some other methods based on energy minimization and active contours have also been studied.

Tupin [12] recently proposed a method well suited to SAR images based on the fusion of two edges detectors [11] with some local post processing. This method gives attractive results and presents good statistical properties, but, as the other proposed methods, it suffers two major drawbacks : firstly, it is very time consuming, and secondly, several thresholds have to be set.

Using morphological operators [10], we propose a simple and fast un-supervised method that can be used over a wide range of images, applications and sensors, and that does not require any threshold. The only parameters that have to be set correspond to the model dimensions of the features we are looking for. To determine these parameters, we decided to characterize the roads in the image as features that are:

• **locally rectilinear** : each road pixel belongs to a segment that is longer than a minimum required length l_0 . We fixed this minimum length l_0 to 21 pixels (corresponding to approximately 280 meters ground lengths for ERS-1 PRI images, which is a realistic minimum straight line length for typical countryside roads and highways),

• **thin** : detected features must not be too wide. We decided to search roads that are up to 3 pixels wide (corresponding up to approximately 40 meters ground widths for ERS-1 PRI images, which is a realistic model too),

• **contrasted** : we will only consider features that are darker than their surroundings (the method could be extended to the detection of clear objects in a very straightforward way by using the dual operators).

Since the features we are looking for are intrinsically characterized by their shape, it seemed natural to us to use a morphological approach to perform the extraction. Section 2 presents the morphological operators we used for the detection, section 3 presents some results. Finally, we conclude and give some future prospects.

nota bene : we used an 8-connectivity square grid to represent the image.

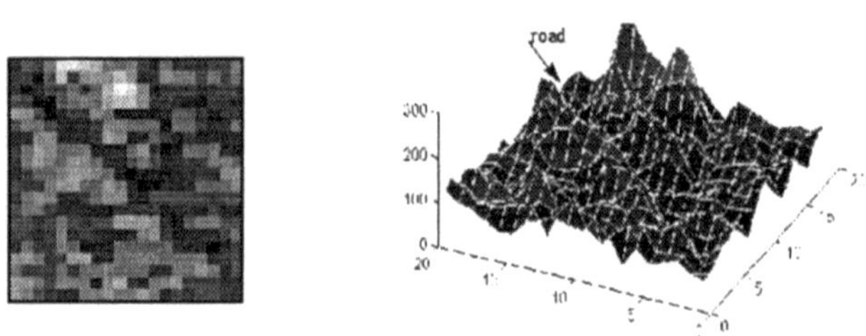

Fig. 2. Part of the original image and the corresponding relief (right).

2. Dark line detection

In the following, we will use the classical topographic analogy for the representation of gray-scale images. In this representation, the image is considered as a topographic relief, the numerical value of each pixel determining the corresponding point elevation.

Figure 2 presents respectively a zoom of the small 20*20 square image taken in the original image (white square in Figure 7(a)) and its corresponding relief. A skilled human observer can notice that a road is crossing this image from the top left corner to the bottom right one.

With this analogy, we are now looking for elongated and narrow valleys in a topographic relief. Three different types of noisy structures have then to be removed, each one violating one of the basic constraint defined in the introduction:

- we will firstly remove the clear structures (corresponding to "**peaks**"), since we only are interested in dark features,
- we will then remove **non-linear valleys** (or linear valleys that are shorter than 21 pixels),
- we will finally remove the remaining linear valleys that are **too wide** (more than 3 pixels).

2.1. Removing peaks

We firstly applied to the original image I_0 a morphological opening by reconstruction, using a square flat structuring element of size 5 [4]. This operator removes the non flat peaks of the image by truncating them. Furthermore, this is a connected operator ; it therefore does not introduce any new discontinuity in the image : the edge information is either totally removed or exactly preserved, which is important for the next operation. Figure 3 presents the result I_1 of this operator on the original image I_0.

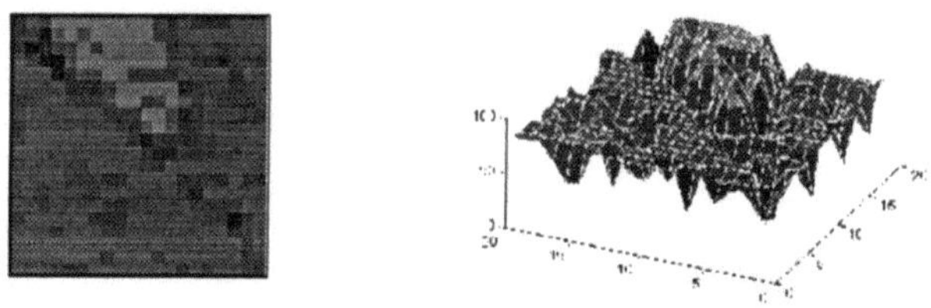

Fig. 3. Filtered image $I_1 = \overline{\gamma}_5(I_0)$.

2.2. Removing non-linear valleys

Morphological closings tend to fill in the valleys and basins of a topographic relief. To preserve the valleys in which 21 pixels can be lined up in at least one direction, and to remove the others, we used the closing obtained by taking the infimum (denoted "$\wedge$") of all the directional closings [8]. It can be computed using 21 pixels long linear structuring elements successively oriented in every possible direction (the infimum of different closing operators still remains a morphological closing).

In a discrete square grid of size $2n+1$, $4n$ symmetrical segments with different directions d_i are usually defined [5]. We therefore used 40 different 21 pixels long linear structuring elements defining 40 different directional closings (Appendix A shows the used segments). Most of the methods found in the literature usually restrict themselves to 8 different directions for computational load reasons. On the contrary, a linear directional closing is very fast. So, taking all the possible directions into account by calculating these 40 closings for each pixel, we nevertheless preserve a reasonable processing time.

Figure 4 presents the result I_2 of this operator applied on I_1.

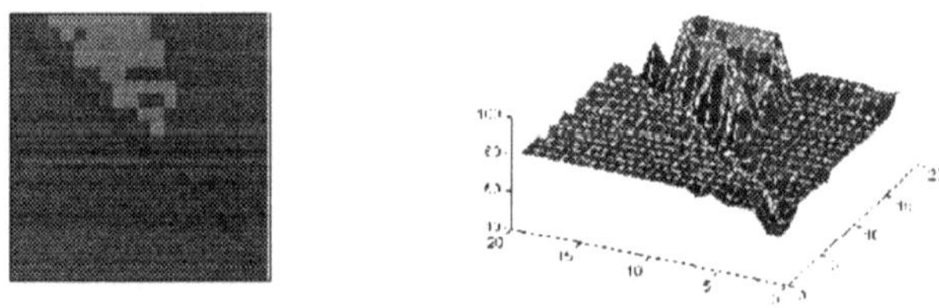

Fig. 4. Filtered image $I_2 = \bigwedge_{d_i} \{\varphi_{d_i}(I_1)\}$, with $i \in \{1, \ldots, 40\}$.

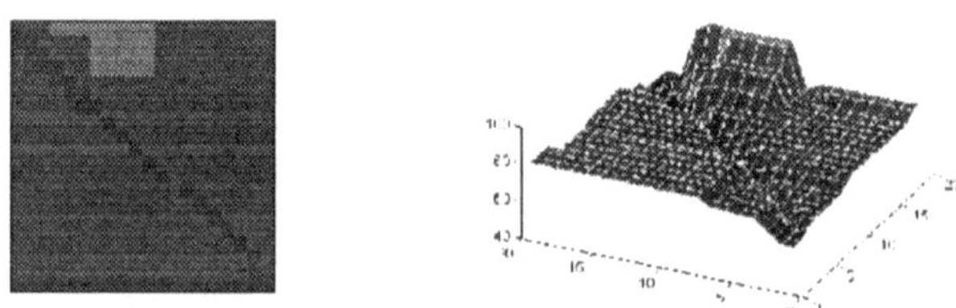

Fig. 5. Filtered image $I_3 = \gamma_5(I_2)$.

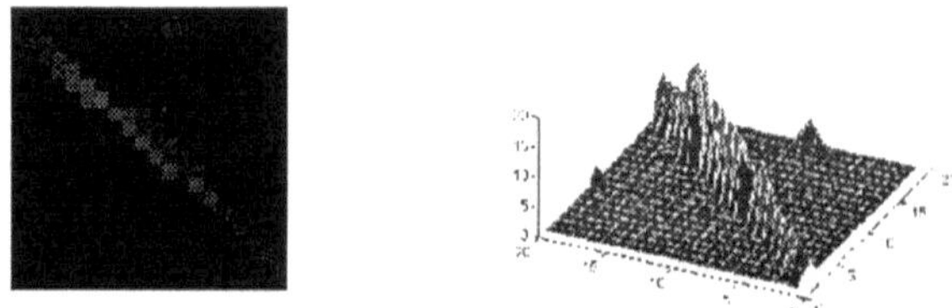

Fig. 6. Filtered image $I_4 = \varphi_3(I_3) - I_3$.

2.3. Removing remaining peaks

Since we firstly used an opening by reconstruction with an 8-connectivity grid, some "isolated" peaks remain that could lead to false detection in the next step. As a consequence, to prepare properly the following extraction operation, we now remove these remaining peaks with a simple opening using a flat square structuring element of size 5 (without any reconstruction).

Behind this operation appears a new constraint : we assume that 2 different roads will be separated by at least 5 pixels, which is a realistic assumption (5 pixels approximately correspond to 60 meters on the ground).

Figure 5 presents the result I_3 of this operator applied on I_2.

2.4. Removing wide valleys and extraction step

We can now apply the last processing to extract the roads. The residue between I_3 and its closing by a flat square structuring element of size 3 is calculated (Top-hat operator). The only remaining structures are the dark features that are up to 3 pixels wide. Any other pixel, belonging either to a plateau or to valley that is more than 3 pixels wide, is put to zero.

Figure 6 presents the result I_4 of this operator applied on I_3.

-a-

-b-

-c-

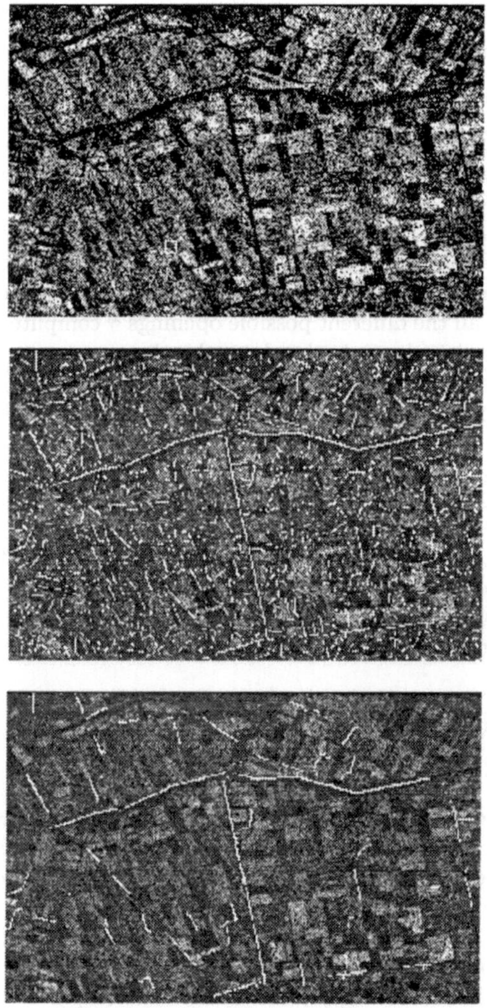

Fig. 7. Original ERS-1 image ©ESA (a); detection result (b); after 'cleaning up' (c).

The final decision is taken by thresholding I_4 at zero. We consider it as quite important not to use any other threshold value since this would introduce a parameter to be set. Furthermore, if I_4 was thresholded at a value $p > 0$, only the roads that are originally at least $p + 1$ gray level darker than their surroundings would be extracted. We do not want to make this kind of assumption on the model ; in particular, we do not want to determine *how* darker the roads have to be, or, equivalently, how deep the valleys have to be, because this widely differs from an image to another, or even in one single image.

3. Results

In this section, the results obtained on a PRI ERS-1 image are presented. The original image is shown in figure 7(a). It represents a flat, agricultural region near Amsterdam and has a 12,5 meters resolution. The small white square superposed on the image defines the small region that has been used along the paper.

Figure 7(b) presents the linear segments detected, superposed in white on the original image. We can see that a lot of small structures are detected. To remove them and to "clean up" the result image, an area opening can be performed. This morphological filter removes all the connected components whose surface is smaller than a given number n_0 of pixels. It can theoretically be seen as the supremum (denoted "$\vee$") of all the different possible openings γ computed with all the different connected structuring elements that have the given area n_0 :

$$I_5 = \bigvee_{\gamma} \{\gamma(I_4)\}$$

For the implementation, we used the algorithm proposed in [13]. Figure 7(c) presents the cleaned result after an area opening of size 50.

Nevertheless, apart from the visual quality of the results, this operator must be used very carefully since all the small removed features are not necessarily noisy features. Some of them really belong to the road network and would highly increase the quality of the reconstruction in the reconnection step (step 2 of Figure 1).

4. Conclusions & future prospects

4.1. Conclusions

This morphological approach for road extraction seems very promising:

• The obtained results are qualitatively comparable to those found in the literature (before the connecting step). The three main roads are perfectly detected and the secondary ones are partially extracted.

• The method is very simple and fast since only sup- and inf- operators are involved (approximately one minute is required for the whole processing on a Sun Ultra Sparc 170 station for a 500*500 image, which is very fast compared with other algorithms).

• The method is independent of the average radiometry of the processed regions.

• Whereas most of the proposed methods are basically based on averaged radiometries differences [7] or, which is more appropriate, averaged radiometries ratios [1,3,

11], our method does not depend on the intensity contrast between the road pixels and their surroundings. Roads just have to be darker : even one single gray level difference is enough for the detection.
• No parameter has to be set, apart from the size of the different used structuring elements :
– size for the openings : 5 * 5
– size of the segments for the closing : 21
– size of the closing for the top hat : 3*3
(– size of the area opening : 50)
But ...
• Some roads are still only partially detected (disconnected segments).
• There are also some spurious detections. For instance, some dark fields are considered as potential roads : the false alarm probability is strictly greater than zero.
• The quantitative influence of the size of the different used structuring elements is very hard to evaluate. The user has to determine them manually, according to his expertise, his aims and the kind of images he has to process.

4.2. FUTURE PROSPECTS

The future prospects of this work include the quantitative study of the performances of this morphological approach and, in particular, the evolution of the false alarm and detection probabilities in function of the structuring elements.

The second major prospect is to complete the reconstruction of the road network from the detected disconnected segments (step 2 of figure 1). Different approaches are conceivable : for instance a morphological method, proposed in [6], consists in directionally dilating the extremities of each detected segment in a restricted cone and in skeletonizing the potential intersection found between two different cones. As explained in [6], this method is well suited to fill in gaps up to 15 pixels long between disconnected segments. We nevertheless believe that for this stage of the processing a more global approach will give better results. For instance, the grouping step proposed by [12], based on a Markov Random Field defined on a set of segments and taking into account the principal properties of the road network, gives very attractive results. The optimal use of this *a priori* knowledge helps to improve the poor local detection performed in the first step (due to the noisy nature of SAR images). This generally poor local detection enables other classical methods to provide good results in the second step.

References

1. M. Adair and B. Guindon. Statistical edge detection operators for linear feature extraction in SAR images. *Canadian Journal of Remote Sensing*, 16(2):10–19, 1990.
2. M. Barzohar and D. Cooper. Automatic finding of main roads in aerial images by using geometrical stochastic models and estimation. *IEEE Trans. on Pattern Analysis and Machine Intelligence*, 18(7):707–721, 1996.
3. A. Bovik. On detecting edges in speckle imagery. *IEEE Trans. on Acoustic, Speech, Signal Proceesing*, 36:1618–1627, October 1988.
4. J. Crespo, J. Serra, and R. W. Schafer. Theoretical aspects of morphological filters by reconstruction. *Signal Processing*, 47:201–225, 1995.
5. M. Gurelli and L. Onural. The adaptive directional median filter. In *Proc. of the 6th Interna-*

tional Symposium on Computer and Information Sciences, Antalya, Turkey, pages 973–979, 1991.
6. B. Kurdy and D. Jeulin. Directional mathematical morphology operations. *Acta Stereologica,* 8/2:473–480, 1989.
7. V. Lacroix. A three module strategy for edge detection. *IEEE Trans. on Pattern Analysis and Machine Intelligence,* 10(6):803–810, 1988.
8. W. Li and J. Ronsin. Directional composite morphological filters in image processing. In *Proc. of the EUSIPCO,* Trieste, Italy, pages 2129–2132, September 1996.
9. N. Merlet and J. Zerubia. New prospects in line detection by dynamic programming. *IEEE Trans. on Pattern Analysis and Machine Intelligence,* 18(4):426–431, 1996.
10. J. Serra. Morphological filtering : an overview. *Signal Processing,* 38:3–11, 1994.
11. R. Touzi, A. Lopes, and P. Bousquet. A statistical and geometrical edge detector for SAR images. *IEEE Trans. on Geoscience and Remote Sensing,* 26(6):764–773, November 1988.
12. F. Tupin, H. Maître, J. M. Nicolas, and E. Pechersky. Detection of linear features in SAR images: application to road network extraction. To be published in *IEEE Trans. on Geoscience and Remote Sensing,* 15 pages.
13. L. Vincent. Grayscale area openings and closings, their efficient implementation and applications. In J. Serra and P. Salembier, editors, *Proc. Workshop 'Mathematical Morphology and its Applications to Signal Processing', Barcelona, Spain, May 12-14,* pages 22–27, 1993.
14. J. Wang and P. Howarth. Use of the Hough transform in automated lineament detection. *IEEE Trans. on Geoscience and Remote Sensing,* 28(4):561–566, July 1991.
15. J. Way and E. A. Smith. The evolution of synthetic aperture radar systems and their progression to the EOS SAR. *IEEE Trans. on Geoscience and Remote Sensing,* 29(6):962–985, november 1991.

Appendix

A. Linear structuring elements

We used 40 different oriented 21 pixels long structuring elements. We only used symmetrical segments, so, we only represented half of the segments (11 pixels long) on figure 8. Furthermore, we used an isotropic angle distribution. So, the 10 segments shown on figure 8 have to be rotated successively by 45, 90 and 135 degrees to give the 40 different used structuring elements.

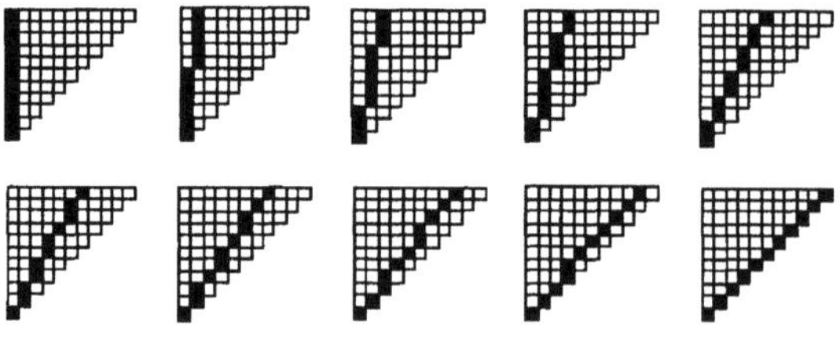

Fig. 8. 10 'seeds' to construct the 40 used segments.

SOFT MORPHOLOGICAL STRUCTURING ELEMENT DECOMPOSITION

A. GASTERATOS, I. ANDREADIS and PH. TSALIDES
Laboratory of Electronics
Section of Electronics and Information Systems Technology
Department of Electrical and Computer Engineering
Democritus University of Thrace
GR-671 00 Xanthi, Greece

Abstract. In this paper a decomposition technique suitable for any grey-scale soft morphological structuring element is presented. A hardware structure implementing this technique is also presented.

Key words: soft mathematical morphology, structuring elements.

1. Introduction

Mathematical morphology is a methodology for image analysis based on set theory and topology [1]. Its two basic operations are erosion and dilation, from which all the other operations are composed. A relatively new approach to mathematical morphology is soft mathematical morphology. This combines mathematical morphology with weighted order statistics. It has been shown that soft morphological operations are less sensitive to additive noise and to small variations in object shape [2].

Applications such as textural segmentation and granulometry require the application of successively larger size structuring elements [1]. However, in morphological processors the size of the structuring element is usually restricted to 3x3 pixels [3], [4]. The handling of larger size structuring elements on existing machines is not a straightforward process. Structuring element's decomposition is required in such cases. This practical problem has concerned many researchers; binary morphology [5], [6] and grey-scale morphology [7], [8]. One decomposition strategy is to represent the structuring element as successive dilations of smaller structuring elements. Algorithms for optimal structuring element decomposition based on the latter strategy have been presented [5] - [7]. Also, several methods for decomposing structuring elements into combined structures of segmented small components have been presented [8].

In this paper, a decomposition technique for any grey-scale soft morphological structuring element is presented for the first time. The difference between a standard morphological structuring element and a soft morphological structuring element is that the domain of the soft morphological structuring element is split into two subsets, i.e. the core and the soft boundary. According to the proposed decomposition technique the domain of a large size structuring element (both the core and the soft boundary) is divided into non-overlapping sub-domains. The small structuring elements which obtain values from these sub-domains are used to form the soft morphological operations locally. The final result is composed from the latter local results. An architecture implementing this technique is also presented.

2. Notations and Definitions

Let $\{k \diamond f(x)\}$ denote the k times repetition of $f(x)$, i.e. $\{k \diamond f(x)\} = \{f(x), f(x), \ldots f(x)\}$ (k times). Also let $\min^{(k)}$ denote the kth order statistic of a multiset with N elements ($N > k$), i.e. the kth smallest element of the multiset and let $\max^{(k)}$ denote the $(N-k+1)$th order statistic of the multiset, i.e. the kth largest element of the multiset. A multiset is a collection of objects, where the repetition of objects is allowed. Soft morphological dilation and erosion of a grey-scale image f by a soft grey-scale structuring element $[\alpha, \beta, k]$, have been defined as follows [9] :

$$f \oplus [\alpha, \beta, k](x) = \max_{\substack{(x-y) \in G_A \\ (x-z) \in G_B}}{}^{(k)} \Big(\{k \diamond (f(y) + \alpha(x-y))\} \cup \{f(z) + \beta(x-z)\} \Big) \quad (1)$$

and

$$f \ominus [\alpha, \beta, k](x) = \min_{\substack{(x+y) \in G_A \\ (x+z) \in G_B}}{}^{(k)} \Big(\{k \diamond (f(y) - \alpha(x+y))\} \cup \{f(z) - \beta(x+z)\} \Big) \quad (2)$$

respectively,

where k is the order index

$x, y, z \in E^{\aleph}$ are the spatial coordinates,

$f : F \rightarrow E$ is the grey-scale image,

$\alpha : G_A \rightarrow E$ is the core of the structuring element,

$\beta : G_B \rightarrow E$ is the soft boundary of the structuring element,

$F, G_A, G_B \subseteq E^{\aleph}$ are the domains of the grey-scale image, the core and the soft boundary of the structuring element, respectively, and $G_B = G \backslash G_A$, where is the domain of the grey-scale structuring element and "\" stands for set difference.

3. Soft Morphological Structuring Element Decomposition Technique

The proposed decomposition technique is described in this section. The domain G of the structuring element is divided into smaller non-overlapping subdomains $G_1, G_2, \ldots G_n$. Also, $G_1 \cup G_2 \cup \ldots G_n = G$. The soft morphological structuring elements obtain values from these domains and they are denoted by $[\alpha_1, \beta_1, k], [\alpha_2, \beta_2, k], \ldots [\alpha_n, \beta_n, k]$, respectively. These have common origin, which is the origin of the original structuring element. Additionally, the points of G which belong to its core are also points of the cores of $G_1, G_2, \ldots G_n$. and the points of G which belong to the soft boundary are also points of the soft boundaries of $G_1, G_2, \ldots G_n$. This process is graphically illustrated in Figure 1. In this figure the core of the structuring element is denoted by the shaded area.

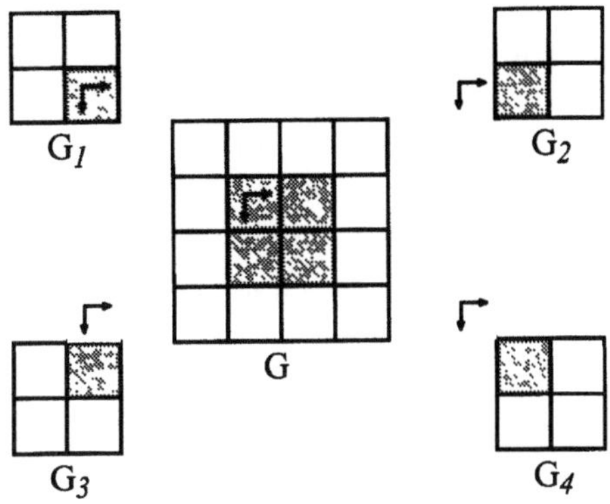

Fig. 1. Example of a 4x4 soft structuring element decomposition

Then, soft dilation and erosion are computed as follows :

$$f \oplus [\alpha, \beta, k](x) = \max_{i=1}^{n}{}^{(k)} \left[\max_{\substack{(x-y)\in G_A \\ (x-z)\in G_B \\ j=1}}^{k}{}^{(j)} \Big(\{k \diamond (f(y) + \lambda_i(x-y))\} \cup \{f(z) + \mu_i(x-z)\} \Big) \right] \tag{3}$$

$$f \ominus [\alpha, \beta, k](x) = \min_{i=1}^{n}{}^{(k)} \left[\min_{\substack{(x+y)\in G_A \\ (x+z)\in G_B \\ j=1}}^{k}{}^{(j)} \Big(\{k \diamond (f(y) + \lambda_i(x+y))\} \cup \{f(z) + \mu_i(x+z)\} \Big) \right] \tag{4}$$

respectively.

Proof:
$\forall (x-y) \in G_A : \alpha(x-y) = \bigcup_{i=1}^{n} \lambda_i(x-y)$
$\Rightarrow f(y) + \alpha(x-y) = \bigcup_{i=1}^{n} [f(y) + \lambda_i(x-y)]$
$\Rightarrow k \diamond (f(y) + \alpha(x-y)) = k \diamond \left(\bigcup_{i=1}^{n} [f(y) + \lambda_i(x-y)] \right) =$
$k \diamond (f(y) + \lambda_1(x-y), f(y) + \lambda_2(x-y), \ldots f(y) + \lambda_n(x-y))$ (5)

Also,
$\forall (x-z) \in G_B : \beta(x-z) = \bigcup_{i=1}^{n} \mu_i(x-z)$
$\Rightarrow f(y) + \beta(x-z) = \bigcup_{i=1}^{n} [f(z) + \mu_i(x-z)] =$
$f(z) + \mu_1(x-z), f(z) + \mu_2(x-z), \ldots f(z) + \lambda_n(x-z)$ (6)

Therefore

$$f \oplus [\alpha,\beta,k](x) =$$

$$\max_{\substack{(x-y)\in G_A \\ (x-z)\in G_B}}{}^{(k)} \left(\begin{array}{c} \{k \diamond (f(y)+\lambda_1(x-y), f(y)+\lambda_2(x-y), \ldots f(y)+\lambda_n(x-y))\} \cup \\ \{f(z)+\mu_1(x-z), f(z)+\mu_2(x-z), \ldots f(z)+\lambda_n(x-z)\} \end{array} \right)$$

$$= \max_{\substack{(x-y)\in G_A \\ (x-z)\in G_B}}{}^{(k)} \left(\begin{array}{c} \{k \diamond (f(y)+\lambda_1(x-y))\} \cup \{f(z)+\mu_1(x-z)\}, \\ \{k \diamond (f(y)+\lambda_2(x-y))\} \cup \{f(z)+\mu_2(x-z)\}, \\ \vdots \\ \{k \diamond (f(y)+\lambda_n(x-y))\} \cup \{f(z)+\mu_n(x-z)\} \end{array} \right)$$

$$= \max_{\substack{(x-y)\in G_A \\ (x-z)\in G_B \\ i=1}}{}^{(k)} \left[\{k \diamond (f(y)+\lambda_i(x-y))\} \cup \{f(z)+\mu_i(x-z)\} \right]$$

Similarly eqn (4) can be proven.

Example : Let us consider the following image f and soft structuring element $[\alpha, \beta]$:

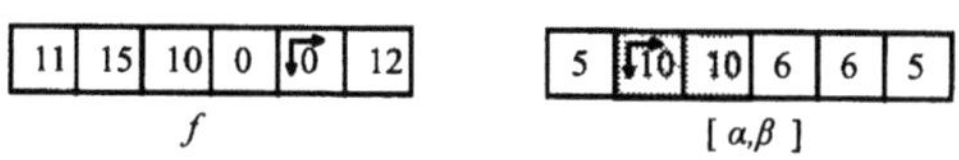

Soft dilation at point (0,0) for $k = 2$, according to eqn. (1) is :
$f \oplus [\alpha,\beta,2](0,0) = \max^{(2)}(\{2 \diamond (10,10)\} \cup \{16,21,16,17\})$
$= \max^{(2)}(10,10,10,10,16,21,16,17) = 17$

According to the proposed technique the structuring is divided into three structuring elements :

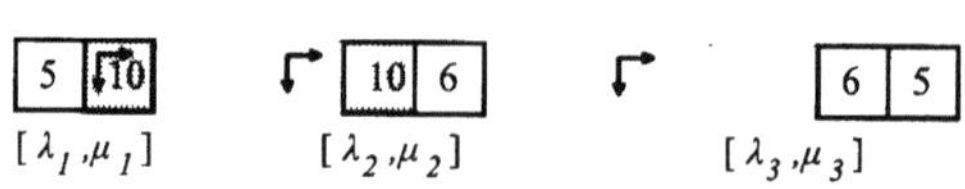

The following multisets are obtained from the above structuring elements :$\{2 \diamond (10), 17\}$, $\{2 \diamond (10), 16\}$ and $\{16, 15\}$, for the first the second and the third stucturing elements, respectively. From these multisets the max and the $\max^{(2)}$ elements are retained : $(\{17,10\}, \{16,10\}$ and $\{21,16\})$. The $\max^{(2)}$ of the union of these multisets, i.e. 17, is the result of soft dilation at point (0,0). It should be noticed that although 17 is the max of the first multiset, it is the $\max^{(2)}$ of the global multiset.

4. Architecture for Decomposition of Soft Morphological Structuring Elements

An architecture for the implementation of decomposition of soft morphological structuring elements is depicted in Figure 2. The structuring element is loaded to the *structuring element management* module. This divides the structuring element into n smaller structuring elements and provides the appropriate one to the next stage. The pixels of the image are imported to the *image window management* module. This provides an image window which interacts with the appropriate structuring element, provided by the *structuring element management* module. Both the previous modules consist of registers and multiplexers (MUXs), controlled by a counter mod$n$, as shown in Figure 3. The second stage, i.e. the *arithmetic unit* (Figure 4) consists of adders/subtractors (dilation/erosion) and of an array of MUXs that are controlled by the order index k. The MUXs provide the multiple copies of the addition/subtraction results to the next stage, i.e. an array of *order statistic modules* (OSMs). An OSM is capable of computing any order statistic of a multiset [4], [10]. Of course, instead of using an array of k OSMs, only one OSM may be used, at the expense of speed. The $\max^{(l)}$ / $\min^{(l)}$ results ($l = 1, \ldots k$) of every multiset are collected by means of an *array of registers* . These registers provide the $n \times k$ $\max^{(l)}$ / $\min^{(l)}$ of the n multisets concurrently to the last stage OSM which computes the final result according to eqns. (3) or (4).

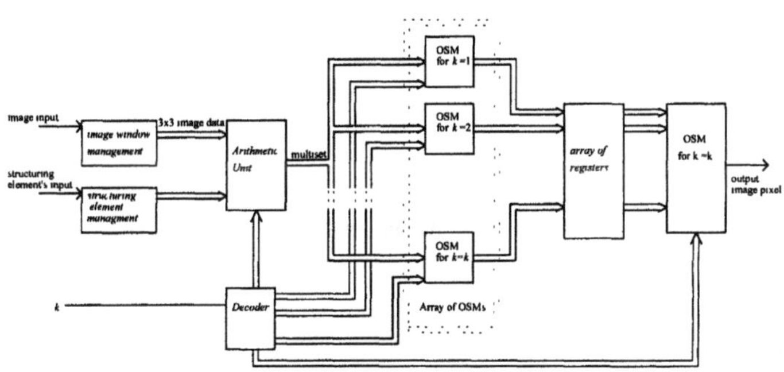

Fig. 2. Architecture for the implementation of the decomposition technique

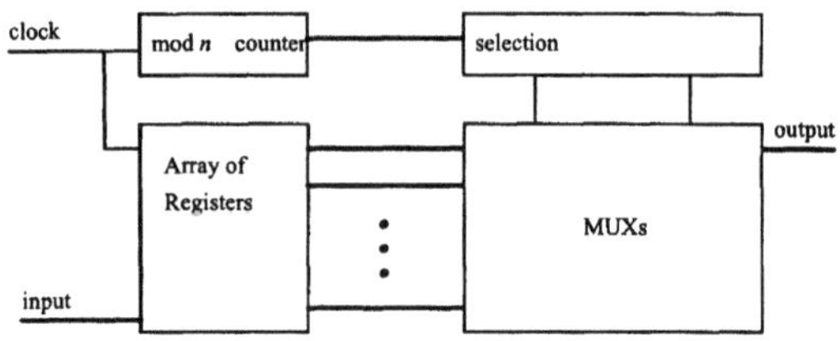

Fig. 3. Data window management

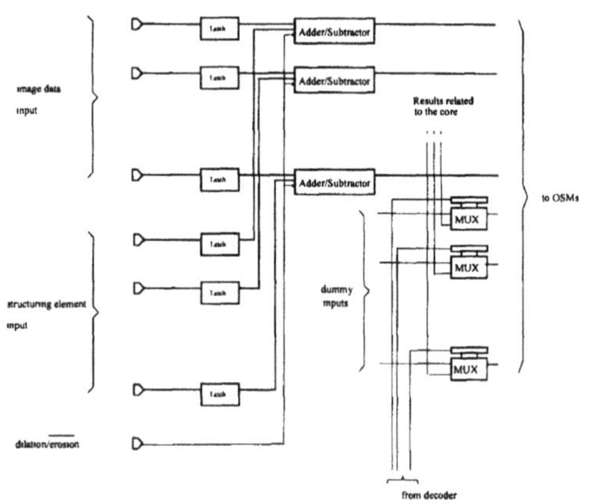

Fig. 4. Arithmetic unit

5. Conclusions

A decomposition technique for any grey-scale soft morphological structuring element has been presented in this paper. According to this technique the domain of the structuring element is divided into non-overlapping sub-domains. These sub-domains are used to compute the soft morphological operations locally. From these local results the global result of the soft morphological operation is composed. An architecture suitable for real time applications implementing this technique has been also presented.

References

1. J. Serra. *Image Analysis and Mathematical Morphology*, volume 1. Academic Press, London, 1982.
2. P. Kuosmanen and J. Astola. Soft morphological filtering. *Journal Mathematical Imag. Vision*, 5:231–262, 1995.
3. L. Luck and C. Chakrabaty. A digit-serial architecture for gray-scale morphological filtering *IEEE Trans. Imag. Proc.*, 4:387–391, 1995.
4. A. Gasteratos, I. Andreadis and Ph. Tsalides. A new hardware structure for implementation of soft morphological filters. , *Lecture Notes in Computer Science* , number 1269, pages 488–494, 1997. Springer-Verlag.
5. X. Zhuang and R. M. Haralick. morphological structuring element decomposition. *Computer Vis. Graph. Imag. Proc.*, 35:370–382, 1986.
6. H. Park and R. T. Chin. Decomposition of arbitrarily shaped morphological structuring elements. *IEEE Trans. Patern Anal. Machine Intell.*, 17:2–15, 1995.
7. O. I. Camps, T. Kanungo and R. M. Haralick. ,Grey-scale structuring element decomposition. *IEEE Trans. Imag. Proc.*, 5:111–120, 1996.
8. F. Y. Shih and O. R. Mitchell. Decomposition of grey-scale morphological structuring elements. *Pattern Recognition*, 24:195–203, 1991.
9. C. C. Pu and F. Y. Shih. Threshold decomposition of grey-scale soft morphology into binary soft morphology. *Graph. Models Imag. Proc.*, 57:522–526, 1995.
10. A. Gasteratos, I. Andreadis and Ph. Tsalides. Realisation of rank order filters based on the majority gate. *Pattern Recognition*, 30:1571–1576, 1997.

MORPHOLOGICAL PROCESSING OF GROUND-TO-GROUND RADAR IMAGE SEQUENCES

ROBERT VOGT, SUSAN STOKELY and GERALD FLISS
ERIM International
P.O. Box 134008, Ann Arbor, MI 48113-4008
email: vogt@erim-int.com

Abstract. This paper discusses the detection and identification of radar scattering mechanisms found in ground-to-ground, inverse synthetic aperture radar (ISAR) data, through the means of analyzing their motion characteristics in the magnitude image domain. The method described uses a morphological watershed algorithm on an inverted radar image to segment the scattering events, and from this segmentation compute attributes and construct a graph of the primary scatterers. When these attributed graphs have been computed for all image frames of a short sequence, cost tables can be computed for matching any pair of scatterers from adjacent frames. From the cost tables, a graph matching process computes the least cost match for each frame pair, with respect to the complete graphs. Given the resulting matches between the nodes in each frame, tracks can then be generated for each major scattering event, and finally, analysis of the motion of the track can be used to classify each scatterer into one of a few major categories. This work demonstrates the possibility of automatically identifying different scattering types based on their motion characteristics, and is helping to contribute to a better understanding of electromagnetic scattering phenomena.

Key words: mathematical morphology, watershed segmentation, radar scattering mechanisms, motion analysis, graph matching

1. Introduction

This paper concerns the analysis of radar image sequences, and the "scattering events" that appear within them. For those not familiar with the principles of imaging radars, we will simply mention here that such radars collect data over two dimensions: a series of angles and a series of frequencies. Depending on the type of radar, a complex-valued "pulse" is sent out, having a certain frequency and amplitude, and its complex-valued "echo" is received shortly thereafter. As this is repeated over many frequencies times multiple viewing angles, a *signal or phase history* dataset is built up (Figure 1), from which one can extract *synthetic apertures.* With ground-to-ground radar imaging, both the radar and the object or scene being illuminated are on the ground. If the radar rotates around the target, we have the standard *synthetic aperture radar* or SAR case; if the radar is fixed and the target rotates in place, we have the inverse SAR or ISAR case, as illustrated in Figure 2. The examples discussed in this paper are all based on the ISAR geometry, where the measurements have been carried out in a special anechoic chamber designed for this purpose.

The phase history itself does not have the characteristics of an image of the illuminated scene; rather, we have to select an aperture or "window" of data from

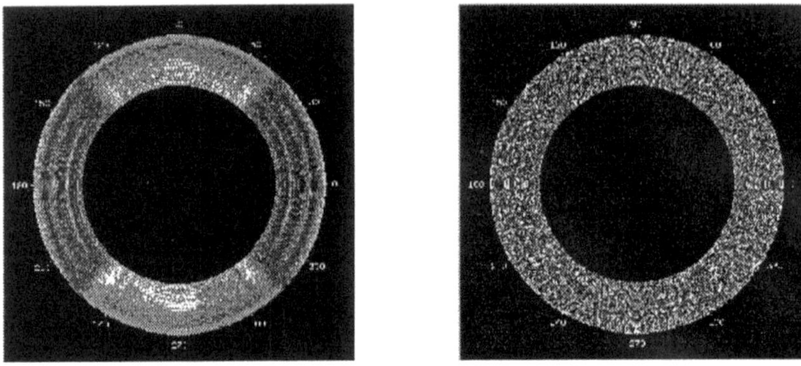

Fig. 1. Example phase history data - Magnitude and Phase

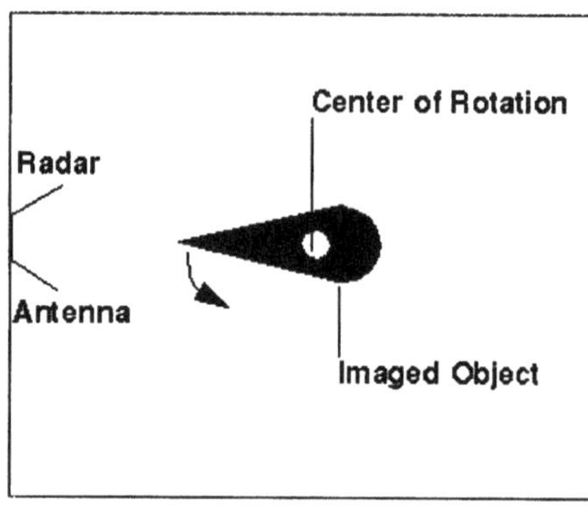

Fig. 2. ISAR measurement geometry

the phase history, and perform a complex fast Fourier transform (FFT) of it. This transform results in an image of the scene, as viewed from the angle at the center of the chosen aperture. The process is somewhat complicated by the fact that the phase history data is collected in a polar coordinate space (essentially within a "donut" of polar angles and a band of radial frequencies) but the FFT has to be carried out on a square grid. In addition, data within an aperture are typically weighted before taking the FFT, to reduce the effects of sidelobes in the resulting image.

The image formed by the FFT of an aperture of the complex-valued phase history data is also complex; that is, it has both real and imaginary parts (the *in-phase* and *quadrature* components, or I and Q, respectively—see Figure 3). Alternatively, it can be considered to have a *magnitude* and *phase*, where the magnitude corresponds to the length of the complex vector and the phase is its angle in polar coordinates. For image viewing purposes, the image phase is typically ignored (it looks very chaotic) and only the magnitude is displayed (Figure 4). Because the dynamic range of radar images is typically very large, the magnitude is usually displayed using a logarithmic or decibel (dB) scale, where dB = 20 * log(Magnitude).

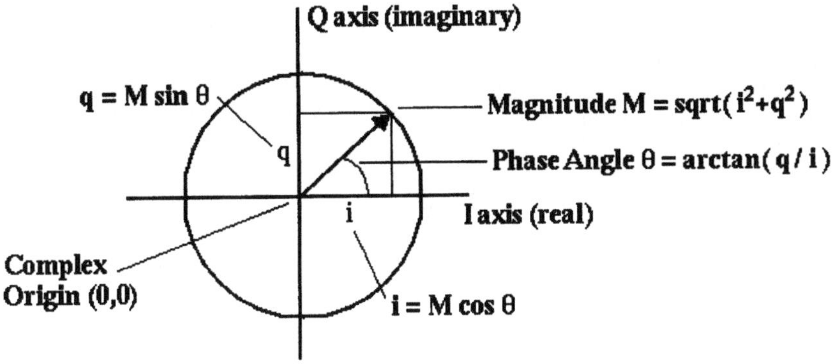

Fig. 3. Complex plane (I & Q, Magnitude & Phase)

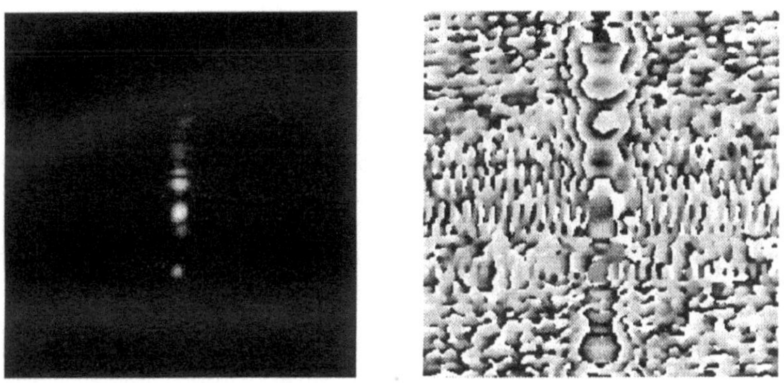

Fig. 4. Example image domain data - Magnitude and Phase

Now that we have explained some of the fundamentals regarding radar images and how they are formed, we can begin to address the automated analysis sequences of such images for the purpose of understanding and classifying particular scattering mechanisms, based on their motion. This work is being carried out as part of scientific studies of radar scattering behavior, with the goal of developing automated tools to perform this kind of analysis [1, 2]. We would like to be able to explain different scattering behaviors based on motion and other criteria, and to isolate particular mechanisms by varying them one at a time, and comparing the results to a reference object.

2. Radar Image Sequence Analysis Process

The following section describes in detail the algorithms we use to analyze the behavior of scatterers in radar image sequences. The overall process to be described here consists of five main parts:

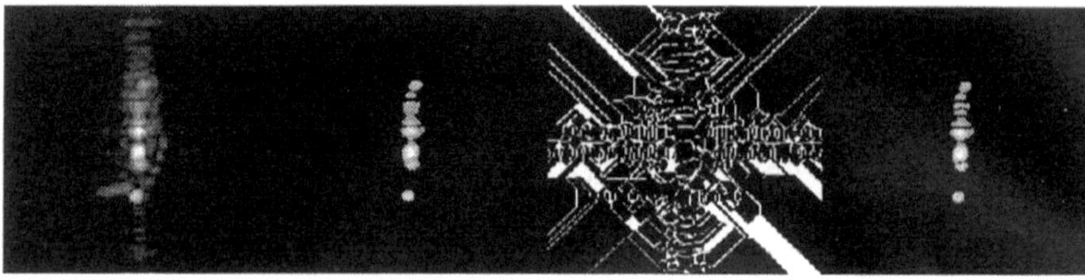

Fig. 5. Segmentation of an ISAR magnitude image using a minimum threshold with an inverted watershed (magnitude image, minimum threshold, minimum energy lines, combined segmentation)

- Image frame segmentation
- Scatterer graph construction
- Node match cost calculation
- Frame-graph matching
- Analysis of scatterer motion tracks

These are each described in the following sections. In some cases, we want to analyze sequences of *target* images directly, such as that shown in Figure 4. In most cases though, we want, for analysis purposes, to look at only the *differences* between the target and corresponding normal or ideal *reference* images, so that we can concentrate on the induced changes or specific anomalous scatterers which have been created on the target (either physically or synthetically). In the latter situation, we have to perform some additional steps to obtain the desired image sequence; i.e., we have to form image sequences for *both* the target and reference data, spatially *register* each pair of frames (using an automatic correlation process on the magnitude images), and then take the (incoherent or non-complex) *positive difference* between each pair of magnitude frames, to obtain the desired *difference frame sequence* for analysis.

2.1. Image Frame Segmentation

Once we have obtained an image frame sequence that we want to analyze, the first task is to segment each frame into different scattering regions. This is done by performing a watershed transformation on each (inverted) frame (selecting peaks instead of basins), following the priority queue approach described by Beucher and Meyer [3]. Radar images that have been upsampled prior to the FFT, and weighted to reduce sidelobes, are extremely well-suited to the use of watershed techniques for segmentation of scatterers, because each component is rounded and smoothed, and cleanly separated by minimum energy lines or "troughs" that occur due to coherent noise/phase-cancellation effects. As illustrated in Figure 5, the watershed, when combined with a minimum energy threshold, does a much better job of segmenting the strong scattering components in such images, than could a threshold operation used alone.

Our modification of the basic watershed technique, for this application, has been to allow the user to extract just the top 'N' scatterers (based on peak values), and only those peaks that are above a certain minimum threshold, 'T'. This allows us to focus on only the strongest scatterers for analysis, which are more stable across multiple frames. The dB-scaled magnitude image (scaled to 8-bits) is used for the segmentation of the scatterers, as it retains more of the minimum energy line information. In addition to computing the *extents* of the main scatterers in this way, we also want to return the *central region* of each scatterer peak, for graph display purposes, and the *centroid* of the regional maximum for each component, for tracking purposes. To obtain the central peak regions, we employ another technique defined in the previously cited paper by Beucher and Meyer. Using the regional maxima or centroids of each desired peak as labeled markers, and the inverted watershed lines (i.e., the minimum energy cuts) as boundaries, we grow the labels out to the maxima of the gradient image (also scaled to 8 bits). The gradient is based on a Sobel kernel, applied to the linearly (not dB) scaled magnitude image. We found that the Sobel gradient had less directional bias, resulting in rounder central regions, and we use the linear-scaled magnitude image here because it is more sensitive to small variations at the upper grey levels, and provides center regions with shapes that appear correct, visually.

2.2. Scatterer Graph Construction

As a result of the steps defined in the previous section, we have segmented and selected the primary scatterer events in each movie frame, and have not only their watershed-based *extents*, but also components representing their *central regions* as defined by the magnitude gradient, and the *centroids* of their regional maxima. The next step in the process is to build the *perceptual graph* of the remaining components in each frame, for use in the graph matching process which follows. The term "perceptual graph" was coined by S. Sternberg [4] due to the fact that it provides a planar graph linkage structure for all the objects in the image, one which appears natural to the eye. The same concept is also referred to in the literature as the *Gabriel graph* [5], originally defined in 1969. To create this type of graph for each frame, we originally planned to use a technique defined by F. Meyer in [6]. A rapid implementation of the basic method defined in that paper consists of the following steps:

- Compute the distance transform of the background
- Find the watershed lines of the distance transform
- Intersect the watershed lines with the distance transform
- Find the regional minima (only within the watershed lines)
- Reduce these minima to single points via pruning
- Follow the swiftest descent paths on the perpendicular sides to the "basins", or foreground components
- The perceptual graph then consists of the union of the original foreground with the swiftest descent paths (including the minima points on the watershed lines)

While the above algorithm performed as desired in generating the required connections, we eventually decided to use a simpler and faster approach to graph generation, based only on the distances of the centroids rather than on geodesic calculations. With this technique, two point nodes are joined if the circle which contains them along a diameter, contains no other point nodes. In such cases, we simply

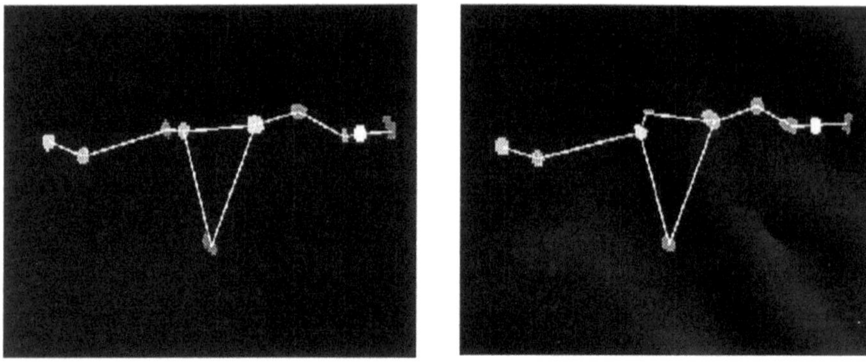

Fig. 6. Example frames showing perceptual graphs of primary scatterers

draw a line in the binary image between the two centroids to generate the image graph. While this algorithm is less general than the geodesically-defined perceptual graph algorithm defined by Meyer, it appears justified in this case—first, because the central regions of the scatterers that we extract are generally convex in shape, and second, because our goal is to generate reasonably stable "neighborhoods" for each node (for the graph matching process), rather than to have a theoretically-correct perceptual graph. Figure 6 provides an illustration of the types of graphs that we obtain using this procedure, for two adjacent frames of another type of measured object. The grey values of the nodes in these images correspond to their peak rank labels in the given frame.

Finally, to complete the construction of the graph structure for each image frame, we also calculate four "attributes" for each node: 1) the peak rank, 2) the area rank, 3) the location of the centroid in terms of the size of a "resolution cell" for the image, and 4) the neighborhood of the node, based on the previously-calculated graph. These attributes are used to match corresponding nodes in adjacent frames, which is the next step in the process.

2.3. Node Match Cost Calculation

Once we have calculated the graph for each frame and the attributes for each node, we need to compute a cost of matching any particular node in one frame, to a given node in the adjacent frame. The costs are defined such that a zero cost corresponds to a perfect match, and higher costs are larger positive numbers. The goal of the graph matching process then is to *minimize* the overall cost of matching one frame graph to the next. To facilitate this process, we calculate a *cost table* for every pair of frames, which contains the individual cost of matching each pair of nodes. The table also contains "drop costs", which are the costs associated with *not* matching a given node with one in the adjacent frame.

There are four components to the cost of matching any two nodes. The first is the absolute difference in the peak value ranks. The second is the absolute difference in the area ranks. The third is the absolute difference in location, measured in terms of resolution cell units in the image. Finally, the fourth is the number of mismatching neighbors in the two node neighborhoods. A neighbor is considered to be mismatched if there is no corresponding node in the other graph, which is within 1 resolution cell and +-30 degrees, of where the neighbor in the first graph is. (This calculation is done in both directions to obtain a more symmetric relation). The four independent cost components (which are all small integers) are simply added together to obtain the overall cost of matching the two nodes; there is currently no weighting applied. A "Maximum Valid Cost" threshold is used in the graph matching process to immediately reject costs which are too high to even be considered. The cost of dropping a node from either graph is inversely related to its best (least) cost match (i.e., Drop Cost = Maximum Valid Cost + c - Best Cost Match). The small positive constant "c", is used here to ensure that nodes which have a very good match available, will not be allowed to drop—i.e., the drop would be considered invalid and thus be rejected outright from consideration.

2.4. Frame Matching of Scatterer Graphs

Once the complete cost tables have been calculated for every pair of frames, the graph matching process is ready to begin. Essentially, the algorithm used to find the graph match with the least overall cost, is an exhaustive search, under certain constraints. First, any node that has already been matched cannot be re-used. Second, any node match that has a cost above the Maximum Valid Cost threshold, will not be considered. Third, the algorithm keeps track of the best overall cost achieved so far, and the current cost of partial matches as they are generated. If the partial cost is ever greater than the best cost achieved so far, that line of the search will be dropped, and a new line will be examined. Two "stacks" keep track of the current best assignment of nodes, and the partial assignment that is currently under consideration. Using these three constraints, despite the fact that the total number of possible matches is quite large (in the millions), the actual number of complete assignments examined is much, much lower, and the graph matching process completes within seconds.

2.5. Analysis of Motion Tracks

The final step in the image sequence analysis process, once the graphs of adjacent frames have been matched, is to generate the track files for individual scatterers, and to analyze the motion represented by these tracks. As the image sequences or movies are typically generated from a central angle (where some significant anomaly was discovered), working outward, we generate the tracks in the same way, starting with the central frame scatterers, and then working outward to each side. In addition, while we typically retain the top 10 or so scatterers for context (in generating and matching the graphs for each frame), when we do track generation, we typically limit this to only the most important *few* scatterers, that are considered to be significant in terms of the overall energy they represent.

The final result of the entire process is a just small number of lists, one for each significant scatterer, of its peak centroid locations in each of the several frames of the image sequence movie. These lists can then be analyzed for their motion characteristics, and be used to classify the scattering event into one of a few types. For example, when expressed in terms of a scene-stabilized (target fixed, radar moving) coordinate system, the most common, "direct specular" scatterers will appear in a constant spot. Extended specular reflections such as the side of the conesphere will "flash" and disappear rather quickly. "Sliding" speculars will move slowly along curved edges. Similarly, in a radar-stabilized (target rotating) coordinate system, resonant scatterers will appear to move horizontally (at constant distance or range), and multiple-bounce returns will appear downrange from the entry point. Using fairly simple rules, most of the more common scattering mechanisms can be identified based on such track characteristics.

3. Results and Conclusions

In this paper we have discussed a process for extracting and analyzing the motion behavior of significant scattering events in ground-to-ground ISAR imagery. After formation of the original sequence of frames to be analyzed, the process begins with an inverted watershed segmentation of the scattering objects in each frame, augmented to extract only the most important objects, along with their peak centroids, and central peak regions. These peaks are then converted to graphs using a "perceptual graph" or "Gabriel graph" approach, and attributes are calculated on each node. The remaining steps of the process involve finding the best overall match of adjacent frame graphs, based on cost criteria calculated from the attributes, and finally, analyzing the track or motion of each object.

The process described has proven quite successful in being able to identify basic scattering mechanisms such as direct, resonant, and sliding specular returns, on a moderate number (several tens) of images. The modified watershed on the inverted image does an excellent job of extracting the most important scattering events or peaks, the perceptual graph algorithm used generates graphs that are quite satisfactory, and the graph matching process based on the attribute match costs that were outlined here, have worked quickly and correctly in matching graphs from adjacent frames, with almost no errors. We are continuing to test this process on larger sets of images, incorporating additional rules for other types of scattering phenomena.

References

1. G. Fliss, J. Steinbacher, S. Stokely, and R. Vogt; 'Feasibility of Automated Analysis of Diagnostic Radar Images', *Proceedings 1997 AMTA Conference*, Greenbelt, Maryland, 1997.
2. R. Vogt, G. Fliss, and J. Mihocka, 'Morphological processing of diagnostic radar images'; *Proceedings of the SPIE Applications of Digital Image Processing XIX*, Vol. 2847, pps. 167-176, Denver, 1996.
3. S. Beucher, and F. Meyer, 'The Morphological Approach to Segmentation: The Watershed Transformation', in *Mathematical Morphology in Image Processing*, (ed.) E. Dougherty, ch. 12, Marcel Dekker, New York, 1993.
4. S. Sternberg, 'Language and Architecture for Parallel Image Processing'; in *Pattern Recognition in Practice*, North-Holland, 1980.
5. K. Gabriel, and R. Sokal, 'A new statistical approach to geographical variations analysis', in *Syst. Zool.*, v. 18, pps. 259-278, 1969.
6. F. Meyer, 'The Perceptual Graph: A New Algorithm'; in *Proc. Congrés ICASSP*, Paris, 1982.

A REGION MERGING ALGORITHM USING MATHEMATICAL MORPHOLOGY: APPLICATION TO MACULA DETECTION

F. ZANA
Centre de Morphologie Mathématique, Ecole des Mines de Paris
35, rue St-Honoré 77305 Fontainebleau, France
zana@cmm.ensmp.fr, fax: (+33) 01.64.69.47.07

I. MEUNIER
Eye University Clinic of Créteil, Paris XII
40 avenue de Verdun 94010 Creteil, France
fax: (+33) 01.45.17.52.27

and

J.C. KLEIN
Centre de Morphologie Mathématique, Ecole des Mines de Paris
35, rue St-Honoré 77305 Fontainebleau, France
klein@cmm.ensmp.fr, fax: (+33) 01.64.69.47.07

Abstract. We present a region merging algorithm based on a watershed cell decomposition and on morphological treatments for macula recognition. After noise removal, a watershed is performed to produce a mosaic image. The resulting decomposition segments the image in cells while maintaining the shape of the Macula. A cell inside the macula is selected as the first step of a merging algorithm. Morphological filters are performed to enhance the capillaries surrounding the macula. A complex criterion based on the gray values and of edges of the filtered image is calculated to merge the cells of the macula while rejecting perifoveal intercapillary zones in order to produce the contour of the macula.

Key words: Region merging, Mathematical Morphology, Watershed, Image understanding, Ophthalmology.

1. Introduction

The macula, as used by the physicians is a term that defines the zone of central vision. Its limits are not precisely defined, therefore we refer to the foveal avascular zone (FAZ). Ideally, the FAZ is defined as the central area of vision delimited by the first continous arcade of capillaries. Fluorescein angiography is required in order to see the smallest capillaries of the perifoveal zone. Fluorescein is a dye introduced with an intravenous injection that enhances specifically the retinal vessels and that permits to distinguish capillaries that can be as wide as 8 to $10\mu m$.
The development of digitized fluorescein angiography [4, 6, 7, 10] combined with image processing systems could modify the knowledge, the classification, and treatment approach in ocular vascular diseases, especially in diabetic retinopathy that remains the first cause of legal blindness in young adults in Western countries.
Previous clinical randomized studies [5] had defined specific criteria for the classification, the severity and the prognosis of ischemic retinopathies (included diabetic

retinopathy, retinal venous occlusion, sickle cell disease). Moreover, these semi quantitative classifications are required for laser treatment and photocoagulation procedures [1]. These classifications are based on vascular modifications analysis, especially microanevrism count and macular non perfusion area measurements. These vascular lesions are the two major features in occurrence and progression of diabetic retinopathy [5]. In addition, these vascular abnormalities are generally correlated with the degree of visual impairment and with a risk factor of progression towards severe proliferative forms of these ischemic retinopathies.
This study was undertaken to design and evaluate image processing programs to detect macular non perfusion area, particularly the foveal avascular zone. Histogram transformations and methods of thresholding have been developed in a former study [7, 8]. Inspired by these techniques, a region merging algorithm using morphological treatments was developed and tested on fluorescein angiographic frames in cases of diabetic retinopathy and venous retinal occlusion.

2. Morphological operators and algorithms

This section is a short reminder of definitions of basic extensive morphological operators (see more details in [9]) followed by the definition of more complex operators that will be used afterwards.

2.1. Basic morphological operators

We define the image whose gray values are contained in $[N_{min}, N_{max}]$ as a functional $S : \mathbb{R}^2 \rightarrow [N_{min}, N_{max}]$, and a structuring element as a functional $B : \mathbb{R}^2 \rightarrow \boldsymbol{B}$ where $\boldsymbol{B}$ is the set of neighbourhoods of the origin.
We then define basic operators, with respect to the structuring element B with scaling factor e, image S and point $M_0 \in \mathbb{R}^2$:
erosion: $\epsilon_B^e(S)(M_0) = MIN_{M \in M_0 + e.B(M_0)}(S(M))$
dilation: $\delta_B^e(S)(M_0) = MAX_{M \in M_0 + e.B(M_0)}(S(M))$
opening: $\gamma_B^e(S) = \delta_B^e(\epsilon_B^e(S))$
closing: $\phi_B^e(S) = \epsilon_B^e(\delta_B^e(S))$
We will not use superscripts for unit scaling factor. Geodesic operators are defined with respect to a norm $||.||$, or a connectivity graph in digitized images (a neighbourhood C of unit radius). They depend on a marker image S_m and on distance d.
The geodesic dilation $S^d = \delta_{S_m}^d(S)$ is defined by:
$S^1(M_0) = inf(\{MAX_{M \in M_0 + C}(S(M))\}, S_m(M_0))$
$S^{d+1}(M_0) = \delta_{S_m}^{d+1}(S)(M_0) = \delta_{S_m}^1(S^d)(M_0)$
The geodesic reconstruction (or opening) is defined by:
$\gamma_{S_m}^{rec}(S) = sup_{d \in \mathbb{N}}(\delta_{S_m}^d(S))$
The geodesic closing is defined by:
$\phi_{S_m}^{rec}(S) = N_{max} - \gamma_{(N_{max} - S_m)}^{rec}(N_{max} - S)$
For our purpose, the most important property of the basic operators of mathematical morphology lies in the fact that the curve variations included in the support of the structuring element are removed. Openings and closings can be viewed as operators that remove undesired elements from the image (replacing them by lower or higher values respectively). As an example, an opening (resp. a closing) with a

1D structuring element of size sz and direction a will only remove local gray level maxima (resp. minima) that are thinner than sz in the direction a. The removed parts are related to noise and to small or thin image elements. Geodesic operators are used to preserve exact shapes.

2.2. Morphological operators with 1D structuring elements

Using the above definitions, we have designed higher level transformations:
- The maximum of geodesic openings with 1D structuring elements of various directions leaves vessels nearly unchanged and removes white spots from the image.
- The maximum of geodesic closings with 1D structuring elements of various directions removes small dark spots, while maintaining shapes.
- The maximum of openings with 1D structuring elements of various directions removes all non-linear components of the image.
- The minimum of closings with 1D structuring elements of various directions enhances some linear components of the image.

2.3. The watershed operator

Let W_0 be the set of minima of S. The watershed is a segmentation tool defined by the algorithm:
$W_{i+1} = \{Flood_{i+1}(Q); Q \in W_i\}$
$Flood_{i+1}(Q) = Set(\gamma_{1_Q}^{rec}(S \leq i+1)) \cap (\cup_{Q' \neq Q})^c$
If we compare the image to a topographic surface, this operator is commonly interpreted as a flooding process. Imagine that we pierce each minimum of this surface. Then we start a flooding process from the minima of the image. During the flooding process, we build a dam on points there the flood merge. The set of catchment basin generated is the watershed image (see [2] for a detailed algorithm). It is commonly used on a morphological gradient, defined as $\delta_B^e - \epsilon_B^e$.
Applied on a gradient image, the watershed segments uniform areas. The result is a cell decomposition of the flat zones of the image.

3. Macula description and identification

3.1. The foveal avascular zone

The FAZ is an histological zone delimited by the first continuous arcade of capillaries that can be as wide as 8 to $10\mu m$. This is just wide enough to permit the movement of a fluorescein molecule. The signal is thus low in intensity, furthermore the background signal is hidden by the xanthophile pigment whose thickness is irregular. Photographs are taken under blue light during the test, at different stages of the diffusion process. We have used images captured with a numeric camera of very high definition (1024x1536). These images contained a global view including the macula, the optic nerve and the vascular arcades. A wide region of interest that included the macula was selected (figure 1, 2, column 1). The algorithm automatically zooms on the FAZ. In those images, the FAZ is defined as the uniform area having the lowest gray level values of the image. However these capillaries which delimit the FAZ are not discernible on a photograph without histogram modification. Furthermore, there can be discontinuities due to the lack of signal. After histogram equalization, the capillaries are hardly discernible and the zone of low gray values is never uniform. Specific treatment is thus necessary.

3.2. Theoretical model of a perfect foveal avascular zone

Our approach is based on the assumption that a perfect FAZ is a flat zone surrounded by capillaries. Under such an hypothesis, the FAZ should be extracted using a watershed on a gradient image: the FAZ should be the cell with the darkest gray level. This algorithm is not successful on its own, for pepper noise, ill-defined boundaries (due to the lack of signal in $10\mu m$ wide capillaries) and non uniform gray values create an over-segmented cell decomposition. The FAZ is particularly noisy since the signal issued from the capillaries is very low in intensity, and can be confused with background fluorescence.

3.3. Realistic modeling of the foveal avascular zone

In order to be closer to the reality, we assume that the previous model is corrupted by the following events:
- A pepper noise is added (caused by the digitization process or the retina texture).
- The FAZ is not uniform in gray values (because the thickness of the epithelium varies), however it is generally delimited by capillaries that have a higher gray value.
- The capillaries surrounding the FAZ may appear as unconnected (because of a low signal or vascular lesions).

4. Treatment for macula identification

Before performing the watershed algorithm we filter the image to remove the pepper noise. First we compute the maximum of geodesic closings with linear structuring elements (with 5-pixel wide elements) in order to remove the small dark holes that are not meaningful minima. This filter has the advantage of enhancing low-contrasted capillaries. Next we remove white spots using the maximum of linear openings (with 5-pixel wide elements). This filter also simplifies the image. Then we compute a cell decomposition using the watershed algorithm on the gradient image. We will refer to this cell decomposition in the following. The cell with darkest gray level is always inside the FAZ. We call it the *kernel-cell.*

At this stage, it is noticeable that the limits of the whole FAZ can be drawn by merging some cells. The model is thus verified. Furthermore, the biggest cells were generally found to belong to an avascular zone. The next step consists in a merging algorithm that starts from the *kernel-cell* and that can be decomposed into five steps:

i) We select the cells that contain at least one pixel whose gray value is smaller than or equal to the maximum value of the *kernel-cell.* We call this set of cells S_1. This corresponds to the fact that the limits of the FAZ are correctly defined by the watershed decomposition, and that a manual threshold (after filtering) can give a reliable though underestimated approximation of the FAZ. Therefore the *kernel-cell* gives a generally underestimated value of the optimal threshold, however the selected cells of S_1 surely belong to the FAZ.

ii) After an histogram equalization used to enhance the display, we perform an alternating filter. It consists of a minimum of linear closing (size 15) followed by

a maximum of linear opening (size 15). The effect is to enhance the surrounding capillaries and to remove some of the central noise with structuring elements. This gives image I_1.
iii) The image I_1 is segmented using M as a threshold value. M is the maximum of the mean gray value on each cell of the set S_1. Cells that have more than 50% of segmented pixels are selected. This gives the set S_2.
iv) A morphological gradient is computed on I_1 to detect surrounding vessels. The gradient image is automatically thresholded at the third of the maximum gray level of I_1 computed on the cells of S_2. The resulting binary image contains vessels and some step edges between inner zones of the FAZ. A closing with reconstruction using a square structuring element of size 4 will fill the perifoveal intercapillary zones since they are often composed of capillaries parallel and close. These zones do not belong to the FAZ and may mislead the algorithm since they are composed of cells having gray values similar to the central area. Inner step edges are reconstructed because they are not close enough. An erosion with a square structuring element of size 2 is performed to give image I_2.
v) Cells of S_2 that contain more than 80% of white pixels of I_2 are excluded. The resulting set of cells corresponds to the FAZ. The FAZ is obtained by filling small holes of the binary image.

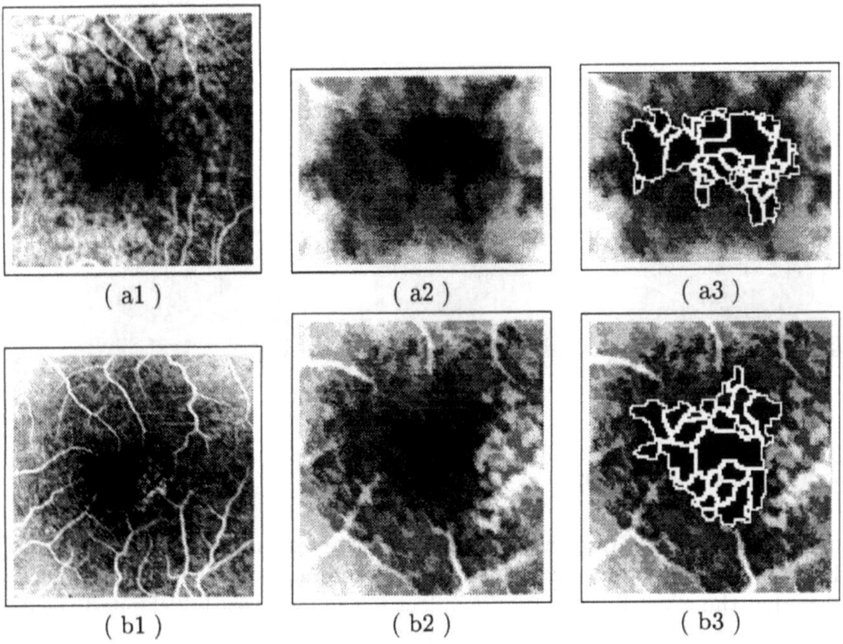

Fig. 1. Column 1: original images with histogram equalization. Column 2: zoom of column 1, based on detection. Column 3: merged cells.

5. Results and discussion

This algorithm has been tested on a set of 9 images.
FAZ have been extracted from a set of six images of macula from different patients with various diseases (4 of whom can be seen in figures 1,2), and three more images from a non-diabetic patient. The exact shape of the FAZ can be determined using the cell decomposition, even though the limits are not always clearly visible. We will thus discuss the reliability of the region merging algorithm with respect to different illnesses and the variability of the segmented zone with respect to images of the same eye.

5.1. Treatment with various abnormalities

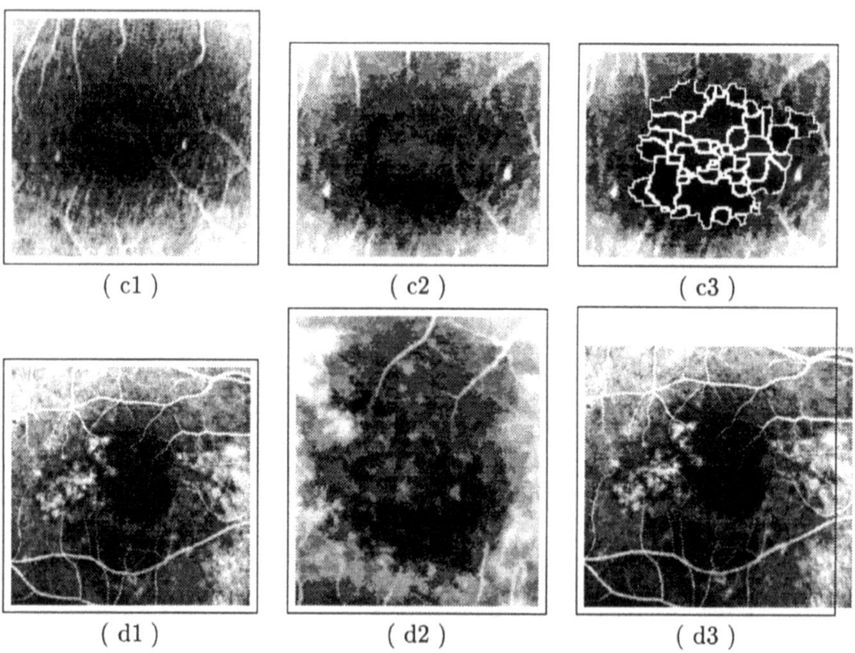

Fig. 2. Column 1: original images with histogram equalization. Column 2: automatic zoom of column 1. Column 3: merged cells.

We have tested our algorithm on a series of six images from different patients affected by diabetic retinopathy or macular lesions. These images can be characterized by the fact that the exact limit of the FAZ does not appear clearly delimited (figures 1,2). It is due to the presence of abnormalities (drusens, microaneurysms or vascular abnormalities) on the border of the FAZ or by a lack of accuracy of the image (capillaries are fuzzy or not visible at this scale). The six images have all been tested successfully except one presenting edema (figure 2 (d)). In case of macular edema with a low contrasted arcade of capillaries delimiting the FAZ, the area surrounding the central zone is darker and the algorithm is misled in the edema zone. Since the signal is very low - the gray level of the macula varies from 10 to 30 - the presence of an edema will always lead to a very tough case. It seems thus that further progress can be done, however the current algorithm is adequate in most cases.

5.2. VARIABILITY OF THE RESULTS

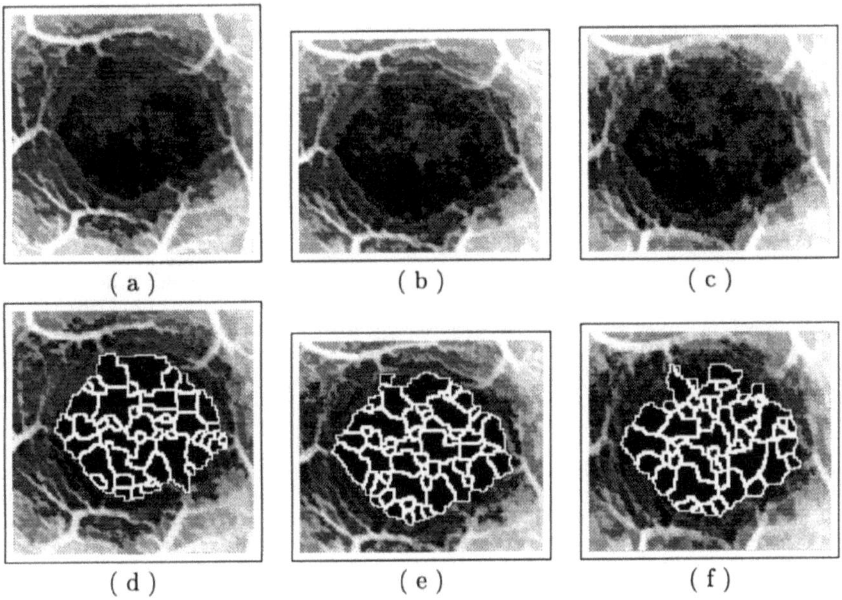

Fig. 3. (a), (b), (c): result of the histogram equalization, after the automatic zoom of the macula. (d), (e), (f): cells selected as belonging to the FAZ.

We have tested our algorithm on a sequence of three images of the same patient during the same angiography test. The patient had no visible abnormalities. All images have been taken during the same test, however they differ in intensity and in the degree of leakage that appears inside the FAZ. The variation in result is very instructive as to the behavior of the algorithm: the first and the second images (figures 3 (a) and (b)) show a very similar and accurate segmentation of the FAZ (figures 3 (d) and (e)), although the cell decomposition is different. The third image (figure 3 (c)), that includes leakage, shows a cell decomposition where the leaking parts are excluded (figure 3 (f)). The variability in surface is maximum between decomposition 3 (f) and 3 (d). Its value, calculated after scale adjustment is 6.5%. The variability between manual drawings made by different physicians ranges from 10 to 20%. From these examples, we deduce that the segmentation is stable and corresponds to the FAZ. Some zones are excluded in later time because they leak (figure 3 (f)), which seems a correct way of delimiting the FAZ. It should be noticed that this image is particularly convenient, especially because capillaries delimiting the FAZ are visible.

6. Conclusion

We have presented a region merging algorithm for macula segmentation. Tests on various abnormalities and a study of the variability of the results have been presented. Therefore we think that the proposed algorithm could be used to detect and follow the evolution of the FAZ (with a registration algorithm [3, 10]), which should give an unbiased measuring tool for clinical studies or an aid to decision making. It gives also means to enhance the quality of images of the FAZ zone that help detecting macular leakage or other abnormalities that are not visible on the original image.

References

1. O. Arend, S. Wolf, F. Jung, B. Bertram, H. Postgens, H. Toonen, and M. Reim. Retinal microcirculation in patients with diabetes mellitus: Dynamic and morphological analysis of perifoveal capillary network. *British Journal of Ophtalmology*, 75:514–518, 1991.
2. S. Beucher and F. Meyer. The morphological approach to segmentation : The watershed transformation. *Mathematical Morphology in Image Processing*, pages 433–481, 1992.
3. L. G. Brown. A survey of image registration techniques. *ACM Computing Surveys*, 24(4):325–376, 1992.
4. M. Goldbaum, S. Moezzi, A. Taylor, S. Chatterjee, J. Boyd, E. Hunter, and R. Jain. Automated diagnosis and image understanding with object extraction, objet classification, and inferencing in retinal images. In *International Conference on Image Processing*, 1996.
5. E. T. D. R. S. R. Group. Classification of diabetic retinopathy from fluorescein angiograms. *Ophtalmology*, 98:807–822, 1991.
6. B. Lay. *Analyse Automatique des Images Angiofluorographiques au cours de la Rétinopathie Diabétique*. PhD thesis, Ecole Nationale Supérieure des Mines de Paris, 1983.
7. I. Meunier, P. Bunel, and G. S. an Maurice Gaudaire. Computerized detection of macular non-perfusion on digitized angiography in retinal vascular diseases. In *International Conference on Biomedical Engineering*, pages 221–222, Tampere, Finland, June 1996.
8. P. K. Sahoo, S. Soltani, A. Wong, and Y. Chen. A survey of thresholding techniques. *Computer Vision, Graphics, and Image Processing*, 41:233–260, 1988.
9. J. Serra. *Image Analysis and Mathematical Morphology*. London, U.K.: Academic, 1982.
10. F. Zana and J. Klein. Robust segmentation of vessels from retinal angiography. In *International Conference on Digital Signal Processing*, pages 1087–1091, Santorini, Greece, July 1997.

Author Index

Subject Index

MIX
Papier aus verantwortungsvollen Quellen
Paper from responsible sources
FSC® C105338

If you have any concerns about our products,
you can contact us on
ProductSafety@springernature.com

In case Publisher is established outside the EU,
the EU authorized representative is:
Springer Nature Customer Service Center GmbH
Europaplatz 3, 69115 Heidelberg, Germany

Printed by Libri Plureos GmbH
in Hamburg, Germany